HANDBOOK OF COMPOSITES
VOLUME 4

HANDBOOK OF COMPOSITES

VOLUME 4

Series Editors

A. KELLY

Vice-Chancellor's Office
University of Surrey
Guildford
Surrey GU2 5XH
England

Yu.N. RABOTNOV

Academy of Sciences of the USSR
Department of Mechanics and Mathematics
Moscow State University
117234 Moscow
USSR

NORTH-HOLLAND – AMSTERDAM • NEW YORK • OXFORD

FABRICATION OF COMPOSITES

Volume Editors

A. KELLY

Vice-Chancellor's Office
University of Surrey
Guildford
Surrey GU2 5XH
England

S.T. MILEIKO

Academy of Sciences of the USSR
Institute of Solid State Physics
142432 Chernogolovka
Moscow
USSR

NORTH-HOLLAND – AMSTERDAM • NEW YORK • OXFORD

ISBN: 0 444 86447 4

First edition: 1983
Second printing: 1986

Publishers:
ELSEVIER SCIENCE PUBLISHERS B.V.
P.O. Box 1991
1000 BZ Amsterdam
The Netherlands

Sole distributors for the U.S.A. and Canada:
ELSEVIER SCIENCE PUBLISING COMPANY, INC.
52 Vanderbilt Avenue
New York, N.Y. 10017
U.S.A.

Library of Congress Cataloging in Publication Data
Main entry under title:

Fabrication of composites.

(Handbook of composites; v. 4)
Bibliography: p.
Contents: High performance composites with resin matrices / N.L. Hancox — Problems of the mechanics of composite winding / Yu.M. Tarnopol'skii and A.I. Beil' — Multidirectional carbon–carbon composites / L.E. McAllister and W.L. Lachman — [etc.]
  1. Fibrous composites. 2. Metallic composites.
I. Kelly, A. (Anthony) II. Mileiko, S.T. (Sergei Tikhonovich) III. Series.
TA481.5.F3 1983   670   82-12528
ISBN 0-444-86447-4

PRINTED IN THE NETHERLANDS

Preface to Series

During 1978 Dr. W.H. WIMMERS, at that time Managing Director of the North-Holland Publishing Company, made arrangements with the Publishers Atomizdat and his own Company, the North-Holland Publishing Company, for the publication simultaneously within the Soviet Union and in the West of Handbooks concerning technical subjects of importance. The idea was that Handbooks consisting of a number of volumes should be edited jointly by an Editor from the Soviet Union and an Editor from the West. These Editors would agree a format for each particular volume and would encourage Authors of individual chapters to write these, the Authors to be chosen from the Soviet Union and from the Western World.

It was hoped, by this initiative, that scientists and technologists in the West and within the Soviet Union would learn something more of one another than they presently glean from the currently published literature.

It would, of course, have been nice to include within this idealistic framework Authors from the so-called Third World to which, of course, the People's Republic of China belongs but a journey of a thousand miles starts with a single step and the first arrangements were possible with the publishing house Atomizdat and with North-Holland.

These volumes are offered by us as a first step along the road towards the idea of publication throughout the world of tracts of importance. There have been a number of difficulties to overcome in reconciling the different cultural systems and the different ways of operating of the scientific communities of the Soviet Union, those in Western Europe and in the United States. At times the interaction has been slow and uncertain. However, we are grateful to all the Authors for their patience and their ready acceptance of Editorial comment.

We are very glad to see this volume appear in print and we hope there will be many others like it.

Yu.N. RABOTNOV

A. KELLY

Preface to Volume 4

This volume covers some aspects of the technology of composite materials. We have assembled here two chapters by authors from the Soviet Union, five from the United Kingdom, one from France and one from the United States.

This is the first volume to go to the press in the series planned in 1978 through the efforts of Dr. W.H. WIMMERS who devoted much care and attention to the idealistic aim of producing a Handbook in an important subject area jointly authored by members of the scientific community throughout the world and published simultaneously in the Soviet Union and in Western Europe. Since the subject was planned, the vicissitudes of international politics have not made the production of such a volume easy. Different cultural systems in the different scientific communities of the Soviet Union and in Western Europe and the United States have made interaction slow and uncertain at times. I am very grateful to all the authors for their patience and their ready acceptance of Editorial comment. To those who submitted their chapters first and have watched the months go by, I can only apologise.

This volume covers a description of the technology of the important fibre composite materials and includes the first comprehensive survey of carbon–carbon composites. In those cases where some theory of how to develop optimum properties has appeared in the literature, this is included. There is also a chapter on biological structures by Professor CURREY, which we hope adds completeness and gives perspective to this volume. It must be admitted, however, that this chapter can only describe natural composite structures and point out their conformance to composite principles without describing how these structures are produced.

The chapters by Russians are more theoretical in nature and those contributed by the American, French and British authors are clear and direct in the descriptions of the technical principles involved.

I am grateful to Dr. T.W. CLYNE for patiently reading two of the Russian chapters and helping me with their production in the form found here.

August 1982

A. KELLY
Worplesdon

Contents

List of Contributors

M.G. BADER, *University of Surrey, Guildford* (Chapter IV)

A.I. BEIL', *Academy of Sciences of the Latvian S.S.R., Riga* (Chapter II)

H. BIBRING, *Office National d'Études et de Recherches Aérospatiales, Châtillon* (Chapter VI)

J.D. CURREY, *University of York, Heslington* (Chapter IX)

N.L. HANCOX, *Atomic Energy Research Establishment Harwell, Oxfordshire* (Chapter I)

D.J. HANNANT, *University of Surrey, Guildford* (Chapter VIII)

T. KHAN, *Office National d'Études et de Recherches Aérospatiales, Châtillon* (Chapter VI)

W.L. LACHMAN, *Materials International, Lexington, MA* (Chapter III)

L.E. MCALLISTER, *Fiber Materials Inc., Biddeford, ME* (Chapter III)

S.T. MILEIKO, *Academy of Sciences of the U.S.S.R., Moscow* (Chapter V)

D.C. PHILLIPS, *Atomic Energy Research Establishment Harwell, Oxfordshire* (Chapter VII)

M. RABINOVITCH, *Office National d'Études et de Recherches Aérospatiales, Châtillon* (Chapter VI)

J.F. STOHR, *Office National d'Études et de Recherches Aérospatiales, Châtillon* (Chapter VI)

Yu.M. TARNOPOL'SKII, *Academy of Sciences of the Latvian S.S.R., Riga* (Chapter II)

CHAPTER I

High Performance Composites with Resin Matrices

N.L. Hancox

Materials Development Division
Atomic Energy Research Establishment Harwell
Oxfordshire
United Kingdom

Contents

HANDBOOK OF COMPOSITES, VOL. 4 – Fabrication of Composites
Edited by A. KELLY and S.T. MILEIKO
© 1983, Elsevier Science Publishers B.V.

1. Introduction

Although natural composites—wood, bones and teeth—have been in existence for many millions of years, fibre reinforced resin matrix materials were not developed until the early 1940's. We should note here that high performance composite materials are taken to be those in which glass, carbon or aramid fibres are used to reinforce a thermosetting or possibly thermoplastic matrix. Consideration of wood flour, inorganic fillers, cotton etc. in, for instance, a phenol formaldehyde matrix is specifically excluded. Since the 1940's the variety of reinforcements and matrices, and end uses, have increased dramatically. Composite materials of many types are now widely employed in the transport, construction, chemical, electrical, marine, leisure, medical and aerospace industries and are a well established class of material. The tonnage consumption of glass reinforced plastic (GRP) in Western Europe and the U.S.A. in 1976 was in excess of 1150 000 tonnes. The present annual world production of carbon and aramid fibres is probably of the order of several hundred and several thousand tonnes, respectively. Among the attractions of resin matrix composites are:

(1) the ability to process easily at low temperatures ($<200°C$),
(2) excellent, specific, thermal and mechanical properties,
(3) the possibility of varying properties with direction,
(4) good resistance to corrosion and other chemical environments.

To produce a composite it is necessary to combine reinforcing fibres or whiskers with a matrix material which can infiltrate the fibres and subsequently be solidified, whilst ensuring, if necessary, the appropriate physical containment of the materials. The aim is to produce a strong, stiff, low density, chemically resistant structure, which can be relatively easily fabricated and have if desired excellent thermal stability. A high performance composite either has the fibres aligned in one or two directions, or it may be composed of an ordered stack of thin laminae to give improved multidirectional mechanical and thermal performance. The fibre provides strength and stiffness in the direction of its long axis, improves the shear modulus of the matrix material, and contributes to the work of fracture of the composite by taking part in failure processes such as debonding and fibre pull-out. Traditionally the role of the matrix is to transfer stress to short fibres, protect the surface of the reinforcement from environmental attack, keep fibres aligned in predetermined directions, and help to avoid catastrophic crack growth through the fibres by providing an alternative failure path along the interface between the fibre and resin. In addition the matrix and interface region largely determine transverse and shear strength properties, and contribute to the flexural strength, longitudinal and transverse compressive strength, torsional fatigue and creep behaviour, dynamic damping, work of fracture, operational temperature range, resistance of the composite to chemical attack and radiation, and assist in the fabricator's ability to manufacture a composite artefact.

The interface is a rather ill-defined but extremely important part of the composite. In a cube of side 10 mm, composed of 60 % (volume percent) of aligned, unidirectional, carbon fibres in a resin matrix, the area of the interface is approximately 3×10^4 mm². The thickness of the interface layer is a matter of conjecture but it may be of the order of 10^{-7} m. The interface may be considered as the region where the matrix is bonded chemically, or keyed mechanically, to the reinforcement plus for a resin, a layer of polymer whose properties differ from those of the bulk resin because of preferential absorption of components of the resin by the fibre, and which may possess a more ordered structure. If the interface is too strong the composite may be overly brittle, while if it is too weak the composite material will be useless as a structural material because of difficulties of diffusing loads into the fibres.

Although we shall describe a range of reinforcements and matrices the composites considered in detail will consist of continuous fibres, or at least a high proportion of these, in a thermosetting resin matrix.

2. Types of reinforcement

Possible starting materials for the reinforcement are shown in Fig. 1. These elements are all from periods 2 and 3 of the Periodic Table. Most are abundant and have a high density of covalent bonds. Unfortunately most, either singly, or in combination, e.g., $A\ell_2O_3$, are brittle and only have a good strength if relatively flaw free. Thus an obvious way to use them is in the form of small diameter fibres or whiskers.

The types of reinforcement usually employed in a composite are glass, carbon, boron and aramid fibre, the latter also known by its trade name of Kevlar. Cellulosic fibres, e.g., jute, are of interest because their very low density $(1.2 \times 10^3 \text{ kg m}^{-3})$ leads to excellent specific properties, comparable in some cases to those of GRP. Several continuous ceramic fibres have been produced recently based on polycrystalline alumina, silicon carbide and a combination of alumina, boria and silicon dioxide respectively. A range of whiskers, which are virtually single crystals with an aspect ratio of 10:1 to $10^4:1$, mainly based on nitrides and carbides of aluminium or silicon, but also

FIG. 1. Possible elements to make fibres or whiskers.

including graphite, magnesia, beryllium and beryllia, have been made. Further details of whiskers are given by MILEWSKI and KATZ (1978).

The commonest and cheapest man-made reinforcing fibre is glass and it has been used in various forms for approaching 40 years. Three grades are of interest for composite work, E, S and R glass. E glass is a borosilicate material with a low alkali content accounting for the great bulk of production, and unless stated to the contrary reference here to GRP will imply E glass reinforcement. S and R glasses contain more alumina and have improved mechanical properties. The former is manufactured in the U.S.A., the latter in Western Europe. Individual glass fibres or filaments are frequently combined into groups of 200 or so untwisted fibres to give a strand. Strands may in turn be combined to form a roving. Yarn is an assembly of twisted fibres or strands used for weaving. Coupling agents, often organic silane complexes are added to the surface of the fibres and strands to improve adhesion between the fibre and resin. Many different types, which can be made compatible with various types of resin, are available. The subject is dealt with at length in BROUTMAN and KROCK (1974) and new developments are often recorded in the Society of Plastics Industry Annual Conference in the U.S.A. For a specific application it is helpful to consult with the glass fibre supplier.

The first practical carbon fibres were developed from textile fibres in the mid 1960's by WATT et al. (1966). Since that time manufacturers in Western Europe, U.S.A. and Japan have developed various grades of fibre. These include medium strength and modulus, high tensile strength, high modulus, and very high modulus fibres. The cost of the fibre is related to that of the precursor, initial oxidation time, final heat treatment temperature, which is greater for high modulus material, and to any special processes such as hot stretching which have to be employed in manufacture. The fibres, which are very anisotropic and have an internal crystallite structure are usually prepared from a polyacrylonitrile precursor, although rayon, or a specially prepared pitch can also be used. Improvements in fibre quality since their introduction coupled with the effect of the composite fabricator's skill on properties, sometimes makes it difficult to compare the results of different workers particularly if these are separated in time by several years. Useful accounts of carbon fibre production and chemistry are given by GOODHEW et al. (1975) and FITZER and HEYM (1976).

To obtain an adequate bond to the resin matrix the fibres, especially the higher modulus variety, must be given an oxidative surface treatment. The mechanisms of adhesion probably involve mechanical keying and chemical bonding, but are not yet fully understood; accounts of work are given by AITKEN et al. (1970), DEITZ and PEOVER (1971), WATT (1972), HORIE et al. (1976) and BREWIS et al. (1979). In fibre nomenclature surface treatment is designated by an S; e.g., HM-S denotes surface treated high modulus fibre.

Carbon fibres are supplied in tows containing from a few hundred to more than 100 000 individual, 8 μm diameter filaments. They are often coated with a small ($\sim$1%) quantity of unreacted epoxide resin size to improve their handling qualities and reduce the risk of fibre damage. Because of the variety of types available, their low density, and excellent specific properties carbon fibres are the preferred reinforcement for most high performance composite applications.

Boron fibres are formed by a chemical vapour technique in which boron is deposited on an incandescent tungsten wire core. The core makes the fibre more dense and of greater diameter ($\sim 100\,\mu$m) than glass or carbon fibre and leaves less potential for a reduced price compared with the others. The use of other core materials, for instance, silica or carbon, has been suggested, and fibres with diameters of 102 and 142 μm consisting of boron deposited on a carbon monofilament core with a coating of pyrolitic carbon are available. Boron filaments have a microcrystalline structure and the core may contain various tungsten borides produced by the boron diffusing into the tungsten during filament manufacture. This causes a residual stress to develop in the fibre leading to a state of biaxial stress at the surface which renders the fibre insensitive to minor scratches and corrosion (KREIDER and PREWO, 1972). The fibre surface is nodular in appearance due to the way that the boron has grown from neighbouring nucleation sites. The greater diameter of boron fibres makes it less easy to form them to a rapidly changing surface profile, though fibre alignment in a laminate is easier. Boron fibres have been more widely used with metal matrices, especially in the U.S.A., than have other types of fibre. Another type of boron based fibre is known as Borsic. This is a boron fibre coated with silicon carbide. This gives improved strength retention at an elevated temperature, but a reduced absolute strength, and increased fibre diameter. A useful account of some of the earlier work on boron filaments is given by LINE and HENDERSON (1969).

Aramid fibres are produced from an aromatic polyamide, poly (p-benzamide). The chemistry and properties are described by HODD and TURLEY (1978), and other interesting background information is given by MORGAN (1979). Two varieties of fibre known as Kevlar 29 and Kevlar 49 are currently produced commercially. They have similar strength but the latter has twice the elastic modulus of the former. The Kevlar 49 material is available in two grades, the one for aerospace applications having a lower tex, or weight per unit length of tow. Aramid fibres are made up of numerous microfilaments (GREENWOOD and ROSE, 1974). This structure causes the compressive and transverse properties to be poor and it is common to use another type of fibre in conjunction with the aramid to take compressive loads.

Some typical properties of glass, carbon, boron and aramid fibres are listed in Table 1. The figures are approximate and will vary, for a given material, from manufacturer to manufacturer, and with the state of development of the material, method of test, and whether the fibre is twisted or not. The tensile strength of virgin E glass fibre is reduced by handling.

Continuous ceramic fibres have much higher operational temperatures than any of the reinforcements described previously. Du Pont's polycrystalline aluminium fibre, known as FP fibre, has according to the company data sheets—DU PONT DE NEMOURS and Co. (1978)—a density of 3.95×10^3 kg m^{-3}, a

TABLE 1.[a] Filament properties.

Material	Density × 10³ Kg m⁻³	Long tensile mod. GPa	Long tensile str. MPa	Trans. tensile mod. GPa	Shear mod. GPa	Comp. str. MPa
E glass	2.54	70	3100	70	28.7	1750
Carbon						
VHM	2.0	517	1860			
HM	1.9	350	2000	12.1	13.7	
HT	1.78	230	2900	20.4	24.0	
A	1.76	215	2400			
Aramid	1.45	63 130	2800	5.38	2.0*	250*
Boron	2.63	420	3400	420	180	2300

Material	Strain at fail. %	Major Poisson's ratio	Diameter μm	Long. coeff. thermal exp. × 10⁻⁶ °C⁻¹	Long. thermal cond. Wm⁻¹ °K⁻¹
E glass	2.5–3.0	0.22	10.0	5	0.89
Carbon					
VHM	0.38	0.25	8.4	−1	140.0
HM	0.5	0.28	11.0	−0.5	100.0
HT	1.3	0.26	8.0	0.5	20.0
A	1.27	0.26	8.5	1.0	10.0
Aramid	2.0–3.0	0.34*	12.0	−2.0	0.57
Boron	0.7	0.13	102 142 203 375	2.8	–

[a] VHM = very high modulus; HM = high modulus; HT = high tensile (strength); * indicates results for a 60% composite. All figures are approximate and derived from manufacturers' data. Values tend to vary from manufacturer to manufacturer, with the state of development of a material, method of test, etc. The tensile strength of E glass is for virgin fibre, handling reduces the value.

filament diameter of 20 μm and a tensile strength and modulus of 1400 MPa minimum and 380 GPa, respectively. The compressive strength, at 7 GPa, is very high. There is good property retention up to 1400°C, and 80% of the static strength remains after 10^7 cycles. A yarn of fibre consists of 210 filaments and the material is available in tape form and can be woven after overwrapping with rayon. The greater density makes the fibre less attractive than the more common reinforcements. However, apart from using FP fibre with resin or ceramic matrices composites can also be produced with aluminium, magnesium or lead matrices though in some cases it may be necessary to modify the metal to promote wetting.

3M's ceramic fibre AB312 is an alumina boria silicon dioxide material available as continuous yarn or roving, chopped fibre, woven fabric or braided. Its properties are described by EVERITT (1977) and MILEWSKI (1978). The density, 2.5×10^3 Kg m⁻³, is lower than that of other ceramic fibres, while the tensile strength and modulus are 1720 MPa and 152 GPa respectively. It is capable of extended use at 1300°C.

A third type of material, which does not yet appear to be commercially available, is a continuous silicon carbide fibre. This has a density of 3.2×10^3 Kg m^{-3}, a filament diameter of 20 to 30 μm, a best tensile strength of up to 6.2 GPa and modulus of 440 GPa (YAJIMA et al., 1976).

Carbon fibres can be supplied in the form of a continuous tow. Glass and aramid fibres are available as rovings or yarns. The former is used for filament winding and the latter for weaving. All three fibres may be woven to give plain, unidirectional, twill or satin weave products. Plain weave fabrics have equal properties in two mutually perpendicular directions, while unidirectional weave has its principal strength in one direction with little reinforcement in the other. Further details of types of weave are given in a publication by FOTHERGILL and HARVEY Ltd. (1977). A variant of woven material is a hybrid consisting of two reinforcements, for instance carbon and glass, carbon and aramid or aramid and glass fibres. Weaving enhances transverse and shear properties, gives drapability, and the use of two materials can lead to improvements in impact and fracture behaviour.

Glass fibres can be supplied chopped into lengths of up to a few centimetres and randomly distributed in a plane, or as continuous fibres randomly spread in a plane. In both cases a binder is used to hold the fibres together and the resulting material is known as a mat. Chopped carbon fibres are available as random, aligned or aligned and needled felts. The advantages of these materials over woven or unidirectional reinforcement are improved drape, i.e., the ability to follow complex surface contours, and possibly better properties perpendicular to the felt. Clearly with either woven or chopped fibre reinforcement specific properties will be reduced.

3. Matrix materials

The matrix is, as we have already pointed out, a vital constituent of any composite material. Except in a few limited applications such as pure tensile loading, the best fibre reinforcement available is of little or no use unless there is a suitable matrix to transfer stress, protect the fibre, enhance transverse properties and improve impact or fracture performance.

Thermoplastics, including polyethylene, nylon 12, cellulose acetate butyrate, polymethylmethacrylate, polysulphone, polyethersulphone and polyacetal have been used as matrix materials, but because the temperatures at which these polymers have a sufficiently low viscosity to infiltrate the fibres are relatively high, problems arise in composite manufacture due to poor infiltration and wetting, and fibre degradation. Continuous fibre composites were prepared by PHILLIPS and MURPHY (1976) by stacking layers of carbon fibre and polymer film and then hot pressing. The most successful matrices were polymethylmethacrylate, polysulphone and polyethersulphone, which it was claimed gave as good results as could be obtained with an epoxide resin. Unfortunately severe fibre buckling occurred when specimens were deformed parallel to the fibres. There is an attraction in making a continuous fibre reinforced composite which could be moulded to shape repeatedly without a loss of properties on applying heat and it would probably find a ready market in orthotic applications. Thermoplastic pellets for injection moulding, containing short (1–3 mm) reinforcement are available and an acetal containing carbon fibre has been used to make gear wheels for use in satellite or spacecraft mechanisms. Advantages of incorporating carbon fibres in thermoplastics are the large increase in thermal and possibly electrical conductivity obtained and the improved dimensional stability. The chemical structures of some of the polymer systems referred to here are shown in Fig. 2. Normally the number average molecular weight of the complete polymer molecule will lie between 10^4 and 10^6. In some materials branching may occur.

A limited amount of work has been carried out with elastomer matrices. The high extensibility of these materials combined with their very low shear moduli (MPa rather than GPa) leads to composites with interesting and intriguing properties. A bar specimen 6.3 mm square reinforced with 60 °/₀ of unidirectional carbon fibre in a polyurethane matrix, cannot be flexed by hand but can easily be twisted through 30 or 40° about the long axis, while a thin sheet reinforced with glass fibre is very stiff in the direction of the fibres but

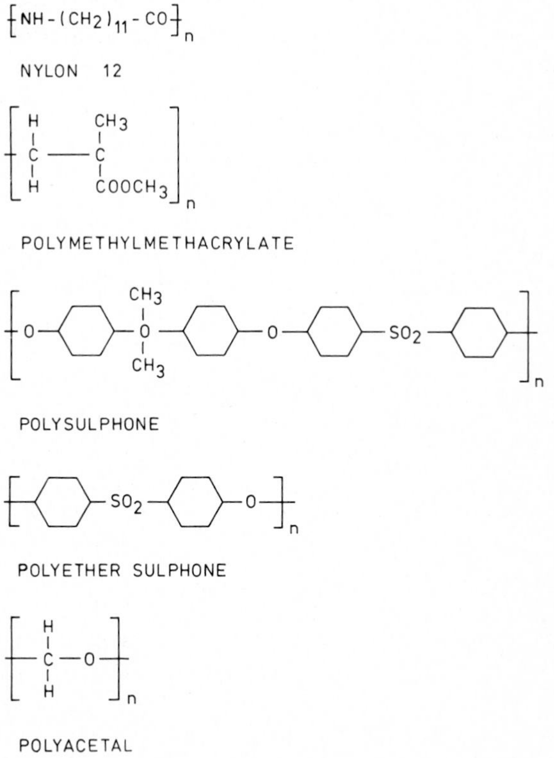

NYLON 12

POLYMETHYLMETHACRYLATE

POLYSULPHONE

POLYETHER SULPHONE

POLYACETAL

FIG. 2. Some examples of thermoplastic matrices.

transverse to the fibres it can be rolled up like a magazine or newspaper. To date urethanes, and elastomers such as Viton and Hytrel, have been employed with continuous carbon and glass fibres. Urethane is a term used to describe a large and diverse family of materials basically produced by a reaction between isocyanates and hydroxyl rich compounds. They contain the NCO group.

The variety of urethane which has been used most widely and with greatest success in our own work is based on a toluene diisocyanate with a number average molecular weight of approximately 2000, containing 4% by weight of isocyanate groups, and cured with diamino diphenyl sulphone. Viton, a copolymer of vinylidene fluoride and hexafluoropropylene, is readily available as a solution in various organic chemicals and impregnates fibres easily. Hytrel is a thermoplastic polyester copolymer consisting of hard and soft segments and impregnation techniques similar to those used for thermoplastics are required. Chemical structures of a type of Viton and Hytrel are shown in Fig. 3. The properties depend on the values of m, n, x, y, z—i.e., the lengths of the different segments in the structural units.

The most convenient and widely used matrices are thermosetting resins. On curing the initially liquid resin goes through a gel stage, and finally becomes a three-dimensionally cross-linked solid. Cross-linking improves modulus, heat and chemical resistance, but may lead to reduced impact strength and elongation at break. The main limitation of thermosets is that few are capable of maintaining their properties if continually exposed to temperatures in excess of 150–200°C. However, at 200°C unprotected carbon and aramid fibres are approaching the limiting temperature at which they are stable, if exposed for any length of time, so that the limitation is not too serious, and is far outweighed by the advantages of ease of handling, processability, and variety of manufacturing methods available with thermosets.

The principal systems employed are epoxides, unsaturated polyesters and polyimides. In addition, vinyl esters, phenolics, furanes, silicones. Friedel Crafts systems and bis dienes have been suggested as possible matrices (JUDD and WRIGHT, 1978a).

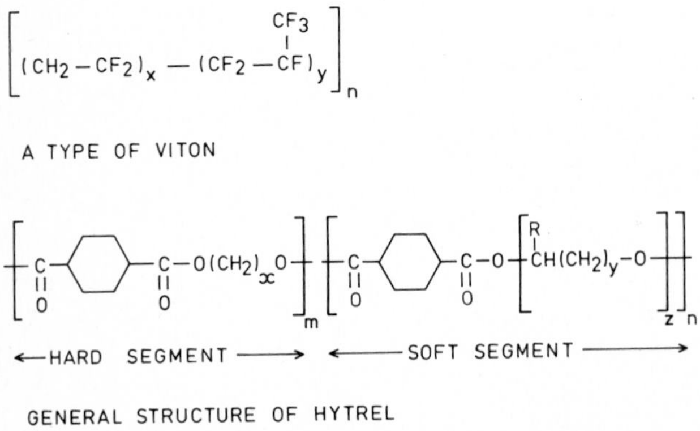

A TYPE OF VITON

GENERAL STRUCTURE OF HYTREL

FIG. 3. The structure of Viton and Hytrel elastomers.

3.1. *Epoxides*

Epoxide resins are the most favoured for use with carbon fibres, and in high performance applications, because of their good mechanical properties, low shrinkage, ability to bond to other materials, environmental stability, and many varieties available. Standard works on the chemistry, cure mechanisms and uses of epoxides are LEE and NEVILLE (1957) and POTTER (1970), while GARNISH (1972) has provided a succinct introductory article, and OSWITCH (1975) has concentrated on high temperature performance. All epoxides are characterised by the presence of the epoxide group. This consists of two carbon atoms and one oxygen atom arranged in a 3 membered ring (see Fig. 4). The group is constrained and reactive. The reactivity depends upon the position of the group in the molecule and on steric factors. The opening of the epoxide ring by a curing agent leads to cross linking and, ultimately, the production of a hard, insoluble, solid. When fully cured all the epoxide groups should have reacted but this probably does not occur in practice.

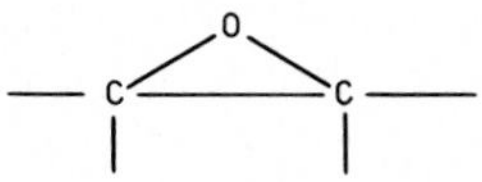

FIG. 4. The epoxide group.

The most common epoxide resin is based on bisphenol A, or 2,2 bis (4 hydroxyphenol) propane, and epichlorhydrin. The simplest member of the diglycidyl ether of bisphenol A series of epoxide resins is shown in Fig. 5. Other members of the series have a higher molecular weight. Other varieties of resin include epoxy novolac, a solid at room temperature, and cycloaliphatic diepoxides. Both of these, because of their multifunctionality produce, on curing, tightly cross linked structures with excellent resistance to heat and chemical environments and improved mechanical properties. Another type with improved heat resistance, faster reactivity and improved bonding properties is based on bisphenol S (SPITZBERGEN et al., 1971).

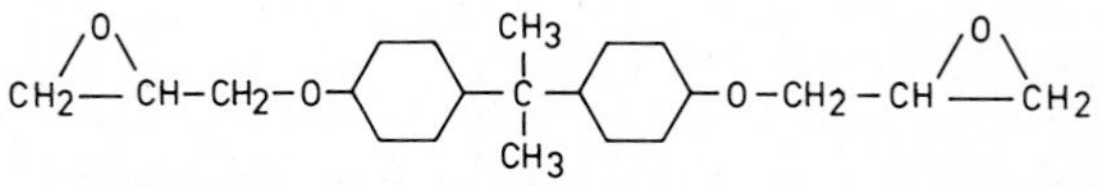

FIG. 5. The simplest epoxide resin molecule based on bisphenol A and epichlorhydrin.

A typical diglycidyl ether of bisphenol A is either a liquid or low melting point solid. For composite work a liquid is preferred because of simpler processing. A typical example consists of a mixture of molecules having a mean molecular weight of 300–400. Apart from viscosity and molecular weight uncured epoxide resins are characterised by the epoxy and hydroxy equivalents, i.e., the weight of resin, in kilogrammes, containing one kilogramme equivalent of either species. The epoxy or hydroxyl content is the amount of either group present as equivalents per kilogramme of resin.

Attempts have been made by SOLDATOS et al. (1969) and CHAMIS et al. (1973) to tailor the matrix to obtain optimum composite properties. The latter authors noted that the area under the matrix stress–strain curve up to 1% is a good measure of the matrix contribution to composite strength, and the initial modulus a useful parameter for relating matrix properties to the longitudinal, compressive and flexural properties of the composite. Matrix ultimate tensile strength, elongation and toughness do not, however, appear to correlate with unidirectional composite properties. BUSSO et al. (1970) have studied the relationship between the chemical structure of an epoxide resin and its mechanical properties, but so far no attempt appears to have been made to use this information to produce optimum performance composite systems.

To cure the epoxide it is necessary to use a hardener and possibly accelerator and, often, heat the constituents, in the correct proportion, for an hour or more at a temperature of 100 to 120°C to gel the resin. For 100 parts by weight of resin, between 10 and 80 parts of hardener and 2 or 3 parts of accelerator are required. To complete the cure heating may be continued for a period of 2 to 24 hours at temperatures up to 180°C. Curing cycles to maximise a given property are usually determined empirically by the resin manufacturers, and will differ not only for different resin types but also for differing final properties.

A wide range of curing agents may be employed with epoxides. Reactive ones, including primary and secondary aliphatic and aromatic amines and acid anhydrides enter into the three dimensional polymer structure, while catalysts such as strong acids and bases promote homopolymerisation. Reactions may be seriously limited by steric hindrance, i.e., non-reactive groups shielding or blocking in space, reactive ones. The final mechanical, thermal, electrical and environmental properties depend on the resin, hardener, catalyst and cure cycle employed. For curing at low temperatures, which is advantageous if it is desired to minimise thermally induced strains in a fibre reinforced composite, multifunctional aliphatic amines or polyamides are useful. Aromatic amines are less reactive and need higher cure temperatures, but give better thermal, mechanical and environmental properties. Liquid anhydrides usually require an accelerator and high curing temperatures but give systems with a long pot life, which wet the fibres well and are excellent for filament winding applications. Table 2 illustrates some of the different hardeners that may be used, pot lives to be expected and the effects of differing cure schedules.

Epoxide resins including those modified with urethanes possess the extremely useful property of being able to be part reacted, or B staged, thus leading to the production of preimpregnated fibre.

The low transverse strain to failure of unidirectional composite materials can be a serious limitation to their use in practice. In filament wound structures high stress magnification can occur at the cross-over point of two fibres, allowing weepage if a fluid is contained, at a stress well below that to cause the

TABLE 2.[a] Details of curing agents, schedules, and pot lives for a diglycidyl ether of bisphenol A epoxide resin.

Hardener	Wt. hardener/ 100 parts resin	Cure schedule Hrs.	Cure schedule °C	Heat distortion temp., °C	Pot life °C	Pot life time	Possible use of system
Diethylamine triamine	11	2 +2	25 100	100	25	$\frac{3}{4}$ hr	room temp. lay-up
4 4′ diamine diphenyl sulphone	30	3 +4 +32	125 175 200	180	25	1 yr	for prepreg. systems
Hexahydro phthalic anhydride + accelerator	65	1 +4	100 150		25	7 days	good high temp. props. and low viscosity
Methyl nadic anhydride + accelerator	90	1 +15	100 260	200	25 90	7 days $2\frac{1}{2}$ hrs	high temp. laminates
BF$_3$ adduct	3	4 +4 +4	120 150 200	62 113 170			
Versamid 125 (a polyamide)	53	24 +1$\frac{1}{2}$	room temp. 150	83			a more flexible resin

[a] The data were kindly supplied by Dr. K.A. Hodd, Brunel University, Middlesex, U.K. The cure schedule affects the properties of the cured resin. The effect is often categorised by quoting the heat distortion temperature. This is the temperature at which a standard test bar deflects 0.254 mm under a stated load.

reinforcement to fail. The use of high strain characteristic resins which adhere well to the reinforcement, though an obvious solution to these problems, has apparently only recently been tried.

It is possible to blend an epoxide with a reactive constituent such as a urethane and either use the same curing agent, e.g., an aromatic amine for both components or, if hydroxyl groups are present in the epoxide, react these with isocyanate groups in the urethane. The two constituents are believed to be intimately cross-linked leading to changes in the properties of the resin and composites prepared from it, including an enhanced transverse tensile strain to failure in unidirectional carbon fibre composites (WELLS and HANCOX, 1978). PENN et al. (1977) and RINDE et al. (1977) have described the use of rubberized epoxide resin prepared from several constituents including a low viscosity epoxide resin, a reactive flexibiliser, a carboxy terminated rubber polymer and a mixed aromatic amine curing agent. The system has been used to produce wound pressure vessels with excellent performance. Recently CHRISTENSEN and RINDE (1979) have suggested that, because of the difference between the fracture modes of restrained and bulk resin, it is the low initial modulus of the resin rather than its ultimate elongation that is important in improving transverse failure strain.

3.2. *Unsaturated polyesters*

Although anyone reading the current literature on high performance carbon, aramid, and boron fibre reinforced plastics might be tempted to think that epoxides or more exotic high temperature performance resins are exclusively employed, most general reinforced, thermosetting, plastics work is carried out using unsaturated polyester resins. These are cheaper than epoxides, easy to work with and can be cured rapidly at ambient or elevated temperatures with, usually, no need for a subsequent post cure. They are extensively employed with glass fibres. The production and properties are described by PARKYN (1972) and BRUINS (1976). The characteristic repeated chemical group, which occurs in the main chain is –[CO–O]–, together with unsaturated aliphatic groups which provide the sites for cross linking reactions with the styrene monomer which is mixed with the polyester to form the resin. Cast resin properties are similar to those of epoxides but the shrinkage on curing is greater and, being non-polar, polyesters do not bond so well to reinforcement. As with epoxides many variants are available.

Polyesters are produced by a condensation reaction between unsaturated and saturated dibasic acids or anhydrides and saturated diols. The properties of the base resin can be altered by changing the nature and ratios of the components. General purpose resins are produced from phthalic or maleic anhydride and propylene glycol, see Fig. 6 for details of these components and the resin produced, while special types are produced by replacing the phthalic anhydride and/or the diol with other components. Orthophthalic acid is used in the majority of general purpose polyesters, isophthalic acid where improved resistance to cracking and chemicals is required, and a bisphenol for a resin with the highest resistance to a wide range of chemicals. To enable the resin to cross link, through the unsaturated bond, and to reduce the viscosity the polyester is dissolved in a monomer such as styrene and stabilised against premature gelation.

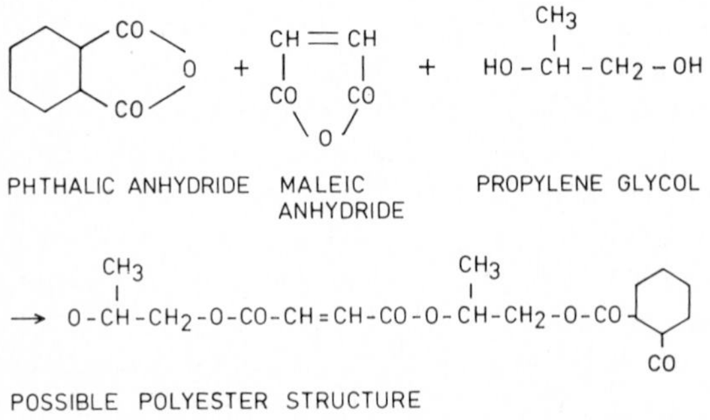

PHTHALIC ANHYDRIDE   MALEIC   PROPYLENE GLYCOL
                       ANHYDRIDE

POSSIBLE POLYESTER STRUCTURE

FIG. 6. Components to give a possible polyester structure.

Other monomers are listed by MILES and BRISTOW (1965). Cure is initiated by a free radical mechanism, and is rapid and exothermic. Free radicals are supplied by a peroxide, e.g., benzoyl peroxide or methyl ethyl ketone peroxide. The former is triggered by the action of heat, the latter by mixing with an accelerator, e.g., cobalt napthate which causes a reaction at room temperature. Approximately 1% of peroxide and 0.1% of accelerator are required.

3.3. Polyimides

The upper working temperature for epoxies, for continuous exposure is about 150°C, with periods of up to a few hours at 170–180°C, and possibly slightly higher for diepoxides (OSWITCH, 1975). For polyesters the same author suggests that a special resin containing aromatic rings in its molecule and cross linked with vinyl toluene monomer would exhibit sustained temperature resistance between 150 and 200°C, and short term stability up to 250°C. To improve on these figures it is necessary to use a compound based on an aromatic or heterocyclic ring structure. Among the compounds of this type which are suitable as matrices polyimides appear the most promising (JUDD and WRIGHT, 1978a; OSWITCH, 1974). These resins can be used at temperatures of 300°C or slightly above. An example of one type of polyimide molecule is shown in Fig. 7.

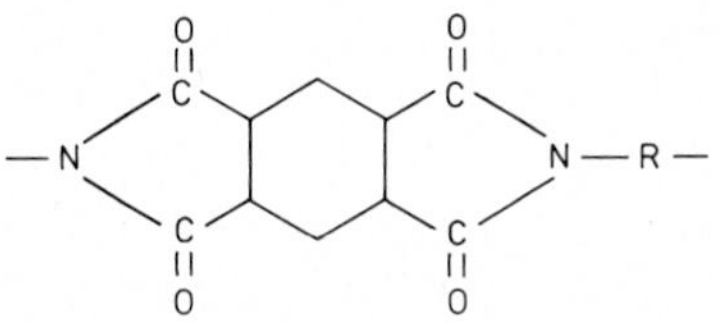

FIG. 7. An example of a polyimide.

Unlike epoxides and polyesters, polyimides may be thermoplastic in nature. They can be prepared by reacting a dianhydride with an aromatic amine at elevated temperature. Careful process control is required. Unfortunately the reaction produces two molecules of water for each polymer molecule, and at the temperatures involved this turns to steam causing castings and composites to have a high void content, as much as $10-15\,^{v}/_{o}$, which significantly reduces mechanical performance. In addition the resins are subject to attack by bases and exhibit relatively great water absorption.

In attempts to overcome these difficulties various imido structures with different terminations which do not produce volatiles when polymerisation takes place, have been devised. If epoxide groups are used for termination the polymer can be cross linked through the agency of a suitable catalyst, such as a Lewis acid. Some of the imido epoxides are soluble in chloroform and can be used to prepare preimpregnated fibre tapes and sheets. Incorporating the epoxy group causes some loss of thermal stability compared with conventional polyimides. The moulding cycle is also lengthy, e.g., 12 hours at 200°C. MARRIOTT (1974) has described additional polyimides formed by combining aromatic diamines with bismaleimides. The products are either stable B staged materials, or on the application of further heat can be fully cross linked.

3.4. Other resin systems

Vinyl esters (JUDD and WRIGHT, 1978a, STAVINOHA and McRAE, 1972, FERRARINI et al., 1979), can be prepared by an addition reaction between an epoxide and a carboxylated monomer, e.g., acrylic acid. Cross-linking is through a styrene monomer using free radical initiation as for a polyester. The epoxide structure in the backbone gives strength and the unsaturation at the ends leads to more even cross-linking. The resins wet glass well. An example of a resin is shown in Fig. 8.

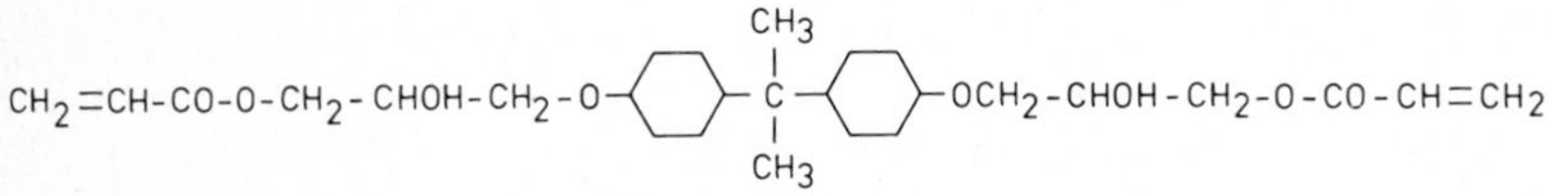

$$CH_2=CH-CO-O-CH_2-CHOH-CH_2-O-\!\!\!\!\bigcirc\!\!\!\!\overset{\overset{\displaystyle CH_3}{|}}{\underset{\underset{\displaystyle CH_3}{|}}{C}}\!\!\!\!\bigcirc\!\!\!\!-OCH_2-CHOH-CH_2-O-CO-CH=CH_2$$

FIG. 8. An example of a vinyl ester resin.

Other resins which are used with glass fibre reinforcement are furanes, silicones, phenolics and Friedel Crafts systems. These are discussed by JUDD and WRIGHT (1978a) and OSWITCH (1975). Cured furanes based on the structure shown in Fig. 9 have excellent abrasion, and chemical resistance in non-oxidising conditions but tend to be brittle. Silicones, which are expensive compared with most other resins, are generally used in solvent form for making preimpregnated fibre sheet. Cross linking, with the elimination of water, is achieved by heating in the presence of a suitable catalyst. The postcure at a temperature of up to 250°C can last for as much as 24 hours. While mechanical properties are lower than those that can be achieved with other resins, due in part to the flexible nature of the polymer chain, the thermal stability is very good and laminates can operate for long periods at temperatures of 250°C and possibly even as high as 500°C.

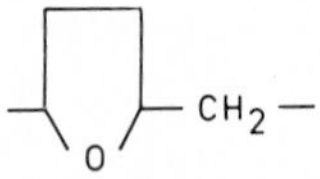

$$-\!\!\!\!\overset{\displaystyle\square}{\underset{\displaystyle O}{\bigvee}}\!\!\!\!-CH_2-$$

FIG. 9. The furane structure.

Phenolic resins are the oldest and some of the most widely used in general, but not in high performance, moulding work. Though they tend to be brittle and exhibit poorer adhesion to glass fibres than polyesters or epoxides they are cheap and phenolic components have good thermal stability being capable of continuous service at 200°C.

Friedel Crafts resins which are produced by the reaction of aromatic and alkyl compounds in the presence of stannic or aluminium chloride have good

TABLE 3.[a] Room temperature properties of bulk cast resins.

Material	Density $\times 10^3$ Kg m^{-3}	Tensile mod. GPa	Tensile str. MPa	Shear mod. GPa	Strain at failure %	Poisson's ratio	Flexural mod. GPa	Flexural str. MPa	Shrinkage %	Coeff. thermal exp. $\times 10^{-6}$ °C^{-1}	Usable temp. range °C
Epoxide	1.1–1.2	2.0–5.0	55.0–120.0	1.5	1.5–8.5	2.5–3.9	2.5–3.9	70.0–130.0	1.0–5.0	55.0–70.0	150+
Unsat. polyester	1.1–1.4	1.2–4.0	42.0–90.0	1.0–2.0	2.0–6.0	0.35–0.36	3.5–5.6	80.0–140.0	5.0–12.0	60.0–70.0	150+
Vinyl ester		3.0–4.0	65.0–90.0	–	1.0–5.0	–	–	100.0–135.0	1.0–6.0	–	150+
Polyimide	1.43–1.89	3.1–4.9	70.0–110.0	–	1.5–3.0	–	–	70.0–120.0	–	–	250+

[a] The values indicate the range of properties and will depend markedly on the exact type of resin and hardener, and cure schedule.

TABLE 4.[a] Chemical and fire resistance unreinforced resins.

Type	Condition				
	Acids	Alkalis	Solvents	Water	Fire
Epoxide	Unaffected except by strong acids	Unaffected except by wet ammonia, caustic soda	Attacked by chlorinated hydrocarbons and ketones	Absorbs up to a few %	Will burn
Polyester	Fairly resistant, except to strong acids	Attacked	May be attacked	Absorbs up to a few %	Will burn
Vinyl ester		Better than polyester		Low absorption	Will burn
Polyimide		Attacked		May absorb a considerable amount	
Silicone	Good resistance	Attacked by strong alkalis	Attacked	Attacked by steam	Resistant
Furane	Attacked by strong acids	Attacked by strong alkalis	Good resistance	Good resistance	Resistant
Friedel-crafts	Good resistance	Good resistance	Good resistance		Resistant
Phenolic	Attacked by strong acids	Attacked by strong alkalis	Good resistance	Absorbs water	Resistant

[a] The comments are general and indicative of performance. Much more detailed information for specific systems is available from manufacturers.

TABLE 5.ª Some laminate properties.

Material	Density × 10³ Kg m⁻³	Long. ten. mod. GPa	Trans. ten. mod. GPa	Shear mod. GPa	Long. ten. str. MPa	Trans. ten str. MPa	Long. comp. str. MPa	Flex mod. GPa	Flex str. MPa	ILSS MPa	Principal Poisson's ratio	Long. coeff. therm. exp. × 10⁻⁶ °C⁻¹
Unid. E glass 60%	2	40	10	4.5	780	28	480	35	840	40		4.5
Bidirectional E glass cloth 35%	1.7	16.5	16.5	3	280	280	100	15	220	60		11
Chopped strand mat E glass 20%	1.4	7	7	2.8	100	100	120	7	140	69		30
VHM CFRP 60%	1.67	290			600	45	520	250	800	52		-1.5
HT CFRP 60%	1.55	140	6.9	5.0	1620	34	1200	124	1680	80	0.33	0.5
A CFRP 60%	1.58	130		4.8				129	1600	96		1.5
Boron 60%	2.1	215	24.2	6.9	1400	63	1760			84	0.3	4.5
Aramid '29' 60%	1.38	50	5	3	1350		238	51.7	535	44		
Aramid '49' 60%	1.38	76	5.6	2.8	1380	30	276	70	621	60	0.34	-2.3

ª In all cases the resin used was an epoxide and the values refer to room temperature. The data is characteristic rather than absolute (see also the footnote to Table 1).

thermal stability, up to temperatures of about 250°C, are easier to process than polyimides and have a good resistance to chemicals. The basic structure, which must be cross-linked to give the finished product, is shown in Fig. 10.

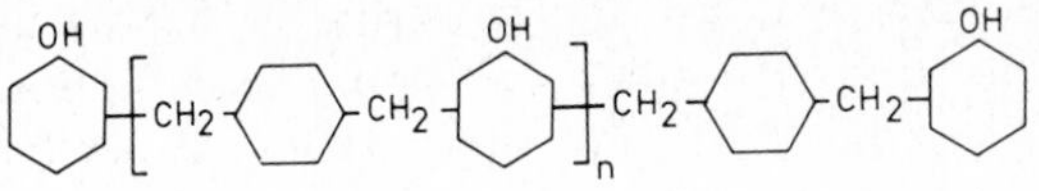

FIG. 10. An example of a Friedel Crafts resin.

Some mechanical and thermal properties of bulk cast resins are shown in Table 3, chemical properties in Table 4 and those of laminates in Table 5. Because of the effects of fabrication and starting materials on the final cast resin or composite properties, the figures should be taken as indicative of what can be obtained rather than absolute values. Resin manufacturers may list the chemical performance against 200 or more specific chemicals, so clearly Table 4 is extremely general, and performance will depend on temperature.

4. Preimpregnated fibre material

Two extremely useful ways of combining fibres and resin to give an inter-mediate material which can be stored and used for laminating or moulding are the techniques for producing preimpregnated fibre and a moulding compound.

In preimpregnation continuous unidirectional fibres, mixtures of fibres, or woven cloths are spread out on a non-stick substrate and impregnated with a metered amount of epoxide resin, hardener and catalyst which is partially cured or B staged. Alternatively an epoxide resin film may be used. Depending on the resin system employed the product may have to be stored at −20°C, although a system containing urethane and a low reactivity hardener can be kept at room temperature for many weeks with no deterioration. The sheets or tapes produced are upwards of a fraction of a millimeter thick, depending on the type of fibre feedstock and how it is spread out. To produce a laminate layers of preimpregnated material, with the backing paper removed, are stacked in or on a mould with the fibre alignment of successive layers arranged in a definite order. This is usually symmetric about a centre plane (e.g., 0°, 90°, 90°, 0°) though it need not be so. The layers are consolidated and cure effected as described later in Sections 7.1.2 and 7.2.1. Preimpregnated material is ideal for preparing laminates and is extensively used in high performance products.

5. Moulding compounds

Polyesters cannot be B staged, but are used to make moulding compounds, which can be regarded as analogous to epoxide preimpregnated fibre material.

The compounds are very widely used in the mass production of a variety of components up to several metres in dimension and possibly containing very complex detail. In many cases this method of manufacture is cheaper and better than fabricating in metal.

There are four types of moulding compound; bulk or dough (BMC, DMC), sheet (SMC), high performance sheet (HMC), and very high performance sheet (XMC). All contain short, randomly dispersed or continuous glass fibre reinforcement in a matrix of polyester resin (possibly vinyl ester), with catalyst, accelerator, up to 50% of particulate filler, 1–3% of chemical thickening agent, release agent, pigment, and shrink resisting compounds. The latter are thermoplastics added to give a smoother surface finish with less shrinkage. The reinforcing fibres are 6–30 mm long in BMC, and 25–50 mm long in SMC and HMC. BMC and SMC contain from 20 to 50 $^{w}/_{o}$ of fibre (MAAGHUL and POTKANOWICZ, 1976), and XMC up to 80 $^{w}/_{o}$ of continuous fibre (ACKLEY, 1976, ACKLEY and CARLEY, 1979). It is customary to quote glass fibre content in terms of weight percentage but carbon fibre content in terms of volume percentage. Glass fibre weight percentages of 20, 50 and 80% are approximately equivalent to volume percentages of 10, 32 and 65% respectively. Up to 25 $^{w}/_{o}$ of chopped glass is sometimes added to XMC to improve transverse properties, while a small quantity of continuous rovings may be used to upgrade SMC (EVANS and NEMETH, 1979). SMC is prepared by a callendering process as a strip a meter or so in width and about 5 mm thick. Layers of reinforcement and resin mix are laid on a polythene sheet, compacted and wound onto a drum. A variety of grades with differing glass contents and made with various types of resin is available. The method of preparation of HMC is similar except that all or most of the filler is omitted from the resin mix. With XMC the continuous fibres are handled by a filament winding technique. The material must be stored under controlled conditions for a period ranging from a few days to several weeks to allow the resin to thicken. This process, known as maturation, continues slowly and indefinitely, and there is a specific time, the moulding window, during which the compound can be successfully used. This may be of 2 weeks or more in duration. Thickening has been reviewed by FEKETE (1970, 1972). It is believed to involve the reaction of alkaline earth (Ca, Mg) oxides or hydroxides with polyester molecules, thus,

$$2(X-\overset{\displaystyle O}{\overset{\displaystyle \|}{C}}-C-OH) + MgO \rightarrow X-\overset{\displaystyle O}{\overset{\displaystyle \|}{C}}-C-O-Mg-O-\overset{\displaystyle O}{\overset{\displaystyle \|}{C}}-C-X + H_2O$$

where X is the polyester and C a divalent carbon atom (NUSSBAUM and CZARMOMSKI, 1970). However, the exact mechanism is not fully understood (ESPENSHADE and LOWRY, 1971). An interesting variation of SMC has been described by MAGRANS and FERRARINI (1979). A specially formulated, styrenated, isophthalic polyester resin with a vinyl ester resin and urethane, in addition to the other additives is used. This gives a system which matures in 24

TABLE 6.[a] Properties of moulding compounds.

Material	Density ×10³ Kg m⁻³	Long. tensile mod. GPa	Trans. tensile mod. GPa	Long. tensile str. MPa	Trans. tensile str. MPa	Flex. mod. GPa	Flex. str. MPa	Shear mod. (in plane) GPa	Inter. lam. shear str. MPa	Principal Poisson's ratio	Coeff. thermal exp. ×10⁻⁶°C⁻¹
BMC	1.93	10.5	8	45–70	70	8–12.6	100–175	6	41	0.31	24
SMC	1.75–1.80	11–16		60–112		8–14.7	245		62		16.8
HMC	1.9	18.2–21		210–280		14.7–16.1	385–420				
XMC		35–42		560–630			1050–1085		59–63		

[a] These are typical figures taken from MAAGHUL and POTKANOWICZ (1976), ACKLEY (1976), MAGRANS and FERRARINI (1979), DENTON (1979), and various commercial sources. They refer to BMC and SMC with 20–50 w/o chopped glass, and HMC/XMC with 60–80 w/o chopped and continuous glass.

hours and has apparently an indefinite life. Processing conditions for moulding compounds are usually of the order of 1–3 minutes press time at a temperature of 100°C plus. Very thick moulding may require a substantially greater time. Some properties of moulding compounds are listed in Table 6.

6. Costs

An important parameter is deciding on which composite material to use in a particular application is that of cost. The majority of resins and two of the fibres are based on petrochemicals, while all types of fibre require a considerable energy input. This does not necessarily mean that the energy required to produce a unit quantity of finished product will be higher for a composite than for a traditional material, such as a metal, especially if the improved specific properties of the fibre reinforced resin are taken into account. CHEREMISINOFF and CHEREMISINOFF (1978) note that in terms of energy input and feedstock per unit volume, a polyester is half as costly as steel and one third as costly as magnesium. Nevertheless, because of the scarcity of petroleum products and inflationary pressures in the world today costs are subject to rapid increases. The figures presented in Table 7 were correct as of 01/12/1979, and in most cases the relative figures would be expected to remain approximately constant with time. The one exception is possibly carbon fibres which may drop in price relative to other reinforcements as the scale of production is increased. The prices quoted for all materials will vary from supplier to supplier, will depend

TABLE 7. Cost of reinforcement and resin.

Material	Cost £/kg	Material	Cost £/kg
VHM carbon fibre	203	Epoxide, med. functionality	1.65
HM carbon fibre	85–163	Epoxide, high functionality	7.40
HT carbon fibre	41–65 [a]		
A carbon fibre	50	Anhydride hardener for epoxide	3.0
		Amine hardener for epoxide	2.49–7.75
		Polyester	0.9–1.37
Aramid fibre	10	Vinyl ester	1.53–1.79
E glass	1.17 [b]	Polyimide	11.21
R glass	8–9 [b]	Silicone	4.00
BMC (20 w/o glass)	0.7–0.75 [b]	Furane	1.4
SMC (25 w/o glass)	0.93–1.16 [b]	Phenolic	0.69–0.87

[a] This is for 3000 filament fibre, 1000 filament fibre is 138–154 £/kg, while 15 000 or 30 000 filament HT-S fibre may cost as little as 35 £/kg when in production in 1980.
[b] Based on 1 tonne minimum. Resin figures are usually based on 1 tonne minimum quantity. Adding 5 w/o glass fibre increases the price of SMC by about 0.05 £/kg. Costs were approximately correct on 01/12/1979.

on whether the material has to be imported, the quantity purchased, the number of fibres per tow, exact grade of resin, etc.

Preimpregnated fibre costs from 1.5 to 3.0 times that of the raw materials, and increases in cost as the thickness of the sheet is reduced. The materials cost of a finished article will be that of the prepreg or moulding compound, or for a wet lay-up, the cost of the reinforcement plus that of the resin and hardener. The contributions of catalysts and accelerators to the price are less important because these are used in much smaller quantities than the other components.

7. Manufacturing methods

To fabricate fibre reinforced composite articles it is necessary to impregnate fibres with a resin or put preimpregnated fibres or moulding compound in a mould and, often, apply heat and pressure. The types of material and forms these take, and manufacturing processes used, are chosen to give or approach as nearly as possible the required performance in the finished article, at the minimum cost. For success, the design should not be a copy of one used for a similar metal product, but a fresh approach based on a full realisation of the anisotropic or pseudo-isotropic nature of the material and its failure characteristics. In addition, the resin system must be carefully chosen since this is the component that is initially exposed to the environment, and manufacturing should be undertaken by properly trained and supervised personnel using the correct equipment. Failure to pay attention to these points can easily lead to a disappointing and inferior product.

All fabrication methods require a mould, be it only a sheet of cellophane laid on a piece of wood, which enables a specific shape to be formed and retained during curing. Open moulds consists of a male or female part. Minimum or zero pressure is applied during lay-up or cure. The finished product has one side with a smooth finish. The moulds are easy, or relatively so, to produce and can be made from simple materials including plaster, casting compounds, filled resins and wood. Closed moulds consist of a male and female component and so the product has a good overall finish. Pressure must be used to close the mould and consequently the mould must be made to the requisite tolerance from a robust material which can withstand any temperatures applied without distortion. Some form of steel is frequently used. The artefact need not necessarily be produced in one operation. For instance an antenna dish for use on a spacecraft might be made by moulding the skins separately and then attaching them to a honeycomb core with adhesive film.

With all types of mould provision must be made to free the product when it is cured. A silicone or PTFE release agent, or even grease, is applied to all mould surfaces and baked on if necessary. In addition the mould should be constructed so that it can be easily disassembled. In the following discussion it will be assumed that the mould has been properly designed and treated so that it can always be released without causing damage to the moulded component.

7.1. Open moulding

7.1.1. Hand lay-up

This is the simplest and oldest method of moulding fibre reinforced plastic (FRP). Fibres in the form of a mat of short or continuous material are laid on, or in, the mould and the resin system worked in using simple hand tools. Woven cloths and unidirectional fibres are more difficult to handle on contoured surfaces and consequently are not used so frequently. The resin used is a room temperature curing polyester or possibly epoxide. The styrene monomer is possibly carcinogenic and hence with polyester resin care must be taken to provide adequate ventilation to minimise the amount of styrene inhaled by operatives. At present in the U.K. the permitted level of styrene monomer is 100 parts per million, but it is intended to reduce this level to 50 parts per million in 1981. Similar comments apply to the U.S.A. Epoxide systems may cause dermatitis and proper precautions, detailed by the resin manufacturers should be taken when handling these materials. Though low cost moulds are suitable and there is no limit on the size of article produced, the hand lay-up method is labour intensive, quality may be variable and dependent on the skill of the operator, and the volume fraction of fibre incorporated is not as high as in press mouldings. To lower the void content fine glass beads or other fillers may be worked into the laminate. Polyester resins exotherm on curing and this limits the thickness of material that can be laid-up in one operation to a few millimeters. The surface finish may be improved by a gelcoat. In this process a sub-millimeter thickness of unreinforced resin is applied to the mould and allowed to gel before the reinforcement is added.

A variant on hand lay-up is the spray-up process. Here glass rovings are chopped into short lengths and sprayed onto the mould surface together with a suitable resin system, such as a low viscosity room temperature curing polyester, and cured 'in-situ'. The method is capable of a high degree of automation, though manual consolidation of the fibre and resin may still be required. The capital outlay is greater than for hand lay-up and the product may again depend on the skill of the operator.

Both methods are very widely used because of their simplicity but neither is really capable of producing high performance composites in the sense in which the term is used in this article.

7.1.2. Vacuum bag, pressure bag, and autoclave moulding

In these methods overall pressure and possibly heat are applied to the material laid-up in, or on, the mould. In the vacuum bag process chopped or woven glass fibres and resin or preimpregnated glass, carbon or aramid fibre, or a mixture of one or more of these, are applied to the mould which is then sealed from the atmosphere by means of a rubber or plastic sheet. A vacuum is produced causing the sheet to press evenly on the fibre thus consolidating it. Heat can be applied by means of steam or electrical or radiant heating of the mould or bag. The application of heat reduces the resin viscosity causing it to flow before gelling occurs, and it is necessary to insert a porous bleeder cloth, to absorb excess resin, between the preimpregnated fibre and non porous release film. Sometimes a pressure spreading plate is used to ensure that the vacuum bag presses down evenly on the mould. Because of the longer time required for lay-up and preparation a resin or preimpregnated fibre system with a reasonable lifetime at room temperature is required. The method is suitable for producing high performance, flat or curved, composite sheets and sandwich panel sections with cores of plastic or aluminium honeycomb or end grain balsa wood. A schematic diagram of a vacuum bag mould lay-up is shown in Fig. 11.

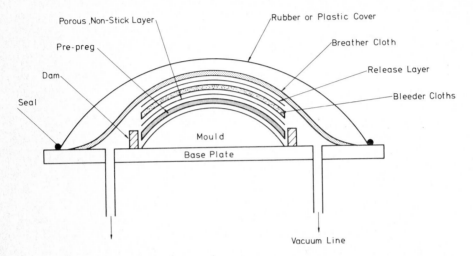

FIG. 11. A vacuum bag moulding assembly.

Pressure bag moulding is an alternative method in which the materials are laid up in a female mould. This is sealed with a rubber bag which is pressurised with steam or air. In the latter case heat is supplied separately. An autoclave or heated, cylindrical, oven which is capable of being pressurised may be used in conjunction with a pressure bag moulding facility.

7.2. Closed moulding

7.2.1. Matched die

The methods described previously use an open mould. To improve the fibre loading and hence performance of a composite article, and obtain, if desired, uniform thickness sections, and good reproducibility a closed mould consisting of a male and female part must be employed. In addition both surfaces of the item will be smoothly finished. It should be made from metal with tolerances consistent with those allowed on the finished article. Gaps or vent holes should be left to allow excess resin to be squeezed out during the moulding process. The mould has to be closed in a press or filled under pressure. While the resin may be cold cured it is usual to heat the mould directly or through contact with the press platens. Fibre mat, cloth or preimpregnated fibres can be used for reinforcement. One problem is that fibres tend to move from their predetermined positions when the mould is heated and closed. It is for this reason that it is difficult to make unidirectional, aligned fibre products with a cross section of more than 5×5 mm using a dry fibre and liquid resin by this method unless special precautions are taken. Closed moulding techniques are used extensively in the production of items from moulding compounds. Large units with considerable detail, and dimensions of several meters can be produced at the rate of one every two or three minutes in this way, and considerable effort is being made to reduce the dwell time to seconds. A simple example of a matched die mould is shown in Fig. 12. The top of the mould is closed to stops and excess resin escapes at either end. A gel coat could be added if desired.

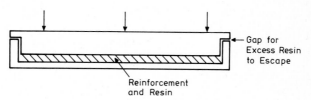

FIG. 12. Matched die moulding.

7.2.2. Resin injection

This technique, also known as resin transfer moulding, can be used for large articles including boat hulls, smaller components, and for long production runs. The reinforcement, chopped strand mat or a fibre cloth is placed between the two parts of the mould. This is then closed and an epoxy, vinyl ester or polyester resin of suitably low viscosity injected into the mould under pressure or induced to flow in under vacuum. Heat may be supplied for curing if required. A schematic diagram of the method is shown in Fig. 13. A health advantage of this, as of all closed mould methods, is reduced styrene monomer emission. Further details of the resin injection technique are given by COUDENHOVE (1979), MARSH et al. (1979) and VACCARELLA (1979).

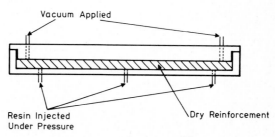

FIG. 13. Resin injection moulding.

7.3. Pultrusion

Unlike other methods pultrusion is a continuous forming process. Glass, carbon or aramid fibres are wetted in a resin bath and pulled through a suitably shaped die. The temperature and pulling speed are arranged to be such that the resin gels on leaving the die, and the product may be post cured later. The profile produced can be cut into lengths, or if sufficiently pliable, wound onto a drum. Epoxides and polyesters have been used. It is possible to make hollow profiles, and by incorporating a cloth or braid add hoop reinforcement to an otherwise unidirectional fibre composite. Pulling speeds vary from a few centimeters to a meter or more per minute, and multiple head dies may be used to speed up production. It is possible to produce sections with a width or depth dimension of 50 cm plus. A schematic diagram of a pultrusion machine is shown in Fig. 14. The post cure oven may be omitted. Earlier work on the method has been described by MEYER (1970) and more recent developments by MARTIN (1978).

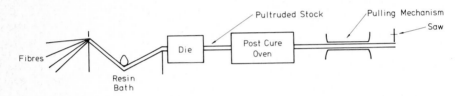

FIG. 14. A pultrusion rig.

7.4. Filament winding

Although we have devoted a complete chapter (Chapter II) in the present volume to this important manufacturing technique, it is mentioned here for completeness.

The method consists of winding continuous glass, carbon or aramid fibres onto a rotating mandrel, from a moving carriage, in a specific predetermined pattern which is calculated to satisfy the loads experienced by the finished article. Filament winding is employed for the production of pipes, storage tanks, pressure vessels, etc. The fibre is usually wetted before winding and laid down under tension. Then the artefact is cured and the mandrel, which may be a solid reusable one, inflatable or sacrificial, removed. Epoxide and polyester resins are used but whatever type is chosen must have a sufficiently low viscosity to wet out the reinforcement and a pot life, at the working temperature, of several hours.

It is possible to wind with preimpregnated fibre, use more than one type of reinforcement, and to combine preimpregnated fibre layers (for instance laid lengthwise along the mandrel) with wound fibres.

8. Faults in composite materials

8.1. Moulding and forming

With the materials and manufacturing techniques previously described it is possible to produce a very wide range of articles and structures including components for artificial satellites and sections of supersonic aircraft, yacht hulls and minesweepers, car components, textile machinery, chemical plant, radomes, artificial limbs, sports goods, etc. Not all would qualify as high performance composites, but in all cases there is great interest in knowing what may go wrong in manufacture and how this influences the subsequent performance of the product.

Let us recapitulate on the sequence of events occurring in a forming or moulding operation. The details will vary somewhat with the type of reaction causing polymerisation. For a wet lay-up process the resin may be worked into the reinforcement by hand or a mechanically applied pressure, or be heated to reduce its viscosity and then forced between the fibres by pressing. This should cause all fibre surfaces to be wetted and in addition ensure the correct fibre volume fraction. The resin may then start to gel and cure at room temperature, with a considerable exotherm, or on the application of external heat. When using BMC, SMC, HMC or XMC the application of pressure and heat causes the charge to flow filling the mould, increasing the volume fraction of reinforcement and causing the resin to polymerise. With preimpregnated fibres heat causes the resin viscosity to fall allowing the material to be consolidated by the application of pressure, making it conform to the mould profile and assisting in the release of solvent. Finally the resin gels and cross links. In all cases, the curing temperature should be in accordance with the manufacturer's instructions and the duration of cure sufficient to ensure that cross linking is complete or virtually so. In certain cases this may mean leaving the article for up to 24 hours at 200°C.

Though the dependence of bulk resin viscosity on temperature, the onset of gelation, the duration of the cure cycle and relationship between curing time and cross-linking are known for many resin systems, many finer details, such as why a particular cure cycle causes a given set of properties, are not so well documented. Far less is understood about the behaviour of resins in thin layers (1 μm) that occur between fibres in composite materials. It is known that large differences can occur between modified resin systems in bulk and thin films. Hancox and Wells (1973) and Scott and Phillips (1975) noted that carboxy terminated butadiene elastomer, which markedly increased the toughness of bulk epoxy resin, had very little influence in composites or thin glue lines. McCullough (1971) and Hancox (1977) have suggested that the physical state of the resin surrounding fibres will be different from that of the bulk because the presence of fibres, especially carbon, reduces the exotherm of the resin. Preferential absorption of the components of epoxide resins may occur at the fibre surface, and a resin might cure in a layered fashion around the fibres giving a more ordered structure. However, these points do not appear to have been pursued experimentally or theoretically.

Various attempts have been made to devise simple methods of determining the degree of cure of a resin (see ARRIDGE and SPEAKE (1972) for further details), and to determine the optimum time for applying pressure in fabrication (MARTIN, 1976, 1977). The results are usually correlated with resin characteristics, e.g., strength, or T_g, the second order glass transition temperature, on an empirical basis and the work throws little light on how things happen.

8.2. Types of fault in manufactured articles

Assuming that the raw materials are of suitable quality, the resin properly mixed, that there is no contamination, and that the artefact and mould have been correctly designed and are used by skilled personnel, the following may occur leading to reduced or modified mechanical, thermal, chemical or electrical performance:

Fibre misalignment,   Errors in fibre and
Voids,   resin volume fraction,
Fibre damage,   Discontinuities,
Poor bonding,   Built-in thermal stresses.

8.2.1. Fibre misalignment

If a fibre set in a resin matrix is oriented at an angle θ to the loading axis, the variation of failure stress with θ (KELLY, 1966) is as shown in Fig. 15. An increase in θ from, say, 0 to 10° results in the strength of a unidirectional composite being reduced by half or more. This reflects the very different properties that laminae or composite bars possess along and at right angles to the long fibre axis. Similarly it can be shown (CALCOTE, 1969) that the longitudinal tensile modulus decreases very rapidly with increasing angle between the fibre and loading axis. ARGON

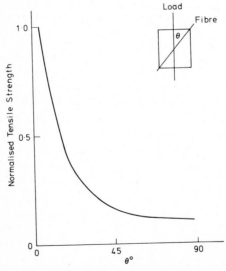

FIG. 15. Variation of normalised tensile strength with orientation of the fibre to the loading axis.

(1972) has used a fibre misalignment model to explain the initiation of compressive failure in unidirectional composites. If fibres in unidirectional composite or laminate sheet are out of alignment by even a few degrees the expected strength and modulus will be reduced markedly from anticipated values. An average laminate or composite bar consists of from several thousand to several hundred thousand fibres and clearly it will not be possible to tension and align every fibre when preparing a moulding. Misalignment also occurs in systems made from woven cloth or preimpregnated fibre sheet. In the latter case it can be due to misalignment in individual sheets or occur when the sheets are stacked to give a laminate. Equally serious and more difficult to control is the movement of fibres that may occur in fabrication when the resin melts or its viscosity is reduced, due to an internal exotherm or external heating, and pressure is applied.

Fibre disorder is visible to the naked eye in the surface of most unidirectional fibre composites and is obvious in polished longitudinal sections of material. In some cases the misalignment of a group of fibres may be as much as 45°. It is difficult to be exact about the proportion of fibres involved but it could be as high as 10%.

Hand lay-up fabrication routes using dry fibre and a liquid resin are most likely to suffer from misalignment because of lack of fibre constraint, the working of the resin into the fibres and the labour intensive nature of the process. However, the effect is also noted in matched die mouldings using preimpregnated fibre. It may be more serious in this case because the performance sought is nearer to the optimum. The fault certainly occurred in the early days of the mass production of high performance CFRP, such as turbine blades. A related effect, the alignment of short fibres which may happen when injection moulding a short fibre reinforced thermoplastic, has been studied by DARLINGTON et al. (1976). The resulting reduction in isotropy can seriously reduce strength and stiffness in certain directions leading to failure. This phenomenon can occur in BMC and SMC.

Fibre misalignment is one of the reasons advanced for the failure of unidirectional components to reach the modulus and strength calculated on a basis of single filament testing.

8.2.2. *Voids*

Voids are probably the most common fault in fibre reinforced plastic. They are due to air trapped between fibres, or incorporated into the resin during mixing, the presence of solvents or other volatiles in preimpregnated material or unreacted styrene in polyester systems, and interstices produced when preimpregnated fibres or moulding compounds fail to coalesce properly under processing. An example of a high void content in a carbon fibre specimen prepared from preimpregnated fibre is shown in Fig. 16. There is a tendency for the voids to be along the boundaries of individual sheets of prepreg. The void may act as a stress raiser or contribute to premature failure because of the presence of unsupported regions of fibre and the exposure of the reinforcement to chemical attack. The presence of voids in the bulk matrix leads to an overall reduction in cured resin properties, which may be reflected in lower transverse, flexural, shear and

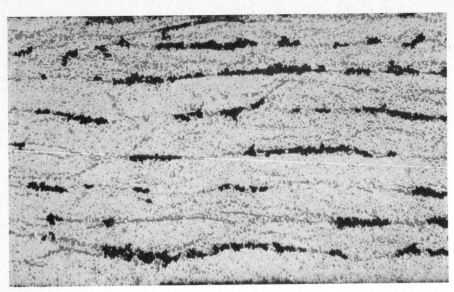

FIG. 16. An example of voids in a carbon fibre composite containing approximately 60 $^v/_o$ of carbon fibre, × 60.

compression performance in the reinforced composite, increased susceptibility to chemical attack and reduced electrical properties of GRP. Void content commonly ranges from less than 1 to 5 $^v/_o$ or greater. Some types of polyimide (see Section 3.3) are particularly prone to voids and this vitiates to some extent their improved temperature capability. Often voids tend to concentrate between tow boundaries or laminae causing a line or planar fault. With hand lay-up using glass cloth or chopped or randomly oriented fibres it is possible to reduce the void content by adding to, and working into, the material fine fillers including small glass beads. It is not practicable to degas large structures to encourage bubbles to rise to the surface and so in these cases the resin system, curing cycle and any pressing conditions should be chosen to minimise air and solvent entrainment and make it as easy as possible for the air and volatiles to escape. With small liquid resin fabrications using matched die moulds and an epoxide resin it is relatively simple to degas the resin and resin plus fibre, though care should be taken if one of the resin components, say the accelerator, has a high vapour pressure and is liable to be preferentially removed. With preimpregnated fibre sheet the trend is to alternative methods of resin impregnation to avoid the solvent that contributes to voidage.

Some examples of the effects of voids on composite properties have been given by, among others, HANCOX (1975, 1977) and JUDD and WRIGHT (1978b). In the former two papers it was noted that flexural modulus was least affected by voids, flexural and compressive strength more so, and that the effects on flexural properties did not appear to ease off as the void content increased. The compression strength decreased by approximately 6% for each 1 $^v/_o$ increase in void content up to 5 $^v/_o$ voids. The shear modulus and strength showed, initially, a slow decline up to 1.5 $^v/_o$ voids, then an accelerating fall, until at 3 $^v/_o$ voids they had been halved; and then a much slower reduction. The latter authors noted that the interlaminar shear strength of CFRP fell by about 7% for each 1 $^v/_o$ increase in void content up to 4 $^v/_o$ voids. Although the results are for unidirectional materials and subject to considerable error due to the difficulty of obtaining a uniform, constant, void content in a specimen, they do illustrate the serious effects voids produce.

Theoretical treatments of the link between void content and a given property have been given by FOYE (1966), for cylindrical voids, GRESZCZUK (1967), for spherical and cylindrical ones, and CORTEN (1968), who assumed that the effective crack length was proportional to the cube root of the void volume fraction. FOYE's more sophisticated analysis for cylindrical voids, GRESZCZUK's strength of materials approach for spherical voids and CORTEN's approach are all fairly well supported by data on shear modulus up to 1 $^v/_o$ voids but then indicate a much larger remaining modulus than is noted experimentally (HANCOX, 1977). GRESZ-CZUK's cylindrical model most nearly approaches the experimental results. In any quality manufacturing process the void content should be regularly monitored and if too high steps taken to reduce it by improving the process control.

8.2.3. Fibre damage

All reinforcing fibres contain surface or internal flaws or both which may be the origin of failure in, for instance, a tensile test. The tensile strength of virgin glass fibre is substantially reduced by handling and individual carbon and boron fibres may be easily damaged because of their brittle nature. Failures at adjacent regions may link up and cause a crack to propagate across the specimen.

It has to be expected that in any handling operation weaker fibres will be broken, appearing for instance as fluff or hairyness on the collimated input to a pultrusion die, or be crushed in pressing, if too high a pressure is applied before the resin is sufficiently liquid to allow adequate fibre movement. One of the most aggressive fabrication processes is hand lay-up where the resin is worked in between fibres and fibre compaction achieved by repeated rolling of the components. However, the reduction in strength properties brought about by this is usually allowed for in the values taken for the fibre properties employed in the calculation of the expected performance. Extremely high pressures in bulk and sheet moulding operations may break fibres into shorter and shorter lengths, and if the length is reduced below l_c, the critical stress transfer value, both the strength and modulus of the composite will be reduced as it is now not possible to utilise the full potential of the individual fibres.

8.2.4. Poor bonding

This is a common and obvious defect in all types of FRP structure. The resin may have failed to wet the fibre because of surface contamination, insufficient resin, incorrect processing temperature, inability to bond to the fibres, or might not have penetrated between individual fibres. The results of any of these are

particularly noticeable on the surface of a composite structure. Fibres may stand
out from the surface giving a rough appearance, and it may be possible to
remove fibres or groups of fibres very easily with a knife blade or any sharp
point. Internally it may be difficult to decide whether the damage is due to poor
bonding or excessive voidage. The strength and stiffness of the composite,
particularly the interlaminar shear strength, are reduced because it is no longer
possible to transfer stress into and between fibres effectively, but the work of
fracture will be increased since extra energy is now absorbed in failure by
pull-out and fibre debonding.

There has been considerable argument over what constitutes the optimum
bond strength between fibre and matrix. There appears to be no point in
having this exceeding the resin or fibre transverse strength, but it may be
advantageous to have it reduced somewhat so that total failure is less brittle in
nature. If the reduction is too great, although the impact resistance is increased, it is difficult to diffuse loads into structures.

For many years various surface agents have been applied to glass fibres to
improve the bond strength to polyesters and epoxides and also to avoid excessive degradation of the bond in wet conditions. Surface treatments for carbon
fibres have been mentioned previously (see Section 2). Suitable agents increase
the interlaminar shear strength of glass from 7 to 35 MPa and that of carbon
fibre composites from 17 to at least 40 MPa. Aramid fibres absorb water and
unless they are dry before the resin is added (either in bulk or in a preimpreg-
nating process) the bond strength is seriously affected.

It should be apparent that poor bonding can occur in any fabrication process,
but that it can be avoided or at least minimised by using fibres with surface
coatings or treatments compatible with the resin system employed, by ensuring
that processing conditions are correct, and that sufficient resin is present.

8.2.5. *Errors in fibre and resin volume fraction*

High performance composites normally contain between 40 and 75 $^v/_o$ of fibre
which is usually evenly distributed throughout the composite. The higher figure
will be approached with unidirectional fibres while the lower one is associated
with mats and woven reinforcements. When forming a composite the volume
fraction of fibres is, initially, usually below the figures quoted, and temperature
to increase the flow properties of the resin, and pressure, are needed to
increase the amount of fibre relative to the matrix. In vacuum bag and
autoclave moulding it is customary to put bleeder plies around the preimpregnated fibre to absorb excess resin, and control the volume fraction of fibres in
the finished composite by the number of bleeder plies used. In other fabrication methods—matched die moulding, resin injection and pultrusion—
knowing the volume of the mould, the amount of fibre to give the desired
packing fraction can be calculated beforehand. Deviations still occur, particularly when using preimpregnated fibre, if pressing is undertaken before the
resin can flow or after it has gelled. The result is reduced mechanical properties
because of the reduced number of fibres in any specific cross sectional area. If

too little resin is used in a wet lay-up moulding or if an excessive amount is bled off, resin starvation may result at the surface leading to an effect similar to that mentioned in the previous section.

It is possible to correct volume loading errors by carefully checking the mould dimensions, amount of fibre and resin or moulding compound charge used, and tuning the heating and pressing cycle to ensure that the pressure is applied when the resin viscosity has the correct value for the required degree of consolidation.

8.2.6. Discontinuities

Discontinuities are lines or areas of excessive weakness in mouldings. They can act as cracks and lead to reduced mechanical properties due to stress concentration at the crack tip and possibly diminish the chemical or environmental resistance of a composite. This type of feature may be formed by poor bonding or inadequate fibre wetting, but in this section we consider splits or cuts in preimpregnated fibre and weld lines in BMC or SMC structures.

Splits in preimpregnated material may be caused by attempting to make too thin a sheet, in which case the fault will be parallel to the fibres, or by cutting at some angle to the fibres. In a multiply laminate a fault in one layer can act as a source of weakness and initiate unexpected failure in the complete structure. OWEN (1976) has noted in experimental fatigue studies on glass fibre composites that a cut in one ply of a specimen, while not affecting the static properties will grow and cause catastrophic failure in a fatigue situation. Such events can be avoided by careful inspection and handling of all materials, but the problem still remains of the effects of deliberate joints in large structures made without adequate fibre bridging of the region. This however is a design problem, not a manufacturing fault, and will not be pursued here.

A weld line is a region of reduced strength and stiffness, rather than a physical gap, which may be present in structures made from moulding compounds. It differs from the joints mentioned in the previous paragraph because it is not incorporated by design. Weld lines arise when two or more separate charges are used to load a mould. Under the action of temperature and pressure the material can be made to flow and fill the mould and coalesce to give to the naked eye a continuous volume. However, the interface areas between charges consist of resin with no bridging fibres and hence will represent a weakened volume of material. Provided the weld line is not in a principal direction of stressing the effect may not be too serious, just as resin rich areas in a unidirectional fibre composite specimen are unlikely to influence the longitudinal performance. BMC and SMC are usually regarded as having pseudo isotropic properties (in a plane in the case of SMC) and so a discontinuity in these materials is potentially serious. The fault can be largely avoided by using one charge per moulding, or if more than one is required, taking care to see that there is fibre continuity across the interface before pressing.

8.2.7. Built-in thermal stresses

The coefficients of thermal expansion (CTE) of fibres and resins differ widely. The CTEs of fibres, in the longitudinal direction, range from -2 to $5 \times 10^{-6} \, ^{\circ}C^{-1}$. Unfilled, bulk, resins are essentially isotropic and have CTEs between about 40 and $60 \times 10^{-6} \, ^{\circ}C^{-1}$. Resin gelling takes place at temperatures of up to 200°C. Consequently stresses are induced in the components of a composite material when this cools from its gelling temperature. In the simplest case of a unidirectional fibre surrounded by resin the fibre will be put into compression and the resin into tension on cooling. The effect can be eliminated by curing at room temperature, but it whould be remembered that systems which nominally cure at room temperature may exotherm raising the temperature significantly, while for better performance and ease of processing, especially of preimpregnated materials, it is desirable and may be necessary to work at a raised temperature.

Thermally induced stresses can be easily demonstrated by bonding a CFRP strip to an aluminium one, or by making a laminate consisting of two sheets of aligned fibre material at right angles to one another, and, in either case, curing at an elevated temperature. On cooling the artefacts will bow or distort, while on heating up again the distortion will gradually disappear. If the built in stresses are too severe delamination may occur. Careful studies indicate that the temperature at which distortion is no longer apparent is not exactly the gelling temperature of the resin and this should be allowed for in any calculation of the effects of thermally induced stresses. In practice unbalanced structures of the types described are not usually used, but the stresses are still present in balanced structures manufactured and cured under the same conditions. If one of the components or plies in a balanced lay-up is damaged, the resulting redistribution of thermally induced stress could cause the component to fail. The matter is further discussed by FAHMY and EL-LOZY (1974).

It is easy, using a strength of materials approach, to calculate the tensile and compressive stresses set up in a two component composite. The result is

$$\sigma_h = \frac{\Delta T (\alpha_h - \alpha_\ell) E_\ell V_\ell}{V_\ell (E_\ell / E_h - 1) + 1} \tag{1}$$

where σ is the stress, ΔT the difference in temperature, α the CTE, E the modulus, V the volume and subscripts h and ℓ refer to the components with the higher and lower CTEs respectively.

In addition,

$$\sigma_h V_h + \sigma_\ell V_\ell = 0 , \tag{2}$$

as there is no nett force on the composite. This relationship allows σ_ℓ to be calculated knowing σ_h. The substitution of typical values for a fibre and resin, or composite and metal, indicate that the stresses involved can be high. For instance in a CFRP aluminium CFRP sandwich containing 11% of composite, with $\Delta T = 130°C$, the compressive stress in the CFRP is approximately 380 MPa (HANCOX et al., 1979). The use of an elastic or rubber interface between the fibres and resin or composite and metal, reduces the stress considerably, so that provided the layer can withstand the stress gradient the situation may be improved.

Most of the defects discussed here can be eliminated or at least reduced to a very low level by careful materials selection and handling, mould design and process control, but in-built thermal stresses cannot be eliminated in this way unless room temperature curing resin systems are used and care taken to contain any exotherm. This has other disadvantages as was explained earlier.

8.3. The effects of faults—Summary

A study of the earlier literature concerning composite materials (for instance, see CHAMIS and SENDECKYJ, 1968) shows the variety and amount of work put into the problem of calculating the thermomechanical properties of composites from a knowledge of the properties and geometry of the components. The assumptions made usually include a homogeneous matrix and fibres, both free of voids, perfectly bonded, initially stress free, with fibres regularly spaced and aligned. Since few, if any, of these assumptions are completely complied with in practice our ability to predict thermal and elastic properties is sometimes surprising; though as ALLRED and GERSTLE (1975) have pointed out for transverse properties the knowledge of basic fibre and resin data is sometimes so poor that it is not possible to distinguish among the theories developed.

The faults we have described, and possibly others, do occur with some of the effects detailed, but they should not be thought of as so serious that they hamper the development and use of composite materials of all types.

To substantiate this let us briefly consider the work of RHODES (1979), who has deliberately incorporated a series of faults, including resin rich areas, inclusions of foreign matter, misaligned tows, broken tows, and voids in a carbon fibre laminate and a carbon fibre sandwich panel. HT carbon fibre with a multifunctional epoxide resin was used. Specimens were tested in the dry condition, after a moisture soak, and after being soaked and freeze cycled. The void content was $3^v/_o$ or above, the maximum inclusion size 8 mm, and a complete tow, representing $7\frac{1}{2}\%$ of the total width of material, was broken. For laminates the flexural modulus was unaffected by the presence of voids, though the flexural and interlaminar shear strengths were slighly reduced especially after moisture soaking. 8 mm long inclusions of silicone paper, equivalent to a debonded area, also led to a reduction in the shear strength after 90 days moisture soak. The presence of a broken tow in the tension skin of a sandwich beam caused a reduced static flexural strength. On fatiguing for 350 000 cycles the broken tow tended to separate from the rest of the laminate, reducing the stress concentration and leading to an increase in the residual strength of fatigued specimens. The effect of broken tows in the compression face of a sandwich construction was much less. RHODES concluded that the design of the carbon fibre components, an engine cowl and section of an aircraft rudder, were such that even allowing for defects of the types outlined above, the structures had an adequate mechanical performance. In practice, strict quality control of the incoming materials, good housekeeping and process control would ensure a much lower defect level than was employed in the test specimens.

9. Hygrothermal degradation

Composites see many environments in use, including a wide range of industrial chemicals, foodstuffs, liquids and particulate matter. There is however one combination, viz. that of water and temperature, that has been more intensively studied over the last decade than the others. The interest in hygrothermal degradation arises for two reasons. Firstly it is relatively easy to study experimentally and to model theoretically, and secondly there is the possibility that it might limit the extensive use of high performance composites in aerospace applications.

The temperature limitation on resin matrix systems was mentioned in Section 3.3, and it is known that the mechanical properties of thermoset and thermoplastic resins, and composites made with these, begin to decrease well before the temperature at which degradation starts. The presence of water can enhance this behaviour and may cause damage that is not reversed when the artefact is returned to room temperature and the water partially removed.

The effects of weathering on polyester glass reinforced composites, in which specimens are subjected to cyclic variations of temperature and moisture, have been reviewed by BLAGA and YAMASAKI (1973). Similar work for epoxy glass reinforced material is described by ROYLANCE and ROYLANCE (1978) who emphasise the role that ultraviolet degradation of the resin can play. SMITH (1977) has stressed the combined effect of radiation attacking overcured regions of the resin and moisture washing the debris away. He believes that this accounts for the much more severe weathering reported in the tropics compared with that in climates which are either dry and sunny or wet but without sunshine. ISHAI and ARNON (1977) have pointed out that glass fibres may be degraded by water while JUDD (1975) has noted that carbon fibres are not affected.

Resins absorb water by volumetric diffusion and SHEN and SPRINGER (1976) and CARTER and KIBLER (1976) have shown that water uptake or loss can be accounted for by simple Fickian diffusion theory. COOK et al. (1977) and SHIRRELL (1978) have noted a more complex behaviour and suggest that it is associated with time dependent microcracking of the matrix. Water swells the resin structure and acts as a plasticising agent, leading to fibre resin delamination and causing a reduction in composite mechanical properties. BROWNING (1973), JUDD (1975) and AUGL and BERGER (1976) found that the effects on mechanical properties such as tensile, compressive and interlaminar shear performance were reversed when the water was removed. However, KAELBLE et al. (1975) and KAELBLE and DYNES (1977) noted irreversible changes in shear strength and work of fracture, and microcracking associated with permanent changes in the properties of epoxide resin matrices. McKAGUE et al. (1975) found that rapid temperature excursions caused a permanent increase in the moisture diffusion behaviour of composites, while LUNDEMO and THOR (1977) reported that the fatigue life of composites is reduced by moisture exposure. In laminates PIPES et al. (1976) and CROSSMAN and WANG (1978) have shown that the non-uniform distribution of moisture causes steep stress gradients near laminate surfaces or free edges which could lead to microcracking or delamination.

The effects of water and temperature, especially on structures for which the design limits are close to the ultimate performance of the material, can be serious. Nevertheless undue pessimism should be avoided. A better appreciation of the many types of resin available and of the fibre resin interface and how its properties can be modified to suit the circumstances, together with studies on methods of protecting the surfaces of composites, should alleviate the problem.

10. Selection of materials for specific applications

It is assumed that the performance of the artifact and the environment it is to experience have been correctly specified and that the design has been carried out in a sound manner with due regard to the anisotropic properties and failure characteristics of fibre reinforced composite materials; it remains to select a suitable reinforcement, resin, and manufacturing technique which will give a product of the required performance at minimum cost.

Carbon fibres have the best combination of elastic modulus, low density, thermal stability and strength but are relatively expensive. They tend to be employed in aerospace, military and medical applications where weight saving can be important. They are also used in sports goods because customers are prepared to pay not only for enhanced performance but also for the prestige of having an article containing carbon fibres. In many of the above applications the fibres would be used as a preimpregnated sheet to make laminates or employed with a filament or tape winding machine.

E glass fibre is extensively used in chemical plant, marine and transport applications either as woven or chopped or random continuous fibre mat, or in the form of continuous material for winding. Its great advantage compared to other reinforcements is its cheapness. In flexural applications it is possible to obtain a stiffness equal to that of carbon fibre laminate by increasing the thickness (and of course the weight) of the article. With either glass or carbon fibres it is important to use materials having a surface treatment or size compatible with the resin matrix.

E glass fibre in the form of moulding compound is widely used for products having considerable detail, and of which large numbers are to be made with a minimum pressing time. These include components for the electrical, building and transportation industries.

Aramid fibres, which have excellent specific properties, are also used in aerospace applications and for sports goods, though care has to be taken in fabrication to keep the fibre from absorbing moisture, and the design may have to be modified to allow for the poor compressive properties of the fibre. Nevertheless the low cost, good modulus and strain to failure make them a useful material.

Boron fibres are principally used in aerospace applications, where their high cost is not such a disadvantage.

An increasingly common method of construction is to use two or more fibre types to make a hybrid. This can lead to reduced cost, for little reduction in properties, or increased performance for a small increase in cost. Examples are carbon fibre beams with a central region of glass fibre composite, glass or aramid fibre structures uprated by the addition of a few percent of carbon fibre, and the use of aramid fibres to improve the impact resistance of carbon fibre composites. Combinations of these types are used in sports goods and aerospace applications. Another type of hybrid, consisting of a honeycomb, metal, or wood core with fibre composite skins is commonly used for antenna dishes, calipers and sports goods.

Having selected the type or types of reinforcement the next step is to choose a class of resin and a specific resin type within that class. Although general guidelines can be given, for detailed information on processing conditions and chemical resistance it is best to consult with the resin suppliers. The resin is often regarded as the 'Cinderella' of a composite material, whereas in reality it is a particularly vital element. It is the first component to experience the chemical/thermal environment and has to protect the reinforcement and allow this to function effectively; it is important in determining fatigue and creep behaviour; and to a large extent it governs the ease with which a composite article can be manufactured.

It is usual to employ epoxides with carbon or boron fibres as the cost and properties of the fibre justify a resin system with superior properties even if it is more expensive than other types of resin. The variety of epoxide will depend on the application of the composite. If a high temperature capability is required a multifunctional epoxide, cured at a higher temperature will be used. For winding applications, low viscosity and adequate pot life are important. Pre-impregnated fibre sheet, used for building up laminated structures, is usually produced with an epoxide resin irrespective of fibre type. Polyimides which are among the most expensive resins and require lengthy processing may be employed with carbon fibres if exceptional heat resistance is called for.

For use with glass fibres it has been traditional to employ polyester resins, though vinyl esters, furanes and phenolics may also be used (e.g., furanes for chemical resistance). In the production of large components cold cure resins will be required due to the limitations on applying heat and pressure to very large structures at reasonable cost.

Hybrids and aramid fibres are usually fabricated with epoxide resins since this type of construction tends to be for high performance uses.

The manufacturing method chosen reflects the size and performance required of the finished article, the availability, cost and skill of labour and number of units to be made. Hand lay-up methods are usual with combinations of glass fibre and polyester resin, where either property optimisation is not required or a large structure (e.g., part of a chemical plant or boat hull) is to be built as a 'one off' using a relatively inexpensive cold curing resin system. Various types of autoclave or pressure moulding systems are used to make carbon, glass or hybrid structures up to several metres in dimension. Preimpregnated fibre can be used with these methods to make laminates with multidirectional properties.

Injection or closed moulding techniques give artifacts with both surfaces well finished and good fibre compaction. Both can be used with any sort of fibre and the latter with preimpregnated material. For moulding compounds a sturdy, closed mould, which may be expensive to produce, is required to enable the material to be shaped and cured under the influence of temperature and pressure. However, the nature of the components made from moulding compounds is such that a large number will be produced, thus offsetting the cost of the mould.

Lengths of section or profile are made from any type of fibre by pultrusion. The sections produced may be used in composite structures later.

Hybrid structures may be produced in one operation using a vacuum or pressure bag mould and an autoclave, or by hot press closed moulding. Alternatively components such as stringers or skins for antenna dishes may be made separately and then bonded to a core.

References

ACKLEY, R.H. (1976), XMC-structural FGRP for matched-metal-die moulding, in: *31st SPI Conf.*, Washington, U.S.A., Society of Plastics Industry, Inc., NY, 16C1–4.

ACKLEY, R.H. and E.P. CARLEY (1979), XMC-3 composite material structural moulding compound, in: *34th SPI Conf.*, New Orleans, U.S.A., Society of Plastics Industry, Inc., NY, 21D1–6.

AITKEN, I.D., G. RHODES and R.A.P. SPENCER (1970), Development of a wet oxidation process for the surface treatment of carbon fibre, in: *7th Internat. Reinforced Plastics Conf.*, British Plastics Federation, London, Paper 24.

ALLRED, R.E. and F.P. GERSTLE (1975), The effect of resin properties on the transverse mechanical behaviour of high performance composites, in: *30th SPI Conf.*, Washington, U.S.A., Society of Plastics Industry, Inc., NY, 9B1–6.

ARGON, A.S. (1972), Fracture of composites, in: H. HERMAN, ed., *Treatise on Materials Science and Technology* Vol. 1 (Academic Press, New York) pp. 79–114.

ARRIDGE, R.G.C. and J.H. SPEAKE (1972), *Polymer* **13**, 443.

AUGL, J.M. and A.E. BERGER (1976), Moisture effect on carbon fibre epoxy composites, in: *8th Nat. SAMPE Tech. Conf.*, California, U.S.A., SAMPE, Azusa, CA, 383–427.

BLAGA, A. and R.S. YAMASAKI (1973), *J. Mat. Sci.* **8**, 654, 1331.

BREWIS, D.M., J. COMYN, J.R. FOWLER, D. BRIGGS and V.A. GIBSON (1979), *Fib. Sci. Tech.* **12**, 41.

BROUTMAN, L.J. and R.H. KROCK, eds. (1974), *Composite Materials* Vol. 6 (Academic Press, New York) Chapters 1, 3, 5, 6.

BROWNING, C.E. (1973), The effects of moisture on the properties of high performance structural resins and composites, in: *28th SPI Conf.*, Washington, U.S.A., Society of Plastics Industry, Inc., NY, 15A1–16.

BRUINS, P.F., ed. (1976), *Unsaturated Polyester Technology* (Gordon and Breach, New York).

BUSSO, C.J., H.A. NEWLEY, T.D. BUCKMAN and H.V. HOLLER (1970), Chemical structure—Mechanical properties studies on pure epoxy resin systems, Part I, in: *25th SPI Conf.*, Washington, U.S.A., Society of Plastics Industry, Inc., NY, 3B1–8.

CALCOTE, L.R. (1969), *The Analysis of Laminated Composite Structures* (Van Nostrand, New York) p. 32.

CARTER, H.G. and K.G. KIBLER (1976), *J. Comp. Mat.* **10**, 355.

CHAMIS, C.C. and G.P. SENDECKYJ (1968), *J. Comp. Mat.* **2**, 332.

CHAMIS, C.C., H.P. HANSON and T.T. SERAFINI (1973), Criteria for selecting resin matrices for improved composite strength, in: *28th SPI Conf.*, Washington, U.S.A., Society of Plastics Industry, Inc., NY, 12C1–12.

CHEREMISINOFF, N.P. and P.M. CHEREMISINOFF (1978), Fibre glass reinforced plastics deskbook, Ann Arbor Sci., Mich., U.S.A., 57.

CHRISTENSEN, R.M. and J.A. RINDE (1979), *Poly. Engrg. Sci.* **19**, 506.

COOK, T.S., D.E. WALRATH and P.H. FRANCIS (1977), *SAMPE Proc. Engrg. Ser.* **22**, 339.

CORTEN, H.S. (1968), Influence of fracture toughness and flaws on the interlaminar shear strength of fibrous composites, in: R.T. SCHWARTZ and H.S. SCHWARTZ, eds., *Fundamental Aspects of Fibre Reinforced Plastic Composites* (Interscience, New York) Chapter 6.

COUDENHOVE, J. (1979), *Reinforced Plastics* **23**, 158.

CROSSMAN, F.W. and A.S.D. WANG (1978), *J. Comp. Mat.* **12**, 2.

DARLINGTON, M.W., P.L. McGINLEY and G.R. SMITH (1976), *J. Mat. Sci.* **11**, 877.

DEITZ, R. and M.E. PEOVER (1971), *J. Mat. Sci.* **6**, 1441.

DENTON, D.L. (1979), Mechanical properties characterisation of an SMC-R50 composite, in: *34th SPI Conf.*, New Orleans, U.S.A., Society of Plastics Industry, Inc., NY, 11F1–12.

E.I. DU PONT DE NEMOURS and CO. INC. (1978), Textile Fibres Dept., Delaware, USA, Data sheets.

ESPENSHADE, D.T. and J.R. LOWRY (1971), Low shrink polyester resins for sheet moulding compound, in: *26th SPI Conf.*, Washington, U.S.A., Society of Plastics Industry, Inc., NY, 12F1–10.

EVANS, D.L. and T.J. NEMETH (1979), Improved SMC for structural electric utility applications, in: *34th SPI Conf.*, New Orleans, U.S.A., Society of Plastics Industry, Inc., NY, 2C1–4.

EVERITT, G.F. (1977), Continuous filament ceramic fibre refractory insulation, in: *Proc. Electrical/Electronic Insulation Conf.*, pp. 236–240.

FAHMY, A.A. and A.R. EL-LOZY (1974), Thermally induced stresses in laminated fibre composites, in: A. BISHAY, ed., *Recent Advances in the Science and Technology of Materials* Vol. 2 (Plenum Press, New York) pp. 245–259.

FEKETE, F. (1970), Thickeners and low-shrink additives for premix and SMC systems, in: *25th SPI Conf.*, Washington, U.S.A., Society of Plastics Industry, Inc., NY, 6D1–24.

FEKETE, F. (1972), A review of the status of thickening systems for SMC, LS-SMC, BMC and LS-BMC compounds, in *27th SPI Conf.*, Washington, U.S.A., Society of Plastics Industry, Inc., NY, 12D1–24.

FERRARINI, J., J.J. MARGANS and J.A. REITZ (1979), New resins for high strength SMC, in: *34th SPI Conf.*, New Orleans, U.S.A., Society of Plastics Industry, Inc., NY, 2G1–7.

FITZER, E. and M. HEYM (1976), *Chemistry and Industry* **16**, 663.

FOTHERGILL and HARVEY Ltd. (1977), Littleborough, U.K., Publication 91B.

FOYE, R.L. (1966), Compression strength of unidirectional composites, in: *3rd Aerospace Sciences Meeting*, American Institute of Aeronautics and Astronautics, NY, AIAA, NY, Paper 66-143.

GARNISH, E.W. (1972), *Composites* **3**, 104.

GOODHEW, P.J., A.J. CLARKE and J.E. BAILEY (1975), *Mat. Sci. Engrg.* **17**, 3.

GREENWOOD, J.H. and P.G. ROSE (1974), *J. Mat. Sci.* **9**, 1809.

GRESZCZUK, L.B. (1967), Effect of voids on strength properties of filamentary composites, in: *22nd SPI Conf.*, Washington, U.S.A., Society of Plastics Industry, Inc., NY, 20A1–10.

HANCOX, N.L. and H. WELLS (1973), AERE R7296.

HANCOX, N.L. (1975), *J. Mat. Sci.* **10**, 234.

HANCOX, N.L. (1977), *J. Mat. Sci.* **12**, 884.

HANCOX, N.L., A.J. HAMMOND and P.R. MARFLEET (1979), AERE R9235.

HODD, K.A. and D.C. TURLEY (1978), *Chemistry in Britain* **14**, 545.

HORIE, K., H. MURAI and J. MITA (1976), *Fib. Sci. Tech.* **9**, 253.

ISHAI, O. and U. ARNON (1977), *J. Test. Eval.* **5**, 320.

JUDD, N.C.W. (1975), The effect of water on carbon fibre composites, in: *30th SPI Conf.*, Washington, U.S.A., Society of Plastics Industry, Inc., NY, 18A1–12.

JUDD, N.C.W. and W.W. WRIGHT (1978a), *Reinforced Plastics* **22**, 39.

JUDD, N.C.W. and W.W. WRIGHT (1978b), *SAMPE J.* **14**, 10.

KAELBLE, D.H., P.J. DYNES, L.W. CRANE and L. MAUS (1975), *ASTM STP* **580**, 247.

KAELBLE, D.H. and P.J. DYNES (1977), *J. Adhesion* **8**, 195.

KELLY, A. (1966), *Strong Solids* (OUP, Oxford, U.K.) p. 152.

KREIDER, K.G. and K.M. PREWO (1972), *ASTM STP* **497**, 539.

LEE, H. and K. NEVILLE (1957), *Handbook of Epoxy Resins* (McGraw-Hill, New York).

LINE, L.E. and U.V. HENDERSON (1969), Boron filament and other reinforcements produced by chemical vapor plating, in: G. LUBIN, ed., *Handbook of Fiberglass and Advanced Plastics Composites* (Society of Plastics Engrg., New York) Chapter 10.

LUNDEMO, C.Y. and S.E. THOR (1977), *J. Comp. Mat.* **11**, 276.

MAAGHUL, J. and E.J. POTKANOWICZ (1976), HMC—A high performance sheet moulding compound, in: *31st SPI Conf.*, Washington, U.S.A., Society of Plastics Industry, Inc., NY, 7C1–6.

MAGRANS, J.J. and J. FERRARINI (1979), Unique electrical properties of ITP SMC, in: *34th SPI Conf.*, New Orleans, U.S.A., Society of Plastics Industry, Inc., NY, 2E1–6.

MARRIOTT, P.J. (1974), Polyamino-bis-maleimides, in: *Reinforced Plastics Congress*, British Plastics Federation, London, Paper 14.

MARSH, H.N., T.E. GRIFFITH and J.V. SPITALE (1979), RTM tooling and moulding for corrosion resistant applications, in: *34th SPI Conf.*, New Orleans, U.S.A., Society of Plastics Industry, Inc., NY, 3A1–10.

MARTIN, B.G. (1976), *Mat. Eval.* **3**, 49.

MARTIN, B.G. (1977), *Mat. Eval.* **6**, 48.

MARTIN, J. (1978), Pultrusion—An overview of applications and opportunities, in: *33rd SPI Conf.*, Washington, U.S.A., Society of Plastics Industry, Inc., NY, 8H1–6.

MCCULLOUGH, R.L. (1971), *Concepts of Fibre Reinforced Composites* (Dekker, New York).

MCKAGUE, E.L., J.E. HALKIAS and J.D. REYNOLDS (1975), *J. Comp. Mat.* **9**, 2.

MEYER, L.S. (1970), Pultrusion, in: *25th SPI Conf.*, Washington, U.S.A., Society of Plastics Industry, Inc., NY, 6A1–8.

MILES, D.C. and J.H. BRISTOW (1965), *Polymer Technology* (Temple Press Books, London) Chapter 5.

MILEWSKI, J.V. (1978), Aluminium oxide and other ceramic filaments, in: H.S. KATZ and J.V. MILEWSKI, eds., *Handbook of Fillers and Reinforcements* (Van Nostrand Reinhold, New York) pp. 583–596.

MILEWSKI, J.V. and H.S. KATZ (1978), Whiskers, in: H.S. KATZ and J.V. MILEWSKI, eds., *Handbook of Fillers and Reinforcements* (Van Nostrand Reinhold, New York) pp. 446–464.

MORGAN, P.W. (1979), *Plastics and Rubber Mat. and App.* **4**, 1.

NUSSBAUM, H.W. and T.J. CZARMOMSKI (1970), Smooth surface premix and sheet moulding compound technology, in: *25th SPI Conf.*, Washington, U.S.A., Society of Plastics Industry, Inc., NY, 6E1–6.

OSWITCH, S. (1974), *Composites* **5**, 55.

OSWITCH, S. (1975), *Reinforced Plastics* **19**, 180, 215.

OWEN, M.J. (1976), Progress towards a safe-life design method for glass reinforced plastics under fatigue loading, in: *Reinforced Plastics Congress*, British Plastics Federation, London, Paper 17.

PARKYN, B. (1972), *Composites* **3**, 29.

PENN, L., B. MORRA and E. MONES (1977), *Composites* **8**, 23.

PHILLIPS, L.N. and D.J. MURPHY (1976), RAE Rept. 76140.

PIPES, R.B., J.R. VINSON and T.W. CHOU (1976), *J. Comp. Mat.* **10**, 129.

POTTER, W.G. (1970), *Epoxide Resins* (Iliffe Books, London).

RHODES, F.E. (1979), Preliminary assessment of the significance of minor manufacturing defects in carbon fibre composites, in: *Significance of Defects in Failure of Fibre Composites* (Institute of Physics, London).

RINDE, J.A., E.T. MONES and H.A. NEWY (1977), Flexible epoxies for wet filament winding, in: *32nd SPI Conf.*, Washington, U.S.A., Society of Plastics Industry, Inc., NY, 11D1–5.

ROYLANCE, D. and M. ROYLANCE (1978), *Poly. Engrg. Sci.* **18**, 249.

SCOTT, J.M. and D.C. PHILLIPS (1975), *J. Mat. Sci.* **10**, 551.

SHEN, C.H. and G.S. SPRINGER (1976), *J. Comp. Mat.* **10**, 2.

SHIRRELL, C.D. (1978), *ASTM STP* **658**, 21.

SMITH, R.B. (1977), Watertown Arsenal, U.S.A., Private communication.

SOLDATOS, A.C., A.S. BURHANS and L.F. COLE (1969), Correlation between high performance epoxy cast resin properties and composite performance, in: *24th SPI Conf.*, Washington, U.S.A., Society of Plastics Industry, Inc., NY, 18C1–6.

SPITZBERGEN, J.C., P. LOEWRIGKEIT, C. BLUESTEIN, J. SUGARMAN and W.L. LAUZE (1971), Improved epoxy resin from dihydroxydiphenyl sulfone (bisphenol S), in: *26th SPI Conf.*, Washington, U.S.A., Society of Plastics Industry, Inc., NY, 19C1–8.

STAVINOHA, R.F. and J.C. MACRAE (1972), Derakane vinyl ester resins, unique chemistry for unique SMC opportunities, in: *27th SPI Conf.*, Washington, U.S.A., Society of Plastics Industry, Inc., NY, 2E1–4.

VACCARELLA, P. (1979), The fabrication of FRP chemical resistant equipment by the resin transfer technique, in: *34th SPI Conf.*, New Orleans, U.S.A., Society of Plastics Industry, Inc., NY, 3B1–4.

WATT, W., L.N. PHILLIPS and W. JOHNSON (1966), *The Engineer* **221**, 815.

WATT, W. (1972), *Carbon* **10**, 121.

WELLS, H. and N.L. HANCOX (1978), *Poly. Engrg. Sci.* **18**, 87.

YAJIMA, S., J. HAYASHI, M. OMORI and K. OKAMURA (1976), *Nature* **261**, 683.

CHAPTER II

Problems of the Mechanics of Composite Winding

Yu.M. Tarnopol'skii

Institute of Polymer Mechanics
Academy of Sciences of the Latvian S.S.R.
23, Aizkraukles Street
Riga 226006
Latvian S.S.R.
U.S.S.R.

A.I. Beil'

Institute of Wood Chemistry
Academy of Sciences of the Latvian S.S.R.
27, Academias Street
Riga 226006
Latvian S.S.R.
U.S.S.R.

Contents

HANDBOOK OF COMPOSITES, VOL. 4 – Fabrication of Composites
Edited by A. KELLY and S.T. MILEIKO
© 1983, Elsevier Science Publishers B.V.

List of Symbols

Coordinates

(a) Cartesian x_1, x_2, x_3; x_1 and x_2 coincide with the orthotropy axis in the reinforcement plane (x_1 is the direction of the maximum stiffness), but x_3 is along the normal to the reinforcement plane.
(b) Cylindrical r, θ, z.

Geometrical characteristics

r – a current radius
r_{in} – inner radius
r_{out} – outer radius
r_{man} – the inner radius of the mandrel
r_i – the outer radius of the ith layer
r_* – position of the moving boundary
h – thickness of a layer or tape
n – number of circuits
H_r – the wall thickness of a ring or cylinder
H_z – width of a ring, cylinder or disc in the axial direction
H_{in} – width H_z on the inner radius
φ – the angle between the radial direction and the wound filament (or tape)
φ_{in} – angle φ on the inner surface of the article
V_i – volume fraction of layers in ith direction
V_f – volume fraction of fibers

Relative linear dimensions

$\rho = r/r_{in}$ – relative current radius
$\rho_* = r_*/r_{in}$ – relative position of the moving boundary
$b = r_{out}/r_{in}$ – relative outer radius of the article
$\Delta b = h/r_{in}$ – relative thickness of a tape
$b_{man} = r_{in}/r_{man}$ – relative dimension of the mandrel
$b_i = r_i/r_{in}$ – relative outer radius of the ith circuit
b_* – the value of b, at which $\rho_* = 1$
$\rho_m (\rho_m^{res})$ – the relative current radius, corresponding to extremal (initial) stresses
c_i – inner-to-outer radii ratio of a layer in a multilayer ring

δb – total relative change in the outer radius due to circuit deformation

Displacements and strains

U_i – absolute values of displacements
$u_i = U_i/r_{in}$ – dimensionless displacements
e_i – strains in contracted notations
e_i^{free} – free strains (of thermal expansion, physico-chemical shrinkage, etc.)
e_i^{ch} – strains of physico-chemical shrinkage

Strength

σ_i^+ – tensile strength in the *i*th direction
σ_i^- – compressive strength in the *i*th direction

Thermoelastic characteristics

t – time
T – temperature
$\Delta T = T_2 - T_1$ – temperature difference
S_{ij} – components of the compliance matrix, $i, j \to 1, 2, 3$; $4 \to 2, 3$; $5 \to 1, 3$; $6 \to 1, 2$ or $r, \theta, z, rz, r\theta, \theta z$
$S_{kj}^{(i)}$ – compliance of the *i*th layer
E_i – Young's modulus in the *i*th direction
E_{man} – Young's modulus of the mandrel material
$E_j^{(i)}$ – Young's modulus of the *i*th layer
$E_3^{(I)} = E_r^{(I)}$ – tangential Young's modulus in the first section of the piecewise-linear $\sigma_r - e_r$ diagram
$E_3^{(II)} = E_r^{(II)}$ – tangential Young's modulus in the second section of the diagram
$\nu_{ij} = -S_{ij}/S_{ii}$ – Poisson's ratio
ν_{man} – Poisson's ratio of the mandrel material
$\nu_{kj}^{(i)}$ – Poisson's ratio of the *i*th layer
$\beta = \sqrt{S_{rr}/S_{\theta\theta}}$ – the anisotropy ratio
β_i – the anisotropy ratio of the *i*th layer
$\lambda = \sqrt{E_\theta/E_r^{(I)}}$ – the anisotropy ratio in the first section of the piecewise-linear $\sigma_r - e_r$ diagram
$\omega = \sqrt{E_\theta/E_r^{(II)}}$ – the anisotropy ratio in the second section of the diagram
μ – the anisotropy ratio before change in temperature
κ – the anisotropy ratio after change in temperature
α_i – coefficient of linear thermal expansion in the *i*th direction
α_{man} – coefficient, characterizing linear thermal expansion of the mandrel; for isotropic mandrels it is equal to the coefficient of linear thermal expansion of the mandrel material

Characteristics of stress state

W – strain energy
σ_i – stresses in contracted notations
σ^0 – stress in the windable tape
$N_0 = \sigma^0 h$ – tensioning per unit width of wound circuits
σ_i^* – the limit of proportionality on the piecewise-linear $\sigma_i - e_i$ diagram
σ_i^{res} – initial technological stresses
P – pressure on the mandrel; internal pressure in the pressing process
q – external pressure in the pressing process
P^- – negative pressure due to mandrel removal
$P_\infty = \lim\limits_{n\to\infty} P$ – pressure on the mandrel after winding of an infinite number of circuits
q_i – external pressure on the ith circuit
q_i^- – negative pressure increment on the radius r_i due to the mandrel removal
σ_i^h – stress after heat buildup
$\bar{\sigma}_i$ – stress averaged throughout the volume
$\sigma_{\theta c}^0$ – coefficient, characterizing the stress level during winding of a ring with alternating tension

Parameters, determined according to the formulae

γ – (3.4)	Y – (4.9)
η – (3.11)	Φ – (4.16)
J – (3.14)	Ω – (4.16)
ξ – (3.28)	δ – (6.3)
ζ – (3.28)	g – (6.3)
D – (4.4)	Z – (6.3)
ψ – (4.9)	

1. Characteristic features of composite winding

1.1. Introduction

A winding operation is the basic fabrication technique for forming load-bearing structural elements made of polymer matrix-based fibrous composites, having the shape of bodies of revolution. The technique allows one easily and with high precision to realize practically any two-dimensional reinforcement scheme. The gist of the winding process is schematically shown in Figs. 1a and 1b. Resin soaked filaments, strands, tapes and fabrics are wound onto the mandrel or a part layer after layer with preset tension. Depending on the type of the winding machine (ROSATO and GROVE, 1964), the mandrel can make one or two

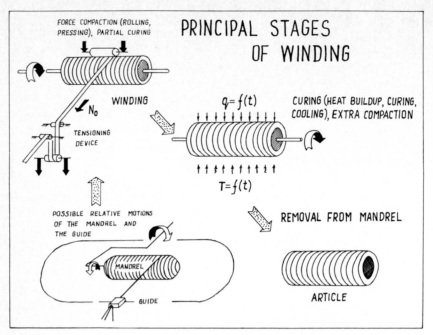

FIG. 1a. The filament winding process. Principal stages of winding.

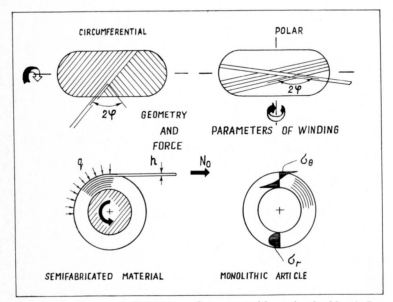

FIG. 1b. The filament winding process. Geometry and forces involved in winding.

rotational movements (around the longitudinal axis—hoop winding, or around a given point—polar winding), and with a thread guide—an inverted translational or rotational movement. This permits one to control the reinforcement laying scheme both within one and the same layer and through the thickness of a packet (or an article) by varying the angle of filament or tape placement, i.e., $\varphi(\rho, z)$. During winding, due to fiber tensioning, the necessary pressure between circuits q is formed, which determines, in the main, the monolithicity (integrity) of the article; in a composite, fiber tensioning contributes to more complete utilization of the strength and stiffness of the reinforcing fibers. If the contact pressure $q = N_0/R$ turns out to be insufficient for compaction of the material, additional layer-by-layer compaction must be employed, for example, by rolling or pressing of the whole packet with extra external or internal pressure. By controlling the filament tensioning and reinforcement scheme, it is possible to reasonably affect strength and stiffness of the material, the system of initial stresses, pressure on a windable part. Thereby, constructional and technological aspects of the fabrication of wound structures are closely interconnected (TARNOPOL'SKII and ROZE, 1969, OBRAZTSOV et al., 1977).

Composite winding differs essentially from that of other materials (wire, electric wires in polymeric insulations, ropes, filaments, paper, fabric, magnetic tapes, etc.) in that the former is a method of converting a semifabricated material into a monolithic article. It should be noted that the semifabricated material possesses extremely low strength and stiffness in the radial direction. Creation of 'a growing body' from such a material by a winding technique is associated with considerable compressibility of circuits and, consequently, with unavoidable loss and redistribution of the preset tension N_0.

Growth in size is accompanied by the appearance and changes in the system of initial stresses, caused by tensioning and compliance of the material.

For conversion of a semifabricate into a monolithic material, the wound article must undergo heat treatment, which includes heat buildup, cure and cooling stages. In all these stages, alteration of the initial scheme of tensioning continues. Upon heat treatment, the temperature and pressure can be constant or vary with time in accordance with the program. The mandrel ensures the necessary shape of the article; it is removed after cure. The winding operation is illustrated in Fig. 1a and the geometrical and force parameters in Fig. 1b. The variation of forces in all stages of the winding process is treated in more detail in Section 1.3.

Three main groups of wound articles may be distinguished: thin-walled shells, thick-walled[1] structural elements and hybrid articles (metal–composite, composite–composite).

For thin-walled shells, it is most important that the optimal design and reinforcing scheme should be found under given loading conditions. The main

---

[1] The conception 'thick-walled' for composites is related not only to the geometrical dimensions, but also the elastic anisotropy.

technological problem of realization of the optimal reinforcing scheme lies in maintaining filaments in an equilibrium state along a preset path in the winding process. For hybrid articles, it is necessary that surface tractions (in particular, pressure) of a definite magnitude should be formed on the interface. This is ensured by stress and time-temperature histories during winding and thermal treatment. For thick-walled structural elements, one more problem arises— that of precise control of the distribution of initial stresses. The problem is made more involved by the fact that the increase in the thickness raises the level of interlaminar stresses, which are negligible for thin-walled shells. Because of the extremely weak transverse tensile strength of composites during processing, it is the interlaminar stresses which become the critical factor. Attention in the present work is focused on the specific features of thick-walled structures. Theory is compared with experiment throughout. Whenever necessary, figure captions give references, containing detailed description of the technique and means of experimental data accumulation.

Wound composites have a layered structure, in which layers form either thin concentric rings or an Archimedes helix in the planar radial section of articles. A multilayer concentric structure is characteristic of angle-ply winding with circuits touching. Helical structures are formed by winding 'in a track'. In the design of wound articles, the helical structure is substituted, as a rule, by a concentrically-circular one due to the fact that the directions of layers in such schemes differ only by a small angle of pitch. Specific features of fibrous composite winding and load-carrying capacity of structural elements may be studied on the simplest model—a ring—and later carried over to cylinders, including even those with complex reinforcing schemes. Test methods for ring specimens are described and summed up in (TARNOPOL'SKII and KINCIS, 1981).

1.2. History in brief

Historically, developments in the winding technology may be traced into the dim past. One of the predecessors of today's wound articles is assumed to be the Egyptian flask (see Fig. 2), made, speaking in modern language, by winding the glass fibers from the melt onto a mandrel, dissolved after finishing the process (PARKYN, 1963). Another instance is wrapping of embalmed mummies (RICHARDSON, 1977) with tapes soaked in resin which were the natural counterparts of polyester resins. Let us note that in the latter case wrappings were performed on a compliance mandrel. For ages pressure vessels have been wrapped with the aim of reinforcement—rope wrappings of bamboo powder rockets in Ancient China, brass string wrappings of wooden cannon barrels, later on steel wire wrappings of steel shotgun barrels, etc.

The advent of fibrous composites marked the beginnings of a new stage. Historically, in the development of fibrous winding technology three stages may be distinguished (ROSATO, 1969).

Stage 1. The adoption of the first patent pertaining to composite winding in 1946.

FIG. 2. A flask, made by winding fiber glass on a soluble mandrel. Egypt XVII dynasty in 1370 B.C., British Museum (PARKYN, 1963).

Stage 2. The creation of the first wound circular sample—a NOL-ring—in 1955, intended for joint technological and strength testing. Close interaction between technological parameters and properties of wound structural elements has been established on a NOL-ring sample.

Stage 3. Publishing of the first monograph on the technology of composite winding (ROSATO and GROVE, 1964) marked the completion of the purely empirical period in the development of composite winding technology.

The mechanics of fibrous composite winding as an important branch of composite mechanics is based on advances made in many fields of science. The mechanics of an elastic filament and, in particular, its equilibrium conditions on a rough surface of an arbitrary shape, are extensively used. In the mechanics of thin-walled wound shells, advances in the mechanics of net shells (in particular, rubbercord) are widely used. In the design calculations of thick-walled wound articles the mechanics of multilayer prestressed structures is of great importance. The winding mechanics of an extendable wire has been treated in SOUTHWELL (1948). The so-called circular model has also been described in the work. The model was later successfully used in descriptions of the winding process for various types of materials.

An important period in the evolution of winding mechanics set in with studying an important step for composite—the turning of a semifabricated material into a finished one. The engineering mechanics of the winding process has been successfully enriched over the last few years. The following contributions must be mentioned: BIDERMAN (1958), BIDERMAN (1979), BIDERMAN et al. (1969), BOLOTIN (1972), BOLOTIN (1975), BOLOTIN et al. (1972), BOLOTIN and NOVICHKOV (1980), ELPAT'EVSKII and VASIL'EV (1972), LIU and CHAMIS

(1965) and OBRAZTSOV et al. (1975). The kinetics of the curing stage have been investigated and described in ROZENBERG and ENIKOLOPYAN (1978). A survey of literature is given in TARNOPOL'SKII (1976), TARNOPOL'SKII and ROZE (1963) and TARNOPOL'SKII et al. (1980).

1.3. The engineering theory

As has been indicated, the winding operation of fibrous composites, in contrast to the winding of traditional materials, is a method of processing a semifabricated material into a monolithic article. The winding operation of a semifabricated material under the given winding tension is inherently a very important stage, but it is only the first technological stage.

Even the winding operation itself differs essentially from the winding operation of traditional materials, firstly, by essentially higher (by orders of magnitude) anisotropy, secondly, by more complex rheological behavior, determined by the high viscosity of the polymeric binder and, thirdly, by the formation of a monolithic material. Here, the monolithicity means that unwinding presents itself as a typical cohesive viscous failure. In general, the winding stage is followed by the stages of heat buildup, curing at an elevated temperature, cooling and removal of the mandrel (if the mandrel is not an integral part of the article). Heat buildup is accompanied by simultaneous thermal expansion of the article and the mandrel, the decrease in radial stiffness, filtration, partial stress relaxation. Upon curing the radial stiffness and transverse strength increase and simultaneous physico-chemical shrinkage occurs. Thick-walled articles are mainly fabricated from hot-cured resins, because polymerization of cold-cured resins is accompanied by uncontrollable self-heating and, as a rule, by considerable physico-chemical shrinkage. Cooling is associated with simultaneous thermal shrinkage of the article and the mandrel and the increase in the radial stiffness and strength. Upon cooling the article may separate from the mandrel at some temperature. If such separation does not occur, it is accomplished by artificial means. The separated mandrel is removed afterwards.

During the entire technological history of the process, the physico-mechanical properties and stress–strain state may undergo a significant change. Therefore, a universal rheological model would be very difficult to set up. The engineering method under consideration involves division of the mechanical history of an article into several stages (including the operational stage), in accordance with the technological stages. In each of these stages, the mechanical behavior of the material follows a specific rheological law. On passing from one stage to another, a discontinuous change in material properties has been observed. Various simplifying hypotheses may be assumed for this case; for example, a hypothesis about inheritance of the state of stress (neglecting differences of strains) or some more complex hypothesis, which may generally be referred to as a hypothesis about inheritance of the stress–strain state. Division into stages is convenient, because it allows a series of simplifications to be applied at each stage, each of which is inappropriate for description of the process as a whole.

Experiments with deformable mandrels have served as an impetus for the development of an engineering theory (BRIVMANIS, 1966). Distinction between the winding process of a tape of semifabricated material and that of an isotropic metallic tape is made by examining an integral force parameter—the dependence of the mandrel winding pressure on the number of circuits being wound under equal conditions (see Fig. 3). A significant part of the pressure is

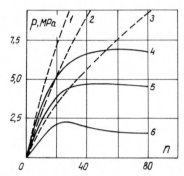

FIG. 3. Pressure on the mandrel dependence on the number of circuits, the tape material and winding tension (BRIVMANIS and GAGANOV, 1971). N_0 = const. – – – a nickel tape; ——— a fiberglass tape (a semifabricated article). Curves 1, 4—N_0 = 50 kN/m; Curves 2, 5—N_0 = 37 kN/m; Curves 3, 6—N_0 = 25 kN/m.

not transferred through the layers to the mandrel, but consumed in the deformation process of the lower lying circuits. This may be attributed to the essential material anisotropy and, in particular, to the high compliance of the semifabricated material in transverse compression. It was possible to evaluate the variation in winding pressure in all stages of the technological process by means of deformable mandrels (see Fig. 4). In the curing process, the constant pressure on the mandrel in different technological regimes is of special interest (cf. Fig. 4). This approach made it possible to develop variants of the theory of initial stresses, in which, by applying the hypothesis about summation of stress states in each of the stages, the polymerization stage had been omitted. On this basis the methods and approaches have been worked out, which made it possible to calculate the variation of stress–strain state in the fabrication process and to compare it with the variation in strength, to calculate initial technological stresses and to consider other features, associated with the winding operation. Furthermore, on the basis of article strength and its service conditions—to propose and describe the optimal technological regimes, contributing to the improvement of the load-carrying capacity of the structures.

Naturally, the first problem required evaluation and description of the anisotropy of the strain and strength properties of a semifabricated and cured material. Attention has been centered on transverse loading.

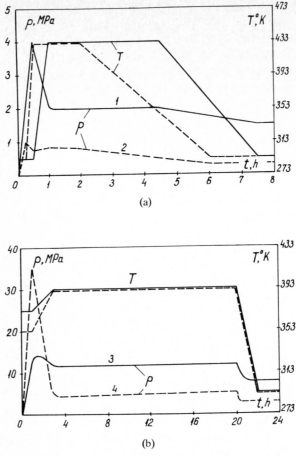

FIG. 4. Variation in the winding pressure with time on a tensometric mandrel in all states of article fabrication. (a) winding without preliminary heating; (b) winding with preliminary heating (BLAGONADJOZHIN et al., 1970); Curve 1—A cold tape, a cold mandrel (——); Curve 2—An impregnated filament, a cold mandrel (– – –); Curves 3, 4—a heated tape ($T = 393°K$), a heated (——) and cold (– – –) mandrel.

2. Anisotropy of wound composites

2.1. A semifabricated material

In order to describe the winding operation analytically, it is necessary to know the strain properties of a semifabricated material. The longitudinal properties of a tape or braid are dependent on the fiber properties; for a rigid reinforcement under maintained pretensioning during processing they may be described by Hooke's law with sufficient accuracy. Transverse deformability is characterized by high compliance. This characteristic is usually determined by compression of a packet of layers of the same thickness as during winding. The boundary conditions in test units should exactly simulate those existing during winding of real articles. The available data, characterizing the behavior of a semifabricated material in transverse static compression, are presented in Fig. 5. It should be noted that the semifabricated materials are more sensitive to the test methodology than the cured materials. This effect not withstanding, test results testify to the essential nonlinearity of the $\sigma_3 - e_3$ diagram, dictated by volumetric compaction of the packet and formation of a monolithic material.

The anisotropy ratio β, characterized by the parameter $\beta = \sqrt{E_1/E_3}$ (where E_1 is the circumferential Young's modulus and E_3 the tangential Young's modulus in transverse compression), of a semifabricated material varies within a wide range, depending on the resin, its composition, temperature and fiber properties in the range $4 \leqslant \beta \leqslant 200$. Consequently, the semifabricated material may be characterized as possessing extremely well defined anisotropy and essentially nonlinear behavior in transverse direction. One needs to know the transverse properties in order to develop a quantitative theory.

During winding, the binder has a characteristic of irreversible creep (more distinct for 'wet' winding and less for winding of a prepreg). Moreover, during winding air bubbles are formed between layers. Therefore, at first glance, characterization of the semifabricated material in terms of the $\sigma_3 - e_3$ diagram

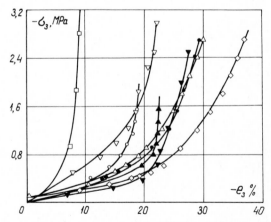

FIG. 5. $\sigma_3 - e_3$ diagrams in transverse compression of a GFRP semifabricate with various reinforcing schemes (PORTNOV and BEIL', 1977).
□—epoxy-thiokol glass-cloth-base laminate, $T = 293°$K;
○—polyester glass-cloth-base laminate, $T = 293°$K;
▽—a unidirectional polyester GFRP, $T = 293°$K;
▲—a polyester glass-cloth-base laminate, $T = 373°$K;
▼—a unidirectional polyester GFRP, $T = 375°$K;
●—a satin polyester glass-cloth-base laminate, $T = 293°$K;
△—a polyester glass mat; $T = 293°$K;
◇—a polyester glass-cloth-base laminate, $T = 293°$K.

in Fig. 5 seems to be inappropriate. However, since the wound articles are designed so that the interlaminar shear stresses should be minimal or completely absent, and circumferential stresses should be carried by the fibers, while in transverse compression volumetric compression mainly occurs, the viscous-flow state of a polymeric binder controls its infiltration from regions of high interlaminar pressure to regions where it is low. In this case, microflow (caused by local gradients of interlaminar pressure due to the presence of voids) and macroflow (caused by a gradient of transverse macrostresses) should be distinguishable. Local pressure gradients are realized in small volumes. They are comparatively high and they cause fast microflows of the binder, resulting in a zero averaged macroflow. Microfiltration stops comparatively soon. The observed short-term creep due to volumetric compression of the semifabricated material is, evidently, attributed to microflow. Isochrones of creep for a polyester prepreg are presented in Fig. 6. Let us note the three characteristic

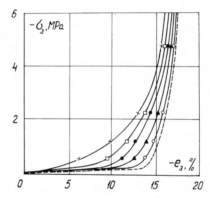

FIG. 6. Isochronic lines of creep in transverse compression of a GFRP semifabricate at 293°K and $t = 45$ (×), 90 ($\square$), 150 ($\bullet$), 240 ($\blacktriangle$), 960 ($\bigcirc$) sec and $t \to \infty$ (---).

traits of a family of isochrones: they are similar, essentially nonlinear and lie within a narrow band. They show that the $\sigma_3 - e_3$ diagrams may be employed in the design calculations. Analogous conclusions follow from the creep data for uncured wound GFRP, made by a wet technique (ABIBOV et al., 1973).

Reverse creep curves testify to the fact that the strains, associated with compaction of the semifabricated material, are irreversible. Residual strains for the majority of semifabricated materials are equal to 8–20%. Repeated loadings and unloadings have nearly a linearly-elastic character. In such a way, the diagram 'loading–unloading' in transverse compression of a semifabricated material may be schematically approximated as a piecewise-linear relationship (see Fig. 7), where σ_3^* is the proportionality limit.

Macroflows proceed comparatively slowly and, in most cases, contribute little to nonuniform deformation of the material due to variation of the reinforcement coefficient. However, since nonuniformity, as a rule, is low in the finished article and mainly attributable to the nonuniform volumetric compaction,

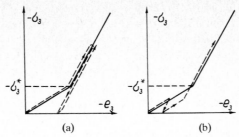

FIG. 7. A piecewise-linear approximation of the diagram for a GFRP semifabricate in transverse compression with unloading. (a) max $|\sigma_3| > |\sigma_3^*|$; (b) max $|\sigma_3| < |\sigma_3^*|$.

preceding the onset of polymerization, consideration of macroflow in the calculation of stress state is not essential.

In such a way, rheological features, manifesting themselves in the winding process of composites, may be conditionally divided into basic ones—the anisotropy of deformability and physical nonlinearity under an active transverse compressive loading, and secondary ones—creep in transverse compression, unlimited interlaminar shear creep, and macroflow of the binder.

The strengths in longitudinal tension σ_1^+ and transverse compression σ_3^- are essential for description of the winding stage. These characteristics do not differ by an order of magnitude from the respective values for cured composites. If the longitudinal loads become compressive, fiber curvatures occur. Therefore, in the design calculations it may be assumed that the strength in longitudinal compression σ_1^- of the semifabricated material is equal to zero.

2.2. Heat buildup and curing

In the heat buildup stage a fall in stiffness of the semifabricated material occurs in transverse compression. A quantitative estimation of the phenomenon may be obtained from the comparison of curves, presented in Fig. 5 for the same materials, but at various test temperatures—20°C and 100°C. The tangential Young's modulus in the initial section of the diagram in transverse compression decreases two-to-three-fold, the tangential modulus in the second section by 40–50%, the proportionality limit two-fold, but the secant Young's modulus by 25–30%. Upon heat buildup, thermal expansion of the material takes place. The value of the longitudinal coefficient of linear thermal expansion, α_1, is nearly equal to that of a cured material. The transverse coefficient of linear thermal expansion α_3 varies greatly and depends essentially on the degree of preliminary compaction. The values of the material after compaction, α_3, may be obtained from binder densities. From the data of measured binder densities (BOLOTIN and BOLOTINA, 1972) in the processes of heat buildup and curing it may be shown that the coefficient of linear thermal expansion is equal to $(15–20) \times 10^{-5}\,°\mathrm{C}^{-1}$. The coefficient is nearly the same as upon cooling of the same cured binders. Consequently, it may be assumed that the coefficient of linear thermal expansion of the semifabricated material in the transverse direction is equal or somewhat higher than the respective coefficient of the cured composite.

Upon polymerization, the stiffness and strength increase in the transverse direction and physico-chemical shrinkage also occurs in the same direction. As a result of polymerization the character of the diagram of transverse compression (see Fig. 8) reverses; the material does not undergo stiffening with the increase in compressive stress, as for the semifabricated material (cf. Fig. 5), but becomes more compliant. The Young's modulus, E_3 (even when compared with the tangential Young's modulus of the semifabricated material in the second section of the σ_3–e_3 diagram), increases by 50–100% as a result of polymerization, but the strength by 50%. After polymerization the material acquires transverse tensile strength.

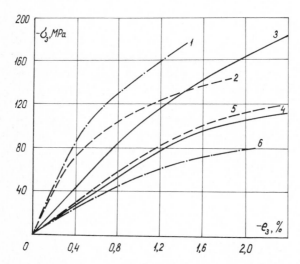

FIG. 8. Diagrams of transverse compression at temperatures 293°K (Curves 1, 2, 3) and 373°K (Curves 4, 5, 6) for three types of unidirectional GFRP.

The magnitude of linear physico-chemical shrinkage in the transverse direction upon polymerization is nearly 0.5%. It should be noted that different data on the physico-chemical shrinkage are available. The differences are caused by various definitions of shrinkage (in particular, definition of the starting point), by different measuring techniques, experimental conditions and different materials properties. Determination of the values of shrinkage is of great significance. The same can be said about the study of the kinetics of increasing the elastic and strength properties. Only the first steps have been made so far in that direction, as a result of which more problems have been posed than ultimate results obtained.

2.3. The cured material

After cooling, stiffening of the material occurs. This is borne out by comparing the diagrams of transverse compression, σ_3–e_3, of the same composites at various temperature (cf. Fig. 8). The diagrams of transverse tension, σ_3–e_3, are practically linear, unless compliant resins with high ultimate strains are used. The Young's modulus in transverse tension, E_3, varies nonlinearly with falling temperature (see Fig. 9); in particular it decreases sharply below the glass transition temperature of the binder. The dependence on the temperature

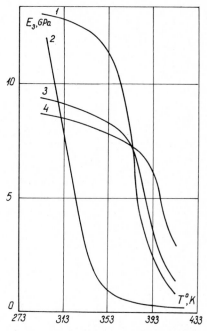

FIG. 9. Transverse modulus of elasticity of epoxy GFRP vs. temperature. Curve 1—0/90; Curve 2—0, winding of a tape (BOLOTIN and BOLOTINA, 1972); Curve 3—0; Curve 4—a glass-cloth-base laminate; Curves 1, 3, 4—wet winding (KOSTRITSKII and TSIRKIN, 1981).

of the transverse tension strength, σ_3^+, is also characterized by nonlinearity (see Fig. 10). The same can be said about longitudinal and transverse coefficients of linear thermal expansion of the material, α_1 and α_3, as well as of their difference, $\alpha_3 - \alpha_1$ (see Fig. 11). Therefore, in studying the cooling process, attention has been focussed on the comparison of strength, σ_3^+, and stress variation with temperature in finished articles.

Mechanical properties of typical unidirectional wound composites are presented in Table 1. By comparing a unidirectional reinforcing scheme with a planar one it becomes obvious that the whole complex of physico-mechanical properties undergoes an essential change, which can be evaluated in terms of

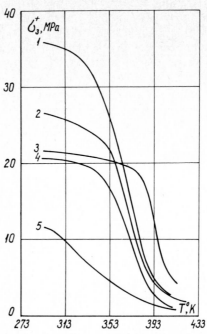

FIG. 10. Transverse tension strength of epoxy GFRP vs. temperature. Curves 2, 5—BOLOTIN and BOLOTINA (1972); Curves 1, 3, 4—KOSTRITSKII and TSIRKIN (1981); Curves 1, 5—0; Curves 2, 3—glass-cloth-base laminates; Curve 4—0/90; Curves 1, 2, 4—wet winding.

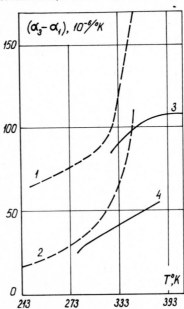

FIG. 11. Difference in the coefficients of thermal expansion $\alpha_3 - \alpha_1$ vs. temperature. Curve 1—OFRP ($V_f = 60\%$) (CLEMENTS, 1977); Curve 2—GFRP ($V_f = 60\%$) (CLEMENTS, 1977); Curve 3—CFRP (TARNOPOL'SKII et al., 1973); Curve 4—GFRP (BOLOTIN and BOLOTINA, 1972).

TABLE 1.[a] Typical characteristics of unidirectional epoxy composites (CHAMIS and KIRALY, 1975; CLEMENTS, 1977; TARNOPOL'SKII and KINCIS, 1981).

Characteristics	Reinforcing fibers			
	Glass	Carbon	Boron	Organic
α_1, $(°K)^- \times 10^{-6}$	3.8	−0.3	6.1	−6.0
$\alpha_2 = \alpha_3$, $(°K)^- \times 10^{-6}$	16.7	40.0	30.4	90.0
S_{11}, 10^{-2}/GPa	1.6	0.5	0.56	1.2
$S_{22} = S_{33}$, 10^{-2}/GPa	4.0	10.0	4.5	23.2
S_{66}, 10^{-2}/GPa	20.0	20.0	18.2	58.8
$S_{13} = S_{12}$, 10^{-2}/GPa	−0.4	−0.2	−0.1	−0.4
S_{23}, 10^{-2}/GPa	−2.0	−4.8	−2.3	−11.6
σ_1^+, MPa	1300	1500	1400	1200
σ_1^-, MPa	850	1800	1600	300
$\sigma_2^+ = \sigma_3^+$, MPa	45	40	55	10
$\sigma_2^- = \sigma_3^-$, MPa	180	250	130	65
τ_6, MPa	45	70	60	30

[a] The plus sign of σ^+ designates tension, the minus sign of σ^- compression.

the mechanics of laminated media. The following changes in transverse properties have been marked: Young's modulus, E_3 (cf. Fig. 9), and the strength in tension, σ_3^+ (cf. Fig. 10), decrease if the unidirectional fiber layup scheme is converted in a crossply scheme, but analogous characteristics in compression increase (see Fig. 12); the coefficient of linear thermal expansion, α_3, increases somewhat also. The above follows from comparison of data for the same material at the same temperature, but having various fiber layups.

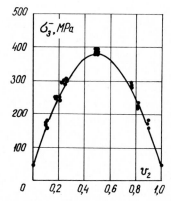

FIG. 12. Strength of a cross-plied laminate in transverse compression. V_2 is the fiber volume fraction in the second direction in a plane (KHITROV and ZHMUD, 1974).

3. Winding stage

3.1. *Basic problems pertaining to the winding process*

During winding, a monolithic composite material is formed from single filaments, layers, etc., having a given reinforcing scheme. In order to ensure integrity it is necessary that a certain interlaminar pressure should be set up. There are two ways of attaining this objective:
  (1) winding operation under controlled winding tension (force winding), and
  (2) the use of extra compaction processes (pressing, rolling, etc.).
  The force winding is less effective in providing the necessary interlaminar pressure with the increase in the absolute dimensions or the radius of the article, because the longitudinal strength, σ_1^+, imposes restrictions on the winding tension. In this case, the extra compaction processes should be widely used. Still, the extra compaction alone is insufficient and the force winding is still desirable, because it allows alignment of the reinforcing fibers and an increase in the longitudinal strength. During force winding as well as extra compaction the previously wound layers undergo some deformation. Therefore, in the process of winding each successive layer, the stress state in the already wound circuits changes. A further change in the stress state occurs in subsequent technological states. In several cases, this leads to a complete drop in pretensioning and development of various structural defects—in particular, local fiber curvatures. Knowledge of the interrelation of parameters in force winding (or extra compaction) and stress and strain fields permits evaluation and effective control of
  – the magnitude of the interlaminar pressure,
  – the pressure on the mandrel or a part being wound,
  – radial distribution of stresses after winding,
  – distribution of initial stresses.
  The interlaminar pressure determines the degree of monolithicity to a large extent. The pressure on the mandrel is the basic characteristic of prestressed hybrid structures; in the winding stage the pressure should not exceed the buckling pressure of thin-walled mandrels or the strength of thick-walled mandrels. At the expense of winding tension, not only a favourable distribution of initial stresses is formed, but also the danger of reinforcing fiber curvature is eliminated; it is desirable that the compressive stresses developed in the lower lying circuits should not reach the critical magnitude. Otherwise the circuit curvatures in compression may result in local strength loss.
  Problems of force winding are more complex than those pertaining to extra compaction (being, in most cases, contact problems, Lame's problem, etc.). Further attention will be focussed on the analysis of force winding.

3.2. *A linearly-elastic circular model*

3.2.1. *Description of the model*
Many problems of winding mechanics have been evolved on the simplest model

of wound articles—a ring. The model makes it possible to consider the basic rheological features of the semifabricated material and the basic peculiarity of the reinforcing scheme—the absence of a radial reinforcement. Analysis of the circular model contributes to the development of an engineering understanding of the phenomena occurring during winding.

According to the circular model, the winding operation involves successive mounting of thin circular layers onto a mandrel and then on to the preceding layer under tensioning equal to the winding tension of a tape (see Fig. 13). The linearly-elastic circular winding model makes allowance for the most important factor—the anisotropy of strain properties of the semifabricated material. The model describes the process qualitatively and, in some cases—under low or very high winding tensions—also quantitatively.

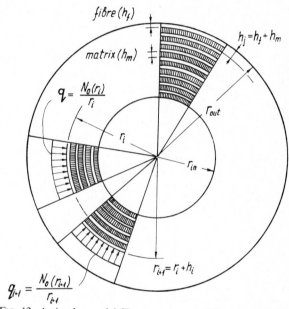

FIG. 13. A circular model (TARNOPOL'SKII and PORTNOV, 1966).

At the moment when the outer radius of the successive circuit being wound lies on the running radius r, the stresses on the outer radius are equal to

$$\sigma_r = 0, \qquad \sigma_\theta = \sigma_\theta^0, \tag{3.1}$$

where σ_θ^0 is the stress, preset by a tensioning device. The stresses vary with the winding pressure during winding of each successive circuit

$$\Delta\sigma_r(r_i) = -\frac{h_{i+1}}{r_i}\sigma_\theta^0(r_{i+1}) = -\frac{\Delta b_{i+1}}{\rho_i}\sigma_\theta^0(\rho_{i+1}) \quad \text{at } \rho_i = b_{i+1} - \Delta b_{i+1} \tag{3.2}$$

where ρ, Δb and b are the radius, the layer thickness and the running outer radius, respectively, expressed in fractions of the inner radius r_{in}. Further, all the linear values, including displacements, will be expressed in fractions of the inner radius. When determining the stress increments $\Delta\sigma_r$ and $\Delta\sigma_\theta$, the already wound ring may be conceived as being a quasi-uniform, cylindrically-orthotropic body and the boundary conditions between the mandrel and the wound body reflect the continuity of radial stresses and displacements; for a ring, it is expressed as

$$\frac{\Delta U_r}{r_{in}} = \Delta u_r = \Delta e_\theta = \gamma\,\Delta\sigma_r S_{\theta\theta}\quad\text{at } \rho = 1 \tag{3.3}$$

where γ is the relative compliance of the mandrel. Physically, it may be explained from (3.3) and the above boundary conditions as follows. Compliance may be characterized as the ratio of the circumferential strain of its outer surface (or, equivalently, the radial displacement in fractions of r_{in}) to the value of the external pressure, causing the displacement. By multiplying the absolute magnitude by E_θ, we shall obtain a dimensionless parameter γ, reflecting the ratio of mandrel compliance to the compliance of the material being wound. For an isotropic mandrel, having characteristics E_{man}, ν_{man} and a relation of outer-to-inner radii b_{man} the value of γ is equal to

$$\gamma = \frac{E_\theta}{E_{man}}\left(\frac{b_{man}^2 + 1}{b_{man}^2 - 1} - \nu_{man}\right). \tag{3.4}$$

The equation of an axisymmetric problem for a cylindrically-orthotropic ring (LEKHNITSKII, 1957), obtained by subsequent substitution of the equilibrium equation

$$\sigma_\theta = \sigma_r + \rho\,\frac{d\sigma_r}{d\rho} \tag{3.5}$$

into Hooke's law in a form of stress–strain relation

$$e_r = S_{rr}\sigma_r + S_{r\theta}\sigma_\theta,\qquad e_\theta = S_{\theta r}\sigma_r + S_{\theta\theta}\sigma_\theta, \tag{3.6}$$

and the result into a strain compatibility equation

$$e_r = e_\theta + \rho\,\frac{de_\theta}{d\rho} \tag{3.7}$$

takes the form

$$\rho^2\frac{d^2\Delta\sigma_r}{d\rho^2} + 3\rho\,\frac{d\Delta\sigma_r}{d\rho} + (1 - \beta^2)\,\Delta\sigma_r = 0. \tag{3.8}$$

Solution of the equation

$$\Delta\sigma_r = C_1\rho^{\beta-1} + C_2\rho^{-\beta-1} \tag{3.9}$$

contains the constants C_1 and C_2, which may be found from the boundary conditions (3.1) and (3.2). The ultimate result takes the form

$$\Delta\sigma_r(\rho, b_i) = -\sigma_\theta^0(b_i)\frac{(\eta\rho^{\beta-1}+\rho^{-\beta-1})\,\Delta b_i}{\eta b_i^\beta + b_i^{-\beta}},$$

$$\Delta\sigma_\theta(\rho, b_i) = -\sigma_\theta^0(b_i)\beta\frac{(\eta\rho^{\beta-1}-\rho^{-\beta-1})\,\Delta b_i}{\eta b_i^\beta + b_i^{-\beta}}, \tag{3.10}$$

$$\Delta u_r(\rho, b_i) = -S_{\theta\theta}\sigma_\theta^0(b_i)[(\beta-\nu_{\theta r})\eta\rho^\beta + (\beta+\nu_{\theta r})\rho^{-\beta}]\frac{\Delta b_i}{\eta b^\beta + b_i^{-\beta}},$$

$$\eta = \frac{\beta + \gamma + \nu_{\theta r}}{\beta - \gamma - \nu_{\theta r}} \tag{3.11}$$

where

$$\nu_{\theta r} = -S_{r\theta}/S_{\theta\theta}, \qquad \beta = \sqrt{S_{rr}/S_{\theta\theta}}.$$

Here S_{rr}, $S_{\theta\theta}$, $S_{r\theta} = S_{\theta r}$ are components of a compliance matrix of the material being wound.

Stresses and displacements in a completely wound ring having the relative outer radius b are summed up as follows:

$$\sigma_r(\rho, b) = \Delta\sigma_r(\rho, \rho+\Delta b) + \Delta\sigma_r(\rho, \rho+2\Delta b) + \cdots + \Delta\sigma_r(\rho, b),$$

$$\sigma_\theta(\rho, b) = \sigma_\theta^0 + \Delta\sigma_\theta(\rho, \rho+\Delta b) + \Delta\sigma_\theta(\rho, \rho+2\Delta b) + \cdots + \Delta\sigma_\theta(\rho, b), \tag{3.12}$$

$$u_r(\rho, b) = \Delta u_r(\rho, \rho+\Delta b) + \Delta u_r(\rho, \rho+2\Delta b) + \cdots + \Delta u_r(\rho, b).$$

As Δb is extremely small, it is reasonable for a large number of circuits to replace summation by integration:

$$\sigma_r(\rho, b) = -(\eta\rho^{\beta-1}+\rho^{-\beta-1})J(\rho, b),$$

$$\sigma_\theta(\rho, b) = \sigma_\theta^0 - \beta(\eta\rho^{\beta-1}-\rho^{-\beta-1})J(\rho, b), \tag{3.13}$$

$$u_r(\rho, b) = -S_{\theta\theta}[(\beta-\nu_{\theta r})\eta\rho^\beta - (\beta+\nu_{\theta r})\rho^{-\beta}]J(\rho, b)$$

where

$$J(\rho, b) = \int_\rho^b \frac{\sigma_\theta^0(y)\,dy}{\eta y^\beta + y^{-\beta}}. \tag{3.14}$$

Thereby as ρ approaches the outer radius b, the number of components in (3.12) reduces and the error from substitution increases.

The integral of the right-hand side of (3.14) for some winding programs $\sigma_\theta^0(\rho)$ may be obtained analytically in a closed form or in a fast-convergent series. Generally, it is obtained numerically and then the direct use of a discrete variant (3.10)–(3.14) is effective.

The energy approach to the design of the winding process was proposed in VASIL'EV (1969). The approach assumes that the stress $\sigma_\theta(b_i, b_i)$ in the running outer ith circuit is not equal to the stress $\sigma_\theta^0(b_i)$, developed by a tensioning device. The value $\sigma_\theta(b_i)$ is obtained from the energy balance equation

$$W_0 = W_1 + W_2 \tag{3.15}$$

where W_0 is the potential energy of the tape of the length equal to that of a current circuit, stretched under the stress σ_θ^0; W_1 is the strain energy increment in the system: the previously wound circuits + the mandrel; W_2 is the energy left over in the last circuit. Expressions for W_0 and W_2 are trivial; W_1 is obtained as

$$W_1 = -\pi \Delta b \Delta u(b_i, b_i) \sigma_\theta(b_i, b_i) r_{\text{in}}^2$$

by using (3.10), where instead of $\sigma_\theta^0(b_i)$, $\sigma_\theta(b_i, b_i)$ is substituted. Eventually, we obtain from (3.15)

$$\sigma_\theta(b_i, b_i) = \frac{\sigma_\theta^0(b_i)}{\sqrt{1 + \dfrac{\Delta b}{b_i}\left(\beta\,\dfrac{\eta b_i^{2\beta} - 1}{\eta b_i^{2\beta} + 1} - \nu_{\theta r}\right)}}. \tag{3.16}$$

The order of deviation of $\sigma_\theta(b_i, b_i)$ from $\sigma_\theta^0(b_i)$ is determined by the order of magnitude $\beta\,\Delta b/b_i$, which may reach 10^{-3} to 10^{-1}. In a physical sense, the two approaches differ in that the force approach (TARNOPOL'SKII and PORTNOV, 1966) corresponds to the winding operation without any friction between the ultimate and penultimate layer, but the energy approach (VASIL'EV, 1969) to the winding operation with an infinitely high frictional coefficient. The actual situation is, evidently, the intermediate state. Let us note that it follows from (3.16) that winding under constant winding tension according to the energy approach corresponds to winding under slowly decreasing winding tension according to the force approach. As $\Delta b \to 0$, both the approaches agree. Therefore, the force approach will be employed further.

Let us note that the displacements of each elementary circular layer do not start simultaneously and they are measured from the moment of mounting of each layer on the lower lying deformed layers. The condition does not allow the concept of pretensioning as an analogue to some volumetric force and the strain, caused by it—an analogue to some shrinkage strain. More exactly, the analogy with thermoelasticity can also be employed, provided that the free strain e_θ^{free}, entering Cauchy's relation $e_\theta = e_\theta^{\text{free}} + u_r/\rho$, allows for the displacement of the lower lying circuits,

$$e_\theta^{\text{free}}(\rho) = S_{\theta\theta}\sigma_\theta^0(\rho) - \frac{1}{\rho}\int_1^\rho e_r(y, \rho)\,\mathrm{d}y\,.$$

Thereby, the solution is not made easier and the problems of winding mechanics may be conceived as an independent branch of the mechanics of deformable bodies.

3.2.2. Winding pressure on the mandrel

The dependence of the relative value of winding pressure on the mandrel $P/\sigma_\theta^0 = -\sigma_r(1, b)/\sigma_\theta^0$ during winding at constant tension on the relative ring dimensions is shown in Fig. 14. The winding pressure increases nonlinearly with

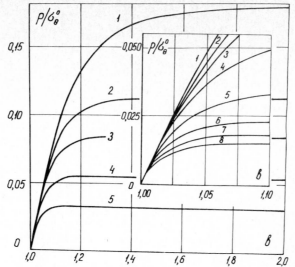

FIG. 14. Pressure on the rigid mandrel P (in fractions of stress σ_θ^0 in a wound tape) vs. the current relative outer radius b of a ring with various anisotropy ratios β. Curves 1–8: $\beta = 10, 15, 20, 30, 50, 70, 90, 110$, respectively.

the thickness of the article, thus asymptotically approaching a certain ultimate value. Therefore the simplest linearly-elastic circular model describes very well the basic qualitative effect during winding of the semifabricated material, observed experimentally (cf. Fig. 3). It has been confirmed within the framework of the model that the main reason for the winding pressure on the mandrel being only a small fraction of the theoretical value

$$\sigma_\theta^0(b-1) = \sigma_\theta^0 \frac{nh}{r_{\text{in}}} = \sigma_\theta^0 n \, \Delta b$$

(calculated under the condition of nondeformability of the type) is the essential ansiotropy of the deformability of the semifabricated material. The ultimate value of winding pressure and the ring thickness, at which the ultimate value is practically reached, decreases sharply with the increase in the material anisotropy ratio β (cf. Figs. 15 and 16). For an absolutely rigid mandrel ($\eta = 1$)

$$\frac{P_\infty}{\sigma_\theta^0} = \frac{2}{\beta} \left(\alpha_0 + \frac{\alpha_1}{\beta} + \frac{\alpha_2}{\beta^2} + \cdots + \frac{\alpha_n}{\beta^n} + R_n \right), \; \beta > 1, \; R_n < 1/[\beta^n(\beta-1)] \quad (3.17)$$

where

$$\alpha_0 < \tfrac{1}{4}\pi < \alpha_1 = G < \alpha_2 = \tfrac{1}{32}\pi^3 < \alpha_3 = 0.98894455 < \cdots < \alpha_4 = \tfrac{5}{1536}\pi^4 < \cdots < 1 \,.$$

Here G is Catalan's constant. By substituting all α_i in (3.17) first by $\tfrac{1}{4}\pi$ and then by 1, we shall obtain the upper and lower estimates of the ultimate winding pressure,

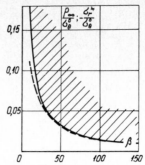

FIG. 15. Relative winding pressure P_∞/σ_θ^0 or the parameter $\sigma_r^*/\sigma_\theta^0$ on the mandrel vs. the anisotropy ratio: $\gamma = 0$ (——); $\gamma = 10$ (– – –). At the value of β, corresponding to the initial section ($\beta = \lambda$) of a piecewise-linear $\sigma_r - e_r$ diagram, the winding of a ring of any thickness in the striated area of parameters is calculated according to the linear theory.

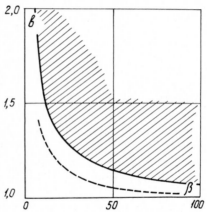

FIG. 16. Relative ring dimension b, at which the pretensioning is still preserved (——) or at which the pressure $P = 0.9 P_\infty$ (– – –) is reached vs. the anisotropy ratio β.

$$\frac{\pi}{2(\beta - 1)} < \frac{P_\infty}{\sigma_\theta^0} < \frac{2}{\beta - 1} \quad \text{at } \beta > 1. \tag{3.18}$$

Mandrel compliance reduces still more the pressure transfer, mainly at the start of the winding process (cf. Fig. 15). Estimates of winding pressure on a complaint mandrel, analogous to (3.18), have the form

$$\frac{\eta + 1}{(\beta - 1)\eta} \left(\tfrac{1}{2}\pi - \text{arctg } \sqrt{\eta}\right) < \frac{P_\infty}{\sigma_\theta^0} < \frac{\eta + 1}{(\beta - 1)\eta}. \tag{3.19}$$

3.2.3. Stress distributions

Since the winding process at constant winding tension is the most widely used, the detailed analysis of stress–strain state is of considerable practical importance. The variation of stresses σ_r and σ_θ as fractions of σ_θ^0 is presented in Fig. 17. Let us note that the circumferential stresses are essentially nonuniform

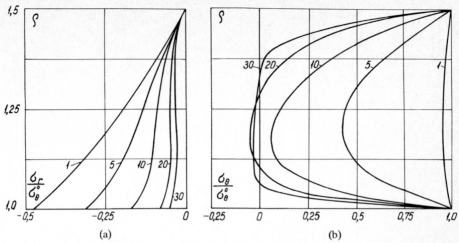

FIG. 17. Form of radial (a) and circumferential (b) stresses during winding at constant winding tension σ_θ^0 on a stiff mandrel for rings having relative dimensions $b = 1, 5$ at various anisotropy ratios β of a semifabricated material ($\beta = 1, 5, 10, 20, 30$).

throughout the section; moreover, there is a zone where they change sign.

Since under processing conditions the fibers may carry only a tensile load, the transition from the state of pretensioning into the state of compression may lead to fiber curvatures. After that the material properties in the directions r and, in particular, θ change and the described design methodology becomes inapplicable. Determination of the range of parameters over which winding at constant tension may result in circuit curvatures, is of practical interest. Conventionally, the level $\sigma_\theta = 0$ may be assumed to be ultimate. By employing the relationships (3.13) and (3.14), it is possible to determine the minimum values of σ_θ for any preset thickness, anisotropy ratio of the semifabricated material and mandrel compliance. For a rigid mandrel, the combination of b and β, at which the initially preset winding tension falls to zero, are presented in Fig. 6. If the winding is accomplished at a combination of parameters, falling within a striated area, fiber curvature is possible and calculation of stresses and displacements according to the above model becomes incorrect.

Let us consider the winding operation at varying winding tension: constant, increasing, decreasing etc. Comparison of stress distributions in circuits for various programs $\sigma_\theta^0(\rho)$, ensuring the same winding pressure on the mandrel, is of interest. Such winding programs are presented in Fig. 18(a) and the respective variation of radial and circumferential stresses in Fig. 18(b), (c), (d) and (e). Here contribution of the winding stresses to initial stresses is shown, i.e., stresses due to mandrel removal for the cured material with $\beta = 2$ are taken into account. For clarity, thermoelastic stresses, independent of the winding program, have been omitted.

The winding program at decreasing winding tension and passing through a minimum, as shown in Fig. 18(c) results in undesirable tensile radial initial stresses, which, together with thermoelastic initial stresses of the same sign,

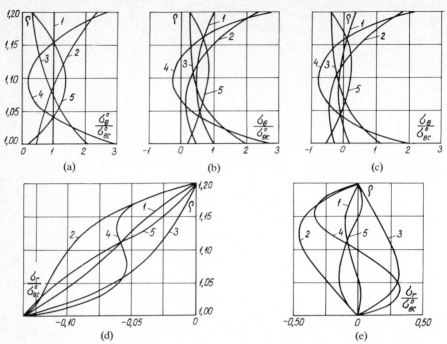

FIG. 18. Winding programs (a) and patterns of circumferential (b) (c) and radial (d) (e) stresses in rings, wound according to the programs. (b) (d) after winding stage, $\beta = 10$; (c) (e) after removal from the mandrel; $\beta = 2$ for the cured material; $\sigma_{\theta c}^0$ is a constant, characterizing the level of tensioning.

may lead to cracking of the wound article. Moreover, during winding according to a program having a minimum, circuit curvature may occur (cf. Fig. 18(d)). The winding operation at increasing winding tension allows, to a greater extent, compensation for radial thermoelastic stresses, and a programmed winding, having a maximum, can prevent the curvature.

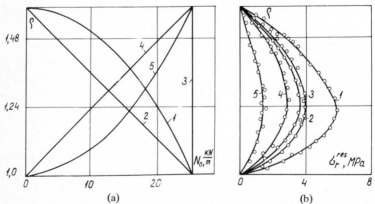

FIG. 19. The effect of variation in the winding tension (a) on the form of σ_r^{res} (b) (VARUSHKIN et al., 1972).

An experimental variation of the winding tension has shown that the extremum values of initial radial stresses (see Fig. 19(b)) in rings, wound according to various programs (see Fig. 19(a)), change in the same manner, as shown by calculations (see Fig. 18(c)).

3.2.4. Evaluation of ring dimensions with allowance for compressibility of a semifabricated material

Due to large radial strains, the thickness of a wound ring is less than $n \Delta b$ and the following question may arise:

> What is the dependence of the actual ring dimensions under winding tension on the number of circuits (or on the dimensions of a ring, containing the same number of circuits, but wound without any winding tension)?

By integrating e_r through the ring thickness, using (3.11), (3.13) and (3.14) and summing up with the expression of $u_r(1, b)$ we may obtain

$$\delta b(b) = -\beta S_{\theta\theta} \int_1^b \sigma_\theta^0(y) \frac{\eta y^{2\beta} - 1}{\eta y^{2\beta} + 1} \, dy. \qquad (3.20)$$

At large values of β and $b > 1.1$ with an error of the order of 1%, (3.20) for the winding at a constant winding tension may be approximated by the following simple expression:

$$\delta b(b) = -S_{\theta\theta}\sigma_\theta^0 \left[\beta(b - 1) - \ln \frac{\eta + 1}{\eta + b^{-2\beta}} \right]. \qquad (3.21)$$

It follows from (3.21) that, if the winding tension is near the ultimate tension $(S_{\theta\theta}\sigma_\theta^0 \sim 0.01)$, the variation in the outer radius is comparable to the ring thickness. In this case, the use of apparatus based on the theory of infinitesimal strains, is incorrect.

3.3. A piecewise-linearly-elastic circular model

3.3.1. Description of the model and calculation relationships

The linearly-elastic circular model, allowing for the most important factor —the anisotropy of strain properties—correctly predicts a number of unique features of composite winding, in particular, the essential nonlinear dependence of the winding pressure (on the mandrel) on the number of circuits. However, the model does not take into account another important factor— nonlinearity of the stress–strain relation in transverse compression σ_r–e_r. The effect of this nonlinear dependence of pressure on winding tension has been experimentally confirmed many times. In all cases, the relative pressure on the mandrel (i.e., in fractions of winding tension) does not remain constant, as for the linearly-elastic model, but increases with the increase in winding tension (Fig. 20).

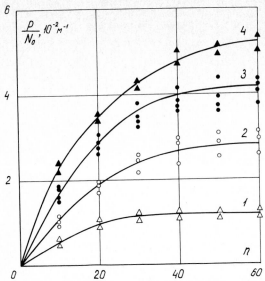

FIG. 20. Dependence of the relative pressure on the mandrel P/N_0 on the number of circuits n during winding at various winding tensions N_0 (BLAGONADJOZHIN et al., 1970). Curves 1–4: $N_0 = 10, 20, 30, 40$ kN/m, respectively.

The effect may be explained only by stiffening of the semifabricated material (cf. Fig. 5) in the transverse direction with the increase in radial stresses. The nonlinear stress–strain relation in transverse compression of a semifabricated material in the first approximation may be expressed by a piecewise-linear relationship (see Fig. 21). The approximation, though it introduces some error, allows the description of a wide spectrum of stress–strain relations, using a minimum number of parameters. For a semifabricated article the parameters fall within the following ranges: the limit of proportionality $\sigma_r^* = 0.5$–2 MPa; the anisotropy ratio $\beta = \sqrt{E_\theta/E_r}$, evaluated in terms of the tangential modulus, E_r, is equal to $\beta = \lambda = 30$–200 up to the limit of proportionality and $\beta = \omega = 4$–50 beyond it. Let us note that the stresses of the order σ_r^* correspond to the mean stresses during winding of large-scale articles. This indirectly indicates the necessity of employing the nonlinear theory in the calculations.

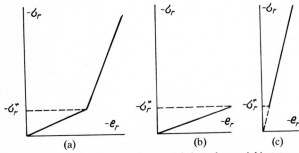

FIG. 21. Approximation of the σ_3–e_3 diagram of a semifabricated material in transverse compression at varying winding tension over different ranges.

At the beginning of the winding process, the radial stresses are low and the process is described by a linearly-elastic model with the anisotropy ratio $\beta = \lambda$, corresponding to the initial section of the $\sigma_r - e_r$ diagram. During winding at low winding tension it may turn out that the radial stresses during the entire process will not reach the limit of proportionality. Since the winding pressure on the mandrel is equal to the maximum value of the absolute radial stress in a ring, then at $|\sigma_r^*| > P_\infty$ the winding process of an article of any thickness is described by the linearly-elastic theory with an anisotropy ratio, corresponding to the initial section of the $\sigma_r - e_r$ diagram. At parameters b and β, falling within the striated area of Fig. 15, the radial stresses in a ring of any thickness do not exceed the limit of proportionality. Another instance, that the limit of proportionality is not reached, is the winding of thin-walled articles. In this case, the upper estimate of the maximum pressure on the mandrel is obtained as $\sigma_\theta^0(b - 1)$. If the value is less than $|\sigma_r^*|$, one has to employ the linearly-elastic model with $\beta = \lambda$.

In a wound ring, the absolute value of radial stresses should monotonically decrease from the surface of a mandrel towards the outer radius. In the opposite case (i.e., appearance of a local extremum or plateau) there is a region of compressive stresses on the curve of σ_θ, related to the variation of σ_r through an equilibrium equation. This is inadmissible because of fiber curvatures. In the piecewise-linear approximation the stress–strain relation in compression is of the type shown in Fig. 21, the ring may be either uniform as to its deformability (if stresses near the mandrel do not exceed the limit of proportionality) or nonuniform (if $|\sigma_r(1, b)| > |\sigma_r^*|$). If the size of the inner circular region with an anisotropy ratio $\beta = \omega$ is comparable to the outer region, having an anisotropy ratio $\beta = \lambda$, the linearly-elastic model is inapplicable. If the regions differ appreciably in size, it is possible to employ the linearly-elastic model with an anisotropy ratio $\beta = \lambda$ or $\beta = \omega$, respectively.

Let at the moment, when the interface between the two regions of a ring lies on the relative radius $\rho = \rho_*$, an infinitesimal change in the external pressure dq be applied. Stress increments, for example $d\sigma_r$, may be formally written as

$$\frac{d\sigma_r}{dq} = \left.\frac{\partial\sigma_r}{\partial q}\right|_{\rho_*=\text{const}} + \left.\frac{\partial\sigma_r}{\partial\rho_*}\right|_{q=\text{const}} \frac{d\rho_*}{dq}. \tag{3.22}$$

This equation means that, upon loading of a ring with pressure, accompanied by moving of the interface towards the outer radius, the stress increment may be conceived as a sum of a stress increment in a multilayer ring with a stationary boundary and a stress increment, resulting from material change in a thin band $d\rho_*$ between the preceding and subsequent boundary positions at finite stresses and under constant boundary conditions. Owing to continuity of strains and stresses at the breaking point the second term in (3.22) is negligibly small.

Stress calculations in a multilayer anisotropic ring are given in LEKHNITSKII (1957). In this case, the problem differs only by the increments and it satisfies somewhat different boundary conditions, analogous to (3.2) and (3.3):

$$d\sigma_r = -\frac{1}{\rho}\,\sigma_\theta^0\,db \quad \text{at } \rho = b,$$

$$(3.23)$$

$$de_\theta = \gamma\,d\sigma_r/E_\theta \quad \text{at } \rho = 1.$$

Formulae of stress and displacement increments in the ith ring have the following form:

$$d\sigma_r^{(i)} = \frac{dq_{i-1}c_i^{\beta_i+1}}{1 - c_i^{2\beta_i}}\left[\left(\frac{\rho}{b_i}\right)^{\beta_i-1} - \left(\frac{\rho}{b_i}\right)^{-\beta_i-1}\right] - \frac{dq_i}{1 - c_i^{2\beta_i}}\left[\left(\frac{\rho}{b_i}\right)^{\beta_i-1} - c_i^{2\beta_i}\left(\frac{\rho}{b_i}\right)^{-\beta_i-1}\right],$$

$$d\sigma_\theta^{(i)} = \beta_i\frac{dq_{i-1}c_i^{\beta_i+1}}{1 - c_i^{2\beta_i}}\left[\left(\frac{\rho}{b_i}\right)^{\beta_i-1} + \left(\frac{\rho}{b_i}\right)^{-\beta_i-1}\right] - \beta_i\frac{dq_i}{1 - c_i^{2\beta_i}}\left[\left(\frac{\rho}{b_i}\right)^{\beta_i-1} + c_i^{2\beta_i}\left(\frac{\rho}{b_i}\right)^{-\beta_i-1}\right],$$

$$(3.24)$$

$$du_r = \frac{dq_{i-1}c_i^{\beta_i+1}b_i}{E_\theta^{(i)}(1 - c_i^{2\beta_i})}\left[(\beta_i - \nu_{\theta r}^{(i)})\left(\frac{\rho}{b_i}\right)^{\beta_i} + (\beta_i + \nu_{\theta r}^{(i)})\left(\frac{\rho}{b_i}\right)^{-\beta_i}\right]$$

$$- \frac{dq_ib_i}{E_\theta^{(i)}(1 - c_i^{2\beta_i})}\left[(\beta_i - \nu_{\theta r}^{(i)})\left(\frac{\rho}{b_i}\right)^{\beta_i} + (\beta_i + \nu_{\theta r}^{(i)})c_i^{2\beta_i}\left(\frac{\rho}{b_i}\right)^{-\beta_i}\right]$$

where b_i is the relative outer radius of the ith ring ($b_0 = 1$, $b_1 = \rho_*$, $b_2 = b$);

$$c_i = b_{i-1}/b_i, \qquad \beta_i = \sqrt{S_{rr}^{(i)}/S_{\theta\theta}^{(i)}}, \qquad \nu_{\theta r} = -S_{r\theta}/S_{\theta\theta}.$$

In the given case, the two elastic characteristics of the rings are equal:

$$E_\theta^{(1)} = E_\theta^{(2)} = E_\theta, \qquad \nu_{\theta r}^{(1)} = \nu_{\theta r}^{(2)} = \nu_{\theta r}.$$

The last equality is necessary for single-valued determination of strains e_r under radial stresses $\sigma_r = \sigma_r^*$ in a planar stress state, when $\sigma_\theta \neq 0$. Pressure increments dq on the surfaces of individual constituent rings are equal to

$$dq_2 = \frac{1}{b}\,\sigma_\theta^0\,db,$$

$$dq_1 = 2\lambda(\eta\rho_*^{2\omega} + 1)b^{\lambda+1}\rho_*^{\lambda-1}\,dq_2/\{[(\lambda + \omega)\eta\rho_*^{2\omega} + (\lambda - \omega)]b^{2\lambda}$$

$$+ [(\lambda - \omega)\eta\rho_*^{2\omega} + (\lambda + \omega)]\rho_*^{2\lambda}\},$$

$$(3.25)$$

$$dq_0 = (\eta + 1)\,dq_1/(\eta\rho_*^{\omega-1} + \rho_*^{-\omega-1}) \quad \text{where } \eta = \eta(\omega).$$

When passing over to small finite differences, it is possible for a given step to employ the given relations for the interface position $b_1 = \rho_*$, borrowed from the earlier step. The increment of a boundary position $\Delta\rho_*$ is obtained on the following grounds. The stresses σ_r on the movable interface are constant and equal to the limit of proportionality. In a mathematical sense, it means that the derivative σ_r along the direction $\rho = \rho_*$ equals zero, whereof by passing over to finite differences, one may obtain

$$\Delta\rho_* = -\frac{\Delta\sigma_r|_{\rho=\rho_*}}{\left.\dfrac{\partial\sigma_r}{\partial\rho}\right|_{\rho=\rho_*}} = -\frac{\Delta\sigma_r|_{\rho=\rho_*}\rho_*}{\sigma_\theta|_{\rho=\rho_*} - \sigma_r^*}. \qquad (3.26)$$

The sequence of calculations is as follows. Up to a certain value of $b = b^*$ the winding operation is accomplished at stresses which do not exceed the limit of proportionality. The value of b^* is obtained by directly employing a discrete form of the circular model (3.10), (3.12) or by the technique of successive approximations—in the case of a continuous form (3.13), (3.14). After the stresses and displacements at $b = b^*$ and $\rho_* = 1$ have been estimated, the first interface position is calculated according to (3.26), and stress and displacement increments—according to the relations for a multilayer ring model. At summed up stresses the new boundary condition is calculated, etc. The idea of the method reminds us of Euler's numerical solution of differential equations; the same comparison shows up in the precision analysis of the method. Precision may be improved by double calculation; at first, the stresses are computed for the preceding boundary position; then $\Delta\rho_*$ and the new boundary position are calculated. Then the stresses at the arithmetic mean value of two ultimate boundary positions are recalculated and $\Delta\rho_*$ is made more precise at refined stresses. It is possible to employ also more refined methods, for example, an analogue of Adams method, constructing the initial section analogously to Krylov's method. Besides, the interface positions in several preceding steps are used. Another method of solution to the winding problem for the material, having a piecewise-linear σ_r-e_r diagram, is treated in PORTNOV and BEIL' (1977). Comparison of numerical results confirmed the equivalence of the two methods.

3.3.2. Numerical results

Let us consider some numerical results, mainly for the winding operation at constant winding tension. In Fig. 22 the winding pressure on the mandrel dependence (in fractions of the stress σ_θ^0) on the outer radius, calculated according to the linear theory with an anisotropy ratios $\beta = \lambda$ and $\beta = \omega$ and nonlinear theory at various winding tensions is presented. At low winding tensions, until the range of radial stresses in an article is beyond the zone of the first section of the piecewise-linear σ_r-e_r diagram, the increase in the winding tension is not accompanied by the change in P/σ_θ^0. On passing out of the zone, the relative winding pressure on the mandrel P/σ_θ^0 increases with the increase in the winding tension. At significant winding tensions, the increase in the relative characteristic is slower and then completely stops, i.e., it becomes possible to pass over to calculations according to the linear model with an anisotropy ratio $\beta = \omega$. Under varying winding tension, an analogous picture has been observed with the increase in the winding tension for one and the same law of its variation.

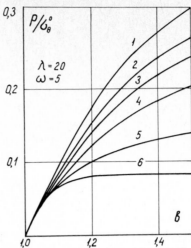

FIG. 22. Dependence of the relative pressure on the mandrel P/σ_θ^0 on the relative ring dimension b at various winding tensions, preset by the parameter $\sigma_r^*/\sigma_\theta^0$. Curve $1—\sigma_r/\sigma_\theta^0 = 0$; Curve $2— = -0.02$; Curve $3— = -0.03$; Curve $4— = -0.04$; Curve $5— = -0.05$; Curve $6— = -\infty$.

Calculation of circumferential stresses during winding at constant winding tension has shown that the value of winding tension significantly affects the distribution. In Fig. 23 an example of distribution of the circumferential stresses (in fractions of the stress σ_θ^0) in identical rings, wound at various winding tensions, is presented. At low winding tensions (within the region of the use of the linear winding model with $\beta = \lambda$), the stress variation σ_θ in fractions of σ_θ^0 is independent of σ_θ^0. With the increase in the winding tension the variation of relative circumferential stresses is disturbed, thereby becoming more uniform throughout the section. At very high σ_θ^0, the effect on distribution of $\sigma_\theta/\sigma_\theta^0$ becomes less and

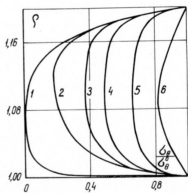

FIG. 23. Forms of the relative circumferential stresses during winding at various tensions: Curve $1—\sigma_r^*/\sigma_\theta^0 \rightarrow -\infty$; Curves 2–6—0.0250, 0.225, 0.02000, 0.00150, 0, respectively ($\lambda = 40$, $\omega = 5$).

eventually completely stops. This corresponds to passing over to the area of use of the linear winding model with an anisotropy ratio $\beta = \omega$.

Accordingly, during winding at high winding tensions, the pretensioning is maintained (cf. curve 6 of Fig. 23), but at low σ_θ^0 a considerable part of pretensioning is lost. In some cases (cf. curve 1 in Fig. 23), compressive circumferential stresses develop, which are associated with fiber curvatures. An analogous picture has been observed at varying winding tensions.

Evaluation of the combination of parameters, at which the pretensioning falls to zero, the range of winding tensions and the ultimate outer radii for a variety of materials are presented in Fig. 24. Curves are of S-shape. The left-hand vertical asymptote refers to winding at low tensions, when the range of radial stresses is not beyond the scope of the first section of a piecewise-linear σ_r-e_r diagram. In the case, the value of the maximum outer radius, at which the compressive stresses in circuits have not yet developed, is independent of the winding tension and it depends merely on the anisotropy ratio $\beta = \lambda$ (cf. Fig. 16). The right-hand vertical asymptote (for some curves of Fig. 24 it is beyond our range and is not shown) corresponds to winding at high tensions. As has been noted, in this case the circumferential stresses are more uniform through the thickness and the minimum of the distribution is zero only at large thicknesses. The right-hand asymptote refers to the linear winding model with an anisotropy ratio $\beta = \omega$ (cf. Fig. 16). The transition area between the two vertical asymptotes corresponds to winding under medium winding tensions, i.e., it may be described only by the nonlinear winding model.

Analogously to the variation in the winding pressure on the mandrel and the circumferential stresses, the variations of radial stresses (see Fig. 25(a)) also change with the winding tension. It is of interest to compare the contribution of the winding stresses to the initial stresses at various winding tensions. To achieve this, it is necessary to subtract the stresses developed as a result of the

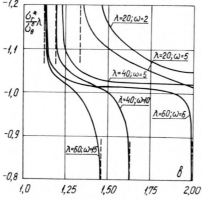

FIG. 24. Dependence of a complex parameter $\sigma_r^* \lambda / \sigma_\theta^0$ on the relative ring dimension b, characterizing danger of layer curvature. The range of parameters on the left from each curve characterize winding without layer curvatures.

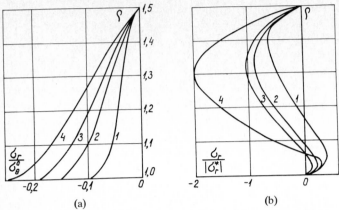

(a)   (b)

FIG. 25. Distributions of radial stresses during winding at four various winding tensions. (a) post-winding; (b) after removal from the mandrel ($\beta = 2$ for the cured material). Winding tensions: Curve 1—$\sigma_r^*/\sigma_\theta^0 = -0.0250$; Curve 2— $= -0.0225$; Curve 3— $= -0.0200$; Curve 4— $= -0.0150$.

removal of the cured article from the mandrel from the curves of Fig. 25. In the calculation for the cured material $\beta = 2$ has been assumed. Calculations have shown that the contribution to the initial radial tensile stresses σ_r^i increases (see Fig. 25(b)) with increasing winding tension.

Experimental studies of the effect of winding tension on initial stresses have revealed that during winding of a cold tape onto a cold mandrel the level of initial stresses increases, but during winding of a heated tape onto a cold mandrel it falls (see Fig. 26). These calculations agree with the experiment of winding of a cold tape onto a cold mandrel.

A series of calculations within a framework of the programs make it possible to numerically estimate the area of use of each of the three winding models: two linear models with an anisotropy ratio λ or ω and the nonlinear one. Calculation results have been compared in terms of the winding pressure on the mandrel; the variations of circumferential and radial stresses, as a rule, differ, slightly. It has been shown that, if the anisotropy ratio λ differs from ω

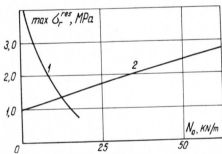

FIG. 26. The effect of winding tension N_0 on max σ_r^{res} in rings made of different GFRP: Curve 1—heated tape, heated mandrel, $b = 1.6$ (VARUSHKIN, 1972); Curve 2—a cold tape, a cold mandrel, $b = 1.185$ (PORTNOV et al., 1969).

by a factor of four or more, the area of use for the nonlinear winding model is controlled by only two parameters, $\sigma_r^*/\sigma_\theta^0$ and λ. In Fig. 27 the respective areas for winding onto a rigid mandrel are shown. It has been assumed that, if the results of the nonlinear theory differ from those of the linear theory by not more than 10%, the calculations should be performed according to the linear theory. The boundary between the areas of use of the linear model with $\beta = \omega$ and the nonlinear model is described by the relation

$$\frac{\sigma_r^*}{\sigma_\theta^0} = -\frac{1}{2(\lambda - 1)}. \tag{3.27}$$

The curves, presented in Fig. 27, are constructed with the parameters set at values, encompassing the majority of actual stress–strain diagrams and practically used tensions.

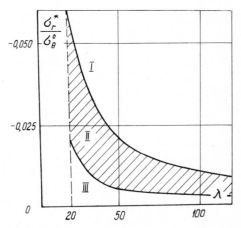

FIG. 27. Areas of use of three variants of theory in description of the winding process: I—a linear theory, $\beta = \lambda$; II—a nonlinear theory; III—a linear theory with $\beta = \omega$. Boundaries are set at $\omega \leqslant 0.25\lambda$.

3.4. A discrete-circular model

3.4.1. Description of the model

In a circular model, the already wound layers are considered as a quasi-uniform cylindrically orthotropic body. For many practical problems—the design of a winding process for complex structures, consideration of the geometrical and physical nonlinearity, etc.—it is unreasonable to pass over to a uniform anisotropic medium, but acceptable to treat the already wound layers as a discrete system of rings. Increments of stresses and strains in the ith circuit may be calculated according to LEKHNITSKII's formulae for a multilayer ring (3.24). Pressure increments on the interfaces between layers are expressed by the following relations:

$$\Delta q_{i+1}b_{i+1}\xi_{i+1} + \Delta q_i b_i \zeta_i + \Delta q_{i-1}b_{i-1}\xi_i = 0 , \tag{3.28}$$

where

$$\xi_i = \frac{2\beta_i c_i^{\beta_i}}{E_\theta^{(i)}(1 - c_i^{2\beta_i})} ,$$

$$\zeta_i = \frac{1}{E_\theta^{(i)}} \left(\nu_{\theta r}^{(i)} - \beta_i \frac{1 + c_i^{2\beta_i}}{1 - c_i^{2\beta_i}} \right) - \frac{1}{E_\theta^{(i+1)}} \left(\nu_{\theta r}^{(i+1)} + \beta_{i+1} \frac{1 + c_{i+1}^{2\beta_{i+1}}}{1 - c_{i+1}^{2\beta_{i+1}}} \right) .$$

The boundary conditions (3.2) and (3.3) are added to the relations and the obtained system of equations relative to Δq_i computed.

3.4.2. Allowance for significant radial strains

It has been shown earlier (Section 3.2.4) that during winding the radial strains may be considerable. Therefore it is necessary to employ finite deformation tensors.

Let us emphasize once more that large strains are allowable only in the radial direction and should not exceed small strains of pretensioning in the circumferential direction; otherwise fiber curvatures will occur.

Strain increments during winding of a current circuit are small. They refer to the current strained state. This leads to referring to Hencky's estimate for e_r in describing the finite deformation state. Physical linearity in this case means that the σ_r–e_r diagram (where e_r is the logarithmic strain) is a straight line. In the design of a multi-ring model with current values of interface radii between layers b_i, the increments of circumferential tensions ΔN_r radial stresses $\Delta \sigma_r$, displacements Δu_r and strains Δe_r and Δe_θ are determined. These increments are summed up with the previous finite values of respective parameters and the new values of b_i are also estimated. Let us note two characteristic features of the calculation: instead of arguments ρ and b_i, the number of the circuit under consideration, i, and the number of the circuit to be wound, j, are used and, instead of $\Delta\sigma_\theta$, summation is over ΔN. This allows one to exclude from consideration the current thicknesses of circuits in the computation, and for determination of σ_θ to use finite values of N and Δb_i. Since in practice all the load in the circumferential direction is resisted by reinforcement layers, the thicknesses of which change slightly, the value of N is more useful than that averaged through the thickness of a circuit σ_θ. Numerical results show that the geometrically linear theory yields somewhat lower values for the winding pressure on the mandrel and a minimum value of N, although the difference is virtually negligible.

3.4.3. Nonlinearly-elastic materials

A piecewise-linear approximation of the σ_r–e_r diagram consisting of two sections is not a necessary requirement for the use of a multi-ring model. The design in a piecewise-linear approximation is completely analogous, consisting of three or more intercepts. A multi-ring model is applicable also in the case of a smooth stress–strain relation. Let the stresses and strains at all points in the already wound ring be known. This means that the tangential moduli and other elastic parameters in each point of the ring are also known. If, during winding of a current circuit, the stress increments are low, it is possible to employ the indicated tangential moduli in the design. This means that, in order to determine the stress increments, the ring must be treated as being linearly-elastic and uniform. A smooth nonuniformity at a large number of layers may be substituted by a piecewise nonuniformity, i.e., (3.24) and (3.28) may be used. After stress and strain increments have been summed up with the preceding ones, the elastic parameters are determined anew, etc. Accuracy may be improved by double calculation, analogous to the one for a piecewise-linear model. Comparison of the results of computation at a smooth stress–strain relation and its piecewise-linear approximations (see Fig. 28) show (see Fig. 29) that the computation according to a piecewise-linear model takes into account the main part of nonlinear effects. Theory-experiment disagreement increases with the increase in tensioning and the number of circuits. This may be attributed to imperfect experimental technique.

Allowance for the finite deformations shows that, for physical nonlinearity, the effect of the strain ultimate increases, compared with the physically-linear variant. However, this is of minor importance compared with the basic factors—the anisotropy of deformability and physical nonlinearity. Therefore a piecewise-linear circular winding model in a geometrically linearized variant makes it possible to satisfactorily assess the effects associated with the non-linearly-elastic compliancy of the semifabricated material in the transverse direction. For a number of problems of winding at low or ultimate winding tensions, the simplest linearly-elastic circular model is applicable, even for quantitative estimates.

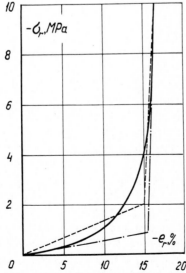

FIG. 28. A diagram of transverse compression of a semifabricated material and two piecewise-linear approximations.

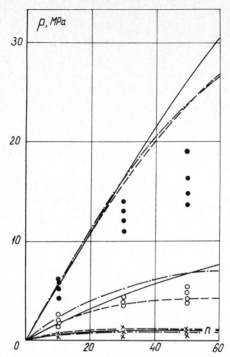

FIG. 29. Dependence of the pressure on the mandrel P on the number of wound circuits n, calculated according to continuous (——), a piecewise-linear (– · –, – – – according to Fig. 28) models, $r_{in} = 0.075$ m, $h = 0.0002$ m, $\gamma = 0$. $\times$—$N_0 = 375$ kN/m; $\bigcirc$—$N_0 = 940$ kN/m; ●—$N_0 = 2340$ kN/m; $\times$, $\bigcirc$, ●—experimental points.

4. Post-winding technological stages

4.1. Initial stresses

As a result of force winding, a system of radial compressive and circumferential tensile stresses is developed in the article. These stresses change during successive technological stages and their sign also changes in many cases. The qualitative change in stress distribution is shown in Fig. 30.

Upon heat buildup, such phenomena as micro- and macroflow of the binder, thermal expansion and simultaneous decrease in stiffness in the transverse direction, partial relaxation of the winding stresses, and change in the winding pressure on the mandrel take place. The latter is caused by the difference in coefficients of linear thermal expansion of the mandrel and the article. Different thermal expansion of a semifabricated material and the mandrel must lead to an increase in the winding pressure on the mandrel. The decrease in stiffness and intensification of dissipation phenomena must lead to the opposite

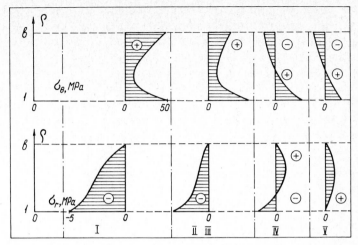

FIG. 30. A diagram, showing the variation in radial and tangential stresses in winding stages, Stage I—winding; Stages II, III—heat buildup and curing; Stage IV—cooling; Stage V—removal from the mandrel.

effect. The second mechanism usually prevails. However, in the literature the opposite cases are also known.

Upon polymerization, the increase in strength and stiffness in the transverse direction and physico-chemical shrinkage take place preferentially in the same direction. Besides, slight relaxation of the winding stresses continues. Here also the two competing mechanisms occur, which, evidently, also result in the winding pressure on the mandrel remaining constant during polymerization (cf. Fig. 4).

Upon cooling, thermal shrinkage and increase in stiffness in the transverse direction take place. Processes of creep and stress relaxation continue. The winding pressure on the mandrel changes due to the difference in coefficients of thermal expansion between the mandrel and the cured article. This situation is complementary to that upon heat buildup. Here, the competitive position of the two mechanisms is different. The free strain caused by the change in stiffness of the cured material upon cooling under conditions of finite stresses is smaller than the thermal shrinkage strain. Therefore, the role of thermal shrinkage is dominant and, upon cooling, the radial compressive stresses decrease, but in many cases zones of tensile radial stresses develop. The winding pressure on the mandrel decreases.

Upon cooling, it is possible that at some temperature separation from the mandrel will take place, because the winding pressure on the mandrel has fallen to zero. However, a variant is possible (which is encountered more often), when the article still preserves some pretensioning after cooling. Removal from the mandrel (mechanical removal, destruction or solution of the mandrel, cooling of the mandrel, etc.) is equivalent to application of some negative inner pressure to the article, i.e., it leads to the growth of tensile radial stresses. In this case, when the parts being wrapped over serve as mandrels, forming an integral part with the windable composite (for example, during winding of bandages), the mandrel is not removed and the pressure on the boundary is not released.

Let us consider for example the distribution of the stresses (see Fig. 31) initially developed in thick-walled GFRP rings. The radial stresses are extremely critical because they are of the same order as the transverse tensile strength. With the increase of the relative thickness of the article the initial stresses sharply increase (see Fig. 32). This leads to the fact that the articles of

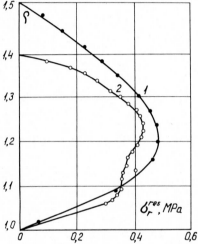

FIG. 31. Patterns of initial technological stresses in GFRP rings (Curve 1) (PORTNOV et al., 1969) and CFRP (Curve 2) (TARNOPOL'SKII et al., 1973).

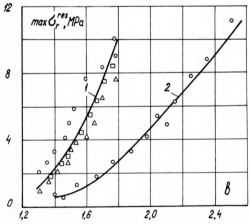

FIG. 32. The effect of relative ring dimension b on max σ_r^{res}. Curve 1—at constant winding tension; Curve 2—at increasing winding tension (VARUSHKIN et al., 1972).

the wall thickness, comparable to the inner radius of the article, as a rule, undergo cracking (TARNOPOL'SKII et al., 1973).

Let us turn to the analysis of the basic reason for cracking—the thermoelastic constituent of initial stresses upon cooling. In order to understand the specific features associated with the processes of heat buildup and cooling, it is useful to analyse in detail the simplest linearly thermoelastic problem for a cylindrically orthotropic ring. This treatment is in line with BOLOTIN and BOLOTINA (1967, 1969).

4.2. A linearly thermoelastic problem for an anisotropic ring

The advent of thermoelastic stresses is mainly caused by the fact that the deformation of thermal expansion does not satisfy the kinematic restrictions. In this connection, additional strains (and corresponding stresses) occur, which are of such magnitude that the summed up strains already satisfy the restrictions. For a ring, such a restriction is the equation of strain compatibility (3.7). Substitution of Duhamel–Neumann's law,

$$e_r = \alpha_r \, \Delta T + S_{rr}\sigma_r + S_{r\theta}\sigma_\theta .$$

$$e_\theta = \alpha_\theta \, \Delta T + S_{\theta r}\sigma_r + S_{\theta\theta}\sigma_\theta, \qquad S_{\theta r} = S_{r\theta} \tag{4.1}$$

into the equation of strain compatibility (3.7) and the equation of equilibrium (3.5) into the result, yields the equation

$$\rho^2 \frac{d^2\sigma_r}{d\rho^2} + 3\rho \frac{d\sigma_r}{d\rho} + (1 - \beta^2)\sigma_r + \frac{(\alpha_\theta - \alpha_r)\,\Delta T}{S_{\theta\theta}} + \rho \frac{d(\alpha_\theta \, \Delta T)}{S_{\theta\theta}\, d\rho} = 0 . \tag{4.2}$$

As a rule, the temperature field at the beginning and at the end of the process is uniform, i.e., $\Delta T = \text{const}$. Solution of (4.2) for a uniform ring has the following form:

$$\sigma_r = C_1\rho^{\beta-1} + C_2\rho^{-\beta-1} + D , \tag{4.3}$$

$$D = \frac{(\alpha_\theta - \alpha_r)\,\Delta T}{S_{rr} - S_{\theta\theta}} . \tag{4.4}$$

By substituting (4.3) into the equation of equilibrium (3.5), one may obtain the formula of circumferential stresses,

$$\sigma_\theta = C_1\beta\rho^{\beta-1} - C_2\beta\rho^{-\beta-1} + D . \tag{4.5}$$

The expression for displacements follows from (4.3), (4.5) and (4.1):

$$u_r = S_{\theta\theta}(\beta - \nu_{\theta r})C_1\rho^\beta - S_{\theta\theta}(\beta + \nu_{\theta r})C_2\rho^{-\beta} + S_{\theta\theta}(1 - \nu_{\theta r})D . \tag{4.6}$$

Let us consider two variants of the boundary conditions:

$$\sigma_r = 0 \quad \text{at } \rho = b, \qquad \sigma_r = 0 \quad \text{at } \rho = 1, \qquad (4.7.1)$$

$$\sigma_r = 0 \quad \text{at } \rho = b, \qquad \sigma_\theta = \sigma_r(\gamma + \nu_{\theta r}) + E_\theta(\alpha_{man} - \alpha_\theta)\Delta T \quad \text{at } \rho = 1, \qquad (4.7.2)$$

corresponding to (4.7.1) free and (4.7.2) constrained with the mandrel heating and cooling. Here, α_{man} is a complex coefficient, taking into account thermophysical and elastic properties and of the dimensions mandrel. It is equal to the relation of the displacement of the outer surface of the mandrel (in fractions of inner radius of the article) to the temperature difference which caused the displacement. For an isotropic circular mandrel, α_{man} coincides with the coefficient of linear thermal expansion of the mandrel material (under a uniform temperature field in the mandrel body). The following values of the constants are obtained from the boundary conditions,

$$C_1 = D\frac{(1 - b^{\beta+1})}{b^{2\beta} - 1}, \qquad C_2 = \frac{(b^{\beta+1} - b^{2\beta})}{b^{2\beta} - 1}D, \qquad (4.8.1)$$

$$C_1 = -\frac{D(\psi + \eta b^{\beta+1}) + Y}{\eta b^{2\beta} + 1}, \qquad C_2 = \frac{D(\psi - b^{1-\beta}) + Y}{\eta b^{2\beta} + 1}b^{2\beta} \qquad (4.8.2)$$

where η is determined from (3.11),

$$\psi = \frac{1 - \gamma - \nu_{\theta r}}{\beta - \gamma - \nu_{\theta r}}, \qquad Y = \frac{E_\theta(\alpha_\theta - \alpha_{man})\Delta T}{\beta - \gamma - \nu_{\theta r}}. \qquad (4.9)$$

Let us analyse the relationships. Upon free cooling or heating, the maximum radial stresses develop on the radius

$$\rho_M = \sqrt{b}\sqrt[2\beta]{\frac{b^\beta - b}{b^{\beta+1} - 1}\frac{\beta + 1}{\beta - 1}}. \qquad (4.10)$$

The level of thermoelastic stresses increases with increase in the thickness. In very thick-walled rings (as $b \to \infty$) the maximum stresses are equal to

$$\sigma_r = D, \qquad \sigma_\theta(1) = D(1 + \beta), \qquad \sigma_\theta(b) = D(1 - \beta). \qquad (4.11)$$

In contrast to an isotropic ring, a uniform temperature change causes stresses. Upon cooling, the radial stresses are tensile, but upon heating they are compressive. The level of radial tensile stresses, as follows from (4.11) and (4.4), may exceed the transverse tension strength. In thick-walled rings, displacements of the inner and outer ring surface upon free cooling–heating have opposite sign.

Analogous ultimate estimates for the case of joint thermoelastic deforming with the mandrel have the following form:

$$\sigma_r(1) = \frac{E_\theta(\alpha_\theta - \alpha_{man})\,\Delta T}{\beta + \gamma + \nu_{\theta r}} + \frac{D(1 + \beta)}{\beta + \gamma + \nu_{\theta r}},$$

$$\sigma_\theta(1) = -\frac{\beta E_\theta(\alpha_\theta - \alpha_{man})\,\Delta T}{\beta + \gamma + \nu_{\theta r}} + \frac{D(1 + \beta)(\gamma + \nu_{\theta r})}{\beta + \gamma + \nu_{\theta r}}, \qquad (4.12)$$

$$\sigma_\theta(b) = D(1 - \beta).$$

Let us note the essential role of the term associated with the difference in coefficients of linear thermal expansion.

Experiments have fully confirmed the basic qualitative theoretical conclusions. The decrease in the curing temperature (see Fig. 33) and the use of mandrels with lower coefficients of linear thermal expansion (see Fig. 34) lead to the decrease in initial stresses.

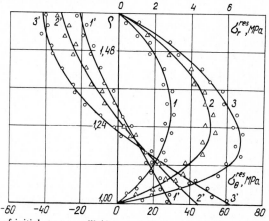

FIG. 33. Variation of initial stresses σ_r^{res} (Curves 1, 2, 3) and σ_θ^{res} (Curves 1', 2', 3') in GFRP rings, cured at various temperatures. Curve 1—383°K; Curve 2—403°K; Curve 3—423°K (VARUSHKIN, 1971).

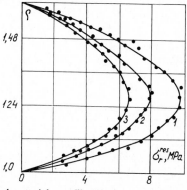

FIG. 34. The effect of mandrel material on σ_r^{res}, Curve 1—Aℓ; Curve 2—steel; Curve 3—a hybrid mandrel (steel + GFRP) (VARUSHKIN et al., 1972).

4.3. A variant of the theory of initial stresses

During winding of a heated tape onto a heated mandrel the stage of heat buildup is excluded from the design scheme. Constancy of the winding pressure on the mandrel during polymerization (cf. Fig. 4) has made it possible to assume that the state of stress does not change in this stage. In such a case, if the hypothesis about inheritance of the state of stress is assumed, the stresses are made up of three constituents: stresses developing during winding (Section 3); stresses due to simultaneous cooling of the article and the mandrel (Section 4.2) and the stresses due to the release of the finite pressure on the mandrel,

$$\sigma_r = \frac{P^-}{b^{2\beta} - 1} (\rho^{\beta-1} - b^{2\beta}\rho^{-\beta-1}),$$

$$\sigma_\theta = \frac{\beta P^-}{b^{2\beta} - 1} (\rho^{\beta-1} + b^{2\beta}\rho^{-\beta-1})$$

(4.13)

where P^- is the negative pressure, equal to the sum of post-winding stress $\sigma_r(1, b)$, and the stress $\sigma_r(1)$ is cuased by simultaneous cooling of the article and the mandrel and computed according to (4.3) and (4.8).

For the most widely used winding process in production—winding of a cold tape onto a cold mandrel—it is necessary to make an allowance for the stage of heat buildup. A theoretical description of the heat buildup process is complicated by the fact that consideration of the effects of elasticity and thermal expansion only leads to results which are directly opposite to the experimental ones. Upon heat buildup, due to a high coefficient of linear thermal expansion of the mandrel, when compared to composites, and because of the anisotropy of linear thermal expansion of composites (the outer radius of thick-walled rings should move outward with rising temperature, but the inner radius inward), the winding pressure on the mandrel should increase, but in practice, a significant fall is observed (cf. Fig. 4).

This contradiction may be explained only by disregarding the sharp growth of deformability of the composite with increasing temperature. In the work of BIDERMAN et al. (1969) a method has been proposed, treating the effect of the change in the radial Young's modulus E_r (radial compliance S_{rr}) on the stress–strain state. The free strain e_r^{free} is registered as the sum of the thermal strain and the strain caused by the change in S_{rr} from its initial value $S_{rr}(T_1)$ to its finite value $S_{rr}(T_2)$,

$$e_r^{\text{free}} = \alpha_r(T_2 - T_1) + \sigma_r(T_1)[S_{rr}(T_2) - S_{rr}(T_1)] \tag{4.14}$$

where $\sigma_r(T_1)$ are post-winding stresses (before the dip in S_{rr}).

The use of Hooke's law (for stress increments, caused by heat buildup),

$$\Delta e_r = \Delta e_r^{\text{free}} + S_{rr}(T_2)\,\Delta\sigma_r, \qquad \Delta e_\theta = \alpha_\theta\,\Delta T + S_{\theta\theta}\,\Delta\sigma_\theta, \tag{4.15}$$

by considering the former expression of a heating problem for a ring, being an integral part of the mandrel, yields

$$\Delta\sigma_r = \frac{(\alpha_{\text{man}} - \alpha_\theta)\,\Delta T}{S_{\theta\theta}(\kappa - \gamma - \nu_{\theta r})}\,\frac{\rho^{\kappa-1} - b^{2\kappa}\rho^{-\kappa-1}}{\eta b^{2\kappa} + 1} + \frac{\Phi(b)(\eta\rho^{\kappa-1} + \rho^{-\kappa-1})}{2\kappa(\eta b^{2\kappa} + 1)} - \frac{\Phi(\rho)}{2\kappa\rho^{\kappa+1}},$$

$$\Delta\sigma_\theta = \frac{(\alpha_{\text{man}} - \alpha_\theta)\,\Delta T\kappa}{S_{\theta\theta}(\kappa - \gamma - \nu_{\theta r})}\,\frac{\rho^{\kappa+1} + b^{2\kappa}\rho^{-\kappa-1}}{\eta b^{2\kappa} + 1} + \frac{\Phi(b)(\eta\rho^{\kappa-1} - \rho^{-\kappa-1})}{2(\eta b^{2\kappa} + 1)} + \frac{\Omega(\rho)}{2\rho^{\kappa+1}}$$

$$(4.16)$$

where

$$\kappa = \beta(T_2), \qquad \mu = \beta(T_1), \qquad \eta = \eta(\kappa),$$

$$\Phi(\rho) = \int_1^\rho [(\alpha_r - \alpha_\theta)\,\Delta T/S_{\theta\theta} + (\kappa^2 - \mu^2)\sigma_r(y)](y^\kappa - \rho^{2\kappa}y^{-\kappa})\,dy,$$

$$\Omega(\rho) = \int_1^\rho [(\alpha_r - \alpha_\theta)\,\Delta T/S_{\theta\theta} + (\kappa^2 - \mu^2)\sigma_r(y)](y^\kappa + \rho^{2\kappa}y^{-\kappa})\,dy.$$

The terms characterizing Poisson's effect have been omitted, because in the problem for a ring they cancel out.

Polymerization and cooling processes may be described by the same relationships as the heat buildup process, only, instead of the strain of thermal shrinkage $\alpha\,\Delta T$, the polymerization relationships will contain strains of physico-chemical shrinkage e^{ch}. The technique described allows the requirements of continuous variation of not only stresses, but also strains in the model of the technological process to be satisfied.

The engineering variant of the theory of initial stresses predicts and allows the analysis of the methods of governing initial stresses. The engineering variant has been laid on the basis of such widely used techniques as the programmed winding, pressing, winding with layer-by-layer curing, the method of artificial temperature gradients. In controlling the reinforcement scheme with the aim of increasing the load-carrying capacity of the article one must consider the interplay between the reinforcing scheme, the initial stresses and the strength fields.

5. Controlling of initial stresses

5.1. A programmed winding

Solution of the problem of linear winding mechanics at varying winding tension (Section 3.2.3) has shown that variation of the winding tension is an effective means of controlling the variation of stresses in wound semifabricated articles and, consequently, also in a finished product. This leads to an opposite problem—how, according to preset post-winding variation of stresses or the form of stress distribution for a finished product, one can find out the law of variation of the winding tension, which ensures this scheme. This opposite problem is the subject of a programmed winding. The solution of the problem is given in the work of TARNOPOL'SKII and PORTNOV (1970).

It is possible to find out the formula directly, interrelating the post-winding distribution of circumferential σ_θ and radial stresses σ_r with programmed winding σ_θ^0 on the basis of the linearly-elastic circular model by using the relations (3.11), (3.13) and (3.14):

$$\sigma_\theta^0(\rho) = \sigma_\theta(\rho, b) - \sigma_r(\rho, b)\beta \frac{\eta\rho^{2\beta} - 1}{\eta\rho^{2\beta} + 1}. \tag{5.1}$$

The following restrictions should be observed: the stresses σ_θ should not be compressive in any of the circuits (to prevent fiber curvatures); the winding tension σ_θ^0 should not exceed the tensile strength of the material being wound. The range of relative and absolute dimensions of wound products over which the winding tension can be effectively employed for controlling initial stresses is limited by these restrictions.

The next step involves the transition to the finished article. The interrelation between the post-winding stresses and the resulting distributions of initial stresses in an article has been found in terms of the relation of Section 3. It is possible to obtain the equation from (3.5), (3.7) and (4.15):

$$\rho^2 \frac{d^2\sigma_\theta}{d\rho^2} + 3\rho \frac{d\sigma_\theta}{d\rho} + (1 - \mu^2)\sigma_\theta = \rho^2 \frac{d^2\sigma_\theta^h}{d\rho^2} + 3\rho \frac{d\sigma_\theta^h}{d\rho} + (1 - \kappa^2)\sigma_\theta^h, \tag{5.2}$$

linking the variation of circumferential stresses before σ_θ and after σ_θ^h heat buildup, where μ and κ designate the anisotropy ratio before and after heat buildup, respectively.

On the assumption that the effects associated with stiffening in the transverse direction at finite stresses and physico-chemical shrinkage in the polymerization process compensate for each other, the stress may be identified with circumferential stresses after polymerization.

The procedure of constructing programs has been completed by taking into account the thermoelastic terms. According to this approach, a winding program designed to set up a preselected system of initial stresses σ_r^{res} and σ_θ^{res} has been proposed. The most widely used program is the program ensuring uniform tensioning in circuits after winding (curve 2 in Fig. 35), which is useful in cases where it is necessary to avoid circuit curvatures. The program is of special interest for conventional materials: fabrics, paper films, magnetic and video-magnetic tapes as well as in fabrication of prepregs. The programmed winding, ensuring uniform tensioning in circuits after heat buildup (curve 3 in

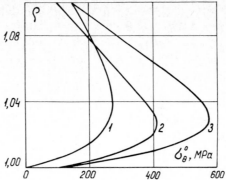

FIG. 35. Winding programs, ensuring: Curve 1—$\sigma_\theta^{res} = 0$ in an article; Curve 2—$\sigma_\theta = 100\,\text{MPa} =$ const. after winding; Curve 3—$\sigma_\theta = 100\,\text{MPa}$ after heat buildup: $\beta = 50$, $\beta E_r\,\Delta T(\alpha_r - \alpha_\theta) = 7500$.

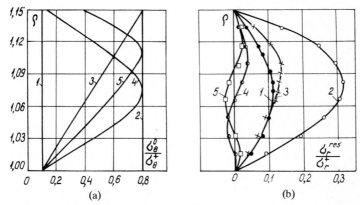

FIG. 36. Variation of initial stresses σ_r^{res} in GFRP rings (b), wound according to programs 1–5 (a) (PORTNOV and SPRIDZANS, 1971).

Fig. 35) makes it possible to avoid circuit curvatures in semifabricated material after heat buildup. The programmed winding, ensuring constant tensioning of circuits after thermal treatment of the product still on the mandrel, is most expedient for producing articles with guaranteed tensioning (bandages, metal-composite structural elements, etc.). The programmed winding ensuring the absence of initial stresses (curve 1 in Fig. 35) is the most widely used. An experimental study of the winding process at a variable winding tension (see Fig. 36a)) has confirmed the fact that, after winding in accordance with the law of variation in tensioning (as shown in Fig. 35, curve 3), having a maximum at the point lying at a distance from the outer radius of 25% of the ring thickness, the initial stresses in the product are nearly zero (see Fig. 36(b)).

5.2. *Pressing and rolling of the semifabricated material*

During fabrication of large-scale products, the pressure developed during force winding is insufficient for obtaining a monolithic article. Extra compaction may be caused by various means, in particular, by post-winding pressing and layer-by-layer rolling. The compaction process is a combination of an active loading with unloading; a schematic diagram of loading–unloading is shown in Fig. 7. Besides, two cases must be distinguished, depending on whether the stress in the loading process is higher or lower than σ_r^*, corresponding to the breaking point on the piecewise-linear σ_r–e_r diagram. It follows from the diagram that it is irrational to develop radial stresses in the product exceeding σ_r^*, because a higher degree of compaction is not achieved than at the stress σ_r^*. On the contrary, the danger of fibers passing from the pretensioning zone into the zone of compression arises, which is connected with the occurrence of curvatures. If the post-winding radial stresses in articles do not reach σ_r^* and the external pressure lower or equal to $-\sigma_r^*$ is applied, then, in a number of cases (for example, at high values of λ), the maximum summed up radial stress will be developed on the outer radius. In this case, the loading process can be described by the linearly-elastic theory, having an anisotropy ratio λ, but the unloading process is described by the same linear theory with the anisotropy ratio ω. The formulae, relating the external pressure q with increments of radial and circumferential stresses are as follows:

$$\sigma_r = -q\,\frac{\eta\rho^{\beta-1} + \rho^{-\beta-1}}{\eta b^{\beta-1} + b^{-\beta-1}}, \qquad \sigma_\theta = -q\beta\,\frac{\eta\rho^{\beta-1} - \rho^{-\beta-1}}{\eta b^{\beta-1} + b^{-\beta-1}}. \tag{5.3}$$

The linear model at high q describes the pressing process only partly, when the sum of radial stresses during winding and pressing does not exceed σ_r^*. After having reached this magnitude, the character of force transfer changes and it may be described only by the nonlinear model, analogous to the one used in the analysis of the winding process. The design methodology of the winding process in the statement involves designing of the pressing process, since the winding operation from the preceding outer circuit until the next one has been treated as a method of developing a certain pressure on the preceding radius.

The pressing process is fraught with danger of producing curvatures. Therefore, calculation of the circumferential stresses during pressing is as essential as the design of compaction degree. The maximum allowable pressure must be lower than the pressure at which the minimum of summed circumferential stress distribution (due to winding and application of the pressure q) is equal to zero. At high anisotropy ratios, λ, the minimum is practically always reached on the outer radius, where the circumferential winding stresses are equal to $\sigma_\theta^0(b)$. Consequently, increments of circumferential stresses in the point $\Delta\sigma_\theta$ due to pressure application must not exceed $\sigma_\theta^0(b)$. In Fig. 37 the dependence of increments of circumferential stresses on the outer radius in fractions of $q\beta$ and increments of the winding pressure on the mandrel P in fractions of q on the outer-to-inner radii b and on the anisotropy ratio $\beta = \lambda$ has been shown, calculated according to the linearly-elastic theory.

In treatment of the problem for a ring of the material, having a piecewise-linear σ_r–e_r diagram during pressing under external pressure, it is necessary to

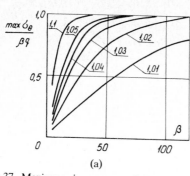
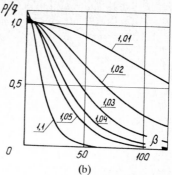

(a)   (b)

Fig. 37. Maximum increments of circumferential stresses max σ_θ (a) and the pressure on the mandrel P (b) during pressing with outer pressure q of a ring, wound onto a rigid mandrel, vs. the anisotropy ratio β at various $b = 1.01$–1.10.

allow for the fact that there are three variants of boundary movement $\sigma_r = \sigma_r^*$, depending on the form of post-winding stresses and the applied pressure:

(i) from the inner radius toward the outer radius,
(ii) from the outer radius toward the inner radius,
(iii) from the outer and inner radii toward the middle.

The two former cases show that the circumferential stresses, although only in one zone, are compressive. It is easy to convince oneself of the fact by considering the equilibrium equation. Consequently, these cases must be excluded. This may be achieved by applying the external pressure, which is lower than $\sigma_\theta^0(b)/\lambda$. Therefore, pressing under external pressure is of practical significance only for articles of small relative thickness, wound onto a stiff mandrel. Besides, the articles being pressed must be fabricated at maximum allowable winding tension (in particular, the outer circuits) and pressed under the pressure which does not exceed $\sigma_\theta^0(b)/\lambda$.

The internal pressure, developed by mandrels of special design, may be applied for compaction of the wound non-cured article. The pressure, applied during pressing, must be low, otherwise the inner circuits will be strongly extended and they might be torn. In thin-walled rings, this may be attributed to average circumferential stresses, being equal to $\bar\sigma_\theta = Pr_{in}/H_r$, where H_r is the ring thickness, and very high. In thick-walled rings, at lower average circumferential stresses their nonuniformity is considerable, particularly on the ring surface, where they approach $\sigma_\theta = \beta P$. Moreover, upon unloading the pretensioning in the middle and outer circuits falls below the level built up during winding. This increases the danger of curvature occurrence still further. During application of the internal pressure, a restricted volume of the material undergoes extra compaction, but the main part of its volume does not undergo significant compaction.

A positive effect for thin-walled articles can be achieved by combining the internal and external pressures during pressing. For thick-walled articles, the effect of combined pressing is reduced to the situation on the inner surface, corresponding to the action of the internal pressure only, but on the outer surface to the external pressure only. The pressing under preset pressure has been treated above. Moreover, there are also other methods of compaction, when displacements are preset, for example, by using special equipment.

The main restrictions of force winding as an effective means of controlling the initial stresses are connected with the high anisotropy of the semifabricated material. Compaction of the material during pressing results in a lower anisotropy ratio and combining of pressing with winding in one process, therefore, seems to be very promising. Two cases may be distinguished in alternating the winding and pressing operations: layer-by-layer pressing, when after a number of layers has been wound, the pressing operation is carried out and then the winding operation is started anew. There is also continuous pressing, when the processes are carried out simultaneously and the pressing operation with a metallic tape takes place. Continuous pressing is reminiscent of the winding technique with a linearly-elastic tape having the anisotropy ratio $\beta = \omega$, corresponding to the second section of the $\sigma_r - e_r$ diagram (cf. Fig. 7).

The significance of the linear analysis should be noted: disregarding the nonlinearity can result in practice in a higher degree of compaction, as compared with the calculated one, and in the reduction of the actual danger of curvatures.

During winding, besides the types of continuous axisymmetric compaction, there are also types of local compaction—for example, rolling etc. Since the pressure in the contact zone in practice always exceeds σ_r^*, the winding operation with simultaneous rolling is accomplished as a linearly-elastic one for the material, having the anisotropy $\beta = \omega$. Moreover, the high local pressure gradient contributes to a more uniform distribution of the binder throughout the volume and to weaker relaxation of the preset tensioning. Solution of the contact problem as well as the experiment (BLAGONADJOZHIN and MEZENTSEV, 1976) indicates the weak influence of the roller radius over the real range of its variation. The basic factors, determining the winding process, accompanied by rolling, are the winding tension and the rolling force. The rolling force should not be too high in order that the fracture of reinforcement should not take place or lose the binder. Therefore, the best result has been obtained for the program at decreasing rolling force with the increase in the current radius of the product.

5.3. Winding with layer-by-layer curing

The idea of the method is the same as for the layer-by-layer pressing: by decreasing the compliance of the wound circuits to achieve the weaker radial strain of successive circuits and thereby a lesser decrease in pretensioning. Upon layer-by-layer winding, the compliance is reduced as a result of partial or complete curing of a number of wound layers, followed by winding of the next layers (FINK, 1965). As the number of layers increases (curing stages), into which the article is divided, the process of layer-by-layer winding approaches some idealized ultimate process, presenting itself as the winding operation of a 'cured' material. It is reasonable to combine the layer-by-layer winding with the programmed winding. The program of variation in the winding tension from layer-to-layer (for a large number of layers) is computed according to (5.1), into which the anisotropy ratio β of the cured material has been inserted. The problem of optimization of the layer-by-layer winding has been treated in TARNOPOL'SKII et al. (1972) and aimed at compensation of the temperature stresses by developing zero initial radial stresses on the boundaries of layers, all the articles has been divided into. For sufficiently thin layers, it means that the intralayer stresses σ_r will also be low.

The contact pressure on the relative radius ρ_i is calculated as a sum of the pressure q_i due to winding, heat buildup and polymerization of the layer from ρ_i to ρ_{i+1}, pressure increments $\Delta q_i(\rho_i)$ on the radius, due to applications of pressure on the radii ρ_j, where $j > i$.

From the obtained sum the pressure increment Δq_i^- due to removal of the mandrel and thermoelastic stresses σ_r, having opposite signs, are subtracted. As a result, we can obtain a system of equations for q_i, which can be easily solved by the computer:

$$q_i + \sum_{j=i+1}^{n} \frac{q_i(\eta\rho_i^{\beta-1} + \rho_i^{-\beta-1})}{\eta\rho_j^{\beta-1} + \rho_j^{-\beta-1}} - \left(q_1 + \sum_{j=2}^{n} \frac{q_j(\eta+1)}{\eta\rho_j^{\beta-1} + \rho_j^{-\beta-1}}\right) \frac{\left(\frac{b}{\rho_i}\right)^{\beta+1} - \left(\frac{b}{\rho_i}\right)^{-\beta+1}}{b^{\beta+1} - b^{-\beta+1}} =$$

$$= D\left[\frac{(1 - b^{\beta+1})\rho_i^{\beta-1} + (b^{\beta+1} - b^{2\beta})\rho_i^{-\beta-1}}{b^{2\beta} - 1} + 1\right]. \tag{5.4}$$

The mandrel compliance over the range of real values does not in practice affect the result of TARNOPOL'SKII et al. (1972). The value q_i increases with the increase in the anisotropy ratio β and the relative dimension b. If the article is divided into more numerous layers, the level of q_i can be essentially reduced. The value of maximum tension σ_θ^0 in developing the system of pressures q_i decreases with the increase in n.

Experimental investigations of layer-by-layer curing have shown that the method of controlling the initial stresses is highly effective. The method of layer-by-layer winding allows the fabrication of articles with compressive radial initial stresses (see Fig. 38). This is especially important in cases where cracking must be prevented. Theoretical relationships were experimentally checked on rings having $b = 1.42$, divided into six equal layers. The winding tension was varied in a similar manner to the theoretically calculated winding tension according to three programs (see Fig. 39(a)). For the sake of comparison, rings were wound at constant winding tension, equal to the minimum and maximum tension (programs 1 and 5 in Fig. 39(a)). The corresponding distributions of initial stresses (see Fig. 39(b)) testify to the fact that the layer-by-layer winding is a method of eliminating the dangers as regards cracking from radial tensile initial stresses.

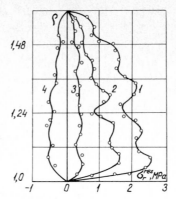

FIG. 38. Curves of initial stresses σ_r^{res} during programmed winding with layer-by-layer curing. Curves 1–4 winding programs (VARUSHKIN et al., 1972); Curve 1 corresponds to the winding program No. 1 in Fig. 19(a), Curve 2 to program No. 2, Curve 3 to program No. 4 and Curve 4 to program No. 5.

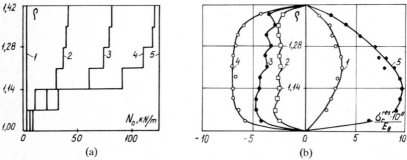

(a)   (b)

FIG. 39. Initial radial stresses in rings (b), wound according to programs 1–5 (a) with layer-by-layer curing (TARNOPOL'SKII et al., 1972).

5.4. *Control of initial stresses in polymerization and cooling stages*

An analysis of stress development during curing and cooling of the article by methods of linear thermoelasticity makes it possible to propose a number of technological means of controlling σ_r^{res} and σ_θ^{res}. They are as follows: pressing during cooling process at the temperature when the radial stress-to-radial tensile strength ratio is maximum, control of the cooling rate, layer-by-layer curing at varying temperatures and a method of temperature gradients.

By comparing the temperature kinetics of transverse strength with the kinetics of growth of thermoelastic stresses, in some cases it is possible to estimate the temperature range over which either the strength is lower than the stresses or their ratio is lower than the allowable safety factor (BOLOTIN and BOLOTINA, 1972; RABOTNOV and EKELCHIK, 1975), while at finite cooling temperature some surplus strength exists. For such a situation, it has been proposed by RABOTNOV and EKELCHIK (1975) to accomplish pressing at some critical temperatures so that the distribution of summed stresses might reach the desired values. External pressure is released at temperatures where the sum of extra tensile (due to pressure release) and thermoelastic radial stresses turns out to be lower than the ultimate strength, taking into account the preset safety factor.

A conclusion has been drawn on the basis of the theoretical study of the relaxation process of thermoelastic stresses by OGIL'KO (1974) that, either by prolongation of the cooling process or by a stepwise cooling with prolonged exposing at each temperature, it is possible to reduce thermo-visco-elastic stresses.

The method of layer-by-layer winding at varying temperatures has been combined with the winding with layer-by-layer curing. The polymerization temperature of each layer is chosen so that, after cooling of each successive layer to one and the same ultimate temperature, a certain tensioning on the previous layers might be developed—which partly compensates for the tensile radial thermal stresses. It is evident that the polymerization temperature (a temperature difference upon cooling) must increase from the inner to the outer radius. The experiment has shown that the level of initial stresses for the proposed technology is lower than for the usual technology.

The method of temperature gradients has been theoretically confirmed in the work of BIRGER (1971) and proposed in AFANASJEV et al. (1980) and BAKHAREV and MIRKIN (1978). The method consists of developing temperature gradients in polymerization and cooling processes so that it is possible to change the stress distribution. In particular, in an idealized case for uniform elastic and thermorphysical characteristics, the temperature distribution $T(\rho)$ in a ring is described by the expression,

$$T(\rho) = T(1)\rho^{(\alpha_r/\alpha_\theta - 1)}, \qquad (5.5)$$

the equation of strain compatibility is satisfied and thermoelastic stresses are absent. It is clear that such a situation cannot be reached in practice, since the temperature distribution in a body is determined by an equation of heat flow and cannot be preset deliberately. The magnitude of the initial gradient is determined by the allowable range for the polymerization temperature, and for thick-walled articles this temperature cannot be high.

5.5. Reinforcement layup

Another method of controlling initial stresses and eliminating the danger of cracking consists in variations of the fiber layup. By varying the fiber layup it is possible to change not only transverse tension strength, but also the distribution of thermoelastic stresses (at the expense of variation of the anisotropy of elastic and thermophysical properties of the article). There are such methods as: introduction of compliant interlayers, an extra radial reinforcement, multilayer structures and a method of winding at alternating angles.

It follows from an analysis of the formulation of thermal radial stresses (BOLOTIN and BOLOTINA, 1967) that the decrease in E_r affects σ_r much more than the respective increase in β. The method of decreasing temperature stresses by introducing thin compliant interlayers by INDENBAUM (1973) has been based on this idea (thereby, E_θ also decreases so that the anisotropy β does not change so abruptly). The method of introducing thermocompensating interlayers is equivalent to dividing a ring into a number of thin, weakly-bonded concentric rings. Since decreasing the ring thickness results in a sharp fall of initial stresses, a definite effect may be achieved. The form of initial radial stresses in two rings of equal dimensions, one of which contains interlayers of compensators, is presented in Fig. 40. In such a way, the experiment has confirmed the theoretical conclusion. The disadvantage of the method is some decrease in load-carrying capacity of the structure (especially, under the effect of concentrated forces).

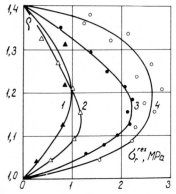

FIG. 40. Curves of initial radial stresses σ_r^{res} in GFRP rings, with compensators (Curves 1, 2) and without them (Curves 3, 4). Reinforcement layup: Curves 1, 3—$[0°, 90°]_m$; Curves 2, 4—$[0°_s, 90°_s]_m$.

It is also possible to reduce the relation between thermal stresses and radial strength in the case when, as a result of variation to the reinforcement scheme, the growth of strength in the radial direction σ_r^+ is more significant than the increase in radial stresses with increasing elastic modulus E_r. An analogous effect has been achieved by extra radial reinforcement with short needles, aligned along the radius in an electrostatic or magnetic field (ZHMUD et al., 1978). Since the fields improve the strength characteristics of composites, the incorporation of small volume fractions of needles results in an essential growth of strength. At the same time, the anisotropy of the wound semifabricated article is reduced, resulting in more favourable distribution of post-winding radial stresses.

In order that the outer part of the article after cooling might exert pressure on the inner part, one can employ not only polymerization at various temperatures, but also the parts can be made from various composites, having different coefficients of thermal linear expansion (BLAGONADJOZHIN et al., 1975; DIMITRIENKO et al., 1976).

The radial thermoelastic stresses may be equal to zero, not only in an idealized nonuniform temperature field, in which the thermal strains satisfy the equation of strain compatibility (3.7), but also at uniform variation in temperature—if the coefficients of linear thermal expansion α_r and α_θ in a cylindrically orthotropic ring change along the radius so that the equation might be satisfied.

It has been proposed in the work of BEIL' et al. (1980) to achieve the necessary variation in the coefficients of thermal linear expansion by variation of the winding angle $\varphi(\rho)$ with increasing the current radius of the article. Let α_r and α_θ be the functions of the winding angle φ (the angle between a circumferential coordinate and one of the directions of a balanced layup); the equation of strain compatibility (3.7) may be rewritten for $\Delta T = \text{const.}$ as follows:

$$\alpha_r(\varphi) - \alpha_\theta(\varphi) = \rho \frac{d\alpha_\theta(\varphi)}{d\varphi} \frac{d\varphi}{d\rho}. \tag{5.6}$$

By employing the initial condition at $\rho = 1$, $\varphi = \varphi_{in}$ one may obtain

$$\rho = \exp\left\{ \int_{\varphi_{in}}^{\varphi} \frac{d\alpha_\theta}{dy} \frac{1}{[\alpha_r(y) - \alpha_\theta(y)]} \, dy \right\}. \tag{5.7}$$

The formula determines the function $\rho(\varphi)$, which is opposite to the one we have searched for $\varphi(\rho)$. Numerical analysis has shown that the winding operation should be performed at an angle φ, increasing with the increase in the current radius of the article. The winding process of a thick cylinder with preset variation of the winding angle along the thickness has made it possible to fabricate an article of high quality without inner cracks. Control of initial stresses by varying the reinforcement scheme simultaneously affects the load-carrying capacity of the article. This fact should be reckoned with in each individual case.

6. Development of a model

6.1. Basic tendencies

The mechanics of composite winding are not restricted to studying an idealized circular model of the article. The next stage involves passing over from basic models to studying the winding operation for bodies of more complex shapes. Along with this, it is useful to further refine the circular model by introducing extra factors—first of all, rheological ones. Although the circular model, described in previous sections, reflects the basic rheological features, under extreme conditions secondary factors may be essential. Statement of the area of application of simpler models is an important task of refined theories.

At present, the two tendencies of the mechanics of composites—generalization applied to bodies of an arbitrary shape and rheological complication of the models—are in the stage of development. As a rule, the methods used are more complex and they lead to more cumbersome results. Let us discuss some individual problems of practical importance.

6.2. *Winding mechanics for cylinders and profiled articles*

Circular models, besides their direct purpose—a qualitative analysis of the phenomena occurring in all stages of winding process—have found application in the treatment of more complex cases than the winding of rings. Let us consider for example the principle of describing the circumferential, circumferentially-axial and helical-balanced (balanced angle-ply) winding of cylinders and circumferential winding of profiled disks by using the circular models.

Circumferential winding of cylinders may be described by means of circular models having some modifications. Depending on the technological procedure of article fabrication, different variants of boundary conditions are possible. In the cases where there are no constraints at cylinder butts and the interlaminar shear compliance S_{rzrz} is high, the state of stress may be considered as planar with small error. Calculation of stresses σ_r, σ_θ and strains e_r and e_θ is based on circular models, but of the strain e_z on the calculated stresses σ_r and σ_θ and the constitutive relation. Let us note that in this case the profile of the butt is more distorted with the increase in the cylinder length.

A technological variant of cylinder winding, in which axial strains are constrained $e_z = 0$ is most often encountered. Passing over from models with a plane state of stress to models with a plane state of strain is accomplished by simple conversion of elastic constants,

$$S_{ij} = S_{ij} - \frac{S_{iz}S_{jz}}{S_{zz}}, \quad i, j \leftrightarrow r, \theta. \tag{6.1}$$

Let us note that in the case of a piecewise-linear model the constants S_{rz}^{I} and S_{rz}^{II} are equal. For composite semifabricated materials, in particular, unidirectional ones, Poisson's ratio ν_{zr} is not so small. Therefore, the anisotropy ratio

$$\beta = \sqrt{E_\theta(1 - \nu_{rz}\nu_{zr})/[E_r(1 - \nu_{\theta z}\nu_{z\theta})]}$$

for a plane state of strain is considerably lower than the value β at a plane state of stress. The experimentally observed fact (see Fig. 41) that the increase of winding pressure on the mandrel with the increase in the number of circuits

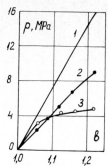

FIG. 41. Dependence of the pressure on the mandrel P on the relative dimension b of the article during winding of an isotropic tape (Curve 1) and GFRP semifabricate onto a cylindrical (Curve 2) and circular (Curve 3) mandrels.

does not attenuate so fast for cylinders as for rings, may be partly attributed to the above conclusion. Calculation of stresses for the winding model, where in each step $\Delta e_z =$ const. and the axial force is equal to zero, is performed by using the solution of a problem for loading an infinite composite cylinder (composite mandrel) under external pressure at zero axial force.

During circumferentially-axial winding, tensioning of layers, laid along the axis does not contribute to the normal component of the reaction. It is necessary to consider only the winding tension $N_0 = \sigma_\theta^0 \Delta b_\theta$ of circumferential layers. In the formula of ultimate transition a characteristic trait appears (cf. (3.12)); instead of Δb the smallest distance min Δb between the repeating elements of the packet must be inserted (for example, for a layup $(0_1/90_1)$ it will be min $\Delta b = 2 \Delta b$; for a layup $n_z : n_\theta$, min $\Delta b = (\Delta b_z n_z + \Delta b_\theta n_\theta)/n_\theta$, where n_z is the number of layers, laid along the axis, but n_θ is the number of layers laid in the circumferential direction, Δb_z and Δb_θ are thicknesses of the respective layers). Therefore, the circumferentially-axial winding is accomplished as if it were at the winding tension $\sigma_\theta \Delta b_\theta / \text{min} \Delta b$, but not at $\sigma_\theta^0 \Delta b$. The anisotropy of the packet may be computed according to the usual formula of the theory of laminated media or experimentally determined in transverse compression and longitudinal tension of a sample of respective structure.

In the case of helicoidal balanced winding, $\sigma^0 \cos^2 \varphi$ should be inserted in the formulae (3.10)–(3.14) instead of σ_θ^0, where φ is an angle between a tangent to the windable tape and a plane, perpendicular to the cylinder axis. Thereby the anisotropy ratio, β, is determined according to the formula of the theory of balanced laminated media, taking into account the formulae of transformation for tensor components S_{ij} upon rotation of the coordinate system or by direct experimenting on a respective layer packet.

During winding of profiled articles (for example, discs), only one change will be introduced into the system of equations and boundary conditions, pertaining to equilibrium equation, in which the variable width of a prepreg $H_z(\rho)$ is enclosed:

$$\sigma_\theta H_z(\rho) = \frac{d}{d\rho}[\rho H_z(\rho)\sigma_z].\tag{6.2}$$

Solution of the linearly-elastic winding problem for a disc of a profile $H_z = H_{in}/\rho$, which is very often used in the fabrication of flywheels, yields

$$\sigma_r = -(g\rho^{\delta-1/2} + \rho^{-\delta-1/2})Z(\rho),$$
$$\sigma_\theta = \sigma_\theta^0(\rho) - [g(\delta - \tfrac{1}{2})\rho^{\delta-1/2} - (\delta + \tfrac{1}{2})\rho^{-\delta-1/2}]Z(\rho),\tag{6.3}$$

where

$$Z(\rho) = \int_\rho^b \frac{\sigma_\theta^0(y)y^{\delta-1/2}\,dy}{gy^{2\delta}+1},$$

$$\delta = \sqrt{\beta^2 + \tfrac{1}{4} + \nu_{\theta r}}, \qquad g = \frac{\delta + \tfrac{1}{2} + \gamma + \nu_{\theta r}}{\delta - \tfrac{1}{2} - \gamma - \nu_{\theta r}}.$$

A problem for programmed winding of profiled articles is stated analogously to the problem for a ring. From (6.3) one can obtain an expression relating the desired distribution of post-winding stresses to a winding program,

$$\sigma_\theta^0(\rho) = \sigma_\theta(\rho) - \sigma_r(\rho)\frac{g(\delta - \tfrac{1}{2})\rho^{2\delta} - \delta - \tfrac{1}{2}}{g\rho^{2\delta}+1}.\tag{6.4}$$

The expression differs from the respective formula for a ring (5.1) by inessential additions.

6.3. Viscous-elastic circular models

Creep in transverse compression is taken into account as follows. By using the constitutive law for transverse compression in a form of the differential equation for the nonlinear rheological model of a typical body (Kelvin model), one can obtain the solution of an axisymmetric problem. The left-hand part in terms of $\dot\sigma_r$ is analogous to a respective equation for σ_r of the nonlinearly-elastic winding problem. The right-hand part, expressed through σ_r, may be considered as preset for a given time moment t. In such a way, the continuous winding operation is substituted by an instantaneous laying of a circuit of the thickness Δb and exposing it in a stationary state for some time Δt, corresponding to the actual time of continuous winding operation of the circuit. The solution of the equation for $\dot\sigma_r$ by a method analogous to the method used in constructing a discrete-circular winding model for nonlinearly-elastic materials, multiplied by time Δt, allows one to determine the new state of stress, preceding the winding operation of the successive layer, etc. The post-winding stress distributions at finite velocity and subsequent relaxation (accelerated upon heat buildup) are between the stress distribution for an instantaneous winding (an instantaneous isochrone σ_r–e_r) and successive relaxation and infinitely slow winding (an isochrone σ_r–e_r as $t \rightarrow \infty$).

Another case of manifestation of the viscous-flow state of the binder is infiltration. The problem, by taking flow into account, is solved in terms of a model analogous to a discrete-circular winding model for nonlinearly-elastic materials, which allows for strain nonuniformity, caused by flow. A numerical solution (see Fig. 42) shows that, under usual conditions, flow does not play an essential role. Another case of manifestation of viscous-flow effects is connected with the structure of the wound articles. A peculiarity of a helical layer layup is the fact that the strain is transferred from circuit to circuit, not only by a compressible interlayer, but also directly along a helix from layer-to-layer (BEIL' and PORTNOV, 1973). The viscosity of the interlayer is a factor counteracting the mechanism. During winding of circuits forming a helical structure, there is a tendency to level out the tightening. This favourably affects the decrease in danger of curvature occurrence. It is important after winding operation to fix the end of the tape firmly, otherwise stress relaxation is possible, associated with movement of the last circuit and occurrence of so-called 'unwinding'.

Solution of the problem may be laid on the basis of a refined statement of a winding problem for bodies from inelastic materials. An advance in the description of the winding process was made by contributions, observed in BOLOTIN and NOVICHKOV (1980), based on the use of a hypo-viscous-elastic idealization.

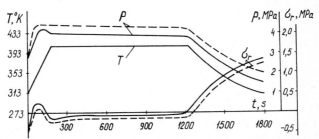

FIG. 42. Time dependence of radial stresses σ_r in a point, where initial radial stresses σ_r^{res} are maximum, and pressure on the mandrel P at preset temperature regime with consideration (——) and without consideration (– – –) of the binder flow (BOLOTIN et al., 1980).

7. Conclusion

Accordingly, division of the fabrication process into separate stages, consideration of the compliance of a semifabricated material, introduction of a circular model and hypothesis about the inheritance of stress–strain state, has allowed us to examine the relationships between the parameters of the process and, particularly, how the winding tension is related to the properties of the finished article, the contact pressure on the part to be wound, and the system of initial stresses. A number of technological procedures follows directly from the already considered models. Among the procedures are: a programmed winding, a winding with layer-by-layer curing and a winding under extra contact pressure, radial reinforcement with short needles, the use of thermocompensating interlayers, a method of temperature gradients, a winding at a varying angle, etc.

The circular models have served as an impetus for researches in the field of engineering winding mechanics. Areas of use for thick-walled wound products are continuously widening. It is necessary to continue researches within the framework of the circular model towards more detailed studies, taking into account the factors which have so far been ignored by the existing theory. In particular, it is necessary to determine the role of creep, physico-chemical shrinkage, the change in resistancy to deformation in all stages of the fabrication process. The researches must be aimed at evaluation of the areas of applicability of the theory, which would allow to pass over from the circular model to the design of products of more complex configuration. Technological methods of controlling the homogeneity and load-carrying capacity of wound products should be further evolved. The focus, as before, is the problem of formation of elastic and strength properties of a composite in the stage of transition from a semifabricated material to a finished article.

References

ABIBOV, A.L., V.A. BUNAKOV, V.N. KOPEIKIN and R.M. KONDRATENKO (1973), *Polymer Mech.* **152** (in Russian).

AFANASJEV, YU.A., V.S. EKELCHIK and S.N. KOSTRITSKII (1980), *Mech. Composite Materials* **651** (in Russian).

BAKHAREV, S.P. and M.A. MIRKIN (1978), *Polymer Mech.* **1118** (in Russian).

BEIL', A.I. and G.G. PORTNOV (1973), *Polymer Mech.* **884** (in Russian).

BEIL', A.I., G.G. PORTNOV, I.V. SANINA and V.A. YAKUSHIN (1980), *Mech. Composite Materials* **1068** (in Russian).

BIDERMAN, V.L. (1958), Determination of tensioning in a steel rope in a load-carrying mechanism, in: *Strength Calculations* Vol. 2 (Mashgiz, Moscow) p. 47 (in Russian).

BIDERMAN, V.L. (1979), State of stress in a helically reinforced ring, in: *Strength Calculations* Vol. 20 (Mashinostroyeniye, Moscow) p. 65 (in Russian).

BIDERMAN, V.L., I.P. DIMITRIENKO and N.A. SUKHOVA (1969), *Polymer Mech.* **892** (in Russian).

BIRGER, B.I. (1971), Proc. USSR Acad. Sci., *Mech. Solids* **92** (in Russian).

BLAGONADJOZHIN, V.L. and N.S. MEZENTSEV (1976), *Polymer Mech.* **1043** (in Russian).

BLAGONADJOZHIN, V.L., G.V. MISHENKOV and V.G. PEREVOZCHIKOV (1970), Investigation of pressure on the mandrel in fabrication process of wound products on a tensometric mandrel, in: *Trans. MEI, Dynamics and Machine Strength* Vol. 74 (MEI, Moscow) p. 133 (in Russian).

BLAGONADJOZHIN, V.L., V.G. PEREVOZCHIKOV, V.D. MERKULOV and V.L. POLYAKOV (1975), *Polymer Mech.* **996** (in Russian).

BOLOTIN, V.V. (1972), *Polymer Mech.* **529** (in Russian).

BOLOTIN, V.V. (1975), *Polymer Mech.* **126** (in Russian).

BOLOTIN, V.V. and K.S. BOLOTINA (1967), *Polymer Mech.* **136** (in Russian).

BOLOTIN, V.V. and K.S. BOLOTINA (1969), *Polymer Mech.* **134** (in Russian).

BOLOTIN, V.V. and K.S. BOLOTINA (1972), *Polymer Mech.* **178** (in Russian).

BOLOTIN, V.V., I.I. GOL'DENBLAT and A.F. SMIRNOV (1972), Structural mechanics, in: *State-of-Art and Future Perspectives* (Stroyizdat, Moscow) (in Russian).

BOLOTIN, V.V. and YU.N. NOVICHKOV (1980), *Mechanics of Multilayered Structures* (Mashinostroyeniye, Moscow) (in Russian).

BOLOTIN, V.V., A.N. VORONTSOV and R.H. MURZAKHANOV (1980), *Mech. Composite Materials* **500** (in Russian).

BRIVMANIS, R.E. (1966), *Polymer Mech.* **123** (in Russian).

BRIVMANIS, R.E. and A.K. GAGANOV (1971), *Winding Structural Elements in Electrical Machines and Apparatus* (Energiya Press, Moscow) (in Russian).

CHAMIS, C.C. and L.I. KIRALY (1975), Rim-spoke composite flywheels: Detailed stress and vibration analysis, in: G.C. CHANG and R.G. STONE, eds., *Proc. 1975 Flywheel Technology Symp.* (University of California, Livermore) p. 110.

CLEMENTS, L.L. (1977), Comparative properties of fiber composites for energy storage flywheels, Part B, in: *Proc. 1977 Flywheel Technology Symp.* (NTIS, U.S. Dept. of Comm., Springfield, Virginia) p. 363.

DIMITRIENKO, I.P., A.A. FILIPENKO and V.D. PROTASOV (1976), *Polymer Mech.* **681** (in Russian).

ELPATEVSKII, A.N. and V.V. VASIL'EV (1972), *Strength of Cylindrical Shells of Reinforced Materials* (Mashinostroyeniye, Moscow) (in Russian).

FINK, B. (1965), Development of reinforced plastics hull structures for deep diving submersible vehicles; in: *Proc. 20th Anniversary Techn. Conf. SPI Reinforced Plastics Division* (SPI, New York, NY) I–A, pp. 1–14.

INDENBAUM, V.M. (1973), Calculation of residual stresses in multilayered cylinders from combined composites, in: *Trans. MEI* Vol. 164 (MEI, Moscow) p. 81 (in Russian).

KHITROV, V.V. and N.P. ZHMUD (1974), *Polymer Mech.* **802** (in Russian).

KOSTRITSKII, S.N. and M.Z. TSIRKIN (1981), *Mech. Composite Materials* **355** (in Russian).

LEKHNITSKII, S.G. (1957), *Anisotropic Plates* (Gostekhizdat Press, Moscow) (in Russian).

LIU, C.Y. and C.C. CHAMIS (1965), Residual stresses in filament wound laminates and optimum programmed winding pension, in: *Proc. 20th Anniversary Techn. Conf. SPI Reinforced Plastics Division* (SPI, New York, NY) 5–D, pp. 1–10.

OBRAZTSOV, N.F., V.V. VASIL'EV and V.A. BINAKOV (1977), *Optimal Reinforcement for Shells of Revolution from Composite Materials* (Mashinostroyeniye, Moscow) (in Russian).

OGIL'KO, T.F. (1974), *Polymer Mech.* **949** (in Russian).

PARKYN, B. (1963), *J. Roy. Soc. Arts* **205**.

PORTNOV, G.G. and A.I. BEIL' (1977), *Polymer Mech.* **231** (in Russian).

PORTNOV, G.G., V.A. GOR'USHKIN and A.G. TIL'UK (1969), *Polymer Mech.* **505** (in Russian).

PORTNOV, G.G. and YU.B. SPRIDZANS (1971), *Polymer Mech.* **361** (in Russian).

PROTASOV, V.D. (1978), *J. All-Union D.I. Mendeleev's Chem. Soc.* **23**, 289 (in Russian).

RABOTNOV, YU.N. and V.S. EKELCHIK (1975), *Polymer Mech.* **1095** (in Russian).

RICHARDSON, M., ed. (1977), *Polymer Engineering Composites* (Applied Science Publishers LTP, London).

ROSATO, D.V. (1969), History of composites, in: G. LUBIN, ed., *Handbook of Fiberglass and Advanced Plastic Composites* (Van Nostrand Reinhold, New York, NY) p. 1.

ROSATO, D.V. and C.S. GROVE (1964), *Filament Winding* (Interscience Publishers, New York, NY).

ROZENBERG, B.A. and N.S. ENIKOLOPYAN (1978), *J. All-Union D.I. Mendeleev's Chem. Soc.* **23**, 298 (in Russian).

SOUTHWELL, R.V. (1948), *An Introduction to the Theory of Elasticity for Engineers and Physicists* (IL, Moscow) (in Russian).

TARNOPOL'SKII, YU.M. (1976), Thick-walled wound composite structures, in: E. SCALA, E. ANDERSON, J. TOTH and B. NOTON, eds., *Proc. 1975 Internat. Conf. Composite Materials* (The Metallurgical Society of AIME, New York, NY) I, p. 221.

TARNOPOL'SKII, YU.M., A.I. BEIL' and G.G. PORTNOV (1980), *Proc. Latvian SSR Acad. Sci.* **80** (in Russian).

TARNOPOL'SKII, YU.M. and T.YA. KINCIS (1981), *Methods for Static Testing of Reinforced Plastics* (Khimiya Press, Moscow, 3rd ed.) (in Russian).
TARNOPOL'SKII, YU.M. and G.G. PORTNOV (1966), *Polymer Mech.* **278** (in Russian).
TARNOPOL'SKII, YU.M. and G.G. PORTNOV (1970), *Polymer Mech.* **48** (in Russian).
TARNOPOL'SKII, YU.M., G.G. PORTNOV and YU.B. SPRIDZANS (1972), *Polymer Mech.* **640** (in Russian).
TARNOPOL'SKII, YU.M., G.G. PORTNOV, YU.B. SPRIDZANS and V.V. BULMANIS (1973), *Polymer Mech.* **673** (in Russian).
TARNOPOL'SKII, YU.M. and A.V. ROZE (1969), *Characteristics of Calculation of Reinforced Plastic Parts* (Zinātne Press, Riga) (in Russian).
VARUSHKIN, YE.M. (1971), *Polymer Mech.* **1040** (in Russian).
VARUSHKIN, YE.M. (1972), *Chem. Oil Industry Engrg.* **8** (in Russian).
VARUSHKIN, YE.M., V.A. POLYAKOV and YU.L. LAPIN (1972), *Polymer Mech.* **75** (in Russian).
VASIL'EV, V.V. (1969), *Polymer Mech.* **1069** (in Russian).
ZHMUD, N.P., V.YU. PETROV and V.N. SHALIGIN (1978), *Polymer Mech.* **226** (in Russian).

CHAPTER III

Multidirectional Carbon–Carbon Composites

Lawrence E. McAllister

Fiber Materials Incorporated
Biddeford
Maine 04005
U.S.A.

Walter L. Lachman

Materials International
Lexington
Massachusetts 02173
U.S.A.

Contents

HANDBOOK OF COMPOSITES, Vol. 4 – Fabrication of Composites
Edited by A. KELLY and S.T. MILEIKO
© 1983, Elsevier Science Publishers B.V.

1. Introduction

Graphite is an attractive material for elevated temperature applications in inert atmosphere and ablative environments based on properties such as high sublimation temperature, improved strength with increasing temperature up to about 2800°C, thermal stress resistance and chemical inertness. In its bulk forms of polycrystalline graphite and pyrolytic graphite, its utility for many applications has been limited by low strain to failure, flaw sensitivity, anisotropy, variability in properties, and fabrication difficulties associated with large sizes and complex shapes.

The advent of carbon fiber technology that emerged in the late 1950's offered the potential for developing graphite materials in truly structural forms. By 1960, under the sponsorship of the Air Force Materials Laboratory (AFML) at Wright-Patterson Air Force Base (WPAFB), a new class of materials that eventually would be designated as carbon–carbon[1] composites was demonstrated by two industrial laboratories in the USA (Schmidt, 1972).

The advancement of carbon–carbon technology was slow for several years. However, by the late 1960's carbon–carbon composites were beginning to emerge as a new class of engineering materials (Stoller and Frye, 1969; Kotlensky and Pappas, 1969). During the 1970's carbon–carbon composites were under extensive development in several laboratories in the USA and Europe (McAllister and Taverna, 1971a, b, 1972, 1974, 1976; Fitzer and Burger, 1971; Bauer and Kotlensky, 1971a, b; Stoller et al., 1971; Mullen and Roy, 1972; Burns and Cook, 1974; Levine, 1975; Chard and Niesz, 1975; Burns and McAllister, 1976; Lamicq, 1977; Fitzer et al., 1978; Girard, 1978; Thomas and Walker, 1978; Lachman et al., 1978).

The original carbon–carbon composites were produced using two-directional (2-D)[2] reinforcements in the form of low modulus rayon precursor carbon and graphite fabrics. The matrix was derived from pyrolyzed high char yield thermosetting resins such as phenolic. Reinforced plastic molding techniques were used to fabricate the precursor composites which were subsequently heat treated to convert the resin matrix to carbon or graphite. These fabric reinforced carbon–carbon composites were higher in strength than polycrystalline bulk graphites in the plane of the fabric reinforcement but exhibited very low properties in other directions. However, because of their improved thermal stress resistance, toughness, and the potential of fabricating large, complex

---

[1] Carbon–carbon (composite) is a generic term referring to a class of materials containing carbon (or graphite) fibers with a carbon (or graphite) matrix.

[2] The capital letter D means directional. For example, 3-D refers to a three-directional composite, i.e., a multidirectional composite with reinforcement in three directions.

shapes, development of 2-D carbon–carbon materials continued. Much of this development was supported by the U.S. Government (SCHMIDT, 1972).

The development of high strength, high modulus graphite fibers during the early 1960's (SCHMIDT, 1972; WATT, 1966) provided the means to produce structural carbon–carbon composites with elevated temperature strength, stiffness and toughness well in excess of other available engineering materials. The ability to tailor properties of these composites also existed. However, it took several years of development to generate sufficient data and to establish the design methodology necessary to take full advantage of carbon–carbon composites as engineering materials.

Even with the improved properties offered by 2-D carbon–carbon composites over polycrystalline bulk graphite, the extreme isotropy of these materials was a serious drawback. Low strength in the unreinforced regions between fabric plies turned out to be a severe limitation. In recognition of the drawbacks of 2-D composite constructions, various techniques were investigated to produce multidirectional reinforced composites. These included needled felts and fabrics, pile fabrics, stitching of fabrics, and multiple warp fabrics (BARTON, 1968). These approaches did not overcome the problem of poor interlaminar strength.

By the late 1960's, techniques for weaving multidirectional structures in block, hollow cylinder, and conical frustum configurations had been developed for use in both resin matrix and carbon–carbon composites (BARTON, 1968; JACOBS et al., 1968). Since that time, multidirectional weaving techniques have been developed and refined to the point where composites can be tailored to meet complex design requirements through the proper selection and distribution of reinforcing elements.

This chapter will discuss multidirectional woven substrates and the densification processes used to produce carbon–carbon composites from these substrates. Resource information for the chapter is from the open literature. Review articles by SCHMIDT (1972) and by STOLLER et al. (1971) provided useful references on the general subject of carbon–carbon composites. A recently published book by ZHIGUN and POLYAKOV (1978) addresses all aspects of multidirectional composites. A chapter in this book which was devoted to a review of the construction, processing and properties of multidirectional carbon–carbon composites was a useful source of references. A paper by LACHMAN et al. (1978) was also used as the basis for much of the discussion on multidirectional structures.

2. Multidirectional substrates

2.1. General

Multidirectional carbon–carbon composites offer the potenti.l of being tailored to meet directional property requirements of an end item. Thermal, mechanical and physical properties of the composite can be controlled by the appropriate design of substrate parameters such as fiber orientation, volume fraction of fibers in required directions, fiber spacing, substrate density, yarn packing efficiency and fiber selection (COOK and CRAWFORD, 1975). Matrix selection and processing technique will also strongly influence final composite properties.

This section of the chapter will address the various types of multidirectional substrates that have been used to produce carbon–carbon composites. Both design of the substrate and fabrication approaches will be presented.

2.2. *Fiber selection*

Fiber selection is based on the application and the environment for which the composite is being designed, and the ability to obtain the fiber in a useful form for a substrate fabrication. The most common form is yarn; a yarn being a continuous length of plied fibers held together by fiber twisting or by an added coating, or both.

Small amounts of yarn coatings or finishes are required for ease of yarn handling during weaving operations. Also, yarn coatings are often required for subsequent impregnation to achieve compatibility between fiber and matrix. The optimum situation occurs when the required compatible coating is also sufficient for handling during the weaving operations.

A variety of carbon fibers in the form of yarns is available for use in multidirectional substrates. High strength, high modulus fibers are usually selected where structural properties are important. However, the lower modulus fibers can be used if lower thermal conductivity is required.

Table 1 summarizes typical properties of various graphite yarn types that may be used to produce multidirectional substrates. These yarns are usually available in the range of 1000 to 10 000 filaments per strand. The size of the yarn bundle will control the fineness of the substrate construction.

It is also possible to select groups of similar or dissimilar yarns and ply or twist them together as required to meet the design objectives of the woven structure. An example of a plied yarn formed by combining dissimilar yarn–fiber types is shown in Fig. 1.

A review of the fiber data in Table 1 reveals that a wide variation in properties can be introduced into carbon–carbon composites depending on the particular fiber type or combination of fibers used. The highest modulus fibers will provide the highest thermal conductivity, density and carbon assay, and the lowest thermal expansion. This property trend relates to the fact that the high modulus fibers have been processed to a high temperature which induces a high degree of graphitization and crystalline orientation along the fiber axis.

In most cases, the fiber properties will change due to various handling and processing steps that occur during the fabrication of carbon–carbon composites. Processing temperature may affect properties of some fiber types especially for carbon fibers that have not been previously subjected to graphitizing temperatures. Fiber–matrix interactions also influence the final composite properties. Some of these process related property effects will be discussed in Section 4.1.

TABLE 1.[a] Typical properties of commercial graphite yarns.

Property/fiber type	Rayon precursor		PAN precursor		Pitch precursor
	Low modulus-strength	High modulus-strength	High strength	High modulus	
Tensile strength (GPa)	0.62	2.2	3.1	2.4	2.1
Tensile modulus (GPa)	41	393	230	390	380
Density (10^3 Kg/m^3)	1.53	1.66	1.73	1.81	2.0
Filament diameter (μm)	8.5	6.5	7	6.5	10
Elongation at break (%)	1.5	0.6	1.3	0.6	0.5
Elastic recovery (%)	100	100	100	100	100
Carbon assay (%)	98.8	99.9	92	99+	99+
Surface area (m^2/g)	<4	1	1	1	1
Thermal conductivity (W/mK)	38	122	2.1	70	100
Electrical resistivity (μ ohm-m)	2	–	18	9.5	7.5
Longitudinal thermal expansion at 21°C (10^{-6}/K)	–	–	-0.5	-0.7	-0.9
Specific heat at 21°C (j/kgK)	–	–	950	925	925

[a] Data from Technical Information Bulletins 465-223, 465-235, 465-204AH and 465-203bb, Union Carbide Corp.

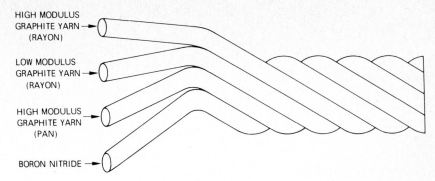

FIG. 1. Yarn plying and twisting (LACHMAN et al., 1978).

In summary, fibers for use in multidirectional carbon–carbon composites must be selected based on a variety of reasons including cost, textile form, properties and stability during processing. Each specific application must be treated as a separate design problem in defining the fiber type best suited for that application.

2.3. *Design of woven structures*

2.3.1. *Fabrics*

A useful form of interwoven yarns for structural reinforcement of composite materials is two-directional (2-D) fabric. Fabrics are characterized by the spacing between adjacent yarns, bundle size, percent of yarn in each direction, yarn packing efficiency and the complexity of the interwoven pattern. Schematics of a plain weave fabric and a five harness satin weave fabric are shown in Fig. 2.

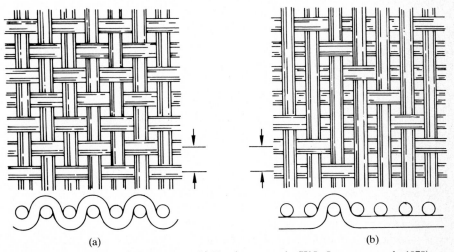

(a)   (b)

FIG. 2. 2-D fabric. (a) Plain weave; (b) Five harness satin (W.L. LACHMAN et al., 1978).

The satin weave construction offers the advantage of the floating yarns which contribute more strength to a composite because of the straight lengths.

Although numerous modifications can be made to fabric weave geometry, the inherent strength of the fabric structure lies in a two-dimensional plane bounded by the thickness of the fabric.

Three-directional (3-D) fabrics can be produced when a third direction of yarn orientation is required. A typical 3-D fabric is shown schematically in Fig. 3. This type of fabric is limited in thickness. However, it does offer the benefit of having straight lengths of yarn in two normal directions with the third direction interlooping to bind the other directions in position. The yarn bundle size, spacing between adjacent yarn bundles, yarn packing efficiency, and the percent of yarn in each direction also characterize the 3-D fabric design. As shown in Fig. 3, the space between adjacent yarns is controlled by the bundle size or diameter of the yarn in each of the three orthogonal directions. This means that the smallest center-to-center yarn spacings are obtained by using yarns with the smallest diameter.

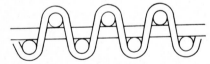

FIG. 3. Single layer 3-D fabric (LACHMAN et al., 1978).

2.3.2. Multidirectional structures

The ideal approach to the tailoring of structural composites is the ability to orient selected fiber types and amounts to accommodate the design loads of the final structural component. Multidirectional weaving technology provides a mechanism to produce such tailored composites.

The simplest type of multidirectional structure is based on a three-directional (3-D) orthogonal construction. As shown in Fig. 4 this type of structure

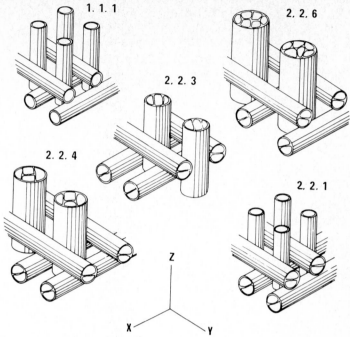

FIG. 4. 3-D orthogonal weave (LACHMAN et al., 1978).

consists of multiple yarn bundles located within the structure on Cartesian coordinates. Each of these yarn bundles is straight in order to obtain maximum structural capability of the fiber.

In 3-D orthogonal preforms, the type of yarn and number of yarns per site can be varied in all three directions. This is shown schematically in Fig. 5. For example, 1.1.1 is a balanced construction with one yarn per site in the X, Y and Z directions. Several unbalanced constructions with various numbers of yarns per site are also illustrated in Fig. 5.

Typical yarn spacings, woven bulk densities, and volume fraction distributions for 3-D structures woven from two types of high modulus graphite yarn are shown in Table 2. These preform characteristics are initially calculated for various constructions by knowing the denier, density and packing efficiency of the yarn. For example, a balanced 3-D orthogonal construction has a maximum packing potential of 75% of the preform volume assuming solid yarns with a square cross section. The remaining 25% of the preform volume is occupied by crossover voids. However, since yarns do not pack to a perfectly square cross section and the filaments within the yarn bundle do not pack solid, the actual fiber volume within a preform will always be less than 75%.

The actual preform characteristics are controlled by the tooling used to fabricate the preforms. The yarn characteristics, number of yarns per site and spacing between sites define the density, fiber volume fraction and distribution

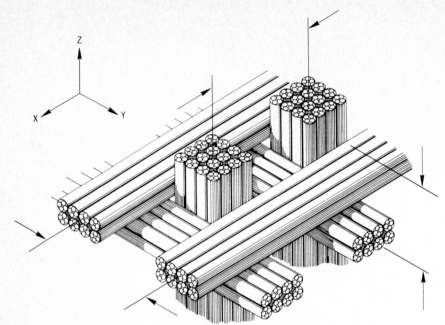

FIG. 5. Typical 3-D block constructions (LACHMAN et al., 1978).

of fibers within the preform. Density of the finished preform can be measured but the actual fiber distribution within the preform cannot be verified.

The volume fraction and distribution of fibers in multidirectional preforms will not change significantly during carbon–carbon processing. Matrix shrinkage during pyrolysis occurs locally within the fiber bundles and voids but does not significantly change the overall dimensions of the preform.

For some applications, an important design parameter is the size and the dispersion of the void pockets. The infiltration of a large void pocket with impregnant will result in localized shrinkage voids during cure and/or pyrolysis. Small and well-dispersed void cells are easier to fill and approach the grain size of solid graphite. To achieve small void cells, yarns with small diameters are spaced closely together and are woven into constructions similar to the 1.1.1 illustrated in Fig. 5, and described in Table 2.

Several weave modifications to the basic 3-D orthogonal design are available in order to form a more isotropic woven structure. Woven preforms with 4, 5, 7 or 11 directions of reinforcement aid in obtaining a composite with isotropic properties (CRAWFORD, 1976). The 5-D design (see Fig. 6) represents the basic 3-D orthogonal structure with the addition of two reinforcement directions in the X–Y plane. This weave configuration will give $\pm45°$ in plane reinforcement with respect to the X–Y yarns along with the axial (Z) direction of reinforcement.

In order to enhance composite properties between the planes of reinforcement of 3-D orthogonal structures, diagonal yarns are introduced into the

TABLE 2. Weave characteristics of 3-D woven structures (LACHMAN et al., 1978).

Thornel[a] 50

Bulk density (g/cm³)	No. of yarn bundles			Center-to-center bundle spacing				Fiber volume fraction[b]		
				X, Y		Z				
	X	Y	Z	(Inch)	(mm)	(Inch)	(mm)	VfX	VfY	VfZ
0.64	1	1	1	0.022	0.56	0.023	0.58	0.14	0.14	0.13
0.75	1	1	2	0.028	0.71	0.023	0.58	0.11	0.11	0.23
0.68	2	2	1	0.040	1.02	0.023	0.58	0.14	0.14	0.12
0.80	2	2	6	0.027	0.69	0.040	1.02	0.12	0.12	0.24

Thornel[a] 75

Bulk density (g/cm³)	No. of yarn bundles			Center-to-center bundle spacing				Fiber volume fraction[b]		
				X, Y		Z				
	X	Y	Z	(Inch)	(mm)	(Inch)	(mm)	VfX	VfY	VfZ
0.70	1	1	2	0.022	0.56	0.023	0.58	0.09	0.09	0.17
0.65	2	2	1	0.033	0.84	0.023	0.58	0.12	0.12	0.09
0.72	2	2	2	0.042	1.07	0.023	0.58	0.09	0.09	0.18

[a] Registered Trademark—Union Carbide Corporation.
[b] Volume fraction of total preform volume occupied by fiber in each orthogonal direction.

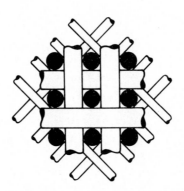

FIG. 6. 5-D construction (LACHMAN et al., 1978).

preform. Diagonal reinforcement across the four corners of the 3-D block structure is shown in Fig. 7. This, combined with the baseline 3-D, produces one type of 7-D structure. Another type of 7-D structure contains diagonal yarns across the faces of the preform as shown in Fig. 8. The diagonal across

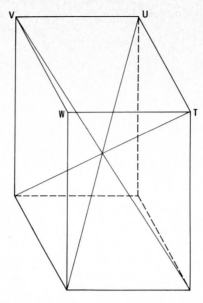

FIG. 7. Typical 'across the corners' diagonals, 7-D construction—type I (LACHMAN et al., 1978).

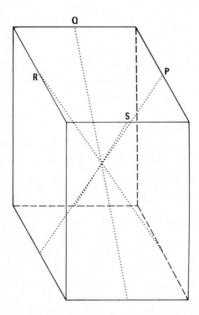

FIG. 8. Typical 'across the face' diagonals, 7-D construction—Type II (LACHMAN et al., 1978).

the corner (see Fig. 7) or across the face (see Fig. 8) constructions can both be utilized to produce 4-D constructions by elimination of the baseline 3-D orthogonal portion of the structure.

7-D structures can also be tailored with respect to preform packing efficiency, preform density, and distribution of directional reinforcement. Table 3 describes examples of available 7-D constructions.

TABLE 3. Various 7-D construction parameters (LACHMAN et al., 1978).

Ends T-50			Volume fraction[b]			Density (g/cm^3)
D^a	X, Y	Z	D^a	X, Y	Z	
3	4	4	0.12	0.17	0.17	0.52
4	4	4	0.14	0.15	0.15	0.62
4	4	6	0.13	0.14	0.21	0.65
4	1	16	0.11	0.03	0.49	0.75

[a] Diagonal directions.
[b] Volume fraction of total preform volume occupied by fiber in each direction indicated.

A combination of the 3-D base structure, diagonal across the corners and diagonal across the face, will produce a third type of isotropic woven structure with 11 directions of reinforcement. The 11-D structure is shown in Fig. 9.

The woven turbine blade configuration shown in Fig. 10 illustrates the types of complex shapes that can be produced with current weave design and fabrication technology. The design of this structure is unique for the combined stresses that the final composite blade must withstand.

The design of cylinders and shapes of revolution, as shown in Fig. 11, with

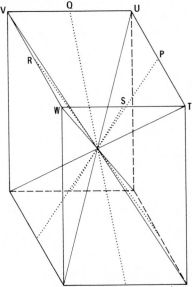

FIG. 9. Typical 'across the corners' diagonals plus 'across the face' diagonals, 11-D construction (LACHMAN et al., 1978).

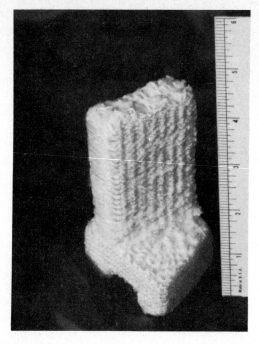

FIG. 10. 3-D woven turbine blade structure (LACHMAN et al., 1978).

reinforcing yarns oriented in three directions is based on the same varying parameters as fabric and blocks, i.e., yarn bundle size, adjacent yarn spacing and the percentage of yarn in each direction. The construction of woven structures of this type is illustrated in Fig. 12. Yarns are oriented in the axial (*A*), radial (*R*) and circumferential (*C*) directions.

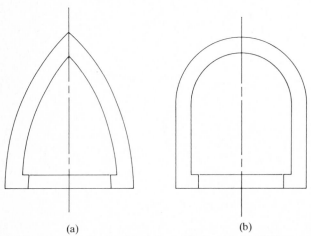

(a)   (b)

FIG. 11. Typical 3-D structure shapes. (a) Ogive with flange; (b) Cylinder with flange and hemispherical end cap (LACHMAN et al., 1978).

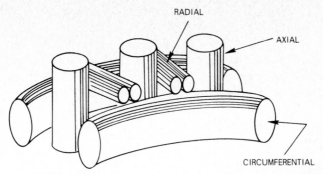

FIG. 12. 3-D cylindrical weave yarn orientation (LACHMAN et al., 1978).

Because of the 'pie' shaped unit cell, an additional design feature, peculiar to cylindrical configurations, must be employed to ensure uniformity of the weave design. As the radial yarns diverge towards the OD of the cylinder, axial yarn bundles must increase in diameter from the ID to the OD to compensate and maintain a uniform density structure (see Fig. 13(a)).

Another design concept (see Fig. 13(b)) to compensate for the 'pie' shaped unit cell is to introduce partial radial elements in a regular pattern from ID to OD. This method also maintains a constant preform density from ID to OD.

Woven structures may also be designed to shapes such as the frustum of a cone. Axial yarns can be oriented to the same half angle as the cone angle. The diametrical change from the aft end to the forward end involves another design

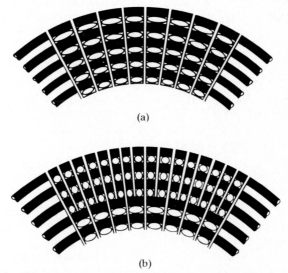

(a)

(b)

FIG. 13. Axial and radial yarn compensation. (a) Axial yarn compensation from ID to OD (axial yarn diameter varies); (b) Radial yarn compensation from ID to OD (axial yarn diameter constant) (LACHMAN et al., 1978).

parameter. In order to maintain constant yarn spacings and uniform density from the aft to the forward end, axial yarns can be tapered. The same construction variables and resulting weave parameters possible for cylindrical weaves can also be designed for frustum shapes. Variations of the cylinder and frustum, namely the ogive and the cylinder with a hemispherical end cap integrally woven, can also be produced with currently available technology (cf. Fig. 11).

2.4. Manufacture of multidirectional structures

2.4.1. General
Section 2.3 described the various multidirectional constructions and shapes that can be designed and fabricated for use in carbon–carbon and other types of composites. Various techniques have been used to produce these structures. Details of the equipment and procedures used to manufacture multidirectional structures are proprietary. However, enough general information is available from the literature to allow some discussion of the subject.

As will be discussed in the following paragraphs, multidirectional structures are fabricated by a variety of methods including weaving of dry yarns, piercing of fabrics, assembly of rigid pultruded rods (from yarns), filament winding, and combinations of the above.

2.4.2. Dry woven structures
Most of the discussion on the design of woven structures in Section 2.3 is related directly to dry woven preform structures. This is the most versatile and widely used approach for the production of carbon–carbon composites. It is also the approach with the least amount of information available on manufacturing techniques.

The process, as described for weaving of 3-D orthogonal block constructions (BARTON, 1968; JACOBS et al., 1968) involves locating spaced, horizontal rows of yarns in alternate X and Y directions in a straight non-interlaced manner. Each adjacent yarn within an X or Y horizontal layer plane is separated by a row of thin tubes. Each of these tubes is replaced with vertical reinforcing yarns (Z direction) after the size of the billet has been established by the horizontal $(X-Y)$ reinforcement layers. Fig. 14 shows a schematic of the weaving process.

The size of the billets that can be fabricated using this process is limited only by the size of the available weaving equipment (JACOBS et al., 1968).

Fig. 15 shows a woven preform representing 3-D orthogonal block construction. The graphite frame is used to prevent distortion prior to processing.

In the weaving of 3-D orthogonal block constructions, several variations can be employed (JACOBS et al., 1968). The distribution of yarns in all three directions can be varied as illustrated in Fig. 5. The fineness of weave can also be controlled. One weave construction which has been designed as 'standard weave' (JACOBS et al., 1968) has yarns located in the Z axis that are ap-

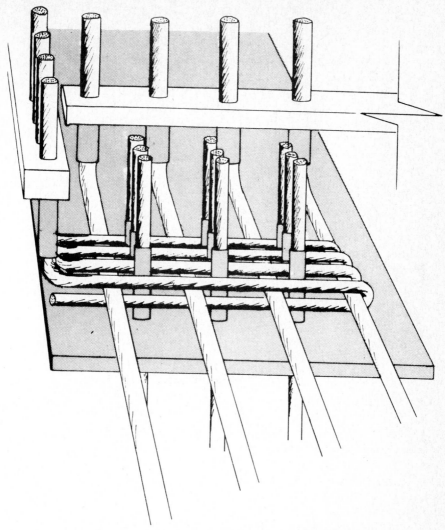

FIG. 14. 3-D weaving schematic (JACOBS et al., 1968).

proximately 1.27 mm (0.050 inches) in diameter and are spaced on 2.54 mm (0.10 inch) centers. Another construction designated as 'fine weave' has Z axis yarns that are approximately 0.64 mm (0.025 inches) in diameter and are spaced on 1.27 mm (0.050 inches) centers. The data in Table 2 show that the center-to-center Z axis yarn spacing as small as 0.58 mm (0.23 inches) can be achieved (LACHMAN et al., 1978).

Cylindrical looms have also been developed to orient fibers in the circumferential, axial and radial directions to produce dry multidirectional structures (BARTON, 1968; JACOBS et al., 1968). These looms are being continually refined and more highly mechanized (GRENIE and CAHUZAC, 1980).

FIG. 15. 3-D orthogonal block preform (photo courtesy Fiber Materials Incorporated).

To operate cylindrical looms, vertical rods are initially inserted into pierced plates. These rods simulate the axial direction of the substrate. The loom then automatically locates yarns between rows of axial rods in the circumferential and radial directions. As a final step the axial metallic rods are automatically replaced by yarns. The resulting construction is represented by the schematic shown in Fig. 12. Weave spacings down to 1.5 mm (radial and circumferential directions) and 0.2 mm (axial direction) can be achieved with fiber volumes ranging from 35 to 55%.

Fig. 16 shows a conical preform that was woven on polar coordinates from graphite yarns.

2.4.3. Pierced fabric structures

The techniques for dry weaving of 3-D structures discussed in Section 2.4.2 and illustrated in Fig. 4 consist of orientating straight non-interlaced yarns in the three orthogonal directions. A modified orthogonal construction in which the $X-Y$ yarns (cf. Fig. 4) are replaced by a 2-D woven fabric of the type shown in

FIG. 16. Conical preform (photo courtesy Fiber Materials Incorporated).

Fig. 2 has also been developed (MCALLISTER and TAVERNA, 1972, 1974, 1976). These preform structures have been called pierced fabric because of their method of manufacture.

To produce pierced fabric structures, woven fabric layers are pierced over an array of metal rods. These metal rods, which represent the Z direction of the structure, are replaced by graphite yarns or pre-cured (rigidized) yarn–resin rods as the final step of the process (MCALLISTER and TAVERNA, 1972, 1974, 1976).

The details of the design and operation of the equipment to produce pierced fabric structures have not been published; however, the process is illustrated schematically in Fig. 17. As with the 3-D orthogonal woven block constructions, the type and distribution of yarns in pierced fabric structures can be varied in all three directions.

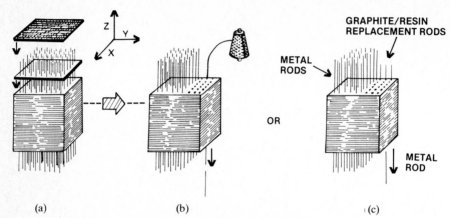

FIG. 17. Pierced fabric block fabrication. (a) Pierce graphite fabric over metal rods; (b) Replace metal rods with graphite yarn; (c) Replace metal rods with graphite–resin rods (MCALLISTER and TAVERNA, 1976).

Various types of fabrics have been used in the $X-Y$ directions of pierced fabric structures (MCALLISTER and TAVERNA, 1974, 1976). Table 4 describes a high modulus satin weave fabric and two low modulus plain weave fabrics that have been evaluated. Table 5 lists various high modulus yarns, tows and pre-cured rods that have been used as Z direction reinforcements in pierced fabric structures. The total fiber cross-sectional area per reinforcing element is also presented in Table 5.

Table 6 gives characteristics of pierced fabric structures prepared from the fabrics described in Table 4. The Z elements of these structures each contained ten high modulus graphite yarns per site on 2.54 mm (0.1 inch) centers. One of the structures was constructed with the $X-Y$ fabric plies rotated every 45 degrees to provide more isotropic properties in the $X-Y$ direction of the final composite.

Table 7 compares a 3-D orthogonal woven block construction with a pierced fabric block construction. Both of these preforms were prepared from the same high modulus graphite yarn. There is a significant difference in the fiber volume and fiber distribution between the two constructions. The pierced fabric block has a higher overall fiber volume and a higher preform density.

As can be seen by the above discussion, pierced fabric construction offers a great deal of versatility in the production of multidirectional block configurations. Various fabric constructions, yarn types and yarn distributions have been demonstrated. The effect of some of these variations on the properties of carbon–carbon composites will be discussed in Section 4.1.

TABLE 4. Comparison of X–Y fabric reinforcements (MCALLISTER and TAVERNA, 1976).

Weave	Thornel 50 [a]	WCA [a]	GSGC-2 [b]
	8 Harness satin	Plain	Plain
Yarn count (Ends/Inch)			
Warp	30	27	26
Fill	34	21	24
Break strength (MPa)			
Warp	1.4	0.6	0.7
Fill	1.4	0.3	0.6
Weight (Oz/Yd^2)	5.6	7.5	7.5
Filaments/yarn	1440	1440	1440
Filament diameter (μm)	6.6	9.1	9.4
Filament density (g/cm^3)	1.6	1.4	1.5
Filament tensile strength (10^3 psi)	2.1	0.7	0.7
Filament modulus (GPa)	379	41	41

[a] Products of Union Carbide.
[b] Product of Carborundum.

TABLE 5. Pierced fabric Z direction variations (MCALLISTER and TAVERNA, 1976).

Z element composition	Total fiber cross-sectional area per element (μm^2)
8 graphite yarns	0.39
10 graphite yarns	0.49
13 graphite yarns	0.63
13 graphite yarns (as precured rod)	0.63
20 graphite yarns	0.98
1 tow (as precured rod)	0.46
2 tows	0.92

Note. All Z elements between 0.89 and 1.4 mm diameter.

TABLE 6. Pierced fabric preform characteristics, $X-Y$ variations
(MCALLISTER and TAVERNA, 1976).

$X-Y$ fabric variation	Preform density (g/cm^3)	Fiber volume [a] (%)	
		$X-Y$	Z
GSGC-2	0.85	47.2	9.3
Thornel 50 8 Harness Satin	0.92	50.8	8.9
WCA rotated every 45°	0.83	48.6	9.3

[a] Percent of total preform volume occupied by fiber in the directions indicated.

TABLE 7. Orthogonal vs. pierced fabric preform comparison
(MCALLISTER and TAVERNA, 1976).

Preform type	Preform density (g/cm^3)	Fiber volume (%)	
		$X-Y$	Z
Pierced fabric	0.9	50	9
Fine weave orthogonal	0.8	32	13

2.4.4. Prerigidized yarn structures

The dry multidirectional structures discussed in Section 2.4.2 are manufactured using textile weaving procedures and equipment. The pierced fabric structures discussed in Section 2.4.3 use woven fabric, but also involve nonweaving approaches such as piercing and the use of prerigidized yarns for some of the preform elements. Multidirectional structures can also be fabricated without the use of weaving or other textile procedures. The basic elements used to produce these structures are prerigidized yarns in the form of rods. These rods are produced from unidirectional high strength carbon fibers and phenolic resins using a pultrusion process (LAMICQ, 1977).

One type of structure produced with these rods is described as a 4-D regular tetrahedral construction with each rod describing an angle of 70.5 degrees with each of the other three. This 4-D construction can be illustrated by the T, U, V, W, 'across the corner' directions as shown in Fig. 7.

The rigidized yarn bundles (rods) used to produce the 4-D structures range from 1 to 1.8 mm diameter with the actual rod shape being hexagonal for maximum packing density. Yarn volumes of up to 75% are claimed because of the packing efficiency of the 4-D tetrahedral structures (LAMICQ, 1977).

Assembly of 4-D structures with rods uses tooling to insure strict conformance with the theoretical geometric pattern (LAMICQ, 1977). The type of tooling used to accomplish this has not been reported. The finished 4-D structures appears to be a simple collection of rigidized graphite fiber rods assembled into a block configuration in accordance with a pre-determined geometric pattern. This type of structure would not be expected to exhibit much integrity prior to introduction of a matrix.

2.4.5. *Cylindrical structures by sub-element fabrication and filament winding*

A non-woven, non-textile approach to producing multidirectional cylindrical structures has also been described by MULLEN and ROY (1972). The resulting structure has reinforcing yarns in the circumferential, radial, and axial directions as shown in Fig. 18. In this process, the radial reinforcement is pre-fabricated as graphite yarn–phenolic sub-element composites. These radial components are assembled into a radial array on a cylindrical substrate with

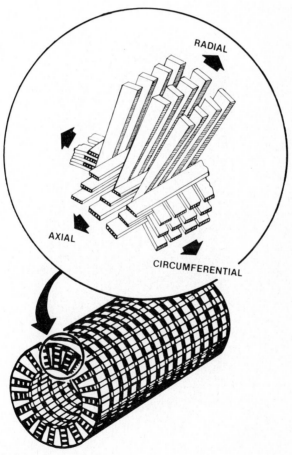

FIG. 18. Schematic of 3-D cylinder construction (MULLEN and ROY, 1972).

vacant corridors in the circumferential and axial directions. Impregnated (pre-pregged) unidirectional high modulus graphite fiber tape is then filament wound into the vacant corridors of the radial array. During this process the circumferential and axial layers are applied in alternating sequence. Phenolic resin is used as the primary matrix binder material throughout all fabrication operations. After completion of the filament winding operation, the structure is cured to form a 3-D graphite–phenolic cylinder that is ready for further processing to convert it to a carbon–carbon composite.

The size and shape of the radial elements can be varied to provide different circumferential and axial fiber volumes.

3. Densification processing of multidirectional substrates

3.1. General

The processing of a multidirectional structure into a fully densified carbon–carbon composite can be accomplished by various procedures. This section will

FIG. 19. Fully processed carbon–carbon billet (photo courtesy Fiber Materials Incorporated).

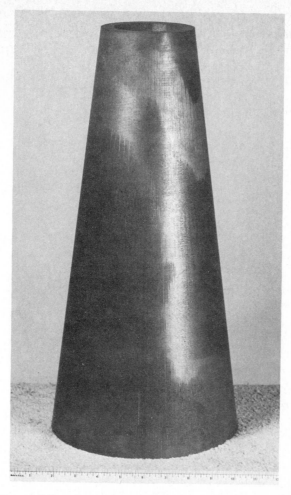

FIG. 20. Fully processed 3-D carbon–carbon conical frustum (photo courtesy Fiber Materials Incorporated).

discuss these procedures in relation to the preform structures, matrix precursor, and process. Fig. 19 shows a fully processed 3-D carbon–carbon billet and Fig. 20 shows a 3-D carbon–carbon frustum. Both of these composites are based on dry woven preforms as shown in Figs. 15 and 16. Their appearance is typical of any of the densification processes discussed in this section.

Several factors must be considered in selecting a matrix precursor and a densification process. Factors relating to the preform structure include:

(a) size and shape of the preform,

(b) yarn type(s) as related to wetting and bonding characteristics, and yarn manufacturing process temperature relative to planned carbon–carbon process temperature,

(c) weave geometry and construction relative to pore size, pore size dis-

tribution, and pore interconnecting channels,
    (d) fiber volume fraction in the preform,
    (e) nature of the preform, i.e., dry woven vs. resin rigidized composite structures.

The process and the matrix precursor must be consistent with the preform characteristics, and must provide properties required for the end item application. In general, high density, high strength carbon–carbon composites are desired.

This section will discuss matrix precursors and processing techniques that are available to densify preform structures into carbon–carbon composites. Much of this technology has been taken from the graphite industry and applied to carbon–carbon processing. However, other industries and technologies have also been drawn upon in establishing carbon–carbon processing techniques. For example, the use of reinforced plastic composites which can be processed into carbon–carbon composites represent an extension of the technology taken from the plastics industry. The use of pressure carbonization as an effective, efficient method to densify carbon–carbon composites utilizes modification of hot isostatic press (HIP) equipment originally developed for metals processing. The use of chemical vapor deposition (CVD) techniques that were originally developed for producing pyrolytic graphite structures and coatings, are used to infiltrate and densify carbon–carbon composites.

The application of these processing techniques to the densification of carbon–carbon composites will be discussed in this section, and the effect of certain materials and process variables on composite properties will be discussed in Section 4.

3.2. Liquid impregnation techniques

3.2.1. Impregnant selection
The number of organic impregnants that could be utilized to process carbon–carbon composites is almost limitless. However, when all of the process and property requirements are considered, the selection is reduced to relatively few materials.

In the selection of a matrix precursor for carbon–carbon densification, the following characteristics must be considered:
    (a) Viscosity.
    (b) Coke yield.
    (c) Coke microstructure.
    (d) Coke crystal structure.

All of these characteristics will be influenced by the time–temperature–pressure relationships encountered during the processing of carbon–carbon composites.

The two general categories of matrix precursors that are commonly used are thermosetting resins such as phenolics and furfuryls, and pitches based on coal tar and petroleum. Characteristics of a typical pitch and resin of the types used to densify carbon–carbon composites are given in Tables 8 and 9.

TABLE 8. Typical properties of coal tar pitch.

Softening point (°C)	94–107
Viscosity at 250° (Cps)	30–50
Benzene insolubles (%)	24–28
Quinoline insolubles (%)	2–7
Coking value	52–62
Specific gravity	1.28–1.31
Sulphur (%)	0.1–0.6
Ash (%)	0.2–0.5

TABLE 9. Typical properties of phenolic resin.

Specific gravity	1.08–1.09
Solids content (%)	60–62
Viscosity at 25°C (cps)	120–200
Refractive index	1.518–1.525
Cure time at 165°C (sec)	85–105
Free formaldehyde (%)	0–0.5
Free phenol (%)	11.5–13.5
Trace elements, Na K, Li, Fe	<5 ppm Ea <10 ppm Total

3.2.1.1. Thermosetting resins. Thermosetting resins are used because they are suitable for impregnation processes and a large technology base exists from their use in plastics processing. Most thermosetting resins polymerize at low temperatures (<250°C) to form a highly cross-linked, thermosetting, non-melting, amorphous solid. Upon pyrolysis, these resins form glassy carbon which does not graphitize at temperatures up to 3000°C (KOBAYASHI et al., 1968; HONDA et al., 1968). Coke yields for thermosetting resins that cyclize, condense, and are readily converted to carbon, range from 50 to 56 percent by weight (MACKAY, 1970). Some resins have been found to give yields as high as 73 percent by weight up to 800°C (McALLISTER and TAVERNA, 1971b). Table 10 presents coke yield and X-ray diffraction data for several thermosetting resin systems. X-ray diffraction data on carbon black and natural graphite are included for comparison. Fig. 21 summarizes characteristics of a furfuryl resin and a coal tar pitch as function of temperature. It can be seen from these data that the thermosetting resin gives a less graphitic, lower density coke than coal tar pitch. The low density of the coke can limit the density of the final carbon–

TABLE 10. Thermosetting resin char characteristics (MCALLISTER and TAVERNA, 1971b).

Resin	Char yield (%) (TGA to 800°C, 5°C/min)	X-ray diffraction data for 2700°C char	
		L_c (Å)	d_{002} (Å)
Phenolic-base cured	57	68	3.43
Phenolic-acid cured (castable)	56	132	3.40
Polyimide	60	75	3.44
Furfuryl ester resin	63	75	3.41
Polybenzimidazole	73	40	3.45
Polyphenylene	71	54	3.44
Biphenol formaldehyde resin	65	83	3.43
Carbon black Not heat treated	–	17	3.63
Carbon black 2700°C heat treat	–	247	3.41
Natural celon graphite	–	[a]	3.36

[a] Too large to measure.

carbon composite. However, there may be applications where a non-graphitic matrix is desired.

There are exceptions to the normal behavior of thermosetting resins which can be used to influence their behavior in carbon–carbon densification processing. For example, if polyfurfuryl alcohol chars are highly stressed at elevated temperature, graphitization can occur (REISIVIG et al., 1968). This stress graphitization effect has also been observed in phenolic precursor matrices in 3-D carbon–carbon composites (MCALLISTER and TAVERNA, 1971a, b). Shrinkage stresses in the vicinity of carbon fibers are believed to have caused the phenolic derived char to graphitize. Matrix in non-reinforced regions of the composite remain glassy.

Another interesting phenomenon that has been observed with a phenolic resin is that application of pressure (48.2 MPa) during carbonization of the resin at temperatures between 400 to 600°C will result in a coke that becomes highly graphitic above 2300°C (KOTOSNONOV et al., 1969). Pressure applied at all other temperatures produced only glassy carbon. It was speculated that external forces, when applied during the 400–600°C temperature range, increased the mobility of the molecular structure resulting in molecular orientation which can be graphitized.

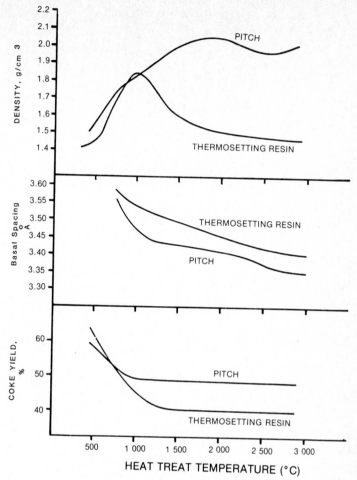

FIG. 21. Characteristics of resin and pitch as a function of heat treatment temperature (BURNS and MCALLISTER, 1976).

    The photomicrograph in Fig. 22 reveals the glassy, amorphous nature of char derived from a furfuryl resin (BURNS and MCALLISTER, 1976). There is essentially no difference in microstructure between resin heat treated at 900 and 2700°C. Resin cured and carbonized under pressure did not exhibit the shrinkage cracks typical of low pressure carbonized resins. It was also found that the pressure-cured and carbonized resin gave a lower coke yield (52 down to 47%) compared to resin cured and carbonized at atmospheric pressure.

    Shrinkage of the matrix precursor during carbonization can be a major problem during processing. Linear shrinkages of up to 20% have been observed with phenolic resins (FITZER et al., 1978). Such high shrinkage tends to cause severe damage and poor properties in 2-D composites that rely on the

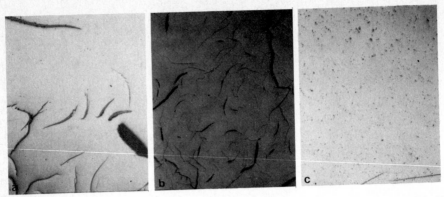

FIG. 22. Microstructure of furfuryl resin char. (a) Atmospheric cure and carbonized to 900°C; (b) Atmospheric cure and carbonized to 900°, heat treated to 2700°C; (c) 5000 psi cure and carbonization, heat treated to 2700°C (photomicrographs courtesy of Fiber Materials Incorporated) (× 100).

matrix to provide integrity to the composite (FITZER and BURGER, 1971). Composite damage due to shrinkage is not as serious a problem with multi-directional structures because of the absence of unreinforced planes within the structure.

The following is a summary of some of the factors which should be considered when selecting a thermosetting resin for densification processing of multidirectional carbon–carbon composites:

(a) Coke yields are in the range of 50 to 70 percent by weight. Limited data indicate that coke yields are not increased by carbonizing under pressure.

(b) Coke structures are glassy and do not graphitize up to 3000°C.

(c) Coke density is very low (<1.5 g/cm³).

(d) High shrinkage occurs during carbonization.

(e) Stresses applied (or induced) during heat treatment can lead to a graphitic coke structure.

3.2.1.2. Pitches. The use of pitches as matrix precursors in carbon–carbon composites is an extension of graphite processing technology. A large data base exists on coal tar and petroleum pitches for processes involving impregnation, carbonization, and graphitization. As shown in Table 8, coal tar pitch has a low softening point, low melt viscosity, and high coking value. As will be discussed, pitches also tend to form graphitic coke structures.

Impregnating pitches of the type used for carbon–carbon composites are a mixture of polynuclear aromatic hydrocarbons. They are thermoplastic in nature as opposed to the highly cross-linked thermosetting resins. From their softening point to about 400°C, pitches undergo various changes including volatilization of low molecular weight compounds, polymerization, cleavage and rearrangement of molecular structure. At temperatures above 400°C it has been discovered that spheres that are initially about 0.1 μm diameter or smaller appear in the isotropic liquid pitch (BROOKS and TAYLOR, 1965).

These spheres, which have been designated as mesophase, exhibit a highly oriented structure with characteristics similar to liquid crystals. On prolonged heating, these spheres undergo coalescence, solidify, and form larger regions of extended order. The lamellar arrangement of the molecular structure in these regions favors the formation of a graphitic structure on further heating to 2500°C or above.

The graphitic nature of pitch precursor chars is illustrated by their high density and small layer spacing, as shown in Fig. 21.

The coke yields from coal tar or petroleum pitches are about 50 percent by weight at atmospheric pressure. This tends to be about the same yield as highly charring resins. However, if pitches are carbonized under high pressure, a significant increase in coke yield can be obtained. Pressure pyrolysis of coal tar pitches at 550°C under nitrogen pressure of about 10 MPa gave coke yields up to about 90 percent (FITZER and TERWIESCH, 1973). Gas pressures greater than 10 MPa were found to have little influence on coke yield for petroleum and extracted coal tar pitches (HÜTTINGER and ROSENBLATT, 1977). The effect of carbonization pressure on coke yield for a petroleum pitch is illustrated in Fig. 23. These data indicate that pressures in the range of 6.9 to 68.9 MPa are sufficient to achieve high coke yields. However, as will be subsequently discussed, factors in

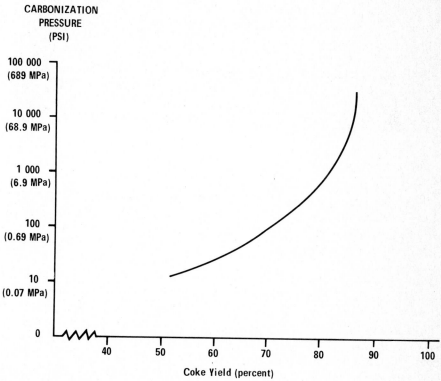

FIG. 23. Effect of pressure carbonization on coke yield for petroleum pitch (LACHMAN et al., 1978).

addition to coke yield must be considered in establishing pressures for carbon–carbon processing.

Microstructures of coke resulting from the carbonization of pitches is also influenced by factors such as pressure and temperature. Fig. 24 shows photomicrographs of petroleum pitch coked at 6.89 and 68.9 MPa followed by heat treatment to 2700°C (BURNS and McALLISTER, 1976). The coke formed at the lower pressure is needlelike, possibly due to deformation of the mesophase by gas bubble percolation. At the higher coking pressure, the microstructure appears to be coarser and more isotropic, possibly due to the suppression of gas formation and escape during coking.

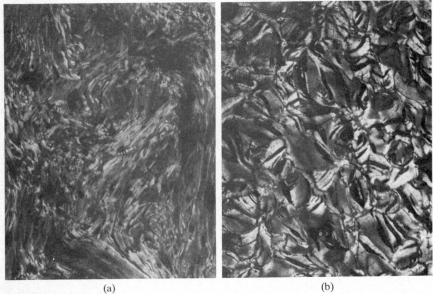

(a)   (b)

FIG. 24. Microstructure formed by petroleum pitch coked at (a) 6.89 and (b) 68.9 MPa and subsequently heat treated at 2700°C (photomicrographs courtesy of Fiber Materials Incorporated) (BURNS and McALLISTER, 1976) (× 100).

Other effects of pressure on coke structure were observed during a study with anthracene (WHANG et al., 1974). Higher pressures tended to result in mesophase formation at lower temperatures. However, when coking was carried out at very high pressures (in the vicinity of 200 MPa), coalescence of the mesophase did not occur.

The following is a summary of some of the factors that should be considered when selecting a pitch for densification processing of multidirectional carbon–carbon composites:

(a) Coke yields at atmospheric pressure are about 50 percent by weight. However, by carbonizing the pitch under pressure of 10 MPa or greater, coke yields will approach 90 percent for some pitches.

(b) Coke structures are graphitic.

(c) Coke density is high ($\sim 2\,\text{g/cm}^3$).

(d) Pressure applied during carbonization affects coke microstructure.

3.2.2. Low pressure processing

The most widely used approach to introducing a carbon matrix into the multidirectional fibrous preforms is through the impregnation of the preform with an organic material and subsequent carbonization of the resultant composite under an inert atmosphere. Conventional densification processing is usually conducted at atmospheric or reduced pressure, and must be repeated several times to reduce the porosity to an acceptable level (McALLISTER and TAVERNA, 1971a).

In this process, the multidirectional structure is usually vacuum impregnated with a resin such as phenolic or a melted pitch such as coal tar, petroleum or synthetic (McALLISTER and TAVERNA, 1971a, 1972, 1976). In some cases, pressure is applied as part of the impregnation cycle to insure penetration of all available porosity in the composite structure. Resin impregnated parts are then cured and post cured. A typical post cure cycle for a phenolic impregnated part is shown in Fig. 25. Pitch impregnated parts are not cured, but are subjected directly to carbonization in a nitrogen atmosphere. Carbonization for resin or pitch processed composites is carried out at a controlled heating rate to temperatures in the range of 650 to 1100°C (McALLISTER and TAVERNA, 1972, 1974; MULLEN and ROY, 1972). A typical carbonization cycle is shown in Fig. 26. The next step in the process is graphitization. This is normally carried out in an induction furnace to temperatures in the range of 2600 to 2750°C (McALLIS-

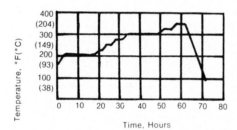

FIG. 25. Post cure cycle for phenolic parts (MULLEN and ROY, 1972).

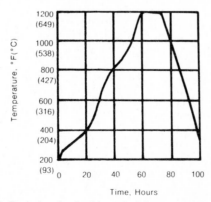

FIG. 26. Typical carbonization cycle (MULLEN and ROY, 1972).

TER and TAVERNA, 1971a, 1972; MULLEN and ROY, 1972). A typical graphitization cycle is shown in Fig. 27. Heating rates for all process cycles may vary depending on the size and shape of the composite being processed.

A schematic of a typical densification process of the type described is shown in Fig. 28 (MCALLISTER and TAVERNA, 1971a). As can be seen, the impregnation–heat treatment cycle is repeated multiple times to complete the process. Graphitization is an optional step in each cycle.

The number of densification cycles required to fully process a carbon–carbon multidirectional composite will vary depending on the impregnant and the process used.

Fig. 29 shows a typical graph of density vs. number of graphitizing cycles for 228.6 mm (9 inch) diameter, 12.7–17.8 mm (0.5–0.7 inch) wall, multidirectional cylinders processed with phenolic resin (MULLEN and ROY, 1972). Six cycles were required to achieve a density of 1.65 g/cm³.

Another example that has been reported (MCALLISTER and TAVERNA, 1976) involved a light phenolic impregnation to rigidize the multidirectional structure. The remaining process cycles utilized cinnamylideneindene, a synthetic pitch, as the impregnant. Multiple cycles consisting of vacuum impregnation at 204°C, carbonization to over 538°C, and graphitization to over 2500°C were

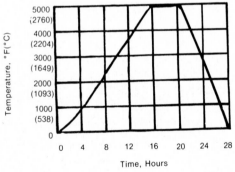

FIG. 27. Typical graphitization cycle (MULLEN and ROY, 1972).

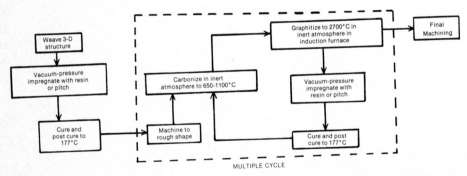

FIG. 28. Typical carbon–carbon process schematic (MCALLISTER and TAVERNA, 1971).

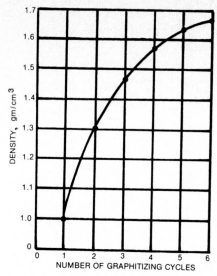

FIG. 29. Densification of multidirectional carbon–carbon cylinder (MULLEN and ROY, 1972).

carried out. The final composite had a density of 1.66 g/cm³, and an open porosity of 12.7 percent.

The key factor in obtaining an efficient densification process for multidirectional carbon–carbon structures is to achieve full impregnation with a high char yield matrix precursor. In order to access the efficiency of the low pressure process, a series of rectangular blocks impregnated with phenolic resins were monitored after selected cycles by removing corings for density and porosity measurements (McALLISTER and TAVERNA, 1972). The experimental data resulting from this study are shown in Fig. 30. These results show that essentially all of the porosity in the composite is open after each cycle, based on the agreement between calculated and experimental data. This condition should allow the blocks to be processed to near theoretical density of 1.8 g/cm³.[3] However, from the standpoint of processing time, this density would not be practical. The actual density of the completed blocks ranged from about 1.60 to 1.65 g/cm³ over an open porosity range of about 8 to 10%. Pore size distribution for a typical composite with 10.7 percent open porosity is shown in Table 11. The pore diameters ranged from 17 to less than 0.08 μm diameter as determined by mercury porosimetry and helium intrusion methods.

Based on the studies discussed, it can be concluded that densification processing of multidirectional carbon–carbon composites can be carried out by low pressure processes utilizing both resin and pitch impregnants. The process is practical for producing composites with a final open porosity of about 10 percent.

---

[3] Theoretical density of 1.8 g/cm³ is based on the fact that both the rayon precursor fiber and the phenolic precursor matrix are glassy (non-graphitic) forms of carbon.

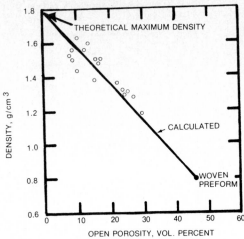

FIG. 30. Experimental density–porosity relationship for a multidirectional carbon–carbon composite (MCALLISTER and TAVERNA, 1972).

TABLE 11. Pore size distribution for a multidirectional carbon–carbon composite (MCALLISTER and TAVERNA, 1972).

Pore volume (%)	Range of pore diameters (Microns)
0.5	10–17
3.3	1–10
3.4	0.1–1
0.1	0.08–0.1
3.4	<0.08
10.7 Total	

3.2.3. High pressure processing

As shown in Fig. 23, pressure carbonization of pitch can increase coke yield from 50% at ambient pressure to over 85% at 68.9 MPa. This pressure effect on coke yield of pitches became the basis of a process for densification of carbon–carbon composites designated as pressure–impregnation–carbonization (PIC). The technique utilizes isostatic pressure to effectively impregnate and densify carbon–carbon composites during the melting and coking stages of the carbonization cycle (BURNS and COOK, 1974; CHARD and CONAWAY, 1974).

All aspects of a densification cycle using PIC processing are the same as for a low pressure cycle described in Section 3.2.2, except for the use of pressure during carbonization. A schematic of a complete PIC densification cycle is shown in Fig. 31.

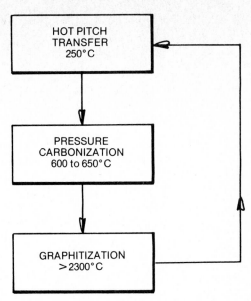

FIG. 31. Schematic of PIC densification process cycle (DIETRICH and McALLISTER, 1978).

In preparation for the PIC process, impregnation of the preform is accomplished by a hot pitch transfer technique (DIETRICH and McALLISTER, 1978). In this process solid pitch is melted under vacuum in a heated (250°C) reservoir tank while the preforms, held in metal canisters, are heated to the same temperature (under vacuum) in an adjacent tank. Hot liquid pitch is transferred to the metal canisters containing the preforms via pipes connecting both tanks. This is accomplished by backfilling the reservoir tank with nitrogen. Once the preforms are fully submerged in hot liquid pitch, the preform tank is backfilled with nitrogen to equalize the pressure in both tanks and stop the transfer of pitch. The canisters containing the impregnated preforms are sealed off under vacuum with metal lids. They are now ready for PIC processing.

The PIC processing step is carried out in specially designed hot isostatic pressure (HIP) equipment (BURNS and COOK, 1974; CHARD and CONAWAY, 1974). A simplified schematic of a HIP system is shown in Fig. 32. The gas-pressure vessel can be either externally heated (hot wall) or water cooled with an internal furnace (cold wall) (BURNS and McALLISTER, 1976). The remainder of the system, as illustrated in Fig. 32, consists of gas storage, high pressure gas transfer lines, compressor, and controls (CHARD and CONAWAY, 1974).

For PIC processing, the sealed metal can containing the impregnated preforms and excess pitch is placed in the pressure vessel. Temperature is raised at a programmed rate to 550–650°C, while pressure is increased and maintained at levels ranging from 6.84 to 103.4 MPa. A typical cycle requires about 24 hours.

Isostatic pressure applied to the thin metal can during the PIC process is

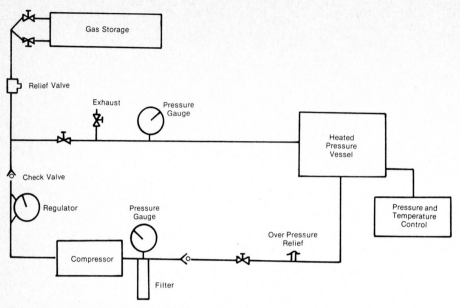

FIG. 32. Simplified schematic of hot isostatic pressure system (CHARD et al., 1978).

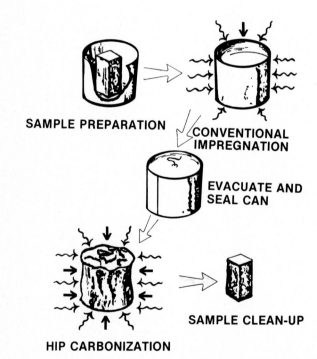

SAMPLE PREPARATION

CONVENTIONAL IMPREGNATION

EVACUATE AND SEAL CAN

HIP CARBONIZATION

SAMPLE CLEAN-UP

FIG. 33. Schematic of sequence of events in a PIC cycle (CHARD et al., 1974).

transmitted to the molten pitch. At elevated temperatures, this metal encapsulating container acts like a 'rubber bag' (CHARD and CONAWAY, 1974).

A schematic illustrating the sequence of events in a PIC cycle is shown in Fig. 33.

After PIC processing, the preforms are removed from the metal container and subjected to graphitization by heating at a controlled rate to temperatures greater than 2300°C (DIETRICH and MCALLISTER, 1978).

The complete process cycle, as shown in Fig. 31, is repeated until the required composite density is achieved. Fig. 34 presents a curve showing composite density vs. the number of PIC cycles for a fine weave multidirectional carbon–carbon composite. A densification curve for a similar composite densified by impregnation and carbonization at atmospheric pressure is included for comparison. Up to a density of about 1.4 g/cm³, PIC processing and atmospheric pressure processing show almost the same efficiency. However, as can be seen from the data, the atmospheric pressure process will not produce a high density composite with a standard pitch impregnant.

Increased coking yield was initially considered the most important advantage of the PIC process. However, as composites were developed with smaller and

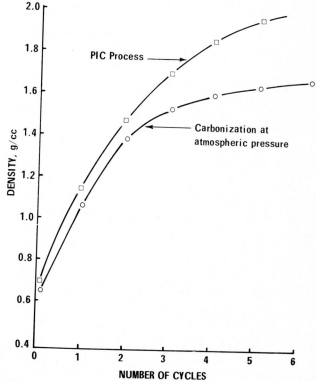

FIG. 34. Densification of fine weave multidirectional carbon–carbon composites (LACHMAN et al., 1978).

more tortuous porosity, other advantages of high isostatic pressure during the impregnation–carbonization phase of the PIC cycle were recognized. During a PIC process, the pitch initially melts and expands within the sealed can and is forced by isostatic pressure into small pores not filled during initial vacuum impregnation (CHARD and CONAWAY, 1974). As the pitch begins to coke, high isostatic pressure maintains the more volatile fractions of the pitch impregnant in a condensed phase. This pressure not only increases coke yield, but also prevents liquid from being forced out of the pores by pitch pyrolysis products (LACHMAN et al., 1978).

The effect of PIC process pressure on densification efficiency can be seen by the data in Table 12. These composites are all in a density range where pitch processing at atmospheric pressure is ineffective (see Fig. 34). A process pressure of 6.9 MPa shows a significant increase in coke yield. However, pressures in the range of 51.7–103.4 MPa appear to be required to obtain efficient densification of the higher density composites.

TABLE 12. Effect of PIC process pressure on the densification of pitch impregnated carbon–carbon composites (BURNS and COOK, 1974).

Carbonization pressure (MPa)	Coke yield (%)	Density (g/cm^3)		Density increase (%)
		Initial	Final	
Atm.	51	1.62	1.65	1.9
6.9	81	1.51	1.58	4.6
51.7	88	1.59	1.71	7.5
51.7	89	1.71	1.80	5.2
103.4	90	1.66	1.78	7.2

In summary, it can be stated that multidirectional carbon–carbon composites can be effectively and efficiently densified by use of PIC processing techniques. The high pressure provides high coke yields from pitches. However, pressure is also required to achieve complete impregnation and to prevent impregnant loss from the composite during pitch pyrolysis.

3.3. Chemical vapor deposition

Chemical vapor deposition (CVD) processing techniques have been used to densify various types of carbon–carbon composites (SCHMIDT, 1972; STOLLER and FRYE, 1969, 1971; BAUER et al., 1971a, b; GIRARD, 1978). However, only limited information has been reported on CVD as a densification technique specifically related to multidirectional carbon–carbon structures (MCALLISTER and TAVERNA, 1971, 1976; GIRARD, 1978). For the purpose of this discussion, it is assumed that general carbon–carbon CVD densification procedures are applicable to multidirectional structures.

The CVD process involves deposition of a hydrocarbon gas such as methane or natural gas onto a carbon substrate (SCHMIDT, 1972; STOLLER and FRYE, 1969, 1971). The process involves diffusion of an active carbon bearing gas into a fibrous substrate in such a manner as to achieve uniform matrix deposition. A number of parameters such as substrate, susceptor design, source and carrier gases, temperature, and pressure will effect matrix characteristics as well as rate, uniformity and efficiency of the process.

Three CVD infiltration techniques are most applicable to carbon–carbon densification. These include isothermal (STOLLER and FRYE, 1971; WARREN and WILLIAMS, 1972; PIERSON, 1968), thermal gradient (STOLLER and FRYE, 1971; PIERSON, 1968), and differential pressure (KOTLENSKY and PAPPAS, 1969) techniques.

A schematric of the isothermal technique is shown in Fig. 35. In this process, the carbon–carbon substrate is heated radiatively by the induction furnace susceptor. Hydrocarbon and carrier gases are introduced into and contained by the susceptor to allow infiltration into the hot substrate. The isothermal process is usually conducted at reduced pressures (STOLLER and FRYE, 1971: WARREN and WILLIAMS, 1972) and tends to produce a uniform deposit through the substrate. This process does produce a surface crust on the preform that must be machined away between cycles to achieve a high density composite. Isothermal processing is amenable to production because a furnace can be loaded with multiple parts.

A schematic of the thermal gradient process is shown in Fig. 36. In this case (STOLLER and FRYE, 1971; PIERSON, 1968), the induction coil and susceptor must be tailored to the geometry of the substrate. The outer surface of the substrate next to the susceptor is the hottest region. Deposition starts at that point and

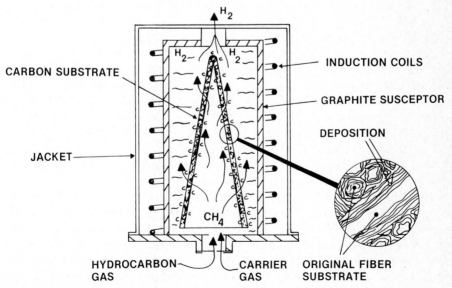

FIG. 35. Schematic of the isothermal CVD technique (STOLLER and FRYE, 1971).

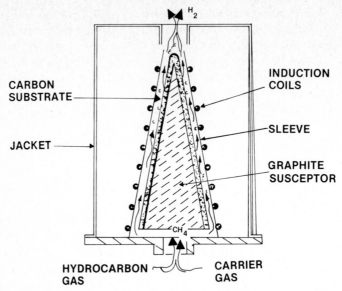

FIG. 36. Schematic of the thermal gradient CVD technique (STOLLER and FRYE, 1971).

progresses radially through the substrate. This process is limited to a single item per furnace, but offers the advantage of short process cycles as compared to the isothermal technique.

The differential pressure technique (KOTLENSKY and PAPPAS, 1969) is a variation of the isothermal technique shown in Fig. 35. In this process the inner portion of the substrate is isolated from the furnace chamber by sealing around the base. Gasses are fed into the inner substrate at a positive pressure with respect to the furnace chamber. This creates a pressure differential across the wall of the preform, forcing gas to flow through the pores.

One of the problems with the use of CVD techniques to densify multidirectional composites is the difficulty in achieving complete infiltration (McALLISTER and TAVERNA, 1971a). This is due to a 'bottleneck' effect in which the CVD matrix closes off small pores, which in turn blocks access to larger pores. A 3-D structure could only be processed to a density of $1.5 \, \text{g/cm}^3$ as compared to $1.65 \, \text{g/cm}^3$ for a low pressure resin densification procedure. However, more recent results have been reported where 3-D structures woven from PAN precursor yarns on 2 mm centers were CVD densified using the isothermal technique (GIRARD, 1978). Densities in the range of 1.5 to $1.7 \, \text{g/cm}^3$ were achieved. The process, as reported, utilized a dry woven preform and isothermal CVD processing with natural gas as the hydrocarbon. An induction furnace was operated at 1100°C, and a pressure of 10 torr. The initial process cycle involved rigidization of the 3-D substrate to a density of 1 to $1.2 \, \text{g/cm}^3$. The substrate was then machined and subjected to another cycle to a density of $1.5–1.7 \, \text{g/cm}^3$. A substrate 120 mm diameter by 250 mm high was processed to a density of $1.6 \, \text{g/cm}^3$ in 27 days using this process.

CVD techniques have also been combined with other densification processes in attempts to tailor the process cycle (McALLISTER and TAVERNA, 1976; MULLEN and ROY, 1972). In one case a 3-D structure was initially subjected to a low pressure densification cycle with phenolic resin. The resulting structure, at a density of $0.98 \, g/cm^3$, was subjected to an isothermal CVD process to a density of $1.43 \, g/cm^3$. Final processing consisted of low pressure densification with phenolic resin. The final carbon–carbon composite achieved a density of $1.53 \, g/cm^3$.

In another case, a combination process consisting of isothermal CVD and low pressure phenolic cycles resulted in a final composite at a density of $1.73 \, g/cm^3$.

In conclusion, it appears that CVD processing techniques are applicable to the production of multidirectional carbon–carbon composites. CVD can be used by itself or in combination with other densification procedures. CVD processing does appear to limit final composite density. However, as will be discussed in Section 4, CVD matrices may offer certain property advantages in the composite.

4. Composite property data

4.1. Effect of composition and process on properties

4.1.1. General

The selection of a particular preform design, fiber type, matrix precursor, and process are necessary steps in defining a multidirectional carbon–carbon composite for a given application. Fiber properties and preform design will generally control the structural properties of the composite. However, carbon–carbon composites by nature are very complex due to the physical and chemical changes and interactions that occur during processing. Some of these process related factors include:

(a) Pyrolysis of an organic impregnant to form a carbon matrix. This conversion involves a 60–65% reduction in volume. This magnitude of shrinkage can create severe process related stresses, and can result in damage to the composite (FITZER and BURGER, 1971).

(b) Change from an organic–carbon fiber interfacial bond to a carbon–carbon fiber interfacial bond. The nature of this bond will depend on a variety of material and process variables.

(c) Change of fiber properties due to extended heat treatment, process induced stresses, and fiber–matrix interactions (McALLISTER and TAVERNA, 1971a, 1972).

(d) Thermal expansion mismatch between fibers and matrix. This condition can cause severe process stresses and composite damage during heat treatment cycles (ADAMS, 1976).

A precise description of how composition and process influence properties of carbon–carbon composites is beyond our present knowledge and understanding. Available data can only indicate trends that have been observed. An attempt will be made to review some of these trends and to shed some light on factors that influence the properties of multidirectional carbon–carbon composites.

4.1.2. Fiber variations

As shown by the data presented in Table 1, fibers with a range of tensile strength and modulus values are available for producing multidirectional structures. These fibers can be used by themselves or be mixed to tailor composite properties. As discussed in Section 2.3, the volume fraction of fibers can also be varied directionally within the structure.

In one study (McAllister and Taverna, 1976), variations in the type and amount of high modulus fibers in the Z direction sites of a pierced fabric multidirectional structure were evaluated. The particular variations included in this study are shown in Table 5. The $X-Y$ direction of these composites, which consisted of low modulus fabric (WCA)[4] as described in Table 4, was not varied.

The densified carbon–carbon composites were evaluated in the Z direction by conducting flexural tests. Fig. 37 summarizes the results of these tests. As expected a general trend of increasing strength and modulus with increasing volume fraction of yarns can be seen. However, when the number of yarns per site was increased from thirteen to twenty, the strength dropped off even though the modulus showed the expected increase. The strength reduction was attributed to fiber damage incurred during preform fabrication. By forcing the yarns into too small a corridor during preform fabrication, fibers were broken, and strength properties were lowered.

When the yarns were prerigidized with resin and inserted in the form of a rod, fiber damage was reduced as seen by the high strength of this composite as compared to the equivalent preform made with dry yarns.

Graphite single tow, as prerigidized rod, also provided acceptable properties. The use of two tows, however, showed the same problems as twenty yarns. The corridors available at each Z site were too small to accept this volume of fibers without damage. As can be seen in Table 5, both thirteen yarns and two tows represent about the same fiber cross-sectional area per site.

Multidirectional pierced fabric constructions as described in Table 6 have been used to evaluate fiber variations in the $X-Y$ direction of multidirectional carbon–carbon composites (McAllister and Taverna, 1976). The preforms described in Table 6 were designed to compare high strength, high modulus with low strength, low modulus fabric as described in Table 4. One preform was also made in which fabrics were rotated within the $X-Y$ planes to provide more isotropic properties in the composite. Mechanical properties measured on carbon–carbon composites prepared from these preforms are presented in Table 13. These data show that the use of high strength, high modulus fabric in

[4] Product of Union Carbide Corporation.

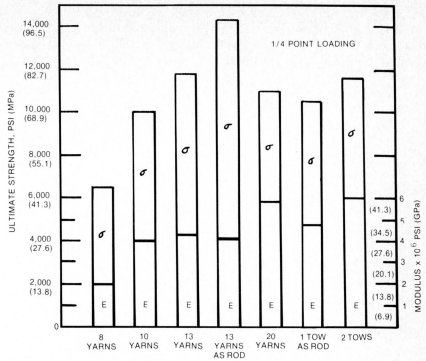

FIG. 37. Flexural properties of pierced fabric carbon–carbon with *Z* direction fiber variations (MCALLISTER and TAVERNA, 1976).

the *X–Y* directions resulted in significant increases in strength and modulus properties as compared to the low strength, low modulus fabrics.

Significant differences have also been shown to exist between these high modulus and low modulus pierced fabric composites in thermal properties.

TABLE 13. Mechanical properties of pierced fabric multidirectional composites with fabric variations in the *X–Y* directions (MCALLISTER and TAVERNA, 1976).

Property	WCA fabric	WCA fabric (rotated)	GSGC-2 fabric	Thornel 50 8 HS fabric
Tensile				
strength (MPa)	35.1	32.4	35.1	104.7
modulus (GPa)	6.9	11.0	11.0	57.9
strain to failure (%)	0.8	0.8	0.6	0.2
Compressive				
strength (MPa)	56.5	48.2	58.6	90.9
modulus (GPa)	7.5	17.9	19.2	70.3
strain to failure (%)	1.2	0.5	0.6	0.2

Figs. 38 and 39 show a comparison of thermal expansion and thermal conductivity behavior. The high modulus fiber, having a degree of crystalline orientation along the fiber axis, contributes to low thermal expansion and high thermal conductivity in the composite.

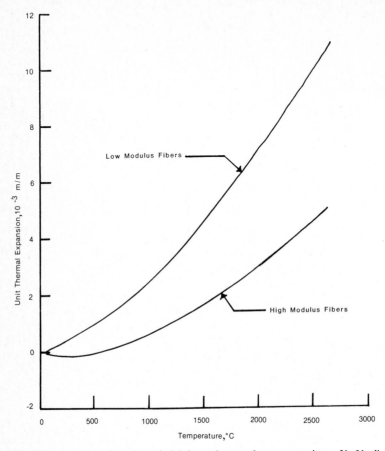

FIG. 38. Thermal expansion for pierced fabric carbon–carbon composites, $X-Y$ directions (McALLISTER and TAVERNA, 1976).

4.1.3. *Multidirectional preform variations*

Very little has been done to make a direct comparison between various types of multidirectional preform structures. In one case, carbon–carbon composites prepared from pierced fabric and fine weave orthogonal constructions were compared (McALLISTER and TAVERNA, 1976). Both preforms used the same high modulus yarns. Table 7 gives comparative data for these preforms. As can be seen, the volume fraction and distribution of fibers is significantly different between the two preforms. However, a qualitative comparison of the $X-Y$ properties can be made for these composites from the data shown in Table 14. These data allow one to compare woven fabric $X-Y$ layers of pierced fabric

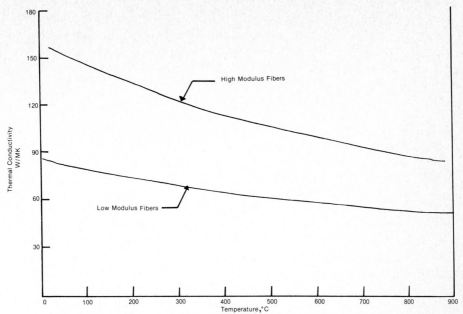

FIG. 39. Thermal conductivity for pierced fabric carbon–carbon composites, X–Y directions (MCALLISTER and TAVERNA, 1976).

TABLE 14. 3-D orthogonal vs. pierced fabric—property comparison in X–Y directions (MCALLISTER and TAVERNA, 1976).

Property	Pierced fabric	Fine weave orthogonal
Tensile		
strength (MPa)	104.7	99.2
modulus (GPa)	57.9	55.8
strain to failure (%)	0.2	0.2
Compressive		
strength (MPa)	90.9	68.9
modulus (GPa)	70.3	42.7
strain to failure (%)	0.2	0.2
Coef. of thermal expansion (10^{-6}/K)		
0–908°K	0	0
908–3019°K	2.0	2.0
Thermal conductivity at 519°K (W/mK)	126.3	114.2
Fiber volume (%)		
X or Y direction	25	16
Composite density (g/cm³)	1.8	1.8

preform to straight non-interwoven $X-Y$ yarns of a fine weave orthogonal composite. Mechanical properties of the two composites are similar indicating that the orthogonal construction with its lower fiber volume may be more effective in translating the fiber properties to the composite. The higher fiber content in the pierced fabric composite contributes to higher thermal conductivity. The coefficient of thermal expansion is the same for both composites.

4.1.4. Matrix variations

As discussed in Section 3, the most commonly used matrix precursors for carbon–carbon composites are resins, pitches, and chemical vapor deposition (CVD) of carbon from hydrocarbon gasses. Only limited studies have been carried out to compare these three matrix precursors in multidirectional carbon–carbon composites. In some cases as discussed below, combinations of these precursors have been used to tailor the process and/or properties of the composite (Mullen and Roy, 1972; McAllister and Taverna, 1971a, 1976).

CVD densification processing may provide a better fiber–matrix interface and/or matrix properties which is reflected in improved composite mechanical properties. This behavior is illustrated in Table 15, which compares properties of a composite processed with resin/pitch to a CVD processed composite. The preforms for this study were a balanced orthogonal 3-D construction with Z direction yarn bundles of 1.27 mm diameter on 2.54 mm centers. A 345 GPa modulus fiber was used.

TABLE 15. Resin/pitch vs. CVD matrix comparison (McAllister and Taverna, 1971a).

	Resin/pitch	CVD
Density (g/cm^3)	1.65	1.5
Tensile strength (MPa)	82.7	120.6
Flexural strength (MPa)	68.9	142.6
Shear strength (MPa)	27.6	51.7

The improved properties obtained by CVD processing may be partially due to the fact that the specimens are held at around 1100°C during the process. If the composite is subsequently heat treated to high temperatures typical of resin or pitch processed composites, the properties may drop off to some extent.

Another study (McAllister and Taverna, 1976) involved several matrix precursors including phenolic resin, cinnamylideneindene (CAI), a proprietary synthetic pitch, and CVD. Pierced fabric preforms with WCA graphite fabric in the $X-Y$ direction and high modulus graphite pre-cured (rigidized) rods in the Z direction were used for this evaluation.

The objective of the study was to identify a matrix precursor that would offer potential improvement over phenolic resin in terms of processing time and mechanical properties.

The referenced material was processed with a thermosetting phenolic using a low pressure process as described in Section 3.2.2.

CAI, a synthetic pitch produced by reacting indene and cinnamaldehyde at room temperature in the presence of an alcoholic base, was evaluated. The CAI also utilized a low pressure densification process as discussed in Section 3.2.2.

Another synthetic pitch included in the evaluation was a proprietary development of an Advanced Technology Center of LTV. The process included an initial impregnation cycle with a copolymer based on 1,4-naphthoquinone and an unidentified aromatic hydrocarbon. This particular copolymer was designed to wet and bond to the graphite fibers. The remaining impregnations used a synthetic pitch formulated from a mixture of 1,4-naphthoquinone and acenaphylene. A low pressure process was used. This consisted of vacuum impregnation at 100°C, followed by pyrolysis to 1000°C, and graphitization to 2300°C. The composite was processed through six cycles to a density of 1.53 g/cm³.

Another process involved a combination of low pressure phenolic cycles and CVD. The composite was processed with phenolic resin followed by carbonization and graphitization. This was followed by isothermal CVD processing to a density of 1.43 g/cm³. Finally, the composite was densified using a low pressure phenolic process, as described in Section 3.2.2, to a density of 1.53 g/cm³.

Flexural (Z direction) and compressive (X–Y direction) testing was conducted to determine the effects of the above matrix variations on properties of the pierced fabric composites. These data are presented in Table 16. There appears to be no major property advantage offered by the CAI synthetic pitch or CVD in combination with phenolic compared to the reference phenolic system. The LTV synthetic pitch, which also included an initial cycle to improve wetting and bonding, showed a significant improvement in Z direction flexural strength.

TABLE 16. Mechanical properties of pierced fabric carbon–carbon composites with matrix variations (MCALLISTER and TAVERNA, 1976).

Property	Matrix precursor			
	Phenolic resin	Phenolic resin + CVD carbon	CAI	LTV synthetic pitch
Flexural, Z direction				
strength (MPa)	89.6	108.9	102.7	153.6
modulus (GPa)	27.56	24.1	32.4	32.4
Compressive, X–Y direction				
strength (MPa)	56.5	73.0	50.3	71.7
modulus (GPa)	7.6	6.9	6.9	10.3

TABLE 17. Hoop tensile properties of 3-D carbon–carbon rings (MULLEN and ROY, 1972).

Matrix precursor	Density (g/cm^3)	Ultimate tensile stress (MPa)	Tensile modulus (GPa)	Strain to failure (%)
Phenolic	1.62	118.5	70.3	0.18
High solids phenolic	0.65	106.8	64.8	0.17
High melt pitch	1.64	94.4	106.1	0.08
Low melt pitch	1.65	128.2	64.1	0.05
Isothermal CVD (not graphitized)	1.59	113.7	77.2	0.15
Isothermal CVD/phenolic	1.73	106.8	77.9	0.13
Differential pressure CVD (not graphitized)	1.35	136.4	68.2	0.20
Differential pressure CVD (graphitized)	1.28	130.2	61.3	0.20
Differential pressure CVD/phenolic	1.58	128.2	64.1	0.20

TABLE 18. Axial compressive properties of 3-D carbon–carbon rings (MULLEN and ROY, 1972).

Matrix precursor	Density (g/cm^3)	Ultimate compressive stress (MPa)	Compressive modulus (GPa)	Strain to failure (%)
Phenolic	1.62	52.9	20.0	0.59
High solids phenolic	1.65	59.9	20.7	0.50
Isothermal CVD (not graphitized)	1.59	103.1	33.8	0.48
Isothermal CVD/ phenolic	1.73	73.9	27.6	0.40
Differential pressure CVD (not graphitized)	1.35	115.1	23.4	0.72
Differential pressure CVD/phenolic	1.58	82.7	18.6	0.65

Another experimental matrix study involved the use of multidirectional cylindrical structures produced by sub-element fabrication and filament winding as described in Section 2.4.5 (MULLEN and ROY, 1972). These were 230 mm diameter cylinders with high modulus yarns in the axial and circumferential directions. Various matrix precursors including phenolic resins, pitches, CVD and CVD/phenolic combinations were evaluated. All matrix processing with liquid impregnants was carried out at low pressure as discussed in Section 3.2.2. Hoop tensile and axial compressive tests were performed on rings. Tables 17 and 18 present the results of this evaluation. The best overall properties in tensile and compression were exhibited by CVD densified composites. It was felt that the CVD process improved the fiber to matrix bonding.

Matrix precursor studies involving high pressure processing such as PIC have not been reported.

4.1.5. Process effects

Processing of multidirectional carbon–carbon composites involves exposure of the structure and its components to extreme temperatures and pressures. These conditions, especially the high temperatures, can produce irreversible changes in the composite. For example, fibers may become more graphitic and change in both properties and dimensions. The large mismatch in thermal expansion characteristics between high modulus graphite yarns and various matrices can produce a very complex stress state in a composite during heating and cooling.

The effect of carbon–carbon heat treatment processing on the properties of graphite fibers has been investigated for rayon precursor yarns (McALLISTER and TAVERNA, 1971a). The objective was to determine the effect of the long graphitization cycles used for carbon–carbon processing on high modulus yarns that had previously been exposed to high temperatures for only a short time. Thornel[5] 25 (172 GPa modulus) and Thornel 40 (276 GPa modulus) were exposed to a temperature of 2600°C for ten hours in an inert atmosphere and under zero stress. In a subsequent experiment, Thornel 50 (335 GPa modulus) was exposed to 2750°C for twenty-four hours. X-ray diffraction studies were carried out on these fibers to determine any changes which may have occurred in crystal structure. As shown by the data in Table 19, the Thornel 25 and Thornel 40 showed some changes in crystal structure, but the Thornel 50 did not. In order to determine the significance of the changes in crystal structure on fiber properties, both control and heat treated Thornel 40 fibers were subjected to tensile testing. Test data obtained on filaments taken from yarns are shown in Table 20. These data show that extended heat treatment had no apparent degradative effect on filament properties. However, it must be recognized that in a composite, in the presence of a matrix, factors other than temperature will affect fiber properties.

The effect of 2700°C heat treatment on low modulus and high modulus fibers, phenolic char and a multidirectional pierced fabric prepared from these

---

[5] Registered Trademark of Union Carbide Corporation.

TABLE 19. Heat treatment effect on the crystal structure of graphite fibers (MCALLISTER and TAVERNA, 1971a).

Fiber	Heat treatment	L_c (Å)	d_{002} (Å)
Thornel 25	As received	25	3.50
	2600°C—10 hrs	50	3.44
Thornel 40	As received	30	3.48
	2600°C—10 hrs	54	3.48
Thornel 50	As received	45	3.48
	2750°C—24 hrs	47	3.47

TABLE 20. Heat treatment effect on graphite filament tensile properties (MCALLISTER and TAVERNA, 1971a).

As received		Heat treated 2600°C—10 Hrs	
Ultimate tensile strength (MPa)	Tensile modulus (GPa)	Ultimate tensile strength (MPa)	Tensile modulus (GPa)
1288.4	314.2	1433.1	178.5
1770.7	266.6	1791.4	308.0
1350.4	269.4	2383.9	313.5
1192.0	232.2	1322.9	239.8
1400.4 AVG	270.6 AVG	1732.9 AVG	260.0 AVG

components has also been studied by X-ray diffraction analysis (MCALLISTER and TAVERNA, 1972). As with the fiber study discussed above, the two diagnostic crystallographic parameters determined were d_{002} (average crystallite dimension or 'stack height' parallel to the crystallographic axis and normal to the basal planes). Increasing L_c and decreasing d_{002} indicates increasing degrees of graphitization.

The data in Table 21 indicate that the high modulus yarn used in the Z direction of pierced fabric structures does not increase in degree of graphitization as a result of heat treatment at 2700°C. The low modulus fabric used in the $X-Y$ direction of the preforms did show a more graphitic structure after heat treatment. Data obtained on the heat treated phenolic coke agreed with other reported X-ray diffraction data indicating that hard carbon is obtained when heat treated to 2800–3000°C (KOBAYASHI et al., 1968).

The interesting aspect of the data presented in Table 21 is that when the resin and fibers are combined into a multidirectional carbon–carbon composite a much more graphite condition is achieved. The existence of three-dimensional order in the crystalline structure of the composite was further substantiated by the presence of hkl reflections and two-dimensional hk lines in all diffraction patterns for the composite.

TABLE 21. X-ray diffraction data relating to carbon–carbon heat treatment (McAllister and Taverna, 1972).

Material	Average basal spacing d_{002} (Å)	Average crystalline size L_c (Å)
Phenolic resin, 2700°C heat treat	3.49	27
High modulus graphite yarn		
as received	3.44	47
2700°C heat treat	3.43	57
Low modulus graphite fabric		
as received	3.52	26
2700°C heat treat	3.41	147
Pierced fabric C–C composite		
after 1st 2700°C heat treat	3.37	453
after 3rd 2700°C heat treat	3.37	453
Natural ceylon graphite	3.36	[a]

[a] L_c crystallite size of celon graphite is probably thousands of angstroms which is too large to be measured by line broadening techniques with the instrumental conditions employed for the other samples.

The reason for achieving such a highly graphitic composite cannot be readily explained. Interpretation of the data is further complicated by the fact that the contribution of fiber and matrix cannot be isolated from the available data. It is obvious that some type of interaction between matrix and fiber within the composite was involved. One explanation was that stress-induced graphitization occurred within the composite during heat treatment. This could be caused by a combination of shrinkage stresses in the matrix and restraint by the fibers.

A rather comprehensive study sponsored by the U.S. Naval Surface Weapons Center (NSWC) has been carried out to examine certain process related behavior patterns and characteristics of a multidirectional carbon composite by in-process testing and micro-analysis (Adams, 1976). In this study, a 3-D orthogonal structure, as described in Table 22, was densified as described in Table 23. As can be seen by the description in Table 22, this multidirectional structure is a very fine weave construction utilizing very high strength (238 GPa), high modulus (517 GPa) fibers. The densification process used resins impregnated at high pressure (34.5 MPa). Carbonization was carried out at ambient pressure.

Tensile and flexural properties were measured after the 7th densification cycle, after the final densification cycle (13th) and after graphitization. These data are presented in Tables 24 and 25. The high fraction of fibers in the Z direction is reflected by higher strength and modulus values in that direction. There is considerable scatter in the data, but the following observations can be made:

TABLE 22. 3-D orthogonal block preform description (ADAMS, 1976).

Density	0.75 g/cm³
Dimensions	16.0 cm (X) × 8.6 cm (Y) × 22.1 cm (Z)
Yarn	– Thornel 75 s, a rayon precursor fiber with high strength (2.38 GPa) and high modulus (517 GPa) properties – 2 ply construction with 720 filaments/ply
Construction	– 1-1-2 orthogonal – X direction—1 yarn/site – Y direction—1 yarn/site – Z direction—2 yarns/site – X–Y sites on 0.41 mm centers – Z sites on 0.76 centers
Preform supplier	Fiber Materials, Inc., Biddeford, Maine

TABLE 23. Densification processing of 3-D orthogonal block preform (ADAMS, 1976).

Initial	– Impregnate with phenolic resin – Cure – Carbonize to 1540°C
Cycles 1–13	– Impregnate with a blend of furfuryl alcohol and epoxy resins using 34.5 MPa pressure – Cure – Carbonize to 1540°C
Final	Some specimens were graphitized to 2650°C after completion of 13 process cycles
Final composite density	1.72 g/cm³

(a) The tensile strength and modulus in the Z direction are nearly the same after the 13th densification cycle as after the 7th cycle.

(b) The Z direction tensile properties are severely degraded after graphitization. Flexural strength, but not modulus, is also degraded.

(c) The Y direction tensile strength is increased after graphitization.

It is apparent from the data in Tables 24 and 25 that significant process induced property degradation is occurring in the composite. In-process SEM studies revealed a ribbon structure in the matrix that was attributed to shrinkage during cure and carbonization. This condition contributes to poor matrix strength and bonding characteristics. After graphitization, the condition became more severe and large gaps were introduced around each Z bundle. This unbonding of the Z bundles explains the degradation of Z direction

TABLE 24. Tensile properties of 3-D carbon–carbon composites (ADAMS, 1976).

Specimen	Process stage after 7th densification			After 13th (final) densification			After graphitization		
	Strength (MPa)	Modulus (GPa)	Elongation (%)	Strength (MPa)	Modulus (GPa)	Elongation (%)	Strength (MPa)	Modulus (GPa)	Elongation (%)
Z-direction									
1	193	84.1	0.23	195	75.8	0.25	109	45	–
2	168	84.1	0.20	211	91.7	0.23	97	44	–
3	170	84.8	0.20	171	95.1	0.18	112	80	–
Average	177	84.3	0.21	192	87.6	0.22	105	57	–
Y-orientation									
1	50.3	24.8	0.20	26	–	–	59	32	0.18
2	39.3	19.3	0.20	23	39	0.08	65	41	0.16
3	24.1	9.7	0.25	33	39	0.08	57	30	0.19
Average	37.9	17.9	0.22	28	39	0.08	61	34	0.18

TABLE 25. Flexural properties of 3-D carbon–carbon composite (ADAMS, 1976).

Z-yarn orientations	Specimen no.	Process stage after 7th densification		After 13th (final) densification		After graphitization	
		Strength (MPa)	Modulus (GPa)	Strength (MPa)	Modulus (GPa)	Strength (MPa)	Modulus (GPa)
Length of beam	1	197	24.1	261	15.8	109	24.8
	2	205	14.4	250	14.4	105	21.3
	3	185	25.5	243	15.1	100	19.3
Average		196	21.3	251	15.1	105	21.8
Width of beam	1	35	15.8	87.5	21.3		
	2	63	11.0	79.9	19.3		
	3	47	6.8	83.4	17.9		
Average		48	11.2	83.6	19.5		
Depth of beam	1	32	4.1	81.3	16.5	93.7	15.8
	2	44	4.8	106.0	19.3	77.2	17.2
	3	30	5.5	75.8	16.5	95.1	17.9
Average		35	4.8	87.7	17.4	88.6	16.9

properties after graphitization. However, this same debonding around bundles did not occur in the X and Y directions. These observations were explained on the basis of a severe thermal expansion mismatch between the highly anisotropic fibers and the surrounding matrix. The high modulus fibers have a coefficient of thermal expansion (CTE) of $-2 \times 10^{-6}/K$ in the axial direction and $18–23 \times 10^{-6}/K$ in the transverse direction. During heat treatment, the Z direction fiber bundles will expand against the matrix which will permanently deform at temperature. Upon cool-down, the bundles will contract, leaving the gap. The fact that the same gaps did not appear around the X and Y fiber bundles was attributed to weave geometry. It was noted that each X and Y bundle was in direct contact at their cross-over points, which reduced the effect of the matrix.

It was concluded that the matrix ribboning effect, in combination with severe process induced stresses, led to low composite mechanical properties. It was suggested that a detailed study of three-dimensional thermal stress effects, in conjunction with changes in fabrication procedures, could lead to improved composite performance.

4.1.6. Fiber–matrix interaction

Interaction between matrix and fiber will be influenced by both composition and process. The nature of this interface in a carbon–carbon composite may determine whether the composite will behave as a brittle, flaw sensitive material, or a tough, thermal stress resistant composite. For resin matrix composites it has been stated that the role of the interface in composite behavior and the ability to control the interface are as important as the understanding and control of the two primary components (SALKIND, 1969). The same is true for carbon–carbon composites. A high strength composite relies on a strong interfacial bond to transfer loads between fiber and matrix. Carbon–carbon composites can be produced with high strength fibers. However, both the fibers and the matrix are very brittle. Because of this, a high strength interfacial bond may not be desired. COOK and GORDON (1964) have predicted that if the interfacial adhesion in a composite produced from brittle solids is too high ($\frac{1}{3}$ to $\frac{1}{5}$ of the cohesion) it will behave as a homogeneous brittle solid. If the interfacial adhesion is in the right range, cracks will be deflected at the interface and toughness and strength will be achieved. An example of this is shown in Fig. 40 which compares typical load–deflection curves for a 3-D carbon–carbon composite and ATJ-S[6] graphite (MCALLISTER and TAVERNA, 1971a). These curves reveal the greater toughness of the 3-D carbon–carbon composite. The carbon–carbon exhibits a 'pseudo-plastic' behavior as illustrated by the shape of the curve. This behavior is believed to be caused by fiber debonding and matrix microcracking at or near ultimate stress. This behavior allows the composite to strain and maintain a load without exhibiting brittle failure because the cracks do not propagate through the three-dimensional

[6] Registered Trademark of Union Carbide Corporation.

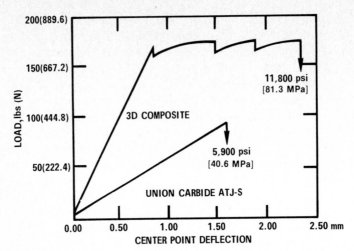

FIG. 40. Typical load–deflection curve for 3-D carbon–carbon compared to ATJ-S graphite (McAllister and Taverna, 1971a).

array of reinforcing fibers. A similar 2-D carbon–carbon material would be expected to fail in interlaminar shear.

The degree of bonding between fibers and matrix will depend very strongly on the nature of the fiber surface, the type of matrix precursor, and processing conditions. Many types of carbon fibers have been activated and/or treated to bond well to resins for use in structural composites. Once such a bond is established, the interface may lose its identity if the composite is processed into carbon–carbon, resulting in a brittle solid.

The importance of fiber surface condition on the mechanical behavior of carbon–carbon composites has been clearly demonstrated in a study involving several commercially available and specially treated fibers (Thomas and Walker, 1978).

The experiments were carried out on unidirectional carbon–carbon composites. However, the general results of this study are applicable to any composite construction because the nature of the matrix–fiber interface will be similar.

The commercial fiber types included in the study are listed in Table 26.

TABLE 26. Commercially available PAN based carbon fiber types (Thomas and Walker, 1978).

High modulus, surface treated—HMS
High modulus, untreated—HMU
High strength, surface treated—HTS
High strength, untreated—HTU
High strain, surface treated—AS
High strain, untreated—AU

Specimens were prepared by compression molding phenolic preimpregnated 10 000 filament tows. These molded composites were processed to carbon–carbon by heat treatment in an inert atmosphere followed by CVD infiltration with pyrolytic carbon.

Table 27 summarizes mechanical properties of resin matrix unidirectional composites prepared from the types of commercial fibers listed in Table 26. In most cases, the surface treated fibers gave slightly higher strength in both the longitudinal and transverse directions, along with significantly higher work of fracture and interlaminar shear strength. The AU fibers, which are manufactured at the lowest temperature, showed good bonding without surface treatment because of the high surface activity associated with low fiber processing temperature.

Table 28 summarizes properties of these composites after processing to carbon–carbon as described above. The difference in properties between the surface treated and untreated fibers is very dramatic. Conversion of the phenolic matrix to carbon has reduced the crack stopping capability of the interface, and drastically reduced strength and toughness. A similar trend would be anticipated with multidirectional carbon–carbon composites.

Pitch, being a thermoplastic, will remain liquid and not establish a bond to the fiber surface until coking occurs. It would therefore be expected to interact differently than a thermosetting resin with an active carbon fiber surface.

Some data have been reported by BRADSHAW (1974) comparing the behavior of phenolic resin and petroleum pitch as matrix precursor for unidirectional carbon–carbon composites. Low temperature processed (1000°C) PAN precursor fibers were used in this particular experiment. These fibers, because of

TABLE 27. Mechanical properties of unidirectional carbon fiber–phenolic composites with various commercial fibers (THOMAS and WALKER, 1978).

	AU	AS	HTU	HTS	HMU	HMS
Longitudinal flexural strength (MPa)	1825	1825	1620	1900	1140	1275
Longitudinal flexural modulus (GPa)	110	115	140	140	210	235
Longitudinal flexural strain (%)	1.7	1.7	1.3	1.1	0.6	0.5
Transverse flexural strength (MPa)	35	36	26	40	20	22
Transverse flexural modulus (GPa)	4.8	5.5	3.5	4.8	2.8	4.0
Transverse flexural strain (%)	0.6	0.6	0.7	0.9	0.7	0.6
Density (g/cm^3)	1.58	1.55	1.54	1.55	1.63	1.64
Interlaminar shear (MPa)	65	80	30	95	20	55
Work of fracture (KJ/m^2)	95	100	85	90	50	45

TABLE 28. Mechanical properties of unidirectional carbon–carbon composites with various commercial fibers (THOMAS and WALKER, 1978).

	AU	AS	HTU	HTS	HMU	HMS
Longitudinal flexural strength (MPa)	470	190	1350	138	1240	240
Longitudinal flexural modulus (GPa)	114	121	142	121	183	175
Longitudinal flexural strain (%)	0.46	0.13	1.17	0.12	0.65	0.15
Transverse flexural strength (MPa)	25	30	16	25	14	21
Transverse flexural modulus (GPa)	8	9	6	8	6	9
Transverse flexural strain (%)	0.25	0.27	0.28	0.29	0.33	0.37
Density (g/cm^3)	1.50	1.45	1.48	1.45	1.58	1.58
Interlaminar shear (MPa)	25	24	20	25	16	27
Work of fracture (KJ/m^2)	9	1	75	1	40	1

their low processing temperature, should have an active surface and provide a good comparison of interface bonding between the two matrices. A composite processed with phenolic resin of 1400°C gave a tensile strength of 1503 MPa. With further heat treatment to 2500°C, strength dropped off to 615 MPa and there was little tendency for crack arrest at the interface. In a composite prepared with petroleum pitch, a tensile strength of 970 MPa was obtained after 2500°C heat treatment and debonding occurred at the fiber–matrix interface.

There appears to be a very complex relationship between the fiber surface, the matrix, and the processing temperature used to produce a carbon–carbon composite. Highly active fiber surfaces resulting from low fiber processing temperatures or from surface treatments may be undesirable depending on the matrix precursor and the manner in which it interacts with the fiber surface.

4.2. Property data for typical multidirectional carbon–carbon composites

Composition, construction and process variations will all influence the properties of multidirectional carbon–carbon composites. These variations along with the associated composite property data have been discussed in previous sections. In this sub-section, property data representative of certain general types of multidirectional carbon–carbon composites will be presented.

Property data typical of early 3-D carbon–carbon block material is presented in Table 29. A typical aerospace grade of polycrystalline graphite is included for comparison. The 3-D carbon–carbon is a pierced fabric construction with low modulus graphite fabric (WCA) in the $X–Y$ directions, and high modulus graphite yarn (Thornel 50) in the Z direction as discussed in Section 2.4.3.

TABLE 29. Properties of 3-D pierced fabric carbon–carbon compared to polycrystalline graphite (SCHMIDT, 1972).

	Temperature (°C)	3-D carbon–carbon pierced fabric		Polycrystalline graphite (ATJ-S)	
		Z direction	X–Y direction	With grain	Across grain
Density (g/cm³)	24	1.65	–	1.83	–
Porosity (%)		10.5	–	9	–
Tensile strength (MPa)	24	103.4	34.5	39.0	30.0
	2485	68.9	64.6	54.4	42.7
Tensile modulus (GPa)	24	41.3	11.0	11.5	7.7
	2485	10.3	6.2	11.0	7.3
Tensile strain-to-failure (%)	24	0.3	0.6	0.5	0.5
	2485	2–7	2–4	2.0	2.2
Flexure strength (MPa)	24	96.5	62.0	42.0	37.6
	2485	103.4	–	72.3	65.5
Compressive strength (MPa)	24	82.7	62.0	88.9	95.6
	2485	158.5	109.6	182.0	192.9
Compressive modulus (GPa)	24	22.7	11.0	6.6	8.7
	2485	12.4	6.2	6.8	1.1
Thermal conductivity (W/mK)	260	55.4	83.0	114.2	90.0
	2485	24.2	27.7	46.7	34.6
Thermal expansion (m/m × 10⁻³)	TO 538	0.2	1.0	3.0	3.5
	TO 2485	4.7	9.4	11.5	13.7

Densification processing utilized phenolic resin in a process similar to that described in Section 3.2.3.

Table 30 presents properties of 3-D orthogonal carbon–carbon blocks densified with CVD and CVD/resin (GIRARD, 1978). The preforms were woven with PAN precursor fibers. The center-to-center fiber bundle spacing was 2 mm in all three directions.

The large difference between flexural and tensile modulus for the CVD densified composite is not an expected result. No explanation for this difference was given by the author.

The highest density and highest strength carbon–carbon composites are woven with high strength, high modulus fibers and densified using PIC processing techniques as described in Section 3.2.3. Table 31 summarizes the preform characteristics and general processing information for a 3-D orthogonal block construction designated as 2.2.6 (CHARD and NIESZ, 1975). Mechanical and thermal properties of the fully densified 3-D carbon–carbon composite are given in Tables 32 and 33 (LEVINE, 1975). The improved properties and high density, as compared to early materials (Table 29), are reflected in these data.

TABLE 30. Properties of 3-D orthogonal carbon–carbon block densified by CVD and CVD/resin (GIRARD, 1978).

	CVD densified		CVD/resin densified	
	Z direction	$X-Y$ direction	Z direction	$X-Y$ direction
Density (g/cm³)	1.6–1.7	–	1.8–1.9	–
Tensile strength (MPa)	70	70	110	70
Tensile modulus (GPa)	50–70	50–70	120	70
Flexural strength (MPa)	50–80	50–80	60	40
Flexural modulus (GPa)	13–50	13–50	–	–
Compressive strength (MPa)	50–80	50–80	100	100
Compressive modulus (GPa)	–	–	70	–
Coef. of thermal expansion, R.T to 2550°C (10^{-6}/K)	3	3	4	4
Thermal conductivity at 20°C (W/MK)	–	10	–	120

TABLE 31. 3-D orthogonal preform and processing information (CHARD and NIESZ, 1975).

Preform characteristics
 – Density—0.75 gm/cm³
 – Construction—2-2-6 orthogonal
   – X direction—2 yarns/site
   – Y direction—2 yarns/site
   – Z direction—6 yarns/site
   – $X-Y$ layers on 0.69 mm centers
   – Z bundles on 1 mm centers
 – Yarn—T-50

Densification
 – Impregnant—coal tar pitch
 – Process
   – Rigidization—impregnation, carbonization and graphitization
   – PIC—15 000 psi—650°C—4 cycles
   – Graphitization—2750°C—after each PIC cycle

Note: FMI preform, PIC processing at Battelle.

TABLE 32. 3-D orthogonal carbon–carbon mechanical properties (Levine, 1975).

	Direction	
	Z	X–Y
Density (gm/cm^3)	1.9	
Tensile		
strength (MPa)		
R.T.	310	103
1900°K	400	124
modulus (GPa)		
R.T.	152	62
1900°K	159	83
Compressive		
strength (MPa)		
R.T.	159	117
1900°K	196	166
modulus (GPa)		
R.T.	131	69
1900°K	110	62

Note. Preform and process information in Table 31.

TABLE 33. 3-D orthogonal carbon–carbon thermal properties (Levine, 1975).

	Direction	
	Z	X–Y
Density (gm/cm^3)	1.9	
Thermal conductivity		
(W/mK)		
R.T.	246	149
1900°K	60	44
Thermal expansion		
(10^{-6}/K)		
R.T.	0	0
1900°K	3	4
3000°K	8	11

Note. Preform and process information in Table 31.

5. Applications for multidirectional carbon–carbon composites

Multidirectional carbon–carbon composites offer a large potential as high performance engineering materials. However, to date their high cost has restricted their use to aerospace and specialty applications. Lack of design data and suitable design methodology have also limited the use of all types of carbon–carbon composites.

Another limitation on the use of carbon–carbon composites is their susceptability to oxidation at high temperatures. This is not a problem for short term applications such as heat shields and rocket nozzles. However, for long term applications reducing or neutral environments are required to prevent oxidative degradation.

The use of carbon–carbon composites for reentry vehicle heatshield applications has been successfully demonstrated (SCHMITT, 1971), and the performance of carbon–carbon materials for nosetip applications has been reported (DiCRISTINA, 1971).

Advanced propulsion systems have generated a need for materials which are more erosion resistant, with increased dimensional stability, and less susceptibility to thermostructural failure during multiple restarts (STOLLER et al., 1971). Representative applications include nozzles, thrust chambers, and ramjet combustion liners. The advantages of carbon–carbon systems lie in simpler design concepts and reduced nozzle weight and volume (LARAMIE and CANFIELD, 1972). A total weight savings of 30 to 50% has been predicted if a rocket nozzle is completely designed using carbon–carbon composites (PARMEE, 1971). Fig. 41 shows the progression from complex tungsten and pyrolytic graphite rocket nozzle designs of the 1960's and 1970's to the simple one piece

1960'S ⟶ 1970'S ⟶ 1980'S

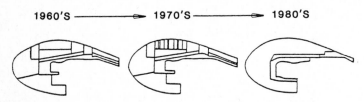

REQUIREMENTS	INCREASING TEMPERATURES AND PRESSURES ⟶		
MATERIAL	TUNGSTEN	PYROLYTIC GRAPHITE WASHERS	CARBON/CARBON
PERFORMANCE	TEMP LIMITED	ACCEPTABLE	REPEATABLE
RELIABILITY	ACCEPTABLE	QUESTIONABLE	HIGH
COST	HIGH	HIGH	MODERATE
DESIGN	COMPLEX	COMPLEX	SIMPLE
WEIGHT	HEAVY	MEDIUM	LIGHT
GROWTH POTENTIAL	VERY LIMITED	LIMITED	HIGH

FIG. 41. Integral throat entrance (ITE) vs. conventional technology (HAWK and KESSLER, 1979).

integral-throat-entrance (ITE) designs of the 1980's (HAWK and KESSLER, 1979). The rocket nozzle throat cross-sectional schematics at the top of Fig. 41 provide a visual comparison between the early multicomponent designs and the current carbon–carbon ITE design.

Other aerospace applications also utilize carbon–carbon composites because of their good thermal shock resistance and high sublimation temperature. Multidirectional constructions would offer advantages for most of these applications. For example, isotopes are used on many space missions to generate heat and electric power (GAVERT, 1969). Carbon–carbon capsules are used to protect these isotopes in the event that reentry into the earth's atmosphere is required. Carbon–carbon composites are also being developed for use as reusable radiative heatshields for manned space vehicles such as space shuttles (CARNAHAN et al., 1971).

Some commercial applications for carbon–carbon composites are beginning to develop. Various carbon–carbon composites are being evaluated for the internal fixation of bone fractures (FITZER et al., 1978). Carbon has been shown to be biocompatible and carbon–carbon composites can be tailored to be structurally compatible with bone. The Young's modulus must be similar to bone and the strength must be high. The specific weight of carbon–carbon is also similar to that of bone. Initial medical experiments with three dogs, involving fixation of the right femur, look promising.

Carbon–carbon hot pressing molds are now a commercial product designated as Filcarb™.[7] These molds can withstand higher pressures and offer a longer use life than polycrystalline bulk graphite.

Disc-brake systems made from carbon–carbon composites have undergone extensive development (BAUER and KOTENSKY, 1971b). The present use of these brake systems includes military aircraft and the Concorde. However, as the technology develops and the economics become more favorable, carbon–carbon composites may be used on other commercial airplanes.

Other applications that have been mentioned for carbon–carbon composites include high temperature ducting systems, components for nuclear reactors, electrical contacts, hot seals, and bearings (STOLLER et al., 1971).

As production techniques become refined and scaled-up to higher volume capacities, multidirectional carbon–carbon composites will become more economically acceptable for most of the applications mentioned above. The availability of low cost carbon fiber will also have a major impact on the future of carbon–carbon products.

References

ADAMS, D.F. (1976), *Materials Sci. Engrg.* **23**, 55.
BARION, R.S. (1968), *SPE J.* **24**, 31.
BAUER, D.W., W.V. KOTLENSKY, J.W. WARREN and W.H. SMITH (1971a), Fabrication and CVD carbon infiltration of carbon and graphite filament wound cylinders, *10th Biennial Conf. on Carbon*, Bethlehem, PA, 1971, Paper FC-36.

[7] Registered Trademark of Fiber Materials Incorporated.

BAUER, D.W., W.V. KOTLENSKY (1971b), CVD carbon infiltration and strength for fabric lay-up carbon–carbon composites, *10th Biennial Conf. on Carbon*, Bethlehem, PA, 1971, Paper FC-37.

BRADSHAW, W.G., P.C. PINOLI and R.F. KARLAK (1974), Large diameter carbon–composite monofilaments, NASA Rept. No. CR-134625, 1974.

BROOKS, J.D. and G.H. TAYLOR (1965), *Carbon* 3, 185.

BURNS, R.L. and J.L. COOK (1974), Pressure carbonization of petroleum pitches, in: M.L. DEVINEY and T.M. O'GRADY, eds., *Petroleum Derived Carbons, ACS Symposium Series* Vol. 21 (American Chemical Society, Washington D.C.) p. 139.

BURNS, R.L. and L.E. MCALLISTER (1976), Densification of multidirectional carbon–carbon composites, Paper presented at *AIAA/SAE, 12th Propulsion Conference*, Palo Alto, 1976.

CARNAHAN, K.R., R.W. KIGER, P.R. DEMPSEY and P.C. PARTIN (1971), Carbon–carbon composites for space shuttle reentry thermal protection, Paper presented at *American Ceramic Society 73rd Ann. Meeting*, Chicago, 1971.

CHARD, W., M. CONAWAY and D. NEISZ (1974), Advanced high pressure graphite processing technology, in: M.L. DEVINEY and T.M. O'GRADY, eds., *Petroleum Derived Carbons, ACS Symp. Series* Vol. 21 (American Chemical Society, Washington D.C.) p. 155.

CHARD, W. and D. NIESZ (1975), High pressure densified carbon–carbon composites, Part I: Processing procedures, *12th Biennial Conf. on Carbon*, Pittsburg, 1975.

COOK, J. and J.E. GORDON (1964), *Proc. Roy. Soc.* **282A**, 508.

COOK, J. and J.A. CRAWFORD (1975), Multidirectional substrates for advanced composites, Paper presented at the *76th Ann. Meeting of the American Ceramic Society*, Chicago, 1975.

CRAWFORD, J.A. (1976), U.S. Patent No. 3 949 126.

DI CRISTINA, V. (1971), Hyperthermal ablation performance of carbon–carbon composites, *AIAA 6th Thermophysics Conference*, Tullahoma, TN, 1971, Paper 71-416.

DIETRICH H. and L.E. MCALLISTER (1978), Properties of ultra high density graphite processed to greater than $2.0 \, g/cm^3$, Paper presented at *American Ceramic Society 80th Ann. Meeting*, Detroit, 1978.

FITZER, E. and A. BURGER (1971), The formation of carbon–carbon composites by thermally decomposing carbon fibre reinforced thermosetting polymers, *Internat. Conf. on Carbon Fibres, Their Composites and Applications*, London, 1971, Paper No. 36.

FITZER, E. and B. TERWIESCH (1973), *Carbon* **11**, 570.

FITZER, E., K.H. GEIGL and W. HÜTTNER (1978), Studies on matrix precursor materials for carbon–carbon composites, in: *Proc. of 5th London Internat. Carbon and Graphite Conf.* Vol. I (Society of Chemical Industry, London) p. 493.

FITZER, E.W., L.M. HUTTNER, L.M. MANOCHA and D. WOLTER (1978), Carbon fibre reinforced composites for internal bone-plates, in: *Proc. of 5th London Internat. Carbon and Graphite Conf.* Vol. I (Society of Chemical Industry, London) p. 454.

GAVERT, R.B. (1969), *SAMPE Quart.* **1**, 56.

GIRARD, H. (1978), The preparation of high density carbon–carbon composites, in: *Proc. 5th London Internat. Carbon and Graphite Conf.* Vol. I (Society of Chemical Industry, London) p. 483.

GRENIE, Y. and G. CAHUZAC (1980), Automatic weaving of 3-D contoured preforms, in: *Proc. of 12th Nat. SAMPE Symp.* (SAMPE, Azusa).

HAWK, C.W. and W.C. KESSLER (1979), A functional approach to the application of carbon–carbon composites to solid rocket nozzles, *30th Internat. Aeronautics Federation Congress*, Munich, 1979, Paper 79-1AF-19.

HONDA, H., K. KOBAYASHI and S. SUGAWARA (1968), *Carbon* **6**, 517.

HUTTINGER, K.H. and V. ROSENBLATT (1977), *Carbon* **15**, 69.

JACOBS, K.M., A.T. LASKARIS and J.W. HERRICK (1968), Three dimensionally reinforced ablative rocket engine components, *AIAA 4th Propulsion Joint Specialist Conf.*, Cleveland, 1968, Paper No. 68-598.

KOBAYASHI, K., S. SUGAWARA, S. TOYODA and H. HONDA (1968), *Carbon* **6**, 359.

KOTLENSKY, W.V. and J. PAPPAS (1969), Mechanical properties of CVD infiltrated composites, *9th Biennial Conf. on Carbon*, Boston, Paper MP-26.

KOTOSONOV, A.S., V.A. VINNIKOV, V. FROLOV and B.G. OSTRONOV (1969), *Doklady Akademii Nauk* **185**, 1316.

LACHMAN, W.L., J.A. CRAWFORD and L.E. McALLISTER (1978), Multidirectionally reinforced carbon–carbon composites, in: B. NOTON, R. SIGNORELLI, K. STREET and L. PHILLIPS, eds., *Proc. of Internat. Conf. on Composite Materials* (Metallurgical Society of AIME, New York).

LAMICQ, P. (1977), Recent improvements in 4-D carbon–carbon materials, *AIAA/SAE 13th Propulsion Conf.*, Orlando, 1977, Paper No. 77-882.

LARAMIE, R.C. and A. CANFIELD (1972), Carbon–carbon composites solid rocket nozzle material processing, design and testing, in: *Proc. of 2nd ASTM Conf. on Composite Materials: Testing and Design, Spec. Tech. Publ.* Vol. 497 (ASTM, Philadelphia) p. 588.

LEVINE, A. (1975), High pressure densified carbon–carbon composites, Part II: Testing, *12th Biennial Conf. on Carbon*, Pittsburg, 1975.

MACKAY, H.A. (1970), *Carbon* **8**, 517.

McALLISTER, L.E. and A.R. TAVERNA (1971a), The development of high strength three dimensionally reinforced graphite composites, Paper presented at the *American Ceramic Society 73rd Ann. Meeting*, Chicago, 1971.

McALLISTER, L.E. and A.R. TAVERNA (1971b), Three-dimensionally reinforced carbon–carbon composites, *10th Biennial Conf. on Carbon*, Bethlehem, PA, 1971, Paper No. FC-40.

McALLISTER, L.E. and A.R. TAVERNA (1972), Development and evaluation of Mod-3 carbon–carbon composites, in: *Proc. of 17th Nat. SAMPE Symp.* (SAMPE, Azusa) p. III-A-3.

McALLISTER, L.E. and A.R. TAVERNA (1974), Composition–construction relationships in 3-D carbon–carbon composites, Paper presented at *Pacific Coast Regional Meeting of the American Ceramics Society*, Los Angeles, 1974.

McALLISTER, L.E. and A.R. TAVERNA (1976), A study of composition—Construction variations in 3-D carbon–carbon composites, in: E. SCALA, E. ANDERSON, I. TOTH and B. NOTON, eds., *Proc. of Internat. Conf. on Composite Materials* Vol. I (Metallurgical Society of AIME, New York) p. 307.

MULLEN, C.K. and P.J. ROY (1972), Fabrication and properties description of AVCO 3-D carbon–carbon cylinder materials, in: *Proc. 17th Nat. SAMPE Symp.* (SAMPE, Azusa) p. III-A-2.

PARMEE, A.C. (1971), Carbon fibre–carbon composites: Some properties and potential applications in rocket motors, in: *Internat. Conf. on Carbon Fibres, Their Composites and Applications*, London, 1971, Paper No. 38.

PIERSON, H.O. (1968), Development and properties of pyrolytic carbon felt composites, in: *Proc. of 14th Nat. SAMPE Symp.*, Cocoa Beach, 1968, II 4B-2.

REISIVIG, R.D., L.S. LEVINSON and J.A. O'ROURKE (1968), *Carbon* **6**, 24.

SALKIND, M.J. (1969), Introductory remarks, in: *Interfaces in Composites, ASTM STP452* (American Society for Testing and Materials, Philadelphia) p. 1.

SCHMITT, H. (1971), Carbon–carbon composites for reentry protection systems, *10th Biennial Conf. on Carbon*, Bethlehem, PA, 1971, Paper FC-57A.

SCHMIDT, D.L. (1972), *SAMPE J.* **8**, 9.

SHINDO, A. (1961), Studies on graphite fiber, Rept. No. 317 of the Government Industrial Research Institute, Osaka.

STOLLER, H.M. and E.R. FRYE (1969), Carbon–carbon materials for aerospace applications, in: *Proc. of AIAA/AIME 10th Structures, Structural Dynamics and Materials Conference*, New Orleans, 1969, p. 193.

STOLLER, H.M., B.L. BUTLER, J.D. THEIS and M.L. LIBERMAN (1971), Carbon fiber reinforced—Carbon–matrix composites, Paper presented at the *1971 Fall Meeting of the Metallurgical Society of AIME*, Detroit.

STOLLER, H.M. and E.R. FRYE (1971), Processing of carbon–carbon composites—An overview, Paper presented at *73rd Ann. Meeting American Ceramic Society*, Chicago, 1971.

THOMAS, C.R. and E.J. WALKER (1978), Effects of PAN carbon–fibre surface on carbon–carbon composites, in: *Proc. of 5th London Internat. Carbon and Graphite Conf.* Vol. I (Society of Chemical Industry, London) p. 520.

WARREN, J.W. and R.M. WILLIAMS (1972), Isothermal CVD processing, *4th Nat. SAMPE Tech. Conf.*, Palo Alto, 1972.

WATT, W., L.N. PHILLIPS and W. JOHNSON (1966), *The Engineers* **221**, 815.

WHANG, P.W., F. DACHILLE and P.L. WALKER (1974), *Carbon* **6**, 137.

ZHIGUN, I.G. and V.A. POLYAKOV (1978), Properties of three-dimensionally reinforced plastics, Y.M. TARNOPOL'SKII, ed. (Svoystva Prostranstvenno – Armirovannykh Plastikov, Riga).

CHAPTER IV

Reinforced Thermoplastics

Michael G. Bader

Department of Metallurgy and Materials Technology
University of Surrey
Guildford
Surrey, GU2 5XH
United Kingdom

Contents

HANDBOOK OF COMPOSITES, VOL. 4 – Fabrication of Composites
Edited by A. KELLY and S.T. MILEIKO
© 1983, Elsevier Science Publishers B.V.

List of Symbols

α – coefficient of thermal expansion
C_0 – fibre orientation parameter
E_f – tensile elastic modulus of fibre
E_m – tensile elastic modulus of matrix
E_c – tensile elastic modulus of composite
E_0 – tensile elastic modulus of composite parallel to fibre direction
E_{90} – tensile elastic modulus of composite normal to fibre direction
E_θ – tensile elastic modulus of composite at angle θ to fibre direction
ε_c – average tensile strain in composite
G_m – shear modulus of matrix
L_f – fibre length
L_t – load transfer length (Kelly model)
L_t^1 – load transfer length (Cox model)
ν – Poisson's ratio
r_f – fibre radius
r_1 – fibre separation
R_{crit} – critical fibre aspect ratio (fracture)
$R_{c(\varepsilon)}$ – strain dependent critical fibre aspect ratio
R_i – sub-critical fibre aspect ratio
R_j – super-critical fibre aspect ratio
σ_f – stress in fibre
σ_m – stress in matrix
σ_c – average stress in composite
σ_{fu} – tensile strength of fibre
$\bar{\sigma}_{f_i}$ – average stress in sub-critical fibre
$\bar{\sigma}_{f_j}$ – average stress in super-critical fibre
T_g – glass transition temperature
T_c – crystalline melting point
τ – interface shear stress
θ – angle between fibre axis and direction of stress
V_f – fibre volume fraction
V_i – fraction of sub-critical fibres
V_j – fraction of super-critical fibres

1. Introduction

Reinforced thermoplastics comprise an important group of mouldable plastics materials which are being utilised to an increasing extent in the manufacture of small load bearing components. Most thermoplastics materials may be reinforced by the incorporation of either particulate or fibrous fillers. Particulate fillers such as chalk, talc and mica are cheap and improve a number of properties, notably dimensional stability, but do not strengthen the material to an appreciable extent. They are really non-reinforcing fillers. Other particulate fillers such as flake glass and glass ballotini and specially treated mineral fillers may give an appreciable reinforcement. However, fibrous fillers consisting of short-discontinuous fibres (usually glass, but carbon and polyaramid textile fibres may also be used) may be utilised more effectively to give a high degree of enhancement of stiffness and strength, together with improved dimensional stability and elevated temperature performance. It is these materials which are the principal concern of this chapter.

The essential characteristic of reinforced thermoplastics materials (RTP) is that they may be shaped by melt fabrication techniques. That is by injection moulding and, to a lesser extent, extrusion, blow moulding and thermoforming. There are also developments in cold mechanical forming processes, although these seem unlikely to be suitable for the more highly reinforced materials. It should be noted, at this point, that traditional thermoplastics processes, such as injection moulding, have been adapted for moulding thermoset based materials, including fibre-reinforced thermosets. The essential characteristic of these processes is that the material is made to flow in the pre-cured state and that in the final stage of the process the material is cured, or cross linked, into an infusable mass in a hot mould. In thermoplastics processing, by contrast, the material is melted or plasticized by heating and then shaped in the plasticized condition and cooled to re-solidify (e.g., by injecting molten polymer into a cold mould). In principle, the thermoplastic suffers no chemical alteration during the processing cycle and may thus be re-processed several times. In practice this ideal is seldom approached since the materials are affected by thermal, mechanical and oxidative degradation.

In Table 1 the relative consumptions of both reinforced and un-reinforced thermosets and thermoplastics are compared. It should be noted that the total thermoplastics consumption is much higher than that of thermosets but this is mainly accounted for by the so-called 'commodity thermoplastics' comprising the polyolefins, polystyrene, polyvinylchloride (PVC) and their derivatives. These materials are utilised mainly in non-reinforced forms, although cheap mineral fillers are used extensively in polypropylene and PVC. A much higher proportion of the 'engineering thermoplastics', (e.g., polyamides (nylons), polyacetals and thermoplastic polyesters) are reinforced, and here the principal reinforcement is discontinuous short glass fibre. The specialized high-performance thermoplastics such as the polysulphones are also often reinforced with fibres. In contrast virtually all thermosets are filled with either reinforcing or

TABLE 1. Annual consumption of selected plastics in the
United Kingdom (1980).

Material	Consumption[d] (1000 t)
Low density polyethylene (LDPE)	400
High density polyethylene (HDPE)	160
Polypropylene[a] (PP)	200
Polyvinyl chloride (PVC)	350
Polystyrene (PS)	150
Polyamides[b] (nylons)	17
Polyoxymethylene[c] (acetals)	6
Amino thermosets	120
Phenolic thermosets	60
Epoxide thermosets	13
Polyesters (mainly thermosets)	45
Polyurethane thermoset	80

[a] Less than 50 000 t in reinforced grades.
[b] Approx. 25% fibre reinforced.
[c] Approx. 25% fibre reinforced.
[d] Figures abstracted from UK Government Statistical Service
Business Monitor PQ 276 (rounded off).

cheap particulate fillers. The unsaturated-polyester, vinyl-ester and epoxy resins are widely used in laminated composites using continuous or relatively long discontinuous fibrous reinforcement and there is an increasing use of short-fibrous reinforcement in the phenolic, amino and melamine-formaldehyde resins which, as mentioned above, may be injection moulded. Another important class of reinforced materials is the class of long-fibre reinforced sheet moulding compounds (SMC) and bulk (or dough) moulding compounds (BMC (DMC)), which are also based on the unsaturated-polyester, vinyl-ester and epoxy resins. These materials are normally compression moulded although BMC may also be injection moulded in a manner similar to that used for conventional thermosets. With these materials the process cycle time is often dominated by the time needed to effect the cure. Typically times of 2–10 minutes are necessary, and this results in production rates which compare unfavourably with sheet metal pressing operations, which are the competitive processes in the automobile and domestic consumer goods markets. This resulted in the emergence of a new class of materials: The fibre reinforced thermoplastic thermo-formable sheet materials. These consist of continuous or chopped glass-fibre reinforcement in thermoplastic matrices. They are produced as sheet materials and are effectively the thermoplastic equivalent of SMC. They are fabricated by preheating sheet blanks and then pressing or stamping between cold dies. This process is much faster than SMC moulding but, at present, is less versatile.

In comparing the roles of these various classes of reinforced materials it should be noted that many applications of composites have originated in hand lay-up mouldings using the unsaturated-polyesters or similar chemical curing resins. These processes are slow and labour intensive and the demand for high production rates has led to competition from thermoplastics on the one hand and the more highly mechanized SMC, BMC and injection moulded thermoset processes on the other. It would seem that thermoplastics compete most favourably in the field of relatively small highly intricate mouldings which are required in very large numbers. In general the most serious limitation of thermoplastics based materials is their lower thermal stability.

In the following sections we will examine some fundamental aspects of the science of reinforced thermoplastics and then discuss in detail compounding and fabrication. This will inevitably concentrate on the injection moulding process as this is by far the most important technique used in fabricating reinforced materials.

2. Fundamental aspects of reinforced thermoplastics

2.1. Thermoplastics

These, by definition, are solid at ambient temperatures but may be melted or softened by heating. In the thermally plasticized state they may be formed by extrusion, injection moulding or thermoforming and then chilled to re-solidify and retain their moulded form. In principle this melt/solidification cycle may be repeated many times but in practice degradation intervenes and limits the length of time that the material may be exposed to elevated temperatures. Thermal degradation usually takes the form of breakdown of the long chain molecules of the polymer into lower molecular weight fractions and there are often also side reactions between the constituents. A few thermoplastics will undergo cross-linking reactions when heated for extended periods and most commercial thermoplastics also suffer from oxidative degradation if exposed to atmospheric oxygen when heated. The initial effects of degradation are usually no more serious than some discolouration, which may or may not be acceptable, depending on the colour and application of the material. Obviously dark coloured compounds will be less affected than white or clear compositions! More severe degradation will alter both the rheology of the melt and the mechanical properties of the moulded material and at this stage is usually unacceptable. The susceptibility of a polymer to degradation is often significantly altered by the presence of additives. These include, of course, mineral and fibrous fillers, coupling agents, pigments, plasticizers, lubricants and stabilisers.

The fabrication of thermoplastics is critically dependent on their melt rheology, which in turn is dependent on the chemical make-up, molecular architecture, molecular weight distribution and the presence of any fillers or additives. For simplicity we may classify thermoplastics into the categories amorphous and semi-crystalline. The former embrace most styrene and vinyl derivatives whilst many of the olefins and the engineering thermoplastics are semi-crystalline. Amorphous polymers do not have a sharply defined melting temperature but soften gradually as the temperature is raised above their glass transition temperature, T_g. Even at temperatures well above T_g the polymer melt is very viscous and exhibits pseudo-plastic rheological properties. That is, the apparent viscosity of the melt decreases as the shearing rate is increased. Crystalline polymers do have a crystalline melting temperature, T_c, but the sharpness of the melting depends on the relative proportions of crystalline and amorphous material. A highly crystalline (>60%) polyamide (nylon) will appear to melt quite sharply when heated a few degrees above T_c. The melt viscosity is controlled principally by the molecular weight distribution of the polymer. Higher average molecular weight results in a more viscous melt but the melt viscosity may be markedly reduced by a relatively small proportion of low molecular weight material. As a general rule lower viscosity melts are easier to injection mould, but high viscosity, and more particularly high melt strength, is needed for extrusion, blow moulding and thermoforming. Likewise higher molecular weights are generally associated with superior mechanical properties, better elevated temperature performance and greater resistance to chemical attack. For injection moulding, in particular, the polymer manufacturer has to strike a very critical compromise between opposed requirements for ease of moulding and optimum mechanical properties.

The addition of a significant fraction of mineral filler or fibre will generally increase the melt viscosity but in many cases the pseudo-plastic behaviour of the melt is enhanced so that the increased viscosity is less apparent at high rates of shear as shown in Fig. 1. This implies that the filled material could be moulded almost as easily as the unfilled provided a very high injection rate was used. However, this is not usually desirable for other reasons which are discussed in subsequent sections. A more practical alternative would be to reduce the viscosity of the filled material by raising the melt temperature (see Fig. 2). This is expedient only if degradation of the polymer can be avoided.

Two other properties of the polymer which are of critical importance in injection moulding are the shrinkage on solidification from the melt and the compressibility of the liquid polymer. Fig. 3 depicts typical specific volume vs. temperature curves for an amorphous and a semi-crystalline polymer. Note that in both cases the solid polymer has a relatively high coefficient of expansion, (typically 2–$5 \times 10^{-5}/°C$ linear or 6–$15 \times 10^{-5}/°C$ volume) and this increases sharply above T_g. When crystallisation occurs there is an isothermal volume shrinkage of perhaps 5%. In practice the amount of shrinkage will depend on the proportion of the material which actually crystallises. This might be between 40 and 80% in typical semi-crystalline thermoplastics, e.g., polypropylene or polyamide 6.6. For any material the extent of crystallisation depends on the rate of cooling through the crystallisation range (normally T_c–$(T_c - 20°C)$). This is important because the extent of crystallisation has an important influence on the mechanical properties of the solid as well as on the 'in mould shrinkage', and variations in thermal history during a moulding

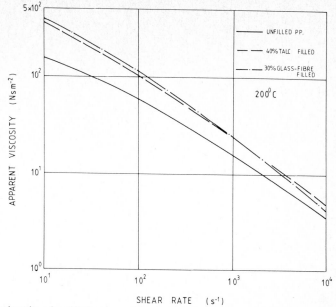

FIG. 1. The viscosity of unfilled polypropylene and similar polymers filled with 40% (w/w) talc and 30% (w/w) glass fibre at 200°C and over 3 decades of strain rate. The data were obtained from capillary rheometer measurements. Note that the viscosity falls by a factor 10 for × 100 increase in strain rate. The curves for the filled materials are steeper, i.e., they are more pseudo-plastic (BADER, unpublished).

campaign will result in variable dimensions of the moulding. In amorphous polymers the problem is not so acute, but T_g itself is dependent on the cooling rate as is the total shrinkage on cooling. Polymer manufacturers often offer controlled crystallinity grades of their semi-crystalline polymers for precision moulding applications. The compressibility of the polymer melt is important because a high hydrostatic pressure can be utilised to force more polymer into the mould and this effect may be used to compensate for shrinkage. There is a large variation in compressibility between different thermoplastics materials. In general, the greater the 'free-volume' the greater the compressibility. Highly branched molecules or those containing large pendant groups are usually more compressible than those with more regular, linear molecules. Effectively this means that semi-crystalline materials tend to have lower compressibility than the amorphous ones.

As noted earlier, polymer melts are viscous liquids and at low shear rates will flow in an essentially laminar manner but at high shear rates various flow abnormalities may occur. Consider the case of pressure induced flow through a circular section passage (see Fig. 4). At low flow rates the Poisuelle assumptions of no flow at the tube wall and isothermal conditions are approximately correct and the velocity profile is of the ideal parabolic form (cf. Fig. 4(a)). At higher flow rates, however, the high shear rates close to the tube walls cause local adiabatic heating, this raises the temperature of the melt close to the wall

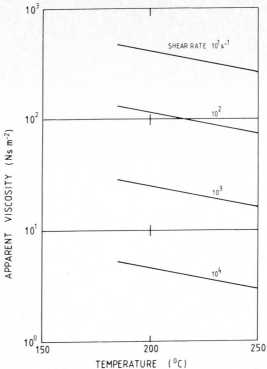

FIG. 2. These data relate to the same glass-filled grade of polypropylene as shown in Fig. 1. At all shear rates the viscosity falls by a factor of ~4 for a 100°C rise in temperature (BADER, unpublished).

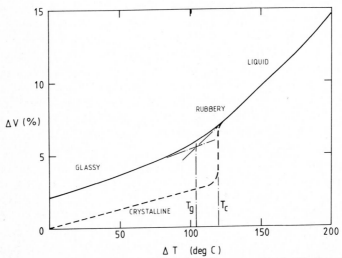

FIG. 3. A schematic representation of the changes in volume associated with the glass/rubber transition and crystalline transition in a typical semi crystalline polymer.

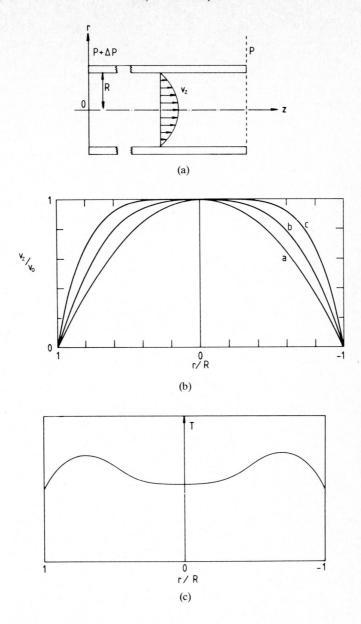

FIG. 4. (a) Representation of pressure induced flow through a circular tube. The velocity profile is parabolic and V_z is zero at the wall; (b) Velocity profiles for Newtonian and 'power law' fluids across a circular tube. Curve a is for a Newtonian fluid—power law index 1. Curve b for a power law fluid index $n = 0.5$ and curve c for $n = 0.25$. Most filled polymer systems approximate to power law fluids with $0.5 > n > 0.25$; (c) This shows a temperature profile for non-isothermal flow of a typical polymer. The adiabatic heating depends on the shear rate (higher nearer to wall) and the volumetric flow rate. The peak temperature is close to the wall. This effect would further distort the velocity profile towards a plug-like flow.

and, thus, lowers its viscosity, so that the shear rate increases. The parabolic velocity contour is then degraded into the 'plug flow' form shown in Fig. 4(b), whilst in Fig. 4(c) a typical temperature profile resulting from adiabatic flow at high strain rate is shown. This effect is important in high shear rate processes like injection moulding and can lead to the 'jetting' effect discussed later. Another effect of high shear rates is that a degree of molecular orientation is induced parallel to the flow direction. If the melt is rapidly cooled from this condition a significant molecular orientation may remain in the solid moulding. This results in a marked anisotropy, especially of shrinkage (which tends to be greater in the direction of orientation), and ductility, which is lower in directions normal to the flow path. This molecular anisotropy will usually be coupled with a high level of internal stress, arising from the rapid cooling. This may give rise to distortion, environmental stress cracking and other undesirable effects during subsequent service.

This brief summary of the properties of thermoplastics is intended to highlight the fact that they are a class of widely disparate materials. In any one class of thermoplastics wide differences in properties, especially melt rheology, may be induced by variations in the molecular weight distribution, the proportions of co-polymers and the amounts of plasticizers and other minor additives. All this before we even consider the effects of mineral fillers or fibres.

2.2. Mineral fillers and reinforcements

Thermoplastics may be filled with either particulate or fibrous fillers which are usually of inorganic mineral origin. Particulate fillers are very extensively utilised and may be added in amounts of up to about 40% by volume. Common fillers are various forms of calcium carbonate (whiting, precipitated chalk, ground limestone, etc), talc, mica, silica and glass in the form of flake or ballotini (small spherical particles). Carbon black is used principally as a pigment in thermoplastics although it acts as a reinforcing filler in rubbers and elastomers. Graphite and molybdenum disulphide are special purpose fillers for improving the tribological characteristics—principally for dry bearing applications. The polymer polytetrafluoroethylene (PTFE, Teflon®, Fluon®)[1] is also used as a filler for increasing the lubricity. The effects of particulate fillers are complex and depend on the nature of the filler, the particle size and shape and the nature of the interface formed between the particle and the polymeric matrix. Mineral fillers are invariably much stiffer than the polymer matrices and will therefore increase the low strain compressive stiffness approximately according to the Voigt bound of the Rule of Mixtures (see Section 2.3). The tensile and shear modulus, strength and ductility are however more dependent on the strength of the particle–matrix interface. Some fillers such as chalk and talc are also comparatively weak and this would limit strength and ductility even if a strong interface were formed. Particle size and shape are important in

---

[1] ® means registered trade name.

this respect since smaller particles and particles of flake or fibrous form have a relatively larger surface area than larger, or more nearly equiaxed ones.

Another important consideration is the behaviour of the particles during the compounding operations when the dry mineral filler is incorporated into the thermoplastic matrix. This operation is generally carried out in some variant of the melt blender (Section 3). Critical considerations are the apparent bulk density, the wetting characteristics, and the ease with which agglomerated particles may be broken up and dispersed in the polymer. Generally, finer particles are more difficult to wet and disperse. It is common practice to subject the filler materials to surface treatments designed to assist wetting and dispersion and also to promote more effective interfacial bonding to the polymer matrix.

Whilst particulate fillers are utilised in both non-reinforcing and reinforcing roles, fibrous fillers are used principally for the enhancement of both stiffness and strength with additional benefits arising from improved dimensional stability and elevated temperature performance. The principal fibrous reinforcing material is E-glass fibre and this is used in well over 90% of all the fibre reinforced thermoplastics (FRTP) produced. Asbestos has been widely used as a reinforcing filler but its usage has been sharply reduced in the past few years on account of the health hazards associated with handling the dry raw material. Current trends indicate that the use of asbestos as a general purpose reinforcement will be virtually abandoned in the near future. This is unfortunate as it can be a cheap and very effective alternative to glass. The other fibres of importance are the carbon fibres which are stiffer and stronger than E-glass, but much more expensive. Nevertheless significant quantities of carbon-fibre filled materials are currently being produced and the indications are that the market will continue to grow. Polyaramid textile fibres (e.g., Dupont's Kevlar® 49) also have potential for reinforcement of thermoplastics. They are stiffer, less dense and as strong as E-glass, but are still comparatively new and little work has been reported of their use in thermoplastics systems. We will therefore confine our discussions to E-glass and carbon fibres.

E-glass is a low-alkali borosilicate glass which is readily drawn into continuous fibres of 5–25 μm diameter. Standard practice is to draw the fibres from a 'bushing' containing 200–2000 holes, or 'tips', so that a *strand* of that number of fibres is drawn simultaneously. This strand is the basic fibre grouping. The individual filament diameter is controlled by the melt temperature, the tip diameter, and the drawing velocity. Glass fibre for reinforcing thermoplastics is usually required in a heavier form than the single strand and this is produced by incorporating a number of strands into a *roving*. A roving is essentially a parallel array of strands without twist. This distinguishes it from a *yarn* which is twisted. In the conventional process the newly drawn strand is sized with a water based emulsion and wound onto a temporary spool to form a 'cheese'. This is dried in an oven, and then the glass strand unwound and several strands combined to form the roving, which is wound up onto another package (spool). Alternatively the strand may be chopped into short lengths (e.g., 4–10 mm) for subsequent incorporation into reinforced plastics. A recent development is the 'direct roving' process. The strands from one or more bushings are combined directly and either wound onto the final package or chopped, by-passing the temporary package and drying stages. The newly drawn glass fibre is extremely strong (~5 GPa) and it owes this strength to the silicate molecular network structure developed by drawing and to the absence of surface flaws and imperfections. If the surface of the fibres is abraded or exposed to the atmosphere for an extended period much of this strength will be lost. It is therefore essential that the fibres be protected immediately they are drawn. This is accomplished by the application of a *size* or *finish*. This consists of a water based emulsion or solution applied by spray, dip or wetted roller to the newly drawn strand. The size consists of a film-forming agent and, usually, lubricating and coupling agents.

The film-forming agent is almost invariably a water based polymer emulsion: polyvinyl acetate, polyvinyl alcohol, polyesters, polyurethane and epoxy resins are commonly utilised. More recently solutions of anionic or cationic polyelectrolytes have been used as film formers. The emulsion is applied to the strand of fibres which are wetted and on subsequent drying the water evaporates to leave behind the polymer droplets. According to the formulation the polymer may spread all over the glass fibre, so that it is virtually encapsulated and the fibres in the strand firmly bound together by the polymer film, or the droplets may maintain their integrity to some extent so that the strand is bound by contact between the fibres and these minute polymer globules. Micrographs of typical glass strands are shown in Fig. 5. The extent to which the fibres are bound in the strand depends on the amount of film former in the size and on the type of film former used. This determines the stiffness of the strand and the manner in which the fibres will separate from the strand during subsequent processing operations. It is not normal practice to apply additional binders when making up a roving from a number of strands, although this is a further option. The direct roving is effectively a single heavy strand. A stiff strand is generally easier to handle and to chop, and less damage is done to the fibres during dry handling operations. However, during the compounding and moulding of GFRTP it is usually required that the chopped or broken strands fully separate into individual filaments, which disperse uniformly in the matrix. (This is in contrast to the practice adopted in SMC and BMC formulation, where the reinforcing element is the chopped strand, and filamentization is considered undesirable.) This requires careful selection of the film-forming agent. The strand integrity must be maintained during the dry handling operations but the film former should melt and/or dissolve in the polymer during compounding and moulding so that the fibres may be dispersed. However, there is some evidence that too rapid a breakdown of the film former leads to a higher level of fibre breakage (into short lengths) and damage during compounding and moulding, especially in screw plasticizing machinery, so that it is desirable to use a size system which maintains some strand integrity until the final moulding operation when complete dispersion should be achieved.

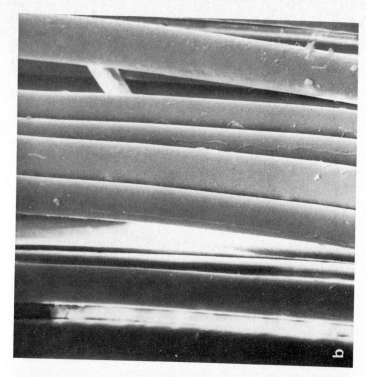

FIG. 5. Scanning Electron Micrograph. (a) Showing a portion of a heavy glass fibre roving and shows typical strands of 13 μm diameter fibres bound together by the size; (b) Taken at a higher magnification shows evidence of the thin coating of size on individual filaments.

This is difficult to arrange in practice since rheological conditions may vary widely between different types of moulding.

The role of the lubricant in the size is to improve the handling characteristics of the dry strand and roving. The coefficient of friction of the strand is reduced and this results in less damage to the glass by abrasion and attrition during handling, and less wear on the machinery. Lubricants are normally waxes or soaps which are compatible with the polymer-fibre system. Care must be taken in selection of lubricants to avoid combinations which might adversely affect the development of a good fibre–matrix interface bond in the final moulding.

The coupling agent is an additive which promotes the development of a strong bond between the fibre surface and the polymer. The exact behaviour of coupling agents is a matter of some controversy (see ISHIDA and KOENIG, 1978). They consist of silane compounds incorporating functional groups which are designed to interact with the polymer matrix. The agent is usually applied to the glass via an aqueous solution (the size) where the silane groups hydrolyse to intermediate silanols of the general form:

$$(HO)_3Si(CH_2)_n\ Y$$

(see PLUEDDEMANN, 1970). The hydrolysed silanol is absorbed onto the glass surface. During compounding the silane molecule is considered to remain intact on the glass surface when the functional group, Y, either reacts with, or is incorporated into the matrix polymer. Essentially the coupling agent is considered to become part of the matrix but remains 'bonded' to the glass via the silanol group. At one time it was considered that the coupling molecules formed a mono-molecular layer on the glass surface but recent studies (e.g., ZISMAN, 1969) suggest that complex multiple layers are formed according to the conditions of application. In addition to the formation of chemical bonds at the interface it is probable that coupling agents improve the wettability of the fibre surface by the molten polymer thus promoting a better physical bond. However, although the beneficial effects of coupling agents are well established in terms of improvements in mechanical properties, and resistance to degradation under wet conditions, of the composite, there is still little understanding of their exact mechanism of action.

The size thus performs a complex and multiple function. The application of the size is most effective immediately the fibres are drawn from the melt, and for this reason are generally applied by the fibre manufacturer. The exact composition of the size is seldom revealed and the manufacturers merely offer a range of proprietary finishes designed for the different matrix polymers and fabrication routes.

Fibre diameter is an important parameter in FRTP. Clearly a smaller diameter will give a greater relative surface area of interface over which the coupling agent may act. Thus, if interface strength is a limiting factor in reinforcement efficiency, a finer fibre would be beneficial. There is considerable evidence that this is the case. However, finer fibres are more expensive due to lower achievable production rates. This is due to the fact that either the tip diameter must be reduced, or the haul off velocity must be increased. The former directly reduces productivity and the latter indirectly through a greater incidence of stoppages due to fibre breakages. In practice 10 μm is considered the lowest viable fibre diameter and 12–13 μm is preferred where ultimate composite performance is not adversely affected. In many instances it is difficult to exploit the benefits of finer fibres as they tend to be broken into shorter lengths during processing so that there is no improvement in fibre aspect ratio which is the critical parameter.

2.3. Principles of reinforcement by discontinuous fibres

In comparison with other composite materials FRTP are amongst the most difficult to model mathematically. This is because they contain short-fibres of variable lengths and orientation dispersed in a non-ideally elastic matrix. The quasi-elastic properties of such a material may be semi-quantitatively modelled by modifying the classical rule of mixtures equations to allow for the short fibres and orientation. Several approaches provide adequate treatments for the practical prediction of low-strain stiffness. This, however, is less than sufficient for design purposes since thermoplastics, even when reinforced, are of lower stiffness in comparison with the more common engineering materials and consequently must be loaded to higher strains in service if they are to compete on a specific strength basis. At higher strains (say $> 0.5\%$) most FRTP materials show pronounced non-linearity in their stress–strain relationships. This may be accounted for, in part, by the non-linear elastic properties of the matrix, and by a progressive reduction in the reinforcing efficiency of the short fibres as the strain is increased. The tensile strength of the material is determined by one or more of a number of complex interactions between fibre, interface and matrix and is difficult to predict from classic micromechanics principles. The same applies to other vitally important properties such as ductility, toughness, creep and fatigue behaviour.

2.3.1. Quasi-elastic properties
Short-fibre reinforced thermoplastics seldom behave in a truly Hookean manner, except at very low strains. This is because the thermoplastic itself is viscoelastic and the fibres are discontinuous. Load can only be carried by a fibre if it is transferred to it from the matrix by shear along the fibre matrix interface. The mechanism of this shear stress transfer has been considered by Cox (1952) and by KELLY and TYSON (1965). Although they approach the problem from different standpoints the results are essentially similar from a practical viewpoint. If we consider an isolated short-fibre embedded in a homogenous matrix, then, as the matrix is extended in tension along the fibre axis, the tensile stress in the fibre builds up from zero at the fibre ends to a maximum in the centre portion of the fibre (see Fig. 6). According to the Cox model the stress distribution along the fibre follows an exponential law (2.1)

M.G. *Bader*

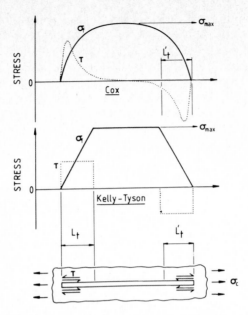

FIG. 6. Stress transfer to a fibre by shear from the matrix according to the models proposed by Cox (1952) and by KELLY and TYSON (1965). The transfer length L_t is defined precisely in the Kelly model but some arbitrary definition is necessary with the Cox model.

and the central portion of a long fibre would be virtually uniformly stressed:

$$\sigma_f = \frac{(E_f - E_m)\sigma_c}{E_m} \left[1 - \frac{\cosh \beta(\tfrac{1}{2}L - x)}{\cosh \beta(\tfrac{1}{2}L)} \right] \tag{2.1}$$

where

$$\beta = [2G_m/(E_f r_f^2 \ln(r_t/r_1))]^{1/2} . \tag{2.2}$$

The simpler Kelly–Tyson concept proposes a constant (frictional) shear stress at the interface from the end of the fibre so that the tensile stress builds up linearly to a maximum which then remains constant over the central portion of the fibre, the transfer length being given by

$$L_t = \sigma_f r_f/2\tau = \varepsilon_c E_f r_f/2\tau . \tag{2.3}$$

Both models show that if the fibre is long, compared with the stress transfer region at the fibre ends, it will reinforce almost as well as a continuous fibre, but if the fibre length is less than about 10 times this transfer length the reinforcement efficiency will be significantly reduced.

The actual transfer length depends on the shear strength of the interface and on the strain. If we assume that the interface strength remains constant, then, as the strain in the composite is increased the transfer region will spread towards the centre of the fibre and the plateau stress will increase to $\varepsilon_c E_f$ (i.e., the strain in the centre of the fibre will be equal to that in the matrix). If the fibre is too short, eventually the two transition zones will merge and no further load may be transferred into the fibre. The concept of a critical length, or more correctly a critical aspect ratio, has been proposed where the criticality is defined in terms of the ultimate strength of the fibre. That is, it is just possible to load a fibre of critical aspect ratio to its failure stress. This condition is given by

$$R_{\text{crit}} = \sigma_{fu}/2\tau.\tag{2.4}$$

In fact there is a critical aspect ratio corresponding to each level of strain given by an equation of similar form:

$$R_{c(\varepsilon)} = \varepsilon_c E_f/2\tau.\tag{2.5}$$

Fibres of lower than critical aspect ratio are loaded to an average stress given by

$$\bar{\sigma}_{f_i} = R_i\tau,\tag{2.6}$$

while, for those of greater aspect ratio,

$$\bar{\sigma}_{f_j} = \varepsilon_c E_f\left[1 - \frac{\varepsilon_c E_f}{4R_j\tau}\right].\tag{2.7}$$

If $R/R_{c(\varepsilon)} > 5$, the reinforcement efficiency of the fibre exceeds 90% of that of continuous fibres. If $R/R_{c(\varepsilon)} = 1$, the efficiency is 50% and below this the efficiency rapidly falls as is shown in Fig. 7. Note that R_c is reduced as the

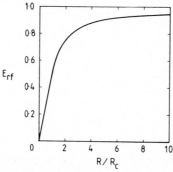

FIG. 7. The fibre reinforcement efficiency E_{rf} (E_{rf} for a continuous aligned fibre system is unity) plotted against the relative fibre aspect ratio R/R_c.

interface strength, τ, is increased. Thus for efficient reinforcement the fibre aspect ratio should be greater than $5R_{c_{(e)}}$ at the failure strain of the fibre.

Now in real FRTP the fibres are not all of the same length but are distributed over a fairly wide range as shown in Fig. 8. The implication of this is that at low strains, when $R_{c_{(e)}}$ is small, all the fibres are effectively super-critical, but as the strain increases a proportion of the shorter fibres will progressively become sub-critical. This means that the stiffness reinforcement efficiency will be high at low strain but will deteriorate as the strain, and hence the stress, increases. This gives rise to the characteristic nonlinear stress–strain curve.

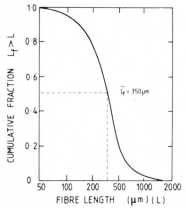

FIG. 8. A typical fibre length distribution for a glass fibre filled nylon. The volume average length is 350 μm but more than 20% of the fibres exceed 500 μm. The length scale is logarithmic (from BADER and COLLINS, 1980).

Superimposed on these effects are those of fibre orientation. A fibre not oriented precisely along the principal tension axis will only be stressed in proportion to the resolved portion of the stress acting along the fibre axis. Thus misaligned fibres will contribute less towards the reinforcement in the principal stress direction. The stiffness of a misaligned fibre composite, E_θ, is given by (2.8) (see Fig. 9). E_0 and E_{90} are the longitudinal and transverse moduli, respectively, for an aligned composite, G_m the shear modulus and ν Poisson's ratio of the matrix.

$$\frac{1}{E_\theta} = \frac{1}{E_0}\cos^4\theta + \left[\frac{1}{G_m} - \frac{2_\nu}{E_0}\right]\sin^2\theta\cos^2\theta + \frac{1}{E_{90}}\sin^4\theta. \tag{2.8}$$

Fig. 9 shows the relationship of E_θ/E_0 to the angle of misalignment, θ.

Therefore, in order to model the elastic behaviour of such a composite it is necessary to utilize a summing or averaging technique which takes account of the fibre aspect ratio and orientation distributions. BOWYER and BADER (1972) used an average orientation factor and summed the contributions of all the fibres according to their aspect ratio. In (2.9) below the two summations refer

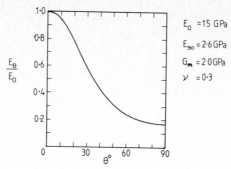

FIG. 9. The relative stiffness versus orientation angle for a hypothetical short fibre composite. The values substituted in (2.8) are typical of a glass fibre filled nylon 6.

to the sub- and super-critical fractions of the fibre respectively. The criterion for criticality is strain-dependent so that the fractions of the fibre distributions lying in each group vary as the strain is increased.

$$\sigma_c = C_0 \left[\sum V_i R_i \tau + \sum \varepsilon_c E_f V_j \left(1 - \frac{\varepsilon_c E_f}{4 R_j \tau} \right) \right] + \varepsilon_c E_m (1 - V_f), \quad R_j > R_{c(\varepsilon)} > R_i.$$

(2.9)

This equation has been shown to accurately model the stress–strain curve for a number of FRTP materials (cf. Fig. 10). The technique, however, suffers from the deficiency that neither the orientation factor C_0 nor the interface strength, τ, are measured independently but are deduced from a curve fitting exercise. With this proviso the fit with experiment has been very good. A further limitation of the technique, (and of the alternatives discussed below), is that it predicts only quasi-elastic behaviour and is of no utility in predicting the actual strength.

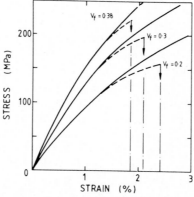

FIG. 10. Experimental stress–strain curves (broken line) and fitted hypothetical curves using (2.9). Note that the fit is good below strains of about 1.5%. Also note the reduction in strain to failure as V_f is increased from 0.2 to 0.38. E-glass fibre filled polyamide 6 (BADER and COLLINS, 1980).

An alternative approach has been proposed by HALPIN and PAGANO (1969) and HALPIN and KARDOS (1978) who model the real material with an 'equivalent laminate' consisting of a balanced symmetric array of continuous fibre laminae of 0°, 90° and ±45° orientation with respect of the principal loading direction. The proportions of the various laminae in the equivalent laminate are adjusted so that it has similar properties to those of the real material. This approach has the merit that the well developed laminate computations may be used to calculate any of the stiffness parameters, at any angle within the plane of lamination. The strength of the material may also be estimated by using the well-known Halpin–Tsai equations (see HALPIN and KARDOS, 1976) with the parameters suitably adjusted to fit the behaviour of a discontinuous fibre composite. This semi-empirical approach procides a model which might be adequate for stiffness critical design calculations but its utility in predicting failure stress is inadequate.

To summarize: Mathematical models for the mechanical properties of fibre reinforced thermoplastics, based on the rule of mixtures equations or on laminate theory, adequately describe their low-strain, quasi-elastic behaviour. The models, however, do not account for the observed reduction in failure strain and therefore do not adequately predict strength or toughness.

2.4. Mechanical properties

As observed in the previous section the stiffness of FRTP may be modelled using either a modified rule-of-mixtures or by an adaptation of continuous fibre laminate theory. In both cases the models have to account for the effects due to the short length and the orientation distribution of the fibres. The stiffness of the composite is then a function only of the fibre fraction (V_f), orientation, aspect-ratio, and the effective interfacial shear strength. In virtually all commercial materials the stiffness reinforcement is less than ideal because the fibres are too short. In principle, shorter fibres could be made more effective if the interfacial shear strength were increased. However, there still remains a limitation in the shear yield strength of the matrix and this, in practice, restricts the possibility of exploiting this principle, as in many cases the strength of the interface appears to approach that of the matrix (see, for example, BADER and BOWYER (1972) and CURTIS et al. (1978)). The only possibilities for improving the stiffness enhancement in current materials are, therefore, to increase fibre aspect ratio or to use a stiffer fibre such as carbon.

The strength of FRTP is controlled by a complex series of interactions between the fibres and the matrix. The general experience is that as the V_f is increased stiffness and tensile strength are both increased, although the strength increases to a lesser extent (see Fig. 11). There is some evidence that strength might actually fall at very high V_f (~0.4) in some materials. The main reason for this relatively lower stength enhancement is that tensile ductility is sharply reduced when fibre reinforcement is introduced. It would appear that all the parameters which enhance the stiffness reinforcement (i.e., higher E_f,

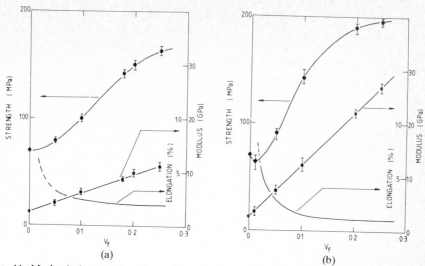

FIG. 11. Mechanical properties of a series of glass filled (a) and carbon fibre filled (b) nylon 6.6 materials. Note the near linear increase of modulus with V_f and the very low ductility of the more highly reinforced materials (taken from CURTIS, 1976).

V_f, R and τ) tend to decrease the tensile ductility or strain to failure. The reasons for this are apparent when we study the manner in which the materials fail.

The unfilled matrix polymers are often extremely ductile, and the semi-crystalline materials, such as the polyamides and polypropylene, often exhibit a yield point at strains in excess of 5%, when they will cold-draw to extension ratios of 3 or more. It should be noted, however, that many unfilled thermoplastics are very notch sensitive. This is apparent from an observation of the difference between notched and unnotched impact test data (see Fig. 12). The presence of a notch introduces a region with a complex triaxial state of stress. The cold-drawing process is inhibited and cracks can be initiated and propagated in an essentially brittle manner.

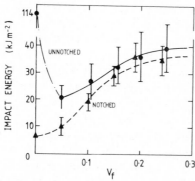

FIG. 12. Charpy impact data for a series of long ($\lesssim \approx 500\,\mu$m) glass fibre reinforced nylon 6.6 mouldings. Note the extreme notch sensitivity of the unfilled material (BADER, unpublished).

In a fibre reinforced plastic the fibres act as internal notches. Regions of locally high internal stress are induced at fibre ends when the composite is stressed and the internal stress state is further complicated by fibre orientation and interaction effects. As the V_f is raised there is less matrix over which to distribute any strain mismatch and interactions between fibres become more probable. The fibre–matrix interface is usually the weakest link in the system. In aligned fibres the ends become debonded at quite low loads and further debonding then propagates along the fibre interface as the loads are increased. This debonding reduces the stiffening efficiency of the fibre, and also constitutes a micro-crack which may propagate into the matrix. Such cracks have been observed to initiate at the ends of well aligned, and along the length of misaligned fibres (BADER and CURTIS, 1977). The composite fails when the local crack density exceeds some critical value and a crack propagates across the section. Highly reinforced materials are observed to fail at strains well below the failure strain of the fibre and, indeed, observations of failed fibres are comparatively infrequent.

Even in materials where the matrix remains quite ductile, plastic or viscous flow is usually restricted to highly localized strain-bands. Flow is intense within the bands but the overall elongation to failure remains very low.

We may therefore conclude that the strength of FRTP materials is limited by the intervention of brittle or semi-brittle failure processes which are initiated by the presence of the fibres.

This reduced ductility also has a strong influence on the low impact strengths associated with FRTP. The filled material is invariably less tough than the unfilled polymer. There is a minimum impact toughness at low V_f ($\sim$0.05) and then an increase as V_f is further increased. Longer fibres appear to confer enhanced toughness. It should be noted that pendulum and drop weight impact tests measure the total energy stored in the specimen at failure, rather than the energy to propagate the fracture and cannot therefore be considered meaningful in terms of elastic fracture mechanics.

The mechanical properties of a selected range of FRTP are summarized in Table 2 which also contains some physical property data which are discussed in the following section. Below we give some remarks on Table 2:

(1) The values quoted have been taken from literature supplied by many manufacturers. Where different manufacturers quote different values for similar materials an average or typical value has been quoted.

(2) Fibre content: Most suppliers quote weight %. In column 2 volume % is also given. This has been calculated from the quoted SG of the unfilled polymer with the SG for E-glass being taken as 2.54 and of carbon fibre (type A) as 1.8.

(3) Melt temperature and mould temperature: Columns 4 and 5 are typical values taken from suppliers recommendations.

(4) Mould shrinkage: This is the shrinkage for a 6 mm thick section in direction of flow. ASTM method D 955 where available; otherwise suppliers data.

(5) Coefficient of thermal expansion: Column 7 ASTM Method D696 in flow direction.

(6) HDT—Heat distortion temperature: Column 8. ASTM Method D648 at 264 psi (1.81 MPa).

(7) Water absorption: Column 9. This is the equilibrium value for immersion at 23°C—ASTM—D570. Values from different suppliers varied widely!

(8) Notched Izod impact strength: Column 13. Most values quoted were of tests of ASTM—D256 on notched $\frac{1}{4}''$ (6 mm) bar. To convert J/m, i.e., Nm/m notch width to ft lb/in multiply values by 0.02 (approx.).

(9) Polymer abbreviations are the following:

Acetal (Homopolymer)—e.g., Delrin® (Dupont),

Acetal (Co-polymer)—e.g., Kematal® (Amcel), Hostaform® (Hoechst),

SAN—Styrene acrylonitrile co-polymer,

ABS—Acrylonitrile–butadiene–styrene, graft co-polymer, e.g., Cyclolac® (Borg Warner), Novodur® (Bayer),

PPO (modified)—Polyphenylene oxide, e.g., Noryl® (General Electric),

PETP—Thermoplastic polyester–polyethylene–terephthalate based, e.g., Hostadur K® (Hoechst),

PBTP—Thermoplastic polyester–polybutylene–terephthalate, e.g., Hostadur B® (Hoechst), Rilsan® (Dupont),

PPS—Polyphenylene sulphide, e.g., Ryton® (Phillips).

A complete discussion of the creep behaviour of FRTP is outside the scope of this chapter, but it should be noted that fibre reinforcement raises the heat distortion temperature (HDT) markedly and there is a comparable reduction in the creep rate at temperatures up to the HDT.

2.5. Physical properties

Fibre reinforcement reduces the in-mould shrinkage very considerably. This facilitates the production of high precision components, and together with the reduction of the coefficient of thermal expansion and the raising of the HDT, combine to confer a much enhanced dimensional stability to FRTP materials. This alone is often a sufficient justification for their selection for moulding small machine parts. The thermal expansion properties are, however, affected by fibre orientation and α is higher in directions normal to any preferred fibre alignment.

The addition of a significant proportion of mineral or fibrous filler generally reduces the specific heat of the compound and reduces the heat input required to melt the material. Glass fibre fillers increase the thermal conductivity by a moderate amount but the more conductive carbon fibre has a much greater effect. Carbon fibre is also electrically conductive and compounds containing more than about $0.05 V_f$ have significant conductance. This can be a useful property where static electricity problems are encountered but, of course, renders these materials unsuitable for many electrical insulating applications where reinforced plastics are widely used.

TABLE 2. Properties of fibre reinforced thermoplastics.

1 Polymer	2 Glass fibre content (W%)	(V%)	3 Specific gravity	4 Typical melt temperature (°C)	5 Typical mould temperature (°C)	6 Typical mould shrinkage (%)	7 Coefficient of thermal expansion (10^{-5}/°C)	8 HDT (°C)
1. Polyethylene (HD)	20	9	1.10	230	40	0.45	5.4	125
2. Polyethylene (HD)	40	20	1.28	230	40	0.35	4.3	125
3. Polypropylene	20	8	1.04	245	40	0.45	4.3	140
4. Polypropylene (chemically coupled)	20	8	1.04	245	40	0.45	4.5	150
5. Polypropylene (chemically coupled)	40	19	1.22	245	40	0.35	2.7	155
6. Nylon 6	40	23	1.46	280	95	0.40	2.2	215
7. Nylon 6.6	20	10	1.28	295	105	0.60	4.1	252
8. Nylon 6.6	40	23	1.46	295	105	0.50	2.5	260
9. Nylon 6.10	40	22	1.41	280	95	0.40	2.2	215
10. Nylon 11	30	15	1.26	250	90	0.50	5.4	173
11. Acetal homopolymer	20	12	1.56	200	95	0.5	3.6	157

12. Acetal co-polymer	30	19	1.63	210	95	0.5	4.3	162
13. Acetal (chemically coupled)	30	19	1.63	210	95	0.3	4.0	165
14. Polystyrene	40	22	1.38	245	65	0.10	3.4	104
15. SAN	40	22	1.40	260	90	0.10	2.7	104
16. ABS	40	22	1.38	260	90	0.15	2.2	107
17. Modified PPO	40	22	1.38	325	105	0.20	1.8	157
18. PETP	30	18	1.61	302	95	0.40	2.7	225
19. PETP	40	26	1.62	240	95	0.35	1.9	232
20. Polysulphone	40	26	1.55	360	150	0.20	2.2	174
21. Polyethersulphone	40	26	1.68	360	150	0.15	2.9	215
22. PPS	40	26	1.65	320	120	0.25	2.7	263
23. Polycarbonate	20	10	1.34	315	120	0.25	2.7	150
24. Polycarbonate	30	17	1.43	315	120	0.20	2.3	150
25. Polycarbonate	40	24	1.52	315	120	0.20	1.8	150
Carbon fibre filled materials								
26. Nylon 6.6	30	21	1.28	295	105	0.25	1.9	257
27. PETP	30	24	1.47	240	95	0.30	0.9	220
28. Polysulphone	30	24	1.37	360	150	0.15	1.08	185
29. PPS	30	24	1.45	320	120	0.15	1.08	263

TABLE 2 (*continued*).

1B Polymer	2B Glass fibre content (W%)	(V%)	9 Water absorption (max) (%)	10 Flexural modulus (GPa)	11 Tensile strength (MPa)	12 Tensile elongation (%)	13 Notched Izod impact[a] (J/m)
1. Polyethylene (HD)	20	9	0.1	4.0	55	2.5	50
2. Polyethylene (HD)	40	20	0.3	7.5	80	2.5	70
3. Polypropylene	20	8	0.02	4.0	63	2.5	75
4. Polypropylene (chemically coupled)	20	8	0.02	4.0	79	4	90
5. Polypropylene (chemically coupled)	40	19	0.09	7.0	103	4	100
6. Nylon 6	40	23	4.6	10.5	180	2.5	150
7. Nylon 6.6	20	10	5.6	9.0	130	3.5	100
8. Nylon 6.6	40	23	3.0	15	210	2.5	136
9. Nylon 6.10	40	22	1.8	9.0	210	2.5	170
10. Nylon 11	30	15	0.4	3.2	95	5	–
11. Acetal homopolymer	20	12	1.0	4.3	60	7	40
12. Acetal co-polymer	30	19	1.8	9.0	90	2	40
13. Acetal (chemically coupled)	30	19	0.9	9.7	135	4	95

14. Polystyrene	40	22	0.1	11.3	103	2.5	60
15. SAN	40	22	0.28	13.4	128	2.5	60
16. ABS	40	22	0.5	7.6	110	3.5	70
17. Modified PPO	40	22	0.09	8.6	135	3.5	80
18. PETP	30	18	0.24	8.3	130	4	85
19. PBTP	40	26	0.4	9.6	150	4	155
20. Polysulphone	40	26	0.6	11.0	138	2	100
21. Polyethersulphone	40	26	–	11.0	205	–	80
22. PPS	40	26	0.06	12.5	160	3	80
23. Polycarbonate	20	10	0.19	5.8	110	6	180
24. Polycarbonate	30	17	0.18	8.2	127	5	190
25. Polycarbonate	40	24	0.16	10.3	145	4	200

Carbon fibre filled materials

26. Nylon 6.6	30	21	2.4	20.0	240	3.5	75
27. PETP	30	24	0.3	13.8	138	2.5	60
28. Polysulphone	30	24	0.4	14.0	158	2.5	60
29. PPS	30	24	0.1	16.9	186	2.5	55

[a] See remark (8) on page 199.

3. Compounding

3.1. Moulding compounds

Most FRTP fabrication is carried out by injection moulding in machines which vary in capacity from a few grams up to perhaps 10 kg per moulding. Screw preplasticizing machinery is now predominant although there has been a resurgence of interest in plunger type or hybrid machines for specialized operations. The general requirement is for a feed stock which is free-flowing homogenous and of high bulk density. This has led to the practice of using a separate compounding operation to combine the polymer and fibre and convert the compound into cylindrical or more nearly spherical moulding granules. The most widely used materials are presented as granules of ~3 mm diameter in which the fibre is fully dispersed and wetted by the polymer, so that there is no requirement for any mixing, or homogenisation, during the moulding operation. At the other end of the scale fine polymer granules, or powder, may be dry blended with chopped glass strand and charged directly into the injection moulder. This process avoids the need for separate compounding but requires that all the necessary blending, wetting, dispersion, and homogenisation is accomplished during the moulding cycle. The general experience is that the potential economies offered by this route are more than offset by the greater difficulty in controlling the moulding operation.

3.2. Extrusion compounding

This is the most widely used of the compounding techniques. In its basic form the equipment is a straightforward single screw extruder, with a screw of about 20:1 length to diameter ratio, and fitted with a 'spaghetti' die (i.e., a die with several apertures of 2–4 mm diameter).

The charge consists of granules of the polymer to be filled and chopped glass strand, usually of 4–10 mm length. The granules and chopped strand are usually dry blended in a tumble blender and the mixture charged into the hopper of the extruder. Alternatively polymer granules and chopped strand may be charged directly into the throat of an extruder, using suitable metering devices which ensure that the screw is 'starve fed' so that there is no build up of charge in the hopper.

The dry mix is then taken up by the screw and conveyed down the barrel where it is melted, compressed into the metering zone and finally extruded through the die apertures. According to the type of polymer the extrudate may either be cut up by a hot face cutter positioned against the face of the die, or the strands may be cooled by air blast or water bath and then chopped cold into granules of a suitable length. The granules formed by the hot face cutter tend to a more spherical shape and flow more readily than the cylindrical pellets formed by cold chopping. The choice of method, however, is determined by the melt rheology of the compound. Materials with very sharp melting points and low melt viscosity are generally chopped cold.

The various processes occurring in the screw extruder have been studied extensively (see, for instance, FENNER (1979) and DONOVAN (1971)). During its passage through the extruder the charge is first compressed into a solid plug which is conveyed into the hot zone of the barrel. The charge is heated by contact with the hot barrel wall and eventually a thin film of melt is established in contact with the barrel wall. The motions of the screw and the mainly solid charge induce the liquid film to build up against the leading edge of the screw flight and eventually to establish a circulatory motion which brings the un-melted material into the melting zone adjacent to the barrel wall. The est-ablishment of this circulatory flow is essential for efficient melting since the very low thermal conductivity of the polymer would otherwise inhibit the flow of heat to the material remote from the barrel wall. This motion, however, results in very high shear rates in the thin liquid film, adjacent to the barrel wall, in the melting zone, and this is sufficient to break the glass fibres as the strands are forced into the zone. Typically the 4–10 mm long strands are broken up to give separated filaments only a fraction of a millimetre long. It has recently been demonstrated (LUNT, 1980) that most of the fibre break-up occurs over just a few screw flights in the melting zone of the extruder.

Relatively little fibre breakage occurs in the metering zone or at the die. This melting operation is, thus, the critical factor which determines the maximum fibre aspect ratio in the moulding compound. (We may also observe that the compound will normally be subjected to a further excursion through a screw melt zone during injection moulding, so that, even if longer fibres were preserved through the compounding process, there is the likelihood of their being broken up during the moulding operation.) The degree of fibre break-up is determined by the shear rate in the melting zone and is also influenced by the constitution of the chopped strand. Higher screw rotation speeds increase the maximum shear rates but also lead to more efficient melting with higher throughput and lower operating energy consumption. Likewise the screws are designed for efficient melting, which requires the establishment of a vigorous circulation of material in the extruder channel. Both of these factors tend to increase the amount of fibre break-up. Another requirement of the compound-ing operation is that the strands should be broken up into individual filaments. This is strongly influenced by the type of size used when drawing the glass strands. A thermoplastic film former which melts at temperatures below those required to melt the polymer to be filled should allow the strands to separate more easily and disperse. However, it is found that less fibre breakage occurs if a stronger film former is used. This maintains the integrity of the sized strand for a longer period during the melting process, because bundles of fibres rather than individual filaments are carried into the high-shear melting zone. Too strong a size would, of course, prevent dispersion altogether. The choice of film former and the amount applied must, therefore, be closely matched to the needs of the particular polymer and compounding/moulding operation. A further requirement of the glass is that the chopped strand should have as high as possible a bulk density and that it should flow freely in the dry state. This is best achieved by using a heavy strand (more filaments, larger diameter fibres) bound with a strong film former. Any filamentization during the dry blending and feeding phases of the process leads to poor flow and clogging in the extruder hopper. A final point concerning extrusion compounding is that at high shear rates glass fibre/thermoplastic blends can be extremely abrasive and this results in very rapid wear of both screw and barrel, especially in the melt zone where the highest effective shear rates occur. It is therefore necessary to use specially hardened components when handling FRTP.

A modification of the technique is to use a screw with a decompression zone and a vented barrel. With these machines the polymer only is charged into the hopper and the glass introduced at the vent usually in the form of continuous roving. Fibre attrition is avoided by keeping the fibre out of the melting zone and adding it directly to the melt. The roving is, of course, broken up to some extent as it enters the second compression zone, but longer fibre lengths may be retained than in the conventional process. The continuous roving, however, can be the source of rapid wear to both screw and barrel, and it is more difficult to control the fibre fraction in the compound.

To summarise: The deficiency of the dry-blend charged, single-screw extruder compounding operation, is that when operated efficiently in terms of output the process causes excessive fibre break-up, to aspect ratios well below the optimum for reinforcing efficiency. This effect may to some extent be controlled by careful selection of the size applied to the glass to give a strand integrity matched to the operating conditions.

3.3. Twin screw compounding extruder

The twin screw extruder is in many ways more suitable for compounding than the single screw machine. The meshing screw design gives a very high throughput and the degree of shear can be controlled over a wider range. The equipment is more expensive and can only be justified when production requirements are sufficiently high. For this reason, although extensively used for processing PVC and other commodity thermoplastics, twin screw machinery is not yet widely utilized for FRTP compounding. Modern twin screw extruders, notably those of Werner Pfliederer design (see STADE (1975) and WOOD (1979)), are built with modular screws consisting of a core on which alternative screw sections may be set up. This allows a basic machine to be modified to adapt it for different materials. The melt zone is designed to introduce less severe shear and special mixing sections may be incorporated at the discharge end to effect fibre dispersion without excessive fibre breakage. It is standard practice to use a vented barrel design and to charge the fibre separately into the melt at vent port, either as chopped strand or continuous roving.

These complex and expensive machines are claimed to offer advantages in productivity and control of fibre aspect ratio in comparison with the single screw extruder. It is probable that they will be more widely exploited when the current generation of single screw machines are replaced.

3.4. Long-fibre compounding processes

A number of processes have been developed to produce compounds containing longer fibres. These produce pellets of cylindrical form with the fibres the full length of the pellet, typically ~6 mm (US patent 3608 033; UK patent 1302 048).

The cross-head die compounder (BADER and BOWYER, 1973) is essentially similar to machines used for producing insulated wire. The polymer only, is melted in a single screw extruder and conveyed to a cross-head die through which a continuous roving is passed. The product is a thin rod of polymer surrounding the continuous fibre roving which may be completely or partially infiltrated by the polymer. This rod is cut into pellets which are suitable for charging directly into the injection moulder. The process is quite attractive but suffers from lower productivity than the conventional extrusion compounding. The roving may not be completely penetrated by the polymer, although this can be remedied by careful die design. The process is most suitable for use with semi-crystalline polymers, such as the polyamides, which have very low melt viscosity. The other problem is that very little filamentization occurs during compounding and full dispersion is effected only during the final moulding operation. There are two variations of this process: The first involves drawing a continuous roving through a bath of molten polymer (US patent 2887 501), thence through a sizing die, the impregnated 'rope' is then chilled and chopped into pellets. Again this process is suitable only with polymer melts of low viscosity and the melt must also be thermally stable. In the second process the roving is drawn through solid polymer powder, in a fluidized bed, and then through a tapering heated tube (UK patents GB 1334 702 and GB 1565 195). The powder is picked up by the roving and melts and impregnates the roving as it passes through the hot tube. After cooling it is again cut into pellets of a suitable length. A further process substitutes a solution of the polymer for the melt. This is clearly of very limited applicability and has not, to the author's knowledge, been used commercially.

The three processes described all give the possibility of charging long fibre containing compounds into the injection moulder. As such they provide the opportunity to retain longer fibres in the final moulding provided that, on the one hand, fibre breakage can be prevented, and on the other full dispersion of the filaments achieved during this moulding operation. In practice although superior mechanical properties may be achieved there have been problems of ensuring homogeneity in the moulding, and fibre orientation and warpage appear to be more difficult to control. As a result long fibre compounds account for only a small proportion of the total FRTP market.

3.5. Direct injection

In this process the compounding stage is by-passed altogether and a dry blend of polymer granules (or powder) and chopped glass strand is charged directly to the moulding machine. This route is inherently cheaper but control of fibre dispersion is more difficult. At one time it was relatively popular for the glass/polypropylene system but appears to have been discarded in favour of separate compounding. It could possibly become viable again for large scale production operations of big mouldings when the economics might prove to be more favourable.

4. Injection moulding

4.1. General principles

The basic design of a modern injection moulding machine is depicted in Fig. 13. It consists of a split mould, with mechanised mould locking and opening and a carriage which holds the screw preplasticizing and injection unit. The screw is designed to reciprocate as well as to rotate in the barrel. A shut-off valve is incorporated into the nozzle.

The operating cycle commences with the mould closed, the carriage retracted (so that the nozzle is not in contact with the mould), and the shut-off valve closed. Rotation of the screw takes in the granular charge from the hopper and transports it forward along the barrel where it is melted. The molten material then enters the metering zone and passes towards the front of the barrel cavity where it builds up a pressure against the closed nozzle valve. This pressure of molten polymer forces the screw backwards against a controlled back pressure in the hydraulic injection cylinder. In this way a quantity of plasticized charge material is built up in the forward end of the barrel. When the quantity of plasticized charge is sufficient the screw rotation is stopped. The carriage is

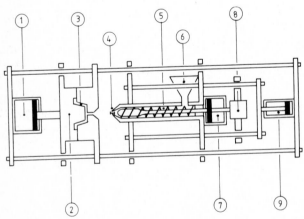

FIG. 13. Injection moulding machine (schematic). (1) Mould locking ram; (2) Split mould; (3) Mould cavity; (4) Nozzle with shut off valve; (5) Barrel and screw; (6) Feed hopper; (7) Injection ram; (8) Screw rotation motor; (9) Carriage positioning ram.

then moved forward so that the nozzle bears against the mould, the nozzle valve is opened and the screw driven forward by the injection ram. This forces the molten charge into the cool mould. The injection pressure is maintained until the charge freezes in the runner system ('dwell' or 'screw-forward' time), when the carriage may be retracted, the nozzle valve closed and the plasticization cycle re-started. The mould remains closed until the charge has solidified sufficiently, it is then opened and the moulding ejected. The mould may then be closed ready to receive the next charge.

The various operations in the moulding cycle are shown in the diagram of Fig. 14. The total cycle time is influenced by three main factors, the plasticization time, the dwell time and the cooling time, and in most practical cases the cooling time is the dominant.

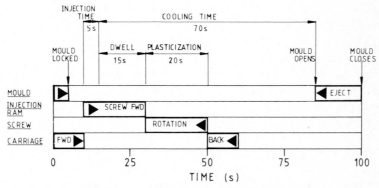

FIG. 14. Typical injection moulding cycle.

The capacity of the machine is rated in terms of the mould locking force, the maximum shot weight, and its plasticizing capacity. The mould locking force must exceed the product of the projected area of the moulding and the injection pressure. It thus determines the maximum dimensions of moulding, within the physical dimensions of the platen size. Since injection pressures are very high, often up to 200 MPa, the mould locking force must be 10–20 kN for each cm^2 of projected area. The largest machines in common usage have mould locking forces of the order of 25 MN, whilst popular machines span the 1–5 MN range. The shot weight (or volume) is determined by the swept volume of the screw stroke (i.e., screw stroke × barrel diameter) and the plasticization capacity is the amount (kg/hour) of charge which may be plasticized by the screw.

The 'dry cycle time' of the machine is the actual time needed for the machine to execute all the functions of the cycle. The cycle time will be the dry cycle time plus the critical portions of plasticization, dwell and cooling times (i.e., allowing for overlaps). Plasticization rate will be critical only if the plasticization time exceeds the cooling time. In general it is considered more satisfactory to operate machines at no more than 75% of their rated shot weight or plasticization capacity. The actual cycle time and production rate are strongly influenced by the characteristics of the material being moulded. Fibre filled thermoplastics generally require high injection pressures and relatively long processing cycles, so that production rates often compare unfavourably with unfilled materials.

4.2. Control of the moulding cycle

The moulding operation is controlled by a number of machine settings. These include the temperatures along the barrel, at the nozzle and the mould. The injection pressure, injection rate, screw back-pressure, screw rotation-speed and the timing of the various cycle operations complete the list of controllable variables in the process.

We may conveniently separate the plasticization and injection operations for the purpose of analysis.

Plasticization is accomplished by the action of the screw in the barrel. This process is virtually identical to that in the compounding extruder, except that, since the plasticization rate is seldom critical, it is unnecessary for the unit to operate at maximum melting efficiency. Likewise most FRTP are moulded from fully homogenized compounds so that the mixing function of the screw plasticizer is not so important. This means that the screw design and operating conditions may be arranged for lower shear rates and, thus, less fibre attrition. The basic need is that the requisite amount of plasticized material be delivered through the metering zone. The plasticization rate is increased when the screw rotation speed is increased, the back-pressure decreased or the barrel temperature raised. In FRTP moulding the back pressure setting is the most influential. Higher pressure settings increase the amount of shear deformation on the melt. This causes further fibre breakage but also ensures good homogenization and disperses any fibre bundles. So, in practice, a compromise must be reached which ensures adequate fibre dispersion with minimum breakage. High back pressure and screw rotation speeds also increase the rate of wear to the screw and barrel.

The injection phase of the operation is critical. The filled polymer melt is highly pseudo-plastic so that the apparent viscosity is lower at high flow rates. In many cases it is advantageous to exploit this effect by the combination of a high injection rate and small gate. There are however a number of dangers which include unpredictable flow patterns (e.g., 'jetting' which is discussed later), degradation of the polymer by the adiabatic heating, induced by the high flow rate, and fibre attrition. The injection pressure has an important influence on mould filling and dimensional control. Most thermoplastics are significantly compressible at the high pressures used for injection moulding, so that high pressure can be used to force more charge into the cavity and help to offset the shrinkage which occurs when the molten polymer freezes. This use of high pressure is effective, only if maintained until the polymer in the gate or runner has frozen, thus, preventing any of the charge from flowing out of the mould cavity when the pressure is relaxed. This effectively determines the dwell time.

The general effects of injection pressure and dwell time for an unfilled thermoplastic are shown in Fig. 15. Whilst very high injection pressures might be advantageous, in practice a limit is imposed by the mould closing force available, the injection ram capacity and the mechanical robustness of the machine.

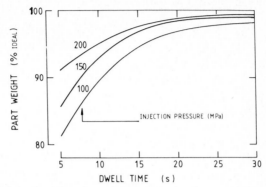

FIG. 15. Effects of injection pressure and Dwell time on the extent of mould filling as reflected in the part weight. In the example chosen an injection pressure of 150 MPa and dwell time of 20 s would give optimum performance.

The cooling phase of the moulding operation is influenced by the temperature and heat capacity of the charge, the dimensions of the cavity and the temperature of the mould. Clearly thin sections and a colder mould would result in fast cooling. But it is usually necessary to maintain the mould temperature at above ambient to avoid various surface defects which occur if the charge is chilled too drastically. Typically polypropylene will be moulded into moulds at 30–50°C and polyamides at 70–100°C. This greatly increases the cooling time, which is the most critical parameter influencing the overall cycle time and hence production rate.

In comparison with unfilled polymers, FRTP are generally more viscous but more strongly pseudo-plastic and require less heat for melting, (owing to the inert filler), likewise there is less heat to abstract during the cooling phase. Flow abnormalities are more significant and for this reason very high shear rates are usually avoided. FRTP exhibit much lower in-mould shrinkage than the unfilled polymers.

We may thus arrive at a very general group of recommendations for moulding:

   (i) avoid overheating;

   (ii) high injection pressure;

   (iii) moderate injection rate;

   (iv) avoid very high shear rates—low back pressure, low screw rotation speed, generous gate dimensions;

   (v) use adequate dwell time.

In the control of the injection moulding operation, conditions for an acceptable moulding might well be achieved by a number of alternative combinations of machine settings. Each combination might, however, result in slightly different moulded dimensions and/or dimensional stability of the moulding. The prime requirement for a moulding campaign is to achieve a high level of consistency in the product. So that once the optimum settings have been established they should not be changed. In the previous generation of injection moulding machines each setting was effected manually by means of adjustments to hydraulic valves, timers, thermostats, etc. This is satisfactory so long as the operation continues at a steady state and no variables due to changes in the charge or ambient conditions intrude. More recently control packages based on micro-computer units have been introduced which implement a system of fully interactive controls. These operate through temperature, pressure and flow sensors set in the mould, runners and barrel of the machine. In this way the machine settings are automatically adjusted to compensate for any change in the charge or in the thermal balance of the machine. Machines fitted with these interactive controls have been shown to produce a more consistent product (MENGES and LUTTERBECK, 1979) and may be operated for long periods without need of manual adjustment. They are particularly suitable for moulding FRTP owing to the greater sensitivity of that material to flow abnormalities.

4.3. Mould cavity design

In designing a component for manufacture in FRTP very careful attention must be directed to ensure that the flow pattern during moulding gives a fibre orientation distribution which optimises the anisotropy of the material with respect to service loads. We may state a number of basic rules and then discuss their further implications:

(i) Maintain a uniform section thickness throughout the component. This ensures a uniform cooling rate.

(ii) Use a generous gate to avoid unnecessary fibre degradation, jetting and overheating of the charge.

(iii) Gates should be positioned so that the main directions of flow are along the most highly stressed directions in the moulding.

(iv) Where multiple gating and/or cored cavities are unavoidable, the flow pattern must be designed so that weld, or knit, lines are formed only in non-critical regions of the moulding.

Ideally there should be a uniform flow path from the gate to the farthest extremity of the cavity. This ideal is seldom possible, since the principal attraction of the injection moulding process is to be *able* to mould complex shapes. This means that section thickness must vary and that cut-outs and other features will induce bifurcated flow patterns, with the inevitable weld lines where the streams subsequently impinge. Cavity design then becomes a compromise which seeks to minimise these deleterious effects.

Consider the alternative ways of gating a simple rectangular plate (see Fig. 16(a)–(e)). In unfilled thermoplastics the edge pin gate would generally be

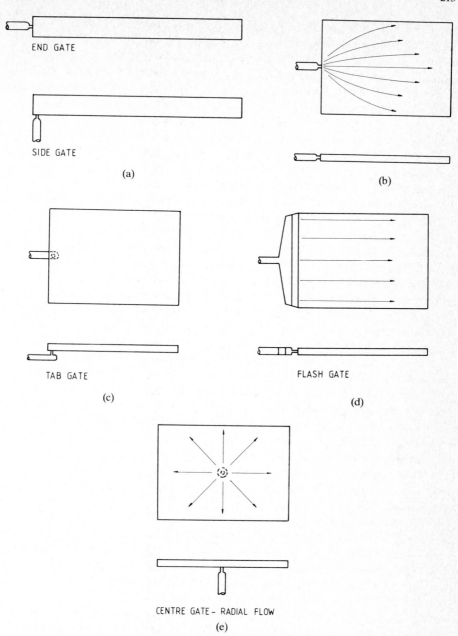

Fig. 16. Illustrations of the consequences of alternative gate geometry on mouldings of simple form. (a) End and side gates on slender moulding. Although the end gate induces a more uniform flow down the section, the side gate eliminates any tendency to jetting and is often preferred; (b) Single edge gate to a rectangular plaque. Fan shaped flow with fibres parallel to flow near the surfaces and normal in the centre; (c) Tab or submarine gate: eliminates any tendency to jetting. Flow pattern similar to (b) except in immediate gate region; (d) Flash gate: uniform parallel flow across the plaque; (e) Centre gate: induces a radial flow pattern. Flow path is shorter.

preferred as it is economic to produce and allows the pseudoplastic properties of the melt to be exploited at high injection rates. It also leaves a minimal witness mark on the moulding when the sprue is broken away. At high injection rates there is the possibility of jetting occurring with the pin gate. This is caused by the fact that at high shear rates the polymer is heated adiabatically with consequent reduction in viscosity. In a narrow gate the highest shear rates are around the periphery of the section and the combination of adiabatic heating and normal pseudoplastic behaviour induces an extreme form of plug flow. When this occurs a narrow jet of material squirts from the gate and impinges on the opposite face of the cavity, which then fills in an irregular manner. The jet tends to fold upon itself and will often form cavities and lines of weakness due to inadequate welding. There is also a possibility that very high temperatures are induced in the high-shear zone with resultant degradation of the polymer. The worst feature of jetting is, however, the totally unpredictable flow pattern. A full flow gate would be preferred for GFRTP so that fibre damage was minimised. This should give a uniform fan shaped flow pattern with the four corners filling last. The fibres would tend to be oriented along the flow direction close to the surfaces of the plate but normal to the flow direction in the centre of the section (DARLINGTON and McGINLEY, 1975; CROWSON and FOULKES, 1980; CROWSON et al., 1980). Ideally as the molten stream flows through the gate it should fill the cavity smoothly from the gate towards the extremities as indicated in Fig. 16. There is a tendency for a line of weakness to develop along the centreline of the flow path. This configuration is also susceptible to jetting if injection rates are too high. This would give a variable and unpredictable flow pattern. In the absence of jetting the plate would be stiffer and stronger along the flow directions and weaker in the transverse direction. Likewise shrinkage is greater in the transverse and less in the flow direction. This may give rise to warping.

The submarine or tab gate has the advantage that any tendency to jetting is eliminated as the jet is immediately quenched and a more regular flow pattern established. The flash gate is more expensive to machine into the mould but when properly designed ensures a uniform flow right across the section. This can help to avoid warping which sometimes occurs as a consequence of differential shrinkage. The radial flow induced by placing the gate in the centre of the plate induces a completely different flow pattern and higher stiffness and strength in the radial direction. Radial flow is often appropriate for components of circular or cylindrical form—e.g., gearwheels.

Where changes of section are essential it is often the best practice to arrange for flow to proceed from thick to thin sections rather than vice versa. This eliminates the possibility of interruption of flow due to the charge freezing prematurely in the thin section.

In designs incorporating long slender passages it may be impossible to adequately fill the cavity from a single gate. Multiple gating is acceptable as long as it is recognised that weld lines are then inevitable. The flow pattern must be designed so that these weld lines are formed in a non-critical part of the component. It must also be recognized that multiple gating inevitably means a more complex runner system with a greater loss of material in the sprue. An example of multiple gate design is shown in Fig. 17(a). Weld lines will also be produced wherever the cavity design induces a bifurcation or

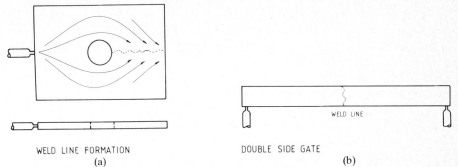

WELD LINE FORMATION   DOUBLE SIDE GATE
(a)   (b)

Fig. 17. Weld line formation. (a) Weld line formed in double gated moulding; (b) Weld line formed downstream of cored out hole.

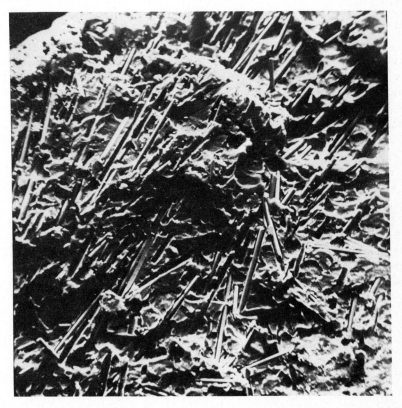

Fig. 18. Fracture surface of weld line failure. The fibre orientation indicates the flow pattern as the two streams impinged (20% glass fibre filled PPO).

branching of the melt stream. This will occur around cored holes, cut outs and inserts (see Fig. 17(b)). It is generally preferable to arrange the gating so that weld lines are formed as close as possible to the gate. This ensures that the impinging melt streams are as hot, fluid and turbulent as possible. This facilitates the formation of a sound weld but, even then, it will still be a source of weakness as no fibres will bridge the surface of contact. Fig. 18 shows the fracture surface formed along a weld line. Note that the fibres all lie in the fracture plane. The weakest weld lines are formed when cool, smooth-flowing streams impinge. This situation must be avoided. Weld line problems may be minimised by raising the melt and/or mould temperatures and by use of higher injection rates (CLOUD and McDOWELL, 1976).

4.4. The influence of the flow pattern on mechanical and physical properties

It will have become apparent, from the previous discussion, that the mechanical and physical properties of GFRTP are strongly influenced by the fibre orientation and, also, that the orientation distribution in injection mouldings is extremely complex. The moulded material tends to be more anisotropic when there is a strongly developed uniaxial orientation. Although it is in principle possible to predict the actual local properties if the orientation is known, in practice the designer needs to work with 'design allowable' values for stiffness, strength, etc. which must not exceed the minimum actual values in critically loaded parts of the moulding. A conservative approach would base these design allowables on the minimum stiffness and strength likely to be encountered in a moulding—i.e., values measured normal to the direction of fibre orientation in a well aligned sample. In this case a fibre reinforced material would give little or no advantage over a particulate filled material, which would, of course, be isotropic. Injection moulded test bars usually show a pronounced uniaxial fibre orientation and test data obtained tends to be higher than that for a typical moulding. Thus, in order to exploit the potential of GFRTP the designer must ensure that the fibre orientation is appropriate to the state of stress envisaged under service conditions. There is still a real problem in obtaining realistic design data. It is acknowledged that the bulk of the available data, obtained from moulded test bars, would give optimistic predictions of performance. Whilst the conservative approach would give design allowables so low as to impair the cost effective exploitation of the material. A new approach (DUNN and TURNER, 1974) is to gather data by testing representative moulded shapes (e.g., discs, plaques, etc.) which incorporate typical orientation variations, weld lines, etc. These tests are hoped to lead to a more realistic data base which will include a practical measure of anisotropy.

In practice there are two main problems associated with GFRTP mouldings which are directly attributable to the flow pattern. These are differential shrinkage, both in the mould and subsequently during storage or service, due to the different fibre orientations within the moulding. This can lead to unacceptable degrees of warping and distortion which cancel out the beneficial effect of the lower coefficient of thermal expansion associated with filled materials. This may usually be controlled by attention to the gating and moulding conditions, expecially mould temperature, but in cases where very high levels of dimensional stability are required it may be necessary to use a particulate rather than a fibre filled material. The other major problem is cracking along weld lines. The formation of weld lines has already been adequately discussed. They are inevitable in complex mouldings but their position can be controlled by careful design and gating so that critically stressed areas are avoided. Weld line problems cannot be avoided by changing to particulate filled or even to unfilled materials, although their relative effect may appear to be less severe, owing to the lower basic properties of those materials.

5. Extrusion

Extrusion of profiled sections accounts for a large proportion of the total output of thermoplastics but the process has not been exploited on a large scale with the fibre reinforced materials. In principle the process lends itself to the production of uniform section, semi finished products, such as tube, pipe, channels and the more complex sections used, for instance, in window frames. Extrusion is a continuous process which is cost effective only when carried out on a large scale. Hence it has tended to be concentrated on the production of articles from the commodity thermoplastics such as polyvinylchloride and its derivatives. Comparatively small proportions of the engineering thermoplastics (e.g., polyamides and polycarbonates) are fabricated by extruding and even less of the fibre filled grades. Many of these engineering thermoplastics are relatively more difficult to extrude than, say, polyvinylchloride but, this is not really the barrier to the use of the process. It is simply that there has been no demand for large quantities of profile in the GFRTP materials. An extruded section might be expected to have a well developed axial fibre orientation and, thus, excellent properties in that direction. There appears to be no technological reason why extruded GFRTP sections should not be produced if an application is identified in which they would be cost effective.

6. Thermoforming

As in the case of extrusion, thermoforming is a major fabrication route for unfilled thermoplastics. It is mainly used for non-structural items fabricated from polystyrene, polyolefins, polyvinylchloride and their derivatives, but some load bearing components are produced from polypropylene and other engineering thermoplastics.

The basic fabrication route involves the production of sheet or foil which is then heated and stamped, vacuum or pressure formed. The high cost of sheet manufacture is offset by die and machine costs which are significantly lower than for injection moulding, so that the two processes are often competitive.

A relatively new development has been the introduction of thermoformable thermoplastics-based fibre composites. There are two distinct areas of activity: The first is the replacement of thermoset resins (e.g., epoxides) in high performance laminates. Here the emphasis has been on utilising thermally stable thermoplastics, such as the polysulphones, silicones and polyamides, in 'pre-preg' materials which are fabricated by press or autoclave moulding. These processes involve minimal flow of the matrix and consist essentially of welding together the layers of pre-preg. They represent a small, highly specialised application of interest only to the high-technology industries. They will not be discussed further in this chapter.

The second group of materials is the group of the hot stampable sheet materials which are being offered as alternatives to the thermoset based sheet moulding compounds (SMC). The materials currently available consist of planar random chopped glass-fibre strands dispersed in either polypropylene or polyamide 6 matrices. The material is supplied as sheet—typically 2–5 mm thick—which is fabricated by melt laminating polymer films with chopped strand mat. In principle any type of mat or cloth could be incorporated, and alternative fabrication processes such as calendering or extrusion used to produce the sheet.

The component is fabricated by first preheating a suitable blank cut from the sheet and then pressing this between cold dies. The hot sheet is squeezed into the die cavity where it is rapidly chilled. The main attraction of the process is that the forming operation is very fast (typically 5–10 s) compared with SMC processing (2–10 minutes), and may use presses similar to those used for metal stamping.

There are two basic variations to the process: solid state and liquid state forming. In the former the preheat temperature is below the melting point of the matrix polymer (but above T_g). This limits the extent of flow possible when stamping and is suitable only for production of shapes of nearly uniform wall thickness (i.e., comparable with metal pressing). In the liquid state process the blank is heated to above the melting point of the polymer, and some sort of support film is usually necessary to transport the heated blank from oven to press. (Typically a polyamide cloth or mesh is used which subsequently becomes incorporated into the part).

During pressing a significant amount of flow may occur and variable thickness mouldings with stiffening ribs, bosses, etc. may be formed. The amount of glass incorporated into the sheet is usually of the order 40–50% by weight (polyamide) which is rather less than in the higher glass content SMC materials and thus, together with the fact that the matrix is less stiff, results in the material having rather lower stiffness. However, the thermoformed composite is generally tougher and has a better surface finish than SMC. These materials have not, as yet, been exploited on a large scale, but it is considered that they have considerable development potential.

References

BADER, M.G. and W.H. BOWYER (1972), *J. Phys. D.* **5**, 2215.

BADER, M.G. and W.H. BOWYER (1973), *Composites* **4**, 150.

BADER, M.G., P.T. CURTIS and R.S. CHATWAL (1975), in: E. SCALA et al., eds., *Proc. Internat. Conf. Composite Materials* Vol. 2 (AIME, New York) p. 191.

BADER, M.G. and P.T. CURTIS (1977), Strength and failure modes in short fibre reinforced thermoplastics, in: F.W. CROSSMAN, ed., *Proc. Fall Meeting Chicago 1977 Failure Modes in Composites* (TMS-AIME, New York).

BADER, M.G. and J.F. COLLINS (1980), The effect of fibre-interface and processing variables on the mechanical properties of glass fibre filled nylon 6, *Proc. IVth Nat. and 1st Internat. Meeting on Composite Materials* (Applied Science Publishers, England) to appear.

BESSEL, T., D. HULL and J.B. SHORTALL (1972), Faraday special discussions, *The Chemical Society* **2**, 137.

BOWYER, W.H. and M.G. BADER (1972a), *J. Mat. Sci.* **7**, 1315.

BOWYER, W.H. and M.G. BADER (1972b), Faraday special discussions, *The Chemical Society* **2**, 165.

CHEN, P.E. (1971), *Poly. Engrg. Sci.* **11**, 51.

CLOUD, P.J. and F. MCDOWELL (1976), Reinforced thermoplastics: Understanding weld line integrity, *Plastics Technology*, August.

COX, H.L. (1952), *British J. Appl. Phys.* **3**, 72.

CROWSON, R.J., M.J. FOULKES and P.F. BRIGHT (1980), *Poly. Engrg. Sci.* **20**(14) 925.

CROWSON, R.J. and M.J. FOULKES (1980), *Poly. Engrg. Sci.* **20**(14) 934.

CURTIS, P.T. (1976), Ph.D. Thesis, University of Surrey, U.K.

CURTIS, P.T., M.G. BADER and J.E. BAILEY (1978), *J. Mater. Sci.* **13**, 317.

DARLINGTON, M.W. and P.L. MCGINLEY (1975), *J. Mater. Sci.* **10**, 906.

DONOVAN, R.C. (1971), *Poly. Engrg. Sci.* **11**, 484.

DUNN, C.M.R. and S. TURNER (1974), in: *Composites Standards, Testing and Design* (IPC Press, Guildford, U.K.) p. 113.

FENNER, R.T. (1979), *Principles of Polymer Processing* (Macmillan, London).

HALPIN, J.C. and N.J. PAGANO (1969), *J. Comp. Mat.* **3**, 720.

HALPIN, J.C. and J.L. KARDOS (1976), *Poly. Engrg. Sci.* **16**(5) 344.

HALPIN, J.C. and J.L. KARDOS (1978), *Poly. Engrg. Sci.* **18**(5) 496.

ISHIDA, H. and J.L. KOENIG (1978), *Poly. Engrg. Sci.* **18**(2) 128.

KELLY, A. and W.R. TYSON (1965), *J. Mech. Phys. Solids* **13**(6) 329.

LEES, J.K. (1968), *Poly. Engrg. Sci.* **8**, 195.

LUNT, J. (1980), Ph.D. Thesis, Liverpool Univ.

MENGES, G. and J. LUTTERBECK (1979), *Plast. Rubber Internat.* **4**(2) 59.

PLUEDDEMANN, E.P. (1970), Adhesion through silane coupling agents, *25th Ann. Techn. Conf.* (The Society of the Plastics Industry Inc.).

STADE, K.H. (1975), New achievements in compounding glass fibre reinforced thermoplastics, *Reinforced Thermoplastics Symp.* (Plastics Institute, London).

WOOD, R. (1979), *Plast. and Rubber Internat.* **4**(5) 207.

ZISMAN, W.A. (1969), *Ind. Engrg. Chem. Product R & D* **8**(2) 98.

Chapter V

Fabrication of Metal–Matrix Composites

S.T. Mileiko

Institute of Solid State Physics
Academy of Sciences of the U.S.S.R.
142432 Chernogolovka
Moscow
U.S.S.R.

Contents

HANDBOOK OF COMPOSITES, VOL. 4 – Fabrication of Composites
Edited by A. Kelly and S.T. Mileiko
© 1983, Elsevier Science Publishers B.V.

1. Introduction

Fibrous composites for structural application are in the first stage of development at the present time. Metal–matrix composites are younger than fibre reinforced plastics. Fabrication processes (technology) of any material are based (or at least are to be based) on results of research in mechanics, physics and chemistry. Combination of such results with the 'black art' leads to a modern technology.

Methods of making and processing of metal materials have been developed for a long time by industrial experience, which has often been subsequently supported by science. Composites arose at a time when technology had already found its scientific base. But composite technology still has no firm scientific foundation which permits strict enough recommendations about choosing the best way of production of a particular composite and the best process parameters.

So any attempt to produce for this book a chapter on metal–matrix composite technology which would have too instructive a nature would not lead to success (at least to full success). Therefore the author sees his aim, firstly, in describing a general and very schematic picture of the state-of-the-art and secondly, to illustrate the present situation with some examples. We do not intend to give a description in detail of existing technological processes, just to make the necessary references. It seems important to understand how the technology predetermines the mechanical behaviour of a composite.

Bearing in mind a possible structure of the chapter let us look at the structure of the field. Of course, Fig. 1 includes only subfields and links between them which seem to be essential. Obviously in this chapter we cannot consider these features. Also note that this scheme reflects the personal scientific interests of the author—the subfield mechanics has its own substructure.

Mechanics of composites is divided into (i) micromechanics, (M1), (ii) macromechanics (M2), and (iii) mechanics of composite structure (M3). Micromechanics is to determine effective mechanical properties of a composite material in terms of structural parameters. Macromechanics is to evaluate constitutive equations of phenomenological nature. These equations are to be used in calculating strength, stiffness, and stability of structural elements made of composite materials (this is connection $M2 \rightarrow M3$). Connection $M1 \rightarrow M2$ reflects an influence of micromechanical models on phenomenological equations as well as on the field of the applicability of the latter.

Connection $Ph \rightarrow T1$ reflects the usage of various physical phenomena in composite technology, examples being diffusion, wetting, sintering, plasma

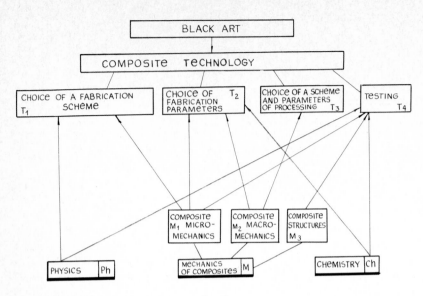

FIG. 1.

spraying, ion plating and so on. Connection Ph → T4 means the usage of some physical phenomena in non-destructive testing of composite structures after fabrication and during service. The examples are X-ray microscopy, acoustic emission and so on. Note that without an effective use of connection M1 → T4 results of such tests will be difficult to understand quantitatively.

Connections M1 → T1 and M1 → T2 determine choosing such a fabrication method and such parameters of a process which lead to a composite with prescribed effective mechanical properties. To understand connection M2 → T2 means sometimes to provide means to optimize a fabrication process; connection M2 → T3 is necessary for choosing proper methods of forming and processing of composite structural components as well as their bonding and fastening.

Connection Ch → T2 helps us to trace limit values of fabrication parameters because of chemical interaction of components. Note that connection M1 → T2 is necessary to estimate an influence of this interaction on macroscopical mechanical properties of a composite. Connection Ch → T4 can supply additional data on how the structure of the composite corresponds to the prescribed one. For example, a structure of an interface layer can be considered here.

Finally, connection M3 → T4 is to be established to evaluate methods of final testing of structural components.

The present chapter is an attempt to give a summary of results in mechanics, physics and chemistry necessary to build up a base for metal–matrix composites technology.

2. Mechanics

Technological problems are traditionally within the sphere of mechanical engineering. In solid state mechanics problems connected to metal cutting, rolling, extrusion, and so on are well known. Such problems are still important from the point of view of composite processing. But they obviously should be revised. For example, problems related to forming by bending or to densifying of a porous semiproduct perhaps demand constitutive equations of a body to be dependent on the state of the body.

But we will concentrate on consequences of micromechanics of composites which are important from the point of view of producing composite materials with desirable mechanical properties. Such a problem looks like a very new one for classical solid state mechanics, but if one considers a composite material as a macrostructure (relative to atomic structure, for example), then it becomes more familiar.

2.1. Micromechanics of metal–matrix composites failure

We intend to give only an expanded summary because some review papers on this subject have been published or are to be published (MILEIKO, 1982; MILEIKO and RABOTNOV, 1980; RABOTNOV and MILEIKO, 1982).

2.1.1. Tensile strength

Two characteristic cases are to be considered, the first being of metal type reinforcement and the second being of brittle type fibre.

2.1.1.1. Metal–matrix/metal fibre composites. This is a case where a maximum on the stress–strain curve of a composite is determined by simultaneous necking of the fibre–matrix system. When the fibre–matrix interface is ideal, then calculations (MILEIKO, 1969) give the dependences of ultimate strength σ_c^+ and ultimate strain ε_{uc} on fibre volume fraction as

$$\sigma_c^+ = \lambda_f \sigma_f^+ V_f + \lambda_m \sigma_m^+ V_m , \tag{2.1}$$

$$V_f = \left[1 + \beta \, \frac{\varepsilon_{uc} - \varepsilon_{uf}}{\varepsilon_{um} - \varepsilon_{uc}} \, \varepsilon_{uc}^{(\varepsilon_{uf} - \varepsilon_{um})} \right]^{-1} . \tag{2.2}$$

Here

$$\beta = \frac{\sigma_f^+}{\sigma_m^+} \, \frac{\varepsilon_{um}^{\varepsilon_{um}}}{\varepsilon_{uf}^{\varepsilon_{uf}}} \, \frac{\exp\{\varepsilon_{uf}\}}{\exp\{\varepsilon_{um}\}} ,$$

$$\lambda_\alpha = (\varepsilon_{uc}/\varepsilon_u)^{\varepsilon_u} \exp\{\varepsilon_u - \varepsilon_{uc}\}$$

where α is either f or m.

A theory for the case of a non-ideal interface has not been evaluated but experiments conducted by OCHIAI and MURAKAMI (1980) show that (2.1) and (2.2) describe the real behaviour of a composite with a weak interface well enough.

In metal–matrix composite an interaction between components does usually occur (see Section 3.4) which leads to interface layers with their own properties. Here it is important to understand the influence of such layers on the behaviour of composites.

Firstly, it should be noted that the plastic properties of the matrix in the vicinity of the more rigid fibre can be different from those of the bulk matrix. This was discovered by KELLY and LILHOLT (1969) who tested tungsten–copper composites. In their experiments the effective tangent modulus $d\sigma/d\varepsilon$ of the matrix at plastic yielding region ($0.05 \leqslant \varepsilon \leqslant 0.4$ per cent) appeared to be one or two orders of magnitude larger than the value of $d\sigma/d\varepsilon$ for the matrix material. The effect is certainly determined by a dislocation pile-up at the fibre–matrix interface (GARMONG and SHEPARD, 1971) and so it should increase with decreasing average distance between the fibres (NEUMAN and HAASEN, 1971).

Secondly, an intermetallic compound formed at the interface can give rise to an increase in the strength of a composite to some extent, if this third component is strong enough (FRIEDRICH et al., 1974).

Thirdly, cracking of a brittle interfacial layer can give rise to an apparent increase in the ultimate stress of a fibre similar to the increase of the limiting stress of a rigid-plastic specimen with a notch (OCHIAI and MURAKAMI, 1976).

2.1.1.2. Composites with brittle fibres. The fracture model described by MILEIKO et al. (1973) and MILEIKO (1979) leads to the following results. If a failure mechanism includes fibres breaking down to the length of critical value, l_c, then the mean ultimate stress of a composite will be

$$\langle \sigma_1^+ \rangle = \alpha \langle \sigma_f^+(l_c) \rangle V_f + \sigma_m^+ V_m \qquad (2.3)$$

where α is a constant factor, $\alpha \approx \frac{1}{2}$. If a composite fails by a weak link mechanism, then its strength should be

$$\langle \sigma_2^+ \rangle = \langle \sigma_f(L) \rangle V_f + \sigma_m' V_m \qquad (2.4)$$

where L is the total length of the fibres within the whole volume of a composite bar of a constant cross-section under tension, σ_m' is the matrix stress at the moment of the first fibre breaks.

Now let us suppose that curves $\langle \sigma_1^+ \rangle$–$V_f$ and $\langle \sigma_2^+ \rangle$–$V_f$ can be drawn as shown in Fig. 2. Now we are to introduce inherent microcracks of length equal to nd_f, where n is the value to characterize the homogeneity of the fibre packing, for the ideal packing $n = 1$. If one fibre breaks, then a microcrack of the length equal to nd_f arises, and on the $\langle \sigma_c^+ \rangle$–$V_f$ plane (Fig. 2) we can draw curve GO_n corresponding to a limit strength in the Griffith–Orowan

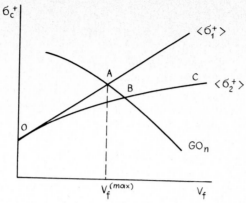

FIG. 2. Dependence of mean strength of a brittle fibre composite on the fibre volume fraction.

sense of an anisotropic material containing a crack of the length nd_f and characterized by effective surface energy equal to $G(V_f)$. The equation of this curve is

$$\sigma = (GC/\lambda nd_f)^{1/2} \tag{2.5}$$

where C is the known combination of the elastic moduli (in the isotropic case $C = E$), λ is a constant.

We have

$$G = G_m^0(1 - V_f) \quad \text{at } V_f < V_f', \tag{2.6}$$

$$G = \frac{V_m^2}{V_f} G_m^0 \frac{d_f}{h_m} \quad \text{at } V_f \geqslant V_f'. \tag{2.7}$$

If to express G_m^0 as

$$G_m^0 = \sigma_m^+ \varepsilon_{um} h_m \tag{2.8}$$

where h_m is a characteristic size, then

$$V_f' = \frac{1}{1 + h_m/d_f}. \tag{2.9}$$

The dependence of the mean strength of a composite on fibre volume fraction is now given by curve OABC (Fig. 2) which is generally non-monotonic. Along part OA failure with the localization of fibre breaks takes place, the strength scatter of the composite is small, the stress–strain curve exhibits large plastic yielding. Along part BC a first fibre break leads to the fracture of the composite, the strength scatter is much larger, the stress–strain curve does not exhibit a plastic yielding part. Along part AB there exists a restricted localization of fibre breaks, macroscopic features of failure combine those of both former mechanisms.

The assumption of ideal bonding at the interface is essential for the model. It permits the contributions to the value of G of energy dissipation due to interface rupture, sliding along the interface, and so on to be neglected. For a real interface, strength decreases might certainly result from these effects. So a possibility of a non-monotonic dependence of the composite strength on the interface strength emerges (OVCHINSKY et al., 1980). When the interface strength goes down, the critical fibre length goes up (and so the effective fibre strength goes down), the contribution of plastic dissipation in the matrix to the effective surface energy goes down, but the contribution of energy dissipation at the interface goes up. But at the present time a tractable model characterized by a small enough number of parameters (that can be determined experimentally) remains to be developed.

Let us now look at conclusions from the model described above which are important from the point of view of requirements to the structure of a composite as well as to the fabrication technology, with a final aim to obtain a composite of high tensile strength. Obviously it is useful to make point A (Fig. 2) to move to the right along the V_f axis. How can this be done?

Firstly, it is clear that improving the homogeneity of fibre packing, i.e., decreasing the value of n in (2.5) makes point A to move to the right, as shown in Fig. 3. There is plenty of experimental data of this type. We show just two sets of evidence to support this idea. The results of testing of two sets of boron–aluminium specimens are shown in Fig. 4, the first being of a normal regular fibre packing and the second being of non-homogeneous fibre packing (see Fig. 5). The strength–fibre volume fraction curve starts to go down (part AB on Fig. 2) at a much lower volume fraction in the case of non-homogeneous packing.

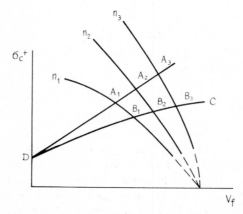

FIG. 3. Influence of fibre packing homogeneity on the strength of brittle fibre composites. Curves DA_1B_1C, DA_2B_2C, DA_3B_3C correspond to values $n_1 > n_2 > n_3$.

Further evidence is supplied by the results of testing silicon carbide–silver matrix composites (TAKAHASHI, 1978) (see Fig. 6). Fibres in these composites were oriented three-dimensionally and randomly packed. So we are to assume such composites to be a superposition of unidirectional composites. If we then compare values of the strength of the composites fabricated on Earth and in Space (where low gravitation provides conditions for uniform distribution of the whiskers and better quality of the matrix), we will see that the difference in the behaviour of these two sets of samples corresponds qualitatively to the prediction of the model.

Note that the well-known maximum on the $\langle\sigma_c^+\rangle - V_f$ curve and a large scatter of the strength of composites at $V_f > V_f^{(max)}$, which have been observed

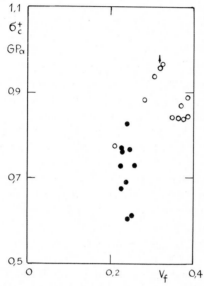

FIG. 4. Strength of the boron–aluminium composite with a D16-matrix. Dark points correspond to specimens with non-homogeneous fibre packing (examples are given in Fig. 5). Experiments by SOROKIN.

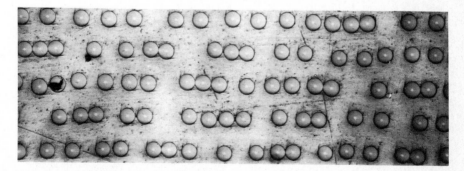

FIG. 5. Examples of non-homogeneous fibre packing in boron–aluminum composites (see Fig. 4).

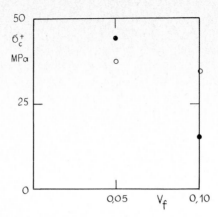

FIG. 6. The strength of the SiC–Ag composites prepared on Earth (●) and in Space (○). The experimental data are from TAKAHASHI (1978).

many times in testing various composites (see Table 1), and especially carbon fibre composites (see Fig. 7), can be also explained sometimes by a poor fibre distribution as a result of using liquid matrix infiltration methods.

So when choosing a fabrication method, we should prefer methods allowing more homogeneous and isotropic fibre packing. Liquid infiltration methods are worse in this respect than solid state methods (unless great care is taken to ensure uniform fibre distribution—e.g., low gravity techniques). But when using solid state methods it is also necessary to take care to produce a homogeneous fibre distribution. At the same time we should remember that this way of improving composites and increasing their strength values has an obvious limit, because $n \geqslant 1$.

Secondly, it is possible to shift point A to the right by using a tougher matrix (with a larger value of G_m^0) as shown in Fig. 8. In this respect solid state fabrication methods look preferable because of the generally low fracture toughness of cast alloys. Again results similar to those shown in Fig. 7 and in Table 1 can partly be explained by low values of fracture toughness of matrix alloys.

TABLE 1. Non-monotonic dependence of the mean strength of brittle fibre composites on fibre volume fraction.

Fibre	Matrix	$V_f^{(max)}$ (Fig. 2)	Sources
B	Aℓ	0.2–0.35	MILEIKO et al. (1973)
B	Mg	0.28	AHMAD and BARRANKO (1973)
C	Aℓ	0.3	JACKSON et al. (1972)
C	Aℓ	0.2	MIMURA et al. (1974)
C	Cu	0.2	KUZMIN et al. (1975)
SiC	Aℓ	0.2	YAJIMA et al. (1981)
Aℓ₂O₃	Mo	0.3	MILEIKO and KAZMIN (1979)

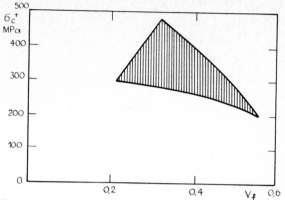

FIG. 7. The area of the strength scatter of the graphite–aluminium composite prepared and tested by JACKSON et al. (1972).

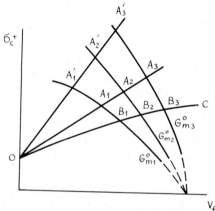

FIG. 8. Influence of the matrix fracture toughness on the strength of brittle fibre composites. Curves OA_1B_1C, OA_2B_2C, OA_3B_3C correspond to one type of fibre and three matrix materials characterized by values of effective surface energy $G_{m1}^0 < G_{m2}^0 < G_{m3}^0$. Parts OA_1', OA_2' and OA_3' correspond to a stronger fibre and the same matrix materials.

If we are to use higher strength fibres in a composite, we have to look for higher values of matrix fracture toughness. Experimental evidence of this can be found in results of testing model composites of the $A\ell_2O_3$–Mo system (MILEIKO and KAZMIN, 1979). Note that the strength of fibres determines the slope of line OA in Fig. 2.

The most effective way to increase the fracture toughness value of a matrix is to use a composite matrix of a metal–metal type (see Section 2.1.2.1). This leads (MILEIKO et al., 1980) to composites with point A located at very high volume fraction of brittle fibres. An example of such composites is boron–steel–aluminium composite. The dependence of the strength of this composite on volume fraction of the brittle fibres is shown in Fig. 9.

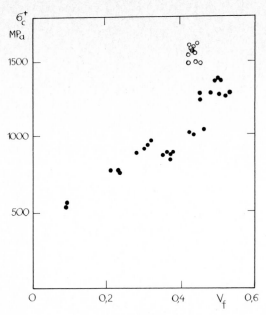

FIG. 9. The strength of boron–aluminium composites (●) and boron–steel–aluminium composites (○). The structure of the latter mainly contains boron fibres with additional steel fibres in the same direction (according to MILEIKO et al., 1980).

Thirdly, a final adjustment of fabrication parameters leading to the highest strength of a composite should be made experimentally. Obviously the fibre–matrix interface strength affects the strength of a composite but we have no quantitative model to plan an experiment at present. This fact is to be taken into account when considering the influence of interface layers on strength. Defects contained in such layers can trigger microcracking and so decrease the strength of a brittle fibre (USTINOV, 1979). But at the same time it seems to be necessary to obtain such layers in some cases to provide bonding to the interface. This point is to be discussed in more detail in Section 3.4.3.

2.1.2. Fracture toughness

The tensile strength of a composite can be very high but it can be used only if the fracture toughness is high enough.

2.1.2.1. Metal–fibre/metal matrix composites.

Let us prescribe to the components of a composite considered above (see Section 2.1.1.1) the values of effective surface energy G_f and G_m. Then we assume that the value of G of an arbitrary homogeneous material depends linearly on the ultimate strain ε_u of the material if we change the value of ε_u and do not change any other characteristic of the material. Then the model considered above leads to the dependence of the effective surface energy of a composite on fibre volume

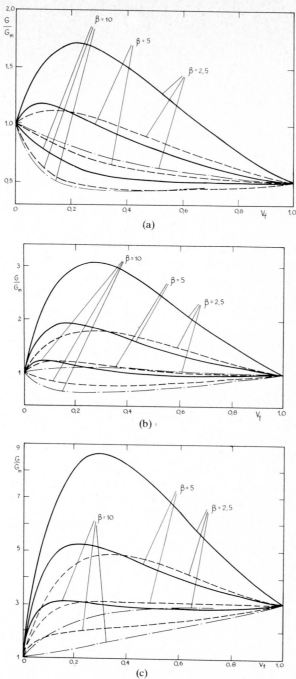

FIG. 10. Dependences of the effective surface energy G of metal–metal composites on fibre volume fraction according to (2.10). $\beta = (\sigma_f^+/\sigma_m^+)\,(\varepsilon_{um}^{\varepsilon_{um}}/\varepsilon_{uf}^{\varepsilon_{uf}})\,(\exp\{\varepsilon_{uf}\}/\exp\{\varepsilon_{um}\})$. $\varepsilon_{uf} = 0.01$ (——), 0.02 (– – – –), 0.05 (– · – · –), $\varepsilon_{um} = 0.25$. (a) $G_f/G_m = 0.5$; (b) $=1$; (c) $=3$.

fraction written as

$$\frac{G}{G_m} = \frac{G_f}{G_m} \frac{\varepsilon_{uc}}{\varepsilon_{uf}} V_f + \frac{\varepsilon_{uc}}{\varepsilon_{um}} V_m \tag{2.10}$$

where the value of ε_{uc} is to be found from (2.2).

Eq. (2.10) was shown to be true by experiments carried out by ARCHANGEL-SKA and MILEIKO (1976). They have shown also that the situation at the interface in a steel–aluminium composite has no essential influence on the value of G. It does not mean of course that this is always true because in these experiments the situation does perhaps accidentally occur where decreasing of the interface strength leads to a decrease of plastic energy dissipation nearly equal to the increase of energy dissipation at the interface.

Some possibilities to control fracture toughness of metal/matrix–metal fibre composites are suggested by the curve $G/G_m(V_f)$ shown in Fig. 10. Obviously such composites can have a very high fracture toughness with respect to cracks normal to a fibre direction. This effect has effectively been used in a composite with brittle fibres and a tough composite matrix (see Section 2.1.1.2). But Fig. 10 shows that such an effect can be obtained only if components with proper characteristics have been chosen.

2.1.2.2. Composites with brittle fibres. An important feature of the behaviour of a macrocrack in a composite with brittle fibres is a formation of a zone of fibre microcracking at the tip of the macrocrack. The evaluation of a size of this zone have been done by MILEIKO and SULEIMANOV (1981). They consider a simple plane model containing two sorts of layers one having regularly spaced defects such that they crack when the local stress reaches a constant value equal to σ^* (see Fig. 11).

To make calculation simple the elastic moduli of the two types of layers are taken to be the same. Then each microcrack is assumed to be influenced by the elastic stress field of the macrocrack and the nearest microcracks only. Then a singular part of the stress at defect (q, m) can be written as

$$\sigma(q, m) = \frac{\sigma_0 \sqrt{l}}{\sqrt{\pi(1+d)}} \, \Phi(q, m, V_f) \tag{2.11}$$

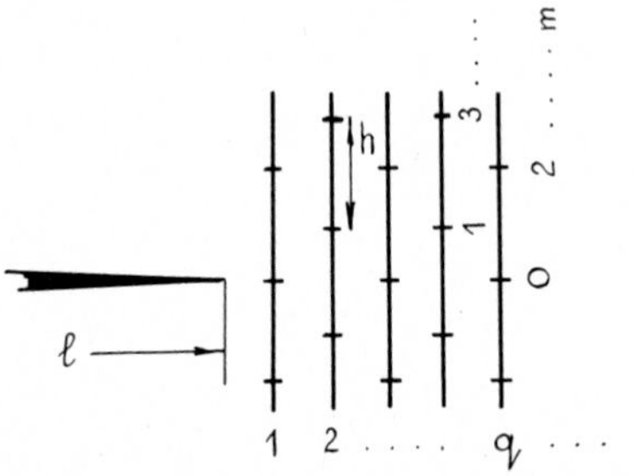

FIG. 11. A macrocrack in a model brittle-fibre composite.

where σ_0 is the stress at infinity, l and d are the dimensionless values of the crack length and the thickness of the brittle layer (the fibre diameter), respectively, and Φ is a known function.

Hence if a configuration of cracked defects around a defect under consideration is known, then the stress at this defect is also known and so we know whether or not it has cracked. Because this configuration is not known a priori we need to analyse all the possible situations. A computer analysis can be done without any difficulty and the result is a size of the microcracking zone.

The size of the zone increases when stress σ_0 increases. It reaches a critical value, say H, after which the microcracking zone starts to move as a whole, and this means the macrocrack starts to propagate (see Fig. 12). To determine the value of H we make four assumptions. Namely, (i) the formation and presence of the cracking zone—which is a zone of lower stiffness—does not change the dependence $K = \lambda \sigma_0 \sqrt{l}$, (ii) Irwin's relationship is true, $K^* = \sqrt{EG}$, (iii) each microcrack dissipates a definite value of energy, for example as a result of matrix (the non-brittle layer) plastic yielding, and (iv) microcracks on a surface of the macrocrack propagation are characterized by a full plastic dissipation, say G_m according to (2.6) and (2.7), all other microcracks by the value of dissipation equal to νG_m.

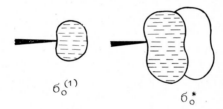

$$\sigma_0^{(1)} \qquad\qquad \sigma_0^*$$

FIG. 12. Growth of microcrack zone with increasing external load; moving of the process zone as a whole.

The equation of this problem can be written as

$$G_m[1 + \nu n(\sigma_0)] = \sigma_0^2 l / E, \tag{2.12}$$

where $n = H/h$. The solution of this equation gives the value of limit stress σ_0^*, which determines the fracture toughness K^* of the composite.

The dependences of K^* on V_f for the model composites (see Fig. 13) lead to the following conclusions.

(i) If plastic yielding of the matrix is the only possibility of energy dissipation in a composite with brittle fibres, the fracture toughness of the composite with respect to cracks normal to the fibre direction can go down monotonically ($\chi = 26$ in Fig. 13) with volume fraction going up as well as go up to infinity, i.e., to a structure which can fail everywhere away the crack ($\chi = 6$). Also the K^*–V_f curve can have a maximum ($\chi = 8$). Such types of behaviour have been observed, for example, by MILEIKO et al. (1976) testing boron–aluminium (the

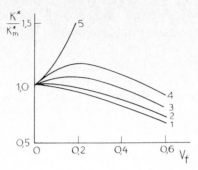

FIG. 13. Dependence of fracture toughness K^* of a model composite on fibre volume fraction, $K_m^* = 9.48\,\text{MN/m}^{3/2}$, $l = 250$, $\chi = 6$ (curve 1), 8 (curve 2), 10 (curve 3), 14 (curve 4), 26 (curve 5) (after MILEIKO and SULEIMANOV, 1981).

χ-values are small) and by COOPER and KELLY (1967) testing tungsten–copper composites (the χ-values are large).

This means that the fibre quality has to be estimated, taking into account defect distributions along a fibre. Fibres with defects should only be used in composites designed to have fracture toughness and strength going up simultaneously. In particular for the model composite material the value of σ_c^+ depends linearly on V_f, so the dependence $K^*-\sigma_c^+$ looks like the dependence K^*-V_f (cf. Fig. 13).

(ii) The fracture toughness of matrix material does essentially influence the fracture toughness of a composite (see Fig. 14). If a composite structure is considered to be an amplifier of matrix toughness, then this amplifier has a nonlinear characteristic.

Therefore, fabrication methods which provide high values of K_m^* should be preferred. Wrought materials look better than cast alloys; among such materials the best are those with high fracture toughness values. Composites having structures of the type patented by MILEIKO et al. (1980) should have high fracture toughness.

(iii) Because the fracture toughness of a composite depends on many structural parameters, even in the simple model, the values of fracture toughness

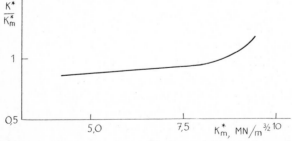

FIG. 14. Dependence of the fracture toughness K^* of a model composite on the matrix fracture toughness K_m^*, $V_f = 0.2$, $\chi = 14$, $l = 50$ (after MILEIKO and SULEIMANOV, 1981).

can be controlled in many ways. But at the same time it leads to a large scatter of experimental data.

It should be noted that in a real metal–matrix composite some processes which contribute to fracture toughness of fibre reinforced plastic and another brittle matrix composites can also have some effect. These processes are well known: fibre–matrix interface debonding, pull out and other localized damage of a structure (KELLY, 1972). An analysis of the relative contribution of each mechanism of energy dissipation to composite fracture toughness has been given by KIIKO (1981); the results can be used to evaluate a possible influence of fabrication parameters on fracture toughness of metal–matrix composites although originally the analysis treats composites without plastically deformed components.

2.1.3. Compressive strength

It was shown by MILEIKO and KHVOSTUNKOV (1971) that when testing metal–matrix composite specimens in compression a buckling mode different from that due to Euler (critical stress σ_e or σ_E) can be observed. When the stiffness of a specimen increases it starts to buckle by shear mode. A shear band is formed in the matrix, and critical stress σ_c^- of a specimen with clamped ends will be

$$\sigma_c^-/\sigma_e = \left[1 + \tfrac{1}{2}\pi \frac{a_0}{h}\frac{h}{l}\frac{\sigma_e}{\sigma_m^*}\frac{1}{V_m}\right]^{-1} \tag{2.13}$$

where σ_m^* is the yield stress of the matrix, h and l are the thickness and the length of the specimen, and a_0 is the initial effective deflection of the specimen.

This buckling mode leads to the dependence of σ_c^- on h/l with a maximum equal to σ_{max}^-. Besides that a maximum on the $\sigma_{max}^- - V_f$ curve occurs, it appears that

$$\sigma_{max}^- \propto E_f^{1/3}\sigma_m^{*2/3} . \tag{2.14}$$

The dependence $\sigma_{max}^-(V_f)$ is shown schematically in Fig. 15, where a typical experimental curve is also plotted. The experimental curve shows no maximum, but this difference has no practical importance. If a rod is designed to have a limit load determined by the shear mode of buckling, then there is no reason to have $V_f > \tfrac{1}{3}$; if the bending mode is to be expected, then increasing of fibre volume fraction is useful in spite of this difference.

The model instructs when and how important the fibre packing can be (see Fig. 16). These instructions should be considered to have only qualitative nature if a structural component differs from the rod considered above. The value of a_0/h describes in a general term a quality of a fabrication method. For example if a boron–aluminium specimen is made out of a plate prepared by hot pressing of a foil-fibres stack, and fibres have been prewound on a cylindrical mandrel of 300 mm diameter, then $a_0/h \sim 0.1$. For a steel–aluminium composite fabricated using a foiled matrix and winding the steel wire

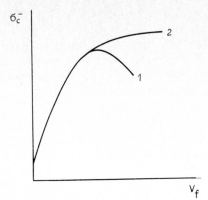

FIG. 15. A comparison of theoretical (curve 1) and schematic experimental (curve 2) dependences of compressive strength of a composite on fibre volume fraction.

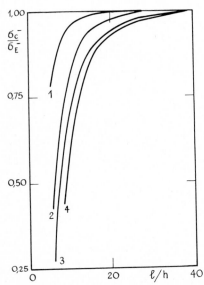

FIG. 16. Influence of the effective initial imperfection of a composite rod on the compressive strength, $E_f/\sigma_m^+ = 10^3$, $E_m/E_f = 0.2$, $V_f = \frac{1}{3}$. Curve 1 corresponds to $a_0/h = 0.01$, curve 2—=0.05, curve 3—=0.10, curve 4—=0.15. A drop of the limiting stress at compression takes place at $l/h \approx 20$. The value of compressive strength at $l/h < 20$ is essentially influenced by the value of a_0/h.

on a mandrel with plane faces, $a_0/h = 0.05$ (MILEIKO and KHVOSTUNKOV, 1971). Note that these values of a_0/h have been obtained comparing experimental data and (2.13).

The choice of a matrix and a fabrication method to provide a high enough value of σ_m^* is also important. Eq. (2.14) shows that if the shear mode of buckling takes place, changing this value is more important than changing E_f.

   Finally, the shear strength of the interface should be as large as possible because any deviation from the ideal value equal to the shear yield stress of the matrix is equivalent to decreasing the effective value of σ_m^*.

2.1.4. Complex stress state

Micromechanical models of composite failure which would give technological recommendations are not known to the author. Hence we are going to give just some experimental data related to the topic.

   WRIGHT and EBERT (1972) studied non-elastic behaviour and failure of a unidirectional boron–aluminium by hydraulic bulge test. The specimens were sheets clamped at their edges, against a die opening (equaxial plane tension is produced when the opening has a circular form). The experimental results show the following:

   (i) Yield locus on the $\sigma_L/\sigma_L^0 - \sigma_T/\sigma_T^0$ plane calculated in micromechanical way using a Finite Element Method deviates essentially to the external direction from the locus corresponding to von Mises–Hill criterion of plastic yielding of an anisotropic body (Hill, 1948). Here σ_L^0 and σ_T^0 are the yield stress in the fibre direction and that in the transverse direction, respectively (see Fig. 17(a)).

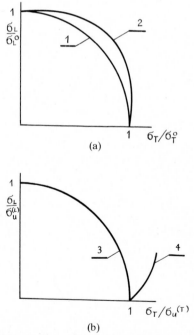

FIG. 17. Schematical illustration of the results by WRIGHT and EBERT (1972). Curves 1 and 3 correspond to the Von Mises–Hill criterion, curve 2 is calculated in a micromechanical fashion and curve 4 is the limiting failure curve.

(ii) Experimental yield locus can deviate from the von Mises–Hill curve still more (see Fig. 17(a)). In fact the experimental evidence of that is given by two points only.

(iii) Experimental strength locus does deviate in the same way from the curve plotted according to von Mises–Hill criterion if one substitutes the values of strength $\sigma_u^{(L)}$ and $\sigma_u^{(T)}$ instead of the values of yield stress σ_L^0 and σ_T^0 (see Fig. 17(b)).

The authors explained the last result assuming the fracture strength of specimens with clamped edges to be higher than that with free edges because in the former case cracks going from the free ends of fibres were excluded. It is important to note that, having tested four specimens, the authors used two points only for their discussion. They noted premature failure as a result of debonding at the interface and/or splitting of fibres. Possible criteria of interface failure and fibre splitting have not been discussed by the authors.

More precise experiments with usual thin wall tube specimens of steel–aluminium composites with circumferential reinforcement have been carried out by KONDAKOV and MILEIKO (1974). The results have been treated by relating them to a model of a composite layer containing two sorts of bands, one being of elastic material and the other of elastic–plastic material with yield stress σ_m^*. Then the limit curve for beginning of plastic yielding is given by the following equation:

$$\psi^2 \left(\frac{\sigma_L}{\sigma_m^*}\right)^2 - \psi \frac{\sigma_L \sigma_T}{\sigma_m^{*2}} + \left(\frac{\sigma_L}{\sigma_m^*}\right)^2 = 1 \qquad (2.15)$$

where

$$\psi = \frac{\kappa \nu_m + (1 + V_f)(E_m/E_f)(1 - \kappa \nu_f V_f)}{1 + V_m E_m/V_f E_f}.$$

Here $\kappa = \sigma_T/\sigma_L$ is the slope of the loading trajectory, indices L and T denote the direction of reinforcement and the transverse direction. The dependences of $\varepsilon_L(\sigma_L, \kappa)$ and $\varepsilon_T(\sigma_L, \kappa)$ can be written as

$$\varepsilon_L = \frac{1}{E_f} \left(\sigma_L' - \nu_f \kappa \sigma_L\right),$$

$$\varepsilon_T = \left[\frac{V_m}{E_m(\sigma_0'')} \left(\kappa - \tfrac{1}{2}\lambda\right) + \frac{V_f}{E_f}\left(\kappa - \nu_f \frac{\sigma_L'}{\sigma_L}\right)\right] \sigma_L \qquad (2.16)$$

where

$$\sigma_L' = \frac{\sigma_L}{V_f}(1 - \lambda V_m),$$

$$\lambda = \frac{\tfrac{1}{2}\kappa + [E_m(\sigma_0'')/E_f](1 - \kappa \nu_f V_f)/V_f}{1 + V_m E_m/V_f E_f},$$

$$\sigma_0'' = \sigma_L \sqrt{\lambda^2 + \kappa^2 - \lambda \kappa},$$

i.e., that σ_0'' is the stress intensity in the matrix, and $E_m(\sigma_0'')$ is the secant modulus.

Experimental data (a typical example is shown in Fig. 18) show that the equations in (2.16) are good enough to describe the behaviour of the composite under study (with a small fibre volume fraction), at least for small enough absolute values of κ.

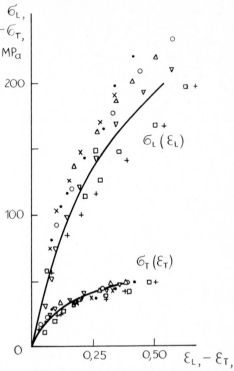

FIG. 18. Stress–strain curves of the steel–aluminium composite ($V_f = 0.1$) obtained in testing thin wall tube specimens loaded along radius $\kappa = 0.25$. Solid lines correspond to (2.16), points represent the results of testing seven specimens (after KONDAKOV and MILEIKO, 1974).

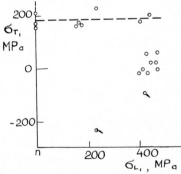

FIG. 19. Failure envelope for the steel–aluminium composite with $V_f = 0.1$ at plane stress state, $\kappa = $ const. (after KONDAKOV and MILEIKO, 1974).

Experimental data on failure (see Fig. 19) show, firstly, that the value of strength $\sigma_u^{(T)}$ is about equal to the matrix strength, which means that the interface bond is good enough. (The specimens were obtained by explosive fabrication methods, described in Section 4.3.) Secondly, if $\kappa > \sigma_u^{(T)}/\sigma_u^{(L)}$, then the composite fails when $\sigma_T = \sigma_u^{(T)}$. These two results contradict to observations made by WRIGHT and EBERT (1972), and this contradiction seems to have important implications for the fabrication process.

2.1.5. Creep and creep-rupture

We are to give just final results obtained by MILEIKO (1970, 1971, 1975).

The evaluation of the creep rate of a composite with short fibres having aspect ratio $\rho = l/d_f$ and a matrix which creeps according to a power law

$$\dot{\varepsilon} = \eta_m(\sigma/\sigma_m)^m \tag{2.17}$$

where m and σ_m are the constants, and η_m is the parameter to be chosen arbitrarily, gives

$$\dot{\varepsilon} = \frac{\eta_m}{m-1}\left(\frac{\sigma}{\sigma_m}\right)^m \frac{1}{\rho^{1+m}}[I(m)]^{-m}\frac{1-V_f^{(m-1)/2}}{V_f^m}. \tag{2.18}$$

Here

$$I(m) = \int_0^1 [z^{-m} + (1-z)^{-m}]^{-1/m}\,dz. \tag{2.19}$$

At $m > 3$, $I(m) \approx \frac{1}{4}$.

Eq. (2.18) is valid if

$$\rho < \rho^* = 2(\eta_m/\dot{\varepsilon})^{1/(1+m)}\left[\frac{1-V_f^{(m-1)/2}}{m-1}\right]^{1/(1+m)}\left(\frac{\sigma_f^+}{\sigma_m}\right)^{m/(1+m)}$$

and

$$\rho < \rho^{**} = 4\left[\left(\frac{\sigma}{\sigma_m}\right)^m\bigg/\left(\frac{\sigma}{\sigma_f}\right)^n\right]^{1/(1+m)}V_f^{(n-m)/(m+1)},$$

where σ_f^+ is the fibre strength, σ_f and n are the constants in a power law for creep of the fibre material, $\eta_m = \eta_f$.

It can be noted that sliding along the interface (KELLY and STREET, 1972) will increase the creep rate of a composite to some degree, but will not change the result qualitatively.

If

$$\rho^{**} \leq \rho < \rho^*,$$

then creep of both components contributes to the result, so that

$$\sigma = \sigma_f \left(\frac{\dot{\varepsilon}}{\eta_f} \right)^{1/n} V_f + \sigma_m \left(\frac{\dot{\varepsilon}}{\eta_m} \right)^{1/m} V_m . \tag{2.20}$$

The process can be accompanied by fibre breaking and corresponding creep rate acceleration.
If

$$\rho^* \leqslant \rho < \rho^{**} ,$$

then creep of a composite starts with stress redistribution, so that the fibre stress σ' goes up and the matrix stress σ'' goes down, and

$$\sigma(t) = \sigma_0'' \left[1 + \eta_m \frac{m-1}{V^{(m)}} \left(\frac{\sigma_0''}{\sigma_m} \right)^{m-1} \frac{E_m}{\sigma_m} t \right]^{1/(1-m)} . \tag{2.21}$$

Here $\sigma_0'' = \sigma''|_{t=0}$, $V^{(m)} = 1 + V_m E_m / V_f E_f$. If the average value of fibre strength $\langle \sigma_f^+ \rangle < \sigma / V_f V^{(m)}$, then at $t = t'$ the first fibre break occurs, followed by a fibre breaking process which is to bring the composite to the kind considered above (with short fibres) or to cause failure according to the weak link mechanism (see Section 2.1.1.2).

Rupture time of a composite with short fibres is given by

$$t_* = \frac{1}{2} \frac{\varepsilon_{uc}}{\dot{\varepsilon}_0} \left[1 + m\psi(m) \right] \tag{2.22}$$

where

$$\varepsilon_{uc} = \frac{\gamma_{ufm}}{\rho} \frac{1 - V_f^{(m-1)/2}}{m-2} , \qquad m\psi(m) \approx 1$$

and γ_{ufm} is the ultimate shear at the interface; the creep rate $\dot{\varepsilon}_0$ is given by (2.18).

The implications for fabrication which follow from these results are obvious. A composite should be obtained with large enough fibre aspect ratio ρ (if short fibres are used as reinforcement), with strong bonding at the interface (the values of γ_{ufm} and ε_{uc}), with uniform fibre distribution, and with a matrix having high enough creep resistance.

2.1.6. Fatigue

No micromechanical model of fatigue failure of metal matrix composites, which would allow recommendation of how to choose and produce a proper structure for given loading conditions, is known to the author. So we are going to discuss very briefly some results of an experiment carried out by MILEIKO and

ANISHENKOV (1980) to make clear some principal features of the fatigue process. These observations can supply a basis to build up a satisfactory model and serve also as a starting point for organization of a fabrication process.

The experiment was cyclic bending of specimens cut out of composite sheets. During loading, the natural frequency of a specimen was recorded and after a given drop of the natural frequency the specimen was said to be failed. After such testing some specimens were studied with optical microscope to observe the system of microcracks.

When fibre volume fraction is low, the fracture process is determined by cracking of the matrix. These microcracks can be arrested by weak interfaces within the matrix volume, if the latter have arisen, for example, because of bonding foils to form the matrix in a fabrication process. The fatigue strength of composites with bonded-foil matrix increases rapidly with an increase of fibre volume fraction (see Fig. 20). It is not important in this case whether reinforcing fibres are brittle or non-brittle and strength characteristics of the fibres are also unimportant. The only essential parameter of the components is the ratio of their Young's moduli. In a composite with a plasma sprayed matrix, the strength value at $V_f = 0$ is lower, it goes up when the fibre volume fraction increases, but the rate of the growth is slower because the matrix has no ordered weak interface to arrest fatigue cracks.

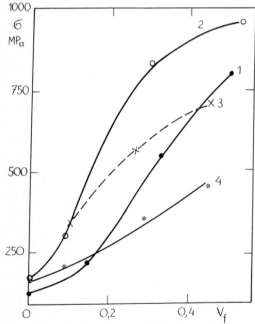

FIG. 20. Dependence of the fatigue strength ($f/f_0 = 0.95$, $N = 10^6$) of composites on fibre volume fraction. Curve 1 corresponds to the boron–aluminium composite with the plasma sprayed matrix, curves 2 and 3 correspond to the boron–aluminium composites with the foiled matrix and the boron fibres of two different reels, and curve 4 corresponds to the steel–aluminium composite with the foiled matrix.

At high fibre volume fractions, fracture of fibres determines the fatigue process of a composite. The choice of a reinforcing fibre becomes important. In brittle fibre composites, a crack leaps from one fibre to another; the useful way to stop such crack leaps is to introduce high-strength non-brittle fibres as additional reinforcement, according to MILEIKO et al. (1980).

2.1.7. On evaluating the interface strength

Each time we have used words like 'the interface strength' to describe a very definite value, if this value has been used both in a model and in a corresponding equation. For example, the critical fibre length is determined in the conventional way (KELLY, 1972) so that

$$l_c/d_f = \langle \sigma_f^+(l_c) \rangle / 2\tau_{ufm} \tag{2.23}$$

where τ_{ufm} is the limiting shear stress on the interface. One more example is given in Section 2.1.5 as a value of the limiting shear strain γ_{ufm} at the interface.

In order to introduce limiting values of stress or strain at the interface, it seems to be natural to evaluate them by measuring macroscopic characteristics to be connected to them. For example, a value of γ_{ufm} determines the ultimate strain ε_{uc} of a composite which can be obtained in an experiment without any difficulty.

But such a procedure is neither always convenient nor leads always to an accurate result. Therefore the problem of independent estimation of the interface characteristics in a plain experiment remains important. It is not easy to solve it because a plain experiment, analogous to the tensile strength test of a homogeneous material, in this case certainly does not exist. The well-known experiments by KELLY (1972) were treated without difficulty because the author assumed the shear stress distribution along the interface to be uniform. GORBATKINA et al. (1981) measured the local shear strength of the fibre–polymer interface, extrapolating experimental data to zero fibre length in a matrix. They obtained a strength value much higher than average strength values. An attempt to account for the stress singularity at the end of a fibre was made by MANEVICH (1981).

The results have been published of more sophisticated studies of stress fields around defects at the interface (DUNDURS and COMNINOU, 1979) as well as at the end of an inclusion of zero thickness (NIKITIN and TUMANOV, 1981). But such results can hardly be used in technological studies at the present time.

The fibre–matrix interface, which can have a number of layers of different stiffness and strength characteristics, is described at present by introducing effective ultimate values according to the scheme discussed above.

2.2. Bonding of solid surfaces

The interface in a composite can originate according to the following scheme. In the first stage, areas of physical contact arise as a result of smoothing of the surfaces due to plastic yielding under applied pressure. At the same time, one or another type of chemical bond starts to form through the contact areas. The rate of surface smoothing decreases as time goes on because the local compressive stress decreases.

In the second stage remaining small voids at the interface evaporate in a sintering process. Also diffusion and formation of chemical compounds at the interface are possible. All these processes lead to formation of an atomic interface structure.

SHIOIRI (1977) has conducted a very instructive experiment observing bond formation between two titanium surfaces. He has been measuring the intensity of an ultrasonic pulse reflected from the interface and has observed two characteristic rates of decrease of the intensity (see Fig. 21), one related to disappearance of long wave roughness of the surfaces and another related to short wave roughness. If the load is removed at $t < t'$ and the specimen kept at high enough temperature, we will never observe the strength of the interface equal to the strength of the bulk material. But if the same is done at $t > t'$, then the strength of interface inevitably reaches the strength of the bulk material.

Generally speaking, if the specimen with small voids at the interface is to be kept at high enough temperature for long enough time, then the voids should disappear due to sintering. The necessary temperature and time conditions can be estimated (see Section 3.3). So the above statement should be practically valid at low enough temperatures when the sintering rate is low. But Shioiri's experiment is still worth mentioning because two important conclusions follow. Firstly, a schematical model of the process of bonding described above is qualitatively true. Secondly, a macroscopic experiment in studying a bonding process is of considerable potential.

Therefore the problem of solid state bonding includes mechanical as well as physical and chemical considerations related to different stages of the process. Here we intend to discuss the mechanical aspect of the problem only.

Plastic smoothing of a surface is usually modelled (KARAKOZOV, 1976) by deformation of a series of wedges by a rigid smooth surface (as shown schematically in Fig. 22(a)). A most comprehensive study of this problem was undertaken by USHIZKY (1978). Briefly his consideration was as follows.

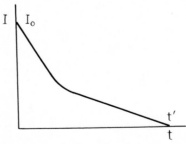

FIG. 21. Dependence of the intensity of a reflected pulse on pressing time for diffusion bonding of two titanium surfaces in SHIOIRI's experiment (a scheme).

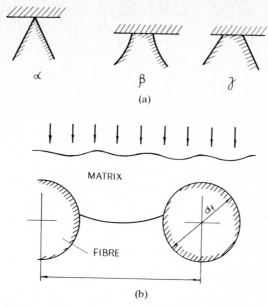

(a)

MATRIX

FIBRE

(b)

FIG. 22. (a) A rigid surface acting on a plastic wedge (α). A real picture of large deformations (β) is replaced by a simpler scheme (γ) to be considered. (b) A matrix material yielding through a lattice of rigid fibres.

Let us represent a surface by a series of wedges which do not interact. The heights of the wedges are $H_1, H_2, \ldots, H_i, \ldots, H_N$, where N is the total number of the wedges, $H_{i+1} \geqslant H_i$. The displacement of the rigid surface is divided into N intervals c_q such that

$$0 < c_1 \leqslant H_N - H_{N-1},$$
$$\vdots$$
$$H_N - H_{N-(q-1)} < c_q \leqslant H_N - H_{N-q} \tag{2.24}$$
$$\vdots$$

This displacement within each interval is divided into n_i steps and the total displacement at the end of the l-step in the q-interval can be written as

$$c_{q,l} = \sum_{i=1}^{q-1} \sum_{j=1}^{n_i} c_{i,j} + \sum_{j=1}^{l} c_{q,j}. \tag{2.25}$$

For a single wedge the solution of HILL's problem can be found (PRAGER and HODGE, 1951) as follows:

$$h = 2c\varphi, \qquad p = 4k\chi \tag{2.26}$$

where h is the contact area between a wedge and the rigid surface, k is the shear yield stress of a wedge material, and p is the force at the wedge-surface contact,

$$\varphi = (1 + \sin \psi) \sec \psi,$$

$$\chi = (1 + \psi)(1 + \sin \psi) \sec \psi$$

(2.27)

where ψ is the angle of the apex of the fan of the slip lines field. Then on each interval c_q we can write the following expressions for the total area of the contact:

$$h_{1,l} = 2c_{1,l}\varphi_N,$$
$$\vdots$$
$$h_{q,l} = 2c_{q,l}\varphi_N + 2[c_{q,l} - (H_N - H_{n-1})]\varphi_{N-1}$$
$$+ \cdots + 2[c_{q,l} - (H_N - H_{N-(q-1)})]\varphi_{N-(q-1)}$$
$$= 2c_{q,l} \sum_{i=N-(q-1)}^{N} \varphi_i - 2 \sum_{i=N-(q-1)}^{N-1} (H_N - H_i)\varphi_i,$$
$$\vdots$$

(2.28)

and for the total force:

$$p_{1,l} = 4kc_{1,l}\chi_N,$$
$$\vdots$$
$$p_{q,l} = 4kc_{q,l}\chi_N + \cdots + 4k[c_{q,l} - (H_N - H_{N-1})]\chi_{N-1}$$
$$+ \cdots + 4k[c_{q,l} - (H_N - H_{N-(q-1)})]\chi_{N-(q-1)}$$
$$= 4kc_{q,l} \sum_{i=N-(q-1)}^{N} \chi_i - 4k \sum_{i=N-(q-1)}^{N-1} (H_N - H_i)\chi_i,$$
$$\vdots$$

(2.29)

After simple transformations (2.28) and (2.29) yield to average values of $h_q(c)$ and $p_q(c)$ on the q-interval as follows:

$$h_q(c) = 2\bar{\varphi} \left[\frac{q}{N} c - \sum_{i=1}^{q-1} \frac{i}{N} \Delta H_i \right],$$

$$p_q(c) = 2k\bar{\chi} \left[\frac{q}{N} c - \sum_{i=1}^{q=1} \frac{i}{N} \Delta H_i \right]$$

(2.30)

where

$$\bar{\varphi} = \frac{1}{q} \sum_{i=N-q}^{N} \varphi_i, \qquad \bar{\chi} = \frac{1}{q} \sum_{i=N-q}^{N} \chi_i, \qquad \Delta H_i = H_{N-(i-1)} - H_{N-i}.$$

If a series of H_i is known, then (2.30) gives the necessary values. But

Ushizky (1978) has obtained asymptotic expressions for the pressure and the contact area assuming a continuous distribution function $F(H)$ to exist. It should be noted that the limiting displacement of the rigid surface is to be found from the condition of constant area of the cross-section of the profile, i.e., $c \leqslant \bar{H}$ where $\bar{H}$ is the average height of the wedges.

Solid state bonding is normally conducted at elevated temperatures when creep of metals is essential. If one uses a power creep law represented by (2.17), then it should be noted that at relatively low temperatures the value of σ_m is high and the value of the exponent m is large. (Examples of temperature dependences of σ_m and m for alloys which can be used as matrix materials, are shown in Fig. 23.) So at low temperatures the rigid-plastic analysis can be considered as a good approximation, if the characteristic stress σ_m for a characteristic time $t \propto \eta_m^{-1}$ is taken as the shear yield stress k. But at high temperatures, which is the usual case, we have $1 < m < 3$ and a creep problem for large deformations has to be solved. At present very approximate approaches are known (see, for example, Karakozov, 1976), and the results can hardly be used without supplementary experimental work.

Up till now we have considered the case of bonding of plane surfaces. When fabricating fibrous composites the situation looks more complex for a number of reasons.

Firstly, a fibre material is usually much more rigid than a matrix material, and the latter yields through a lattice of fibres (cf. Fig. 22(b)). At the matrix/fibre boundary there exist both normal and tangential components of stresses. Such a process can be modelled as yielding of a rigid-plastic matrix in a convergent channel and a result of one such solution is presented by Toth et al. (1972), referring to an unpublished report. They write the dependence of pressure q on the ratio of d_f to distance L between fibres as

$$q \propto \sigma_m \ln(1 - d_f/L)^{-1} . \tag{2.31}$$

The value of the logarithmic term changes with a factor 2 when the fibre volume fraction changes from 20 to 50 per cent.

Secondly, on the metal surface an oxide layer may have been formed. This layer is to be fractured during the mechanical bonding stage to provide conditions for physical–chemical bonding. So the tangential component of the stresses helps to do it but it becomes more difficult to formulate boundary conditions in the corresponding problem.

Thirdly, when a plasma sprayed matrix is used in a composite fabrication method (see Section 4.1.1), the mechanical behaviour of the latter changes both quantitatively and qualitatively during mechanical bonding. At first the material which is to become a matrix looks rather like a powdered medium, then like a metal undergoing creep. The porosity of the matrix and its density changes to a large degree.

Fourthly, boundary conditions at the fibre–matrix interface can change even at the first stage of bonding due to physical–chemical interaction at the interface.

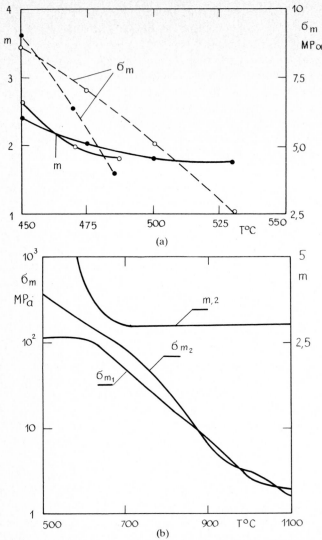

FIG. 23. Temperature dependences of the creep parameters of some aluminium and titanium alloys at high temperatures. (a) Aluminium alloys D16 (●) and Aℓ–6% Mg (○); (b) Titanium alloys OT-4 (Ti–3Aℓ–1.5 Mn)—curve 1; $m = 3$, and VT-14 (Ti–4Aℓ–3Mo–1V)—curves 2 (after RABOTNOV and MILEIKO, 1970).

All this makes proper planning of an experiment very important. It can be done having qualitative estimates as a result of solving a model problem. We will give an example of the usage of some approximate estimate, neglecting some of the factors mentioned above.

In order to evaluate the pressing time t (cf. Fig. 21) necessary to form the complete physical contact, we need to obtain a combination of temperature T and pressure q which provides the result without unwanted effects like fibre degradation (dissolving, breaking) or formation of too thick an interface layer containing products of chemical interaction between the fibre and the matrix. If an optimum combination of fabrication parameters (T_0, q_0, t_0) is known for a composite with one matrix material, then a first approximation to an optimal set of parameters for a composite with other matrix material can be obtained in the following way,

Note that for the power creep law the solution of a creep problem is characterized by an interesting property (RABOTNOV, 1966). Namely, if all external loads increase proportionally to one factor, say λ, and for one value of λ, say $\lambda = \lambda_0$, a solution of a problem has been obtained, and if we know the stress field $\sigma_{ij}^0(\bar{x})$ and the displacement field $u_i^0(\bar{x})$, then for an arbitrary value of λ the stress field will be $(\lambda/\lambda_0)\sigma_{ij}^0$ and the displacement field will be $(\lambda/\lambda_0)^m u_i^0$. Therefore if creep parameters for two matrix materials $\eta_0 = \eta_1 = \eta$, $\sigma_m^0(T)$, $m_0(T)$, $\sigma_m^{(1)}(T)$ and $m_1(T)$ are known, an it is possible to assume $m_0 = m_1 = m$ in a temperature interval of interest, then temperature T_1 is to be chosen such that $\sigma_m^0 = \sigma_m^{(1)}$. If the possibility of changing the temperature is restricted by the chemical interaction, then the following equation has to be satisfied:

$$(q_1/q_0)(\sigma_m^0/\sigma_m^{(1)}) = (t_0/t_1)^{1/m} . \tag{2.32}$$

If $m_0 \neq m_1$, then a consequence of the well-known Calladine–Drucker's (RABOTNOV, 1966) theorem should be used. The only generalized force here is q, so choosing the parameters inequality

$$(q_1/\sigma_m^{(1)})^{m_1}(\sigma_m^0/q_0)^{m_0} \leq 1 \tag{2.33}$$

has to be taken into account if $m_1 > m_0$. This inequality can be especially useful if optimum sets of hot pressing parameters for a number of matrix materials with various values of the exponent m are known.

The temperature dependences of creep parameters for aluminium alloys D16 and AMr6 are presented in Fig. 23. For a boron–aluminium composite with the D16-matrix the set of the parameters of hot pressing close to an optimal one is

$$485°C-25\ \text{MPa}-1.5\ \text{h} .$$

If for the AMr6-matrix the parameters

$$530°C-30\ \text{MPa}-1.5\ \text{h}$$

are taken, then these two combinations will nearly satisfy (2.32). It should be noted that a deviation from these parameters can give composites with worse properties.

An example is the dependence of the tensile strength of boron–aluminium composites with the AMr6-matrix ($V_F = 0.28 \pm 0.02$) on temperature of hot pressing at constant pressure $q = 30\ \text{MPa}$ and time $t = 1.5\ \text{h}$. We have the following:

$$T \, (^{\circ}C) \qquad 520 \quad 530 \quad 550$$

$$\langle \sigma_c^+ \rangle \, (\text{MPa}) \qquad 396 \quad 496 \quad 402$$

The dependence of compressive properties of the same composite on temperature of hot pressing is illustrated in Fig. 24. It can be seen that the difference in critical compressive stresses for the composites fabricated at temperatures 530 and 550°C is most essential at small ratios of thickness h to length l of specimens. It is supposed to be a consequence of decreasing either the yield stress σ_m^* of the matrix or the shear strength of the interface (see Section 2.1.3).

One more example on the influence of technological parameters on mechanical properties of boron–aluminium composites: In the experiments the temperature of hot pressing was kept constant, $T = 530°C$. One set of speci-

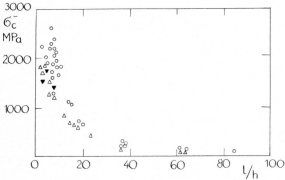

FIG. 24. Critical stresses at compression of boron–aluminium specimens obtained at various temperatures of hot pressing, ▼—520°C, ○—530°C, △—550°C, $q = 30$ MPa, $t = 1.5$ h.

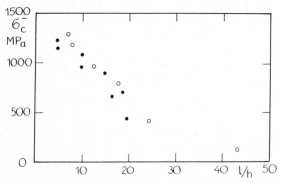

FIG. 25. Critical stresses at compression of boron–aluminium specimens obtained at various fabrication parameters. ●—$q = 30$ MPa, $t = 1.5$ h. ○—$q = 60$ MPa, $t = 0.5$ h. Temperature is 530°C in both cases.

mens with $V_f = 0.25$ was fabricated at the standard conditions (see above for the AMr6-matrix), the other set was fabricated at double the pressure, but the pressing time was 3 times less, which is a bit more than necessary according to (2.32). The mean tensile strength of specimens of the second set was 10 per cent higher, the compressive strength of all the specimens was nearly the same (see Fig. 25).

3. Physics and chemistry in composite technology

A non-homogeneous structure of a fibrous composite provides numerous examples of physical and chemical interactions which determine the very possibility of the existence and stability of such a structure, as well as a number of its mechanical properties.

3.1. Interfaces in composites

The physics, chemistry and mechanics of the surface of solids have been rapidly developed in recent years. This appears to be because, firstly, the importance of the problem has been appreciated, and secondly, many powerful experimental techniques (like a variety of methods for local analysis, surface spectroscopy of various kinds, and so on) have become available. But at present more conventional knowledge of the structure of interfaces and interface bonding is still of some importance in composite technology. The latest results of surface science cannot yet be interpreted in composite technology terms.

A consequence of such a situation is the usage of the knowledge of interface structures by composite engineers only in a qualitative way. Well-known classifications of the interfaces (see, for example, HARROD and BEGLEY, 1966; METCALF, 1974; PORTNOI et al., 1979) can perhaps be considered as too provisional, but nevertheless they are more or less useful.

Obviously purely mechanical bonds which provide shear load transfer can arise as a result of fraction along the interface because of normal compressive stress or/and penetration of the matrix into microcavities on the fibre surface (like those at the surface of boron fibres).

On the other hand atomic bonding can be seen. This is a large or small angle grain boundary of the coherent atomic structure. So in this case the well-developed techniques of analyzing small angle grain boundaries, and also rapidly developing techniques of studying large angle boundaries, can be applied to analyze the structure of the interface.

These two cases (see Fig. 26) are obviously extreme ones. The first can be observed almost always, the second is rather an exception in composites—an example being directionally solidified eutectics. It is also obvious that these two cases are not mutually exclusive.

No doubt that in both cases the strength of the interface is determined by defects of an interface structure rather than an ideal structure. The models of

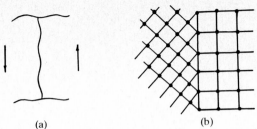

(a)   (b)

FIG. 26. Schematic representation of (a) purely mechanical interface bonding, and (b) atomic coherent bonding.

defects can be macroscopic in the case of mechanical bonding. Defects can be introduced into a coherent atomic structure on atomic level, a well-known example being the dislocation wall model of the small angle (tilt or twist) boundary.

Between these two cases any possible bonding mechanism can work, including Van der Waals bonds, non-coherent crystal boundary, and so on.

The structure of the interface is influenced to a large extent by possible diffusion of fibre and matrix elements. This process leads to solid solution regions in the vicinity of the interface, and to formation of a third phase layer, or a number of layers of new phases. Bonding between the layers and the original components as well as between the layers themselves can be of various types (a schematical example is shown in Fig. 27).

Recently some works have appeared which attempt to connect fabrication parameters to the structure of the interface and parameters of this structure to the interface strength. For example, YUE et al. (1975) have calculated the interface energy assuming a usual potential function of the pair interaction between atoms of two sorts within the diffusion zone at the vicinity of the interface plane. The structure of this zone is determined by temperature,

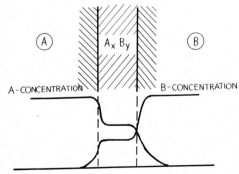

FIG. 27. An example of the structure of the interface region.

pressure, and time of the process of interface formation. A clear physical result has been obtained, namely that the bond strength goes up monotonically with the parameters mentioned above, but it appears to be impossible to connect this result to the choice of an optimal set of fabrication parameters.

It should be noted once more that a calculation of the real strength of the interface can be done only by taking into account defects of an interface structure. This situation is quite similar to the evaluation of the real strength of a homogeneous body which is impossible without consideration of cracks, dislocations and another defects. Also note that words like 'a physical contact' have been used in Section 2.2 to describe a possibility of forming an interface bond which is not only of pure mechanical nature.

3.2. Wetting

Wetting of a solid surface with a liquid provides the next possibility to form an interface bond (first being plastic smoothing of interfaces). An examination of wetting conditions is important not only from the point of view of evaluating bond strength but also because of the necessity of evaluating the possibility of using a liquid infiltration method in producing composites.

3.2.1. Surface energy and wetting conditions

The present understanding of wetting, as well as of liquid–solid interaction (which determines wetting), can be found in a number of books. We just point out two of them. The first one by NAIDICH (1972) contains a lot of experimental data important for evaluating liquid infiltration methods. The second one by KOSTIKOV and VARENKOV (1981) is about liquid metal–carbon interactions, which can be important from the point of view of fabrication of carbon fibre–metal matrix composites.

The main relationships between characteristic parameters to determine wetting are written as follows:

$$\gamma_{LS} \left(\frac{1}{R_1} + \frac{1}{R_2} \right) + \rho g x = 0 \, ,$$

$$\gamma_{LS} - \gamma_{SV} + \gamma_{LV} \cos \theta = 0 \, . \tag{3.1}$$

Here γ is the value of the interface energy related to the particular interfaces by indexes L (liquid), S (solid) and V (vapor), R_1 and R_2 are the main radii of the curvature at a point of the LV interface, x is the vertical coordinate of the point, ρ is here the liquid density, and θ is the angle of contact.

The second equation is illustrated by the equilibrium of the forces of surface tension shown in Fig. 28(a). If the angle of contact $\theta > \frac{1}{2}\pi$, then it is said that the liquid does not wet the solid, and the liquid column does not rise in a capillary tube (see Fig. 28(b)). If $\theta < \frac{1}{2}\pi$, then the liquid wets the solid and the liquid column does rise in a capillary tube (see Fig. 28(c)). The first equation

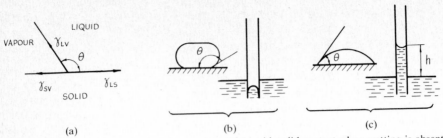

FIG. 28. (a) An illustration of Young's equation; (b) Liquid–solid contact when wetting is absent, $\theta > \frac{1}{2}\pi$; (c) Liquid–solid contact at wetting, $\theta < \frac{1}{2}\pi$.

gives the height of the liquid column rise for $r \to 0$:

$$h = \frac{2\gamma_{LV} \cos \theta}{\rho g r} \tag{3.2}$$

where r is the radius of a capillary tube.

To measure a value of the angle of contact, a sessile drop method is often used, for which the experimential scheme is obvious (cf. Fig. 28). The experimental data obtained in this way are usually taken to estimate the liquid–solid interaction, although theoretical models exist which give values of θ.

3.2.2. Some features of wetting

The above consideration of the solid–liquid interface was of a thermodynamical nature. The results would be useful in a general table giving (i) the values of surface energy of melted metals considered to be a basis for matrix alloys measured in vacuum, (ii) those of typical fibre materials as well as of possible coatings which can be used to provide wetting and to prevent too strong a chemical interaction at the fibre–matrix interface, and (iii) the values of interface energy or the angle of contact.

But the wetting kinetics (which is not analyzed by thermodynamics) can be quite complicated for particular pairs of substances. It includes chemical interaction at the interface and formation of a new phase, mutual solution of elements both in the solid and in the liquid, solution of elements from the vapour in the liquid (if the process is not going on in vacuum), absorption on the solid surface of trace elements contained in the liquid, and so on.

These factors make the use of a simple table for the values of surface energy quite difficult. Nevertheless we present such a table (see Table 2) which can be considered as preliminary information only. It could also be useful as a list of references for more complete information on experimental conditions as well as on features of the wetting kinetics.

One of the many factors which influence wetting is surface roughness. Wetting of a rough surface is determined by the local angles of contact which can differ markedly from the apparent angle. When surface roughness increases, wetting as a rule gets worse. To illustrate the situation we note that the value of $d\theta/d(R_a/\lambda_a)$, where R_a and λ_a are the mean values of amplitude and wavelength of the roughness, for the pair Cu–HfC, for example, is equal to 103° at 1200°C. The value of the true angle of contact is equal to 128° under these conditions. HITCHCOCK et al. (1981), who have obtained these data, have found at the same time that an improvement of wetting can be reached by ultrasonic agitation.

Generally speaking the kinetics of wetting is usually described by a curve such as that shown in Fig. 29. Parameters of this curve depend on temperature and other conditions of the experiment. These can be conditions which provide wetting, i.e., the angle of contact reaches $\frac{1}{2}\pi$ (a case shown in this figure), but there can also be situations when the value of θ does not reach $\frac{1}{2}\pi$, and that is a case of non-wetting.

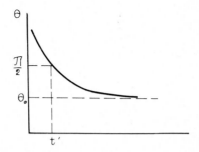

FIG. 29. Schematic dependence of the angle of contact on time of liquid–solid contact. Wetting is achieved at time t'.

3.2.3. Treatment of fibre surface

Fibre surface coating usually has two goals, the first being to provide wetting by a molten matrix alloy if a liquid phase fabrication method is to be used, the second being to provide a diffusion barrier to arrest possible forming of unwanted phases at the interface, fibre degradation, and other effects (to be described partly in Section 3.4.3).

To start with the wetting condition, we look at Table 2 and see that coating of carbon and oxide fibres is of most importance.

The story about carbon fibre coating started quite a long time ago when such fibres were metallized by an electrolytic process. The fibres were coated with nickel (JACKSON, 1969; WATANABE, 1979), copper (JACKSON, 1969), tantalum (SARA, 1974), and other metals.

But some authors noted that it looked preferable to use chemical coating methods, which provide a more uniform surface film. For example, KOSTIKOV et al. (1979) deposited nickel and copper on the fibre surface from solutions of appropriate salts. In their method the fibre surface has to be activated before the metal deposition. The activation essentially provides the surface with catalytical properties by surface oxidation and subsequent treatment in nitric acid, chlorous tin and chlorous palladium.

TABLE 2.[a] Wetting of some possible fibre and fibre coating materials with some molten metals. (The three figures at each point mean: temperature (°C)—the angle of contact, degrees (or surface/interface energy, erg/cm^2, if the figure has an asterisk)—reference to a source given below the table[b].)

Metals	C	SiC	B$_4$C	TiC	ZrC	HfC
Aℓ 800-850-1	800-3180*-1 850-90-2 1000-75-3 1020-90-2	1000-65-3 1100-33-4	1000-145-4 1000-60-3 1100-119-4 1150-33-4 1200-28-3	700-118-4 1150-20-4		
Ni 1550-1700-1	1550-1470*-1	1450-65-4	1460-88-4 1500-87-4 1480-43-5	1380-23-4 1450-17-4 1500-0-4	1380-24-4 1450-32-4	1380-23-4
Cu 1100-1300-1	1100-3400*-1 1285-120-2	90-4	960-137-4 1100-136-4 1150-134-3 1150]1300}-126-5 1500-0-4	1100-108-4 112 1300-$^{70}_{109}$-4	1100-135-4 1200-126-4	1200-132-4

Metals	B	Aℓ_2O	SiO$_2$	ZrO$_2$	TiN	TiB$_2$	ZrB$_2$
Aℓ	1000-0-3	1250-48-4			900-135-4	900-98-8	900-106-8
Ni		1500-1290*-6 1500-150-4 1527-2200*-7	1500-125-4	1500-130-4		$\left.\begin{array}{c} T_m \\ 1650 \end{array}\right\}$-25-8 1480-40-4	$\left.\begin{array}{c} T_m \\ 1650 \end{array}\right\}$-72-8 1500-42-2
Cu	1150-45-3	1500-125-4	1100-134-4		1130-155-4 1500-148-4	1120-142-4 1130-143-8	1100-123-4 1130-135-8 1400-36-4

[a] The data in Table 2 as a rule have been obtained in vacuum but in some cases an inert atmosphere was used.

[b] *References*: 1. KENDALL (1974)[c]; 2. NICHOLAS and MORTIMER (1971); 3. KOLESNICHENKO (1972); 4. NAIDICH (1972)[c]; 5. PANASUK et al. (1981); 6. SUTTON and FEINGOLD (1966); 7. KURKJIAN and KINGERY (1956); 8. SAMSONOV et al. (1972).

[c] Some data given by KENDALL (1974) and NAIDICH (1972) have been taken from the third work.

Metal coatings of carbon fibres allow them to be wetted by metals with a lower melting point, but they appear to be unstable in metal melts, i.e., fibre degradation, leading to a drop of the strength of a composite, is still observed. Hence one has to look for more thermodynamically stable substances for fibre coating. As is known (see, for example, Table 2) many molten metals of practical interest wet some carbides, nitrides and oxides. So mixtures of such substances are widely used for carbon fibres coating.

KASHIN (1981) and RASHID and WIRKUS (1972) have suggested and studied a method of coating carbon fibres with carbides. The fibres are pulled through a molten metal bath which is chemically inert to carbon (examples being copper or tin) and contains elements to react with carbon to form a carbide. The thickness of a titanium carbide layer, for example, depends on the titanium content in the melt, temperature and time of the reaction. At a subsequent stage of infiltration of a fibre bundle by an aluminium alloy, aluminium carbide is formed at the titanium carbide–aluminium interface. The rate of this process can be decreased by doping aluminium with such elements as silicon, magnesium and gallium.

But even carbide coatings do not appear to be stable enough in metal matrices. That is why PEPPER and ZACK (1978) have suggested coating fibres with a mixture of silicon carbide and silicon oxide. This is done by a chemical vapor deposition method; silicon tetrachloride is supplied to the fibre surface in a flow of hydrogen gas and the reduction of silicon tetrachloride takes place in the presence of an oxygen-containing gas. The thickness of the coating is in the range of about 10 nm to 1 μm.

However, coating of carbon fibre by a mixture of titanium and boron is certainly used most widely (MEYERER et al., 1978). Usually the coating procedure involves a CVD-method, a mixture of tetrachlorides of titanium and boron being reduced by zinc vapor. The thickness of the coating is about 20 nm.

Other methods of coating are also known, the main goal usually being to deposit a coating material in an atomized or finely dispersed state.

Some researchers, however, believe that the usage of such refined methods in fibre coating procedures is not the best engineering solution. So GODDARD (1978) and GODDARD and KENDALL (1977) have suggested a procedure which looks simpler. A bundle of fibres, that can be carbon or polycrystalline sapphire, is pulled first through a potassium bath. For carbon fibres, the temperature of the bath should be about 550°C and the time around 10 minutes. Then if a metal to be used as a matrix (for example lead) wets potassium, the coated fibres are pulled through the molten matrix, which wets the coating and replaces it at the fibre surface. When a matrix (for example aluminium) does not wet potassium, an intermediate bath is necessary. It can be a tin bath for example, as tin wets potassium and then aluminium wets tin.

3.2.4. Matrix modification

A definite influence on the fibre surface during fabrication processes can be achieved by matrix alloy modification. It can be important from the point of view of liquid phase techniques of composite fabrication. To promote wetting it is necessary to decrease the surface energy of a fibre. Therefore we are looking for dopants which would react with a fibre material with reaction products which would form a thin surface layer (which is wetted by the molten matrix). Because it is difficult to predict the thickness of the layer and the bond at the layer–fibre interface, the procedure needs extensive trial and error. Hence it is easier to illustrate the situation by a series of examples, a part of which can be obtained in patents.

Let us start with a proposal made by KALNIN (1977) to promote wetting of carbon fibres by a matrix containing large enough amounts of magnesium. The author has noticed that adding of relatively refractory magnesium nitride Mg_3N_2 to a molten matrix markedly enhances wetting, provided that particle size is less than 2 μm. The improvement is such that the infiltration technique (see Section 4.4) becomes possible and high compressive and shear strength of the composites can be obtained. Certainly the result relies on formation of magnesium carbonitride (MgC_xN_y) or magnesium cyanamide ($MgCN_2$) on the fibre surface.

Doping the matrix by fine particles of magnesium nitride can be done in various ways. It is possible to add the particles into the molten matrix alloy in a concentration of about 0.2 to 25 per cent by weight (preferably 1.0 to 10 per cent). But it is better to obtain the particles in situ by providing the reaction between the molten magnesium metal with the nitrogen ambient (preferably at a temperature between 800 and 850°C). In such a process Mg_3N_2 particles tend to be very fine. Still another way of getting the necessary modification of the matrix is to add into the matrix a metallic nitride capable of reacting with magnesium metal to form magnesium nitride. Such nitrides are silicon nitride, aluminium nitride, titanium nitride, etc.

This procedure can be applied to metal alloys based on aluminium, zinc, titanium, chromium, etc, containing at least 10 per cent magnesium, according to the estimate of KALNIN (1977).

A further example is the modification of an aluminium matrix by alloying it with lithium to promote wetting of polycrystalline alumina fibres by an aluminium alloy matrix (CHAMPION et al., 1978). The resulting alloy can be infiltrated into a fibre bundle. X-ray diffraction studies of fibres extracted from a composite show the presence of small quantities of $LiAℓO_2$ on the fibre surface. A minimum concentration of lithium in the matrix and preferred values of technological parameters have been found by the authors after mechanical testing of composites obtained in this way. They found that, when the lithium concentration goes above approximately 3.5–4.0 per cent by weight, the tensile strength of the composites goes down in a nearly linear fashion with the lithium concentration. Increasing the contact time with the molten alloy leads to increasing amounts of $LiAℓO_2$ on the fibre surface, but the tensile strength of the composite with fibre volume fraction equal to 60 per cent appears to decrease by only 20 per cent in the contact time interval from 1 to 10 minutes. (Lithium concentration in these tests was 3.3 per cent, and melt temperature was 700°C.)

Another possibility for matrix modification is to dope the matrix by elements which are adsorbed by the fibre surface, thus decreasing the fibre surface energy. To evaluate the kinetics of wetting in this case, it is also necessary to conduct a special experiment. The angle of contact should depend on time as shown schematically in Fig. 29. Time t' should decrease with increasing temperature.

A review of experimental data on such matrix modification for the case of graphite fibre–aluminium matrix composites has been presented by KENDALL (1974). These data show that the value of θ becomes less than $\frac{1}{2}\pi$ at temperatures between 800 and 925°C (when aluminium starts to wet graphite fibres) if the matrix metal is doped by about 1 per cent of such elements as Ga, Cr, Ni, Co (NICHOLAS and MORTIMER, 1971). The same occurs at temperatures between 900 and 1120°C if doping elements are V, Nb, Ti, Zr. Aluminium does not wet graphite at temperatures of at least 925°C if the doping elements are Si, Mg, Fe.

Wetting of vitreous carbon and sapphire with molten copper is improved by doping the copper with titanium (STANKING and NICHOLAS, 1978), and still better results can be obtained if titanium and tin are employed simultaneously (although tin itself does not improve wetting). Copper with 10% Ti starts to wet vitreous carbon at 1150°C and copper with 1.2% Ti and 2.8% Sn also wets vitreous carbon at the same temperature. To wet sapphire at 1150°C, it is necessary to add 8% Ti or 3.5% Ti and 11.8% Sn.

The role of interfacially active substances in wetting of sapphire by nickel has been investigated in a number of works.

KURKJIAN and KINGERY (1956) measured sapphire–nickel interface energy γ_{LS} when nickel is doped by such elements as Sn, Zn, Cr and Ta. They discovered that tin and zinc do not influence the value of γ_{LS} up to about 2 to 3 weight per cent. Doping by chromium leads to a decrease in the value of γ_{LS}, starting with a chromium content of about 0.1 per cent. At 10 per cent chromium content γ_{LS} decreases by about $\frac{1}{3}$. But the most effective way to decrease γ_{LS} is to dope copper with titanium. A titanium content of about 0.1 per cent decreases γ_{LS} by one half of the original value. This result was confirmed by ALLEN and KINGERY (1959), who also demonstrated the possibility of wetting sapphire by nickel alloys containing more than 1 per cent of titanium.

SUTTON and FEINGOLD (1966) studied in detail nickel matrix modification with relation to wetting of sapphire filaments. They showed that doping nickel with chromium and zirconium increases the angle of contact at 1500°C, while titanium doping decreases the value of θ down to 95° at 1500°C. These experiments were carried out in vacuum.

Sometimes it seems to be useful to dope a matrix material with two elements. The wetting of sapphire by copper doped by titanium is greatly improved if indium is added, slightly improved by adding aluminium, and is not improved by additional doping with nickel and gallium (NICHOLAS et al., 1981).

3.2.5. On the strength of the interface

As was mentioned above, measurements of the values of the angle of contact and/or the interface energy are to be conducted, not only to estimate the feasibility of liquid phase fabrication methods, but also to attempt to evaluate the interfacial bond strength—which is sometimes called an adhesion strength. In the latter case, the interfacial energy is assumed to have no strong temperature dependence, even below the crystallization point. However, such an approach can lead to an upper limit of the bond strength similar to the results of calculation of the ideal strength of a defect-free solid. A real stress state in the vicinity of an interface is perturbated by defects of the interface and it determines a real limiting load of an element with the interface.

The usual way of evaluating the bond strength is very simple. If the interface energy is γ_{LS} and the values of surface energy of the two phases are γ_L and γ_S, then the work to fracture the interface is

$$W_{LS} = (\gamma_L + \gamma_S) - \gamma_{LS}. \tag{3.3}$$

Introducing the angle of contact according to (3.1) we obtain

$$W_{LS} = \gamma_L(1 + \cos\theta). \tag{3.4}$$

Now experimental data for the values of γ_L and θ (see, for example, Table 2) enable us to estimate the work to fracture of some particular interfaces.

Note that besides the already-mentioned reasons for the difference between real values of the interface strength and those given by the values of W_{LS}, there are some other reasons which will be discussed in Section 3.4.3.

3.3. Sintering

At present the physics of sintering has become a very developed field of solid state physics. The main question is that of densification of a porous body, and it has been considered from both the thermodynamic and kinetic point of view. Sintering takes place at the final stage of surface bonding in a composite fabrication process when small voids at the interface remain and have to be removed, and also when powder metallurgy methods are used, i.e., when a matrix is formed from a powdered metal. Densification of a plasma sprayed matrix is also accompanied by sintering processes.

Sintering is well described in a number of books, that by GEGUZIN (1967) being an example. The reader can find all the details there. Here we are to discuss very briefly only such points as the equilibrium of a single void in a solid, bonding of two similar particles, and bonding of two dissimilar particles.

The problem of the behaviour of a single void in an infinite linear-viscous body has been formulated by FRENKEL (1945). He considers a spherical void of radius $R(t)$ decreasing with time. A driving force of the process is the surface tension, while the viscosity of the material resists the decrease in radius. Because of the spherical symmetry, the displacement rate v only has a radial component

$$v_r = \beta/r^2$$

where β is determined by the rate of void diminishing, namely,

$$\beta/R^2 = dR/dt.$$

Then we have

$$\varepsilon_{rr} = dv_r/dr = -2\beta/r^2,$$

and the density of the rate of energy dissipation

$$W = 2\eta \varepsilon_{rr}^2$$

where η is the viscosity coefficient, and the total rate of energy dissipation is

$$\tilde{W}_1 = 8\eta \int_R^\infty \pi \varepsilon_{rr}^2 r^2 \, dr = \tfrac{32}{3}\pi\eta \, \frac{\beta^2}{R^3}.$$

The rate of free energy decrease due to the decrease of the surface area is

$$\tilde{W}_2 = \frac{d}{dt}(4\pi\gamma R^2) = -8\pi\gamma R \, \frac{dR}{dt}$$

where γ is the surface energy.

The condition $\tilde{W}_1 = \tilde{W}_2$ leads to

$$\frac{dR}{dt} = -\tfrac{3}{4}\frac{\gamma}{\eta},$$

and finally the time necessary to remove the void is

$$t_* = \tfrac{4}{3}\frac{\eta}{\gamma} R_0 \tag{3.5}$$

where $R_0 = R\big|_{t=0}$.

To consider the thermodynamic equilibrium between a void and a gas of vacancies in a solid, the treatment will not differ from an analysis of vapour equilibrium around the curved surface of a solid. In both cases the equilibrium gas pressure p increases when the radius of curvature decreases, namely,

$$p = p_0 \left(1 + \frac{2\gamma}{R}\frac{\Omega}{kT}\right) \tag{3.6}$$

where p_0 is the equilibrium pressure around a plane surface and Ω is the atomic volume.

Therefore a void can evaporate into the solid body and this tendency becomes more pronounced as the radius of the curvature decreases. The process will be limited by the rate of vacancy diffusion away from the void. This problem was considered by PINES (1946). We shall not repeat PINES' calculation, which can also be found in GEGUZIN's (1967) book, and just give the final result:

$$t_* = \frac{kT}{6D\gamma\Omega} R_0^3 \tag{3.7}$$

where D is the self-diffusion coefficient.

It is easy to see that applying an external pressure accelerates the process of void evaporation. Moreover, if the process is limited by vacancy diffusion, then a critical void size exists above which the void will not disappear. In fact a real mechanism of void disappearance is much more complicated, but the schemes described above represent a real situation fairly well.

Bonding of two similar spherical particles (see Fig. 30) is a result of a number of processes. They are (i) a viscous flow of the material from a particle volume to the regions of a large curvature at the neck, (ii) the volume vacancy diffusion from the regions of high pressure in the vicinity of the concave surface of the neck to those of a low pressure in a vicinity of the convex surface of the particles, (iii) the volume diffusion of vacancies to the region in between the particles, (iv) the surface diffusion of atoms from the convex parts of the surface to the concave parts, (v) the transport of atoms in the same direction through the vapour phase because of a difference in the vapour pressure, and (vi) plastic yielding due to external forces. All these ways of combining two particles in one are analyzed by GEGUZIN (1967), who gives the estimate of the sintering rate, i.e., the formula for $r(t)$, r being the radius of the neck.

Bonding of dissimilar particles includes more complicated processes. If materials of the particles are mutually insoluble, combining of two particles A and B becomes thermodynamically possible only when

$$\gamma_{AB} < \gamma_A + \gamma_B \tag{3.8}$$

where γ_{AB} is the surface energy of the A–B-interface (PINES, 1956). If one

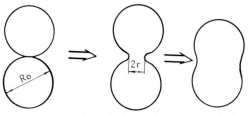

FIG. 30. A scheme of bonding of two similar particles.

assumes that

$$\Delta \gamma_{AB} = \gamma_A - \gamma_B > 0\,,$$

then the kinetics of particle bonding will depend on which inequality is fulfilled:

$$\gamma_{AB} > \Delta \gamma_{AB} \quad \text{or} \quad \gamma_{AB} < \Delta \gamma_{AB}\,.$$

If $\gamma_{AB} > \Delta \gamma_{AB}$, then the equilibrium shape of the final particle should be that of the concentric sphere with particle B in the centre. If $\gamma_{AB} < \Delta \gamma_{AB}$, then the situation will be similar to that shown in Fig. 30.

If the materials of the particles are mutually soluble, then bonding can be accompanied by new events (GEGUZIN, 1967). Those of most interest are determined by a difference of the diffusion coefficients. If for example $D_A > D_B$, then fluxes j_A and j_B of atoms A and B through the surface of contact will be different. Hence extra vacancies will arise in body A near the interface and the rate of vacancy generation will be proportional to $j_A - j_B$. These vacancies will be adsorbed either by dislocations (when creep will occur, moving the interface in the direction of body A), or by immobile defects and inhomogeneities, leading to the formation of macroscopical voids in body A in the vicinity of the interface. The first event is usually called the Kirkendall effect, the second is called the Frenkel effect.[1]

During bonding of dissimilar bodies, new features can arise because of contact melting of a material of higher melting point interacting with a material of lower melting point, and also because of mass transport through a vapour phase if there is a difference in the values of saturation vapour pressure for the two materials.

3.4. Diffusion through a fibre–matrix interface

If there is no thermodynamic equilibrium between the fibre and the matrix, then physical–chemical interaction between them is inevitable. The consequence will be a structure of the interface region which can contain substances with properties different from those of the component materials. In principle the situation is quite clear but there exist some technical difficulties in describing all the details.

The first difficulty arises because the phase diagrams of complex systems for elements presented in matrix and fibre materials are usually unknown. An exact calculation of such diagrams is impossible, unlike that for simple binary systems. An experimental study of the diagrams (SOKOLOVSKA and GUZEI, 1978) is a very time-consuming procedure.

---

[1] In English literature the Frenkel effect is also called the Kirkendall effect.

Secondly, the kinetics of the interaction in complex systems cannot be described exactly without special experiments. This means that it is difficult to evaluate interaction processes in a composite without making specimens and studying them. In fact almost any matrix modification demands a repeat of the whole study. This explains the large number of publications on this point.

Thirdly, the influence of the interface region on the mechanical properties of a composite can be varied, depending on the type of loading. This cannot be described systematically at present.

With regard to the first difficulty it should be noted that a lot of data has been collected and published (see, for example, SOKOLOVSKA and GUZEI, 1978). We have to refer to these publications because it seems impossible to present this experimental information in a compact form. The kinetics of the interaction and an influence of interface zones on mechanical properties of a composite will be discussed very briefly, although it is understood that any brief discussion of these topics will leave aside some important details. Nevertheless we will try to give a brief review of the recommendations about the organization of diffusion barriers.

3.4.1. Kinetics—a simple case

A simple case occurs when in the interfacial zone no new chemical compounds appear. If the fibre volume fraction in a composite is small enough for diffusion in the vicinity of one fibre to be unaffected by concentration gradients around other fibres, then the diffusion kinetics is determined by a solution of the well-studied equation for concentration c, written in cylindrical coordinates (and assuming radial and axial symmetry) as follows:

$$\frac{\partial c}{\partial t} = \frac{1}{r} \frac{\partial}{\partial r} \left(r D(c) \frac{\partial c}{\partial r} \right), \tag{3.9}$$

where the diffusion coefficient D depends on the concentration. Eq. (3.9) is solved usually assuming $D = \text{const}$, because otherwise a numerical analysis has to be applied.

With respect to the situation in a fibrous composite, such as analysis was done by HERRING and TENNEY (1973). The authors neglected non-axisymmetry of the problem and considered the concentration changes only along definite radii in a periodic fibre array aligned parallel to the fibre axis. The next difficulty was to choose a dependence of the diffusion coefficient on concentration of nickel in copper (because the data published previously had the scatter of one order of magnitude). Nevertheless, the solution appeared to correlate with the experimental results from the Cu–Ni-system. The solution procedure was a finite-difference one with the obvious boundary conditions. Also the auxiliary condition

$$\int_0^{R_A} c(r) \, dr = c_0 R_f$$

was to be satisfied. Here R_A is the radius of a cylindrical zone of influence associated with a single fibre, c_0 is the initial concentration of the diffusing element in the fibre.

Note that in the author's experiment with the Cu–Ni-system, the porosity in a reacted zone of the matrix has been clearly observed which corresponds to the Kirkendall–Frenkel effect. Of course it has not been taken into account in the calculations.

3.4.2. Kinetics in the case of chemical reaction

We start with a particular case studied in detail by KIM et al. (1978) to give an illustration of a complex problem. This is a case of interface reaction in the boron–aluminium composites. The first one is that with commercially pure aluminium (1100 alloy) as a matrix. The matrix of the second one is the aluminium–magnesium alloy containing small quantities of silicon, copper and chromium (6061 alloy). Specimens have been investigated after isothermal exposure at temperatures of 350°C and 500°C in air and dry argon.

The study of a layer remaining on the fibre surface after extraction from the 1100-alloy matrix by SEM-observation and X-ray diffraction shows that separate particles of $A\ell B_2$ start to appear after exposure at 500°C for 2.5 h. The number of such particles increases with time and after exposure for about 30 h a continuous layer of $A\ell B_2$ covers the fibre surface.

In the case of the aluminium–magnesium alloy, X-ray analysis of the reaction products reveals $A\ell B_{12}$. The particle size reaches about 5 μm after exposure at 500°C for 2.5 h and the surface layer becomes continuous after exposure for about 7 h. The exposure at 350°C for about 14 h corresponds (in terms of appearance of the particles) to that at 500°C for 2.5 h.

In both cases only aluminium and boron are present in the final reaction products, but the compositions of the products and the rates of reactions are quite different. The authors note that in the binary $A\ell$–B-phase diagram both carbides are present, $A\ell B_2$ being stable up to a relatively low temperature (975°C) and $A\ell B_{12}$ being stable up to 2070°C. Then the ternary $A\ell$–Mg–B-system shows, besides the two borides mentioned above, three aluminium–magnesium intermetallics and four magnesium borides. Because of the presence of alloying elements in the matrices the possibility of formation of aluminium and boron oxides as well as complex oxides of $A\ell$–Si–O and $A\ell$–Mg–O types, and another compound also exists. A thermodynamic analysis of the stability of the possible compounds gives the following chain of compounds, in order of growing stability at temperature 527°C: $A\ell B_2$–MgB_2–$A\ell B_{12}$–BN–$A\ell N$–MgB_4–B_2O_3–$A\ell_2O_3$–$A\ell_2MgO_4$–$A\ell_2SiO_5$.

To start the analysis of the kinetics, the authors assume that the diffusion of various elements in less stable $A\ell B_2$ proceeds faster than in $A\ell B_{12}$ at comparable temperatures ($J_{A\ell B_2} > J_{A\ell B_{12}}$), although a real situation can be complicated by defects of the structure, diffusion mechanism and ternary additions. Then, on the basis of the comparison of ionic and atomic radii they order the elements of interest in the following sequence of increasing diffusion coefficient: Mg–Aℓ–Si–B.

If one considers these assumptions and the experimental data together, the following conclusions can be produced. In the case of the 1100 aluminium alloy, the growth of the interface layer starts with the formation of $AℓB_2$ on the matrix side and $AℓB_{12}$ on the fibre side. Then, because the diffusional flux of aluminium through $AℓB_2$ is larger than that of boron through $AℓB_{12}$, the $AℓB_2$–$AℓB_{12}$ interface moves in the direction of the fibre surface. It leads to disappearance of the $AℓB_{12}$ phase. The whole process can be influenced by the formation and fracture of oxides of aluminium and boron. This decreases the growth rate of the boride layers.

To explain the opposite result in the case of the 6061-matrix, the authors have to assume $J_{AℓB_{12}} > J_{AℓB_2}$ because of the influence of ternary additions. In fact preliminary Auger electron analysis indicates the presence of magnesium and silicon at the extracted fibre surface.

Therefore a solution of the problem of the formation of a chemical compounds layer at the interface, although simple in principle, seems to be too complex because of the absence of much of the necessary data. These are mainly the diffusion coefficients and their dependence on defects in the structure. So investigations at the present stage, which is one of data accumulation, seem to be too time-consuming, not too effective but still unavoidable. Note that the data are presented usually in the form of the constants in the equations of purely phenomenological nature, namely

$$h = kt^{1/2},\tag{3.10}$$

$$k = K \exp\{-Q/RT\}.\tag{3.11}$$

Here h is the interface zone thickness, K and Q are constants depending on the content of alloying elements in a matrix, the fibre surface conditions, and the atmosphere. These factors are shown in order of decreasing influence.

To present here even a brief table of the results published by many authors would be hardly beneficial and the reader can find such data in review papers by METCALF (1974) and books by PORTNOI et al. (1979a) and SOKOLOVSKA and GUZEI (1978).

3.4.3. *The effect of component interaction on mechanical properties of composites*

Physical and chemical interaction at the interface occurred at the fabrication stage and also during service under some conditions (high temperatures and corrosive environments) and can influence the mechanical properties of composites in various ways. In particular the interface strength can be changed, a new phase can contribute to the composite strength, the mechanical properties of the fibre can be changed, and the mechanical behaviour of the matrix can also be altered. Some consequences of these changes can be predicted on the basis of results presented in Section 2.1.

3.4.3.1. The interface strength changes. Note firstly that observation of the structure of the interface (Section 3.1) does not bring any quantitative information about interface strength, and neither do thermodynamic considerations (Section 3.2.5). We should also remember that to measure the interface strength (Section 2.1.7) is quite a complex and contradictory problem. Finally it should be stressed that any consideration of this problem has to be connected to a particular problem or type of loading.

When a metal–fibre composite is to be loaded in the fibre direction, the problem has got a very simple solution (see Sections 2.1.1.1 and 2.1.2.1); namely, increasing the interface strength always leads to a better result— although quite good results (high strength and fracture toughness) can be obtained with a relatively weak interface. The choice of technological parameters does not appear to be very critical. Formation of a brittle inter-metallic layer on the interface does not lead to a decrease of the composite strength and some increase of the effective fibre strength can be observed.

A brittle fibre composite (Sections 2.1.1.2 and 2.1.2.2) can be affected by interface strength changes in a more complicated fashion. If its failure is accompanied by fibre breakage at weak points (part OA in Fig. 2), then increasing the interface strength leads to decreasing the critical fibre length and so to increasing fibre stress contribution to the ultimate composite stress (because of the scale dependence of the fibre strength). It also leads to increases in the plastic dissipation contribution to the effective surface energy of a composite and to decreasing energy dissipation at the interface. The latter should be favourable in the case of a tough matrix, and less favourable in the case of a less ductile matrix—cast alloys and plasma sprayed matrices being the examples. Certainly in this case an optimal interface strength and thus an optimal set of fabrication parameters are expected to exist. An example is supplied by bending tests of graphite fibre–aluminium composite with various contents of the aluminium carbide phase at the interface (see Fig. 31). The strength of such composites go up with the carbide content at small volume fractions of the carbide when obviously the interface strength can go up. But then the composite strength goes down, certainly because of the formation of a brittle layer decreasing the effective fibre strength (see Section 3.4.3.3).

A similar result was obtained when testing composites with the tungsten fibre and the copper matrix doped by manganese, which is soluble in tungsten (see Fig. 32).

Obviously the interface strength strongly affects the transverse strength of a composite. This property seems to be very sensitive to fabrication parameters. To give an example, we refer to the results of testing boron–aluminium composites by HOOVER and ALLRED (1974). The transverse strength of the composite, with commercially pure aluminium as a matrix and fibres coated with silicon carbide, changes from about 22 to 92 MPa depending on fabrication parameters. Hot pressing in an argon atmosphere gives better results than in air.

The influence of the interface strength and therefore of fabrication

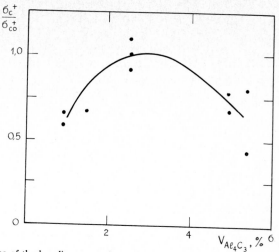

FIG. 31. Dependence of the bending strength of the graphite–aluminium composite on the content of the carbide phase at the interface. The matrix is commercially pure aluminium ($V_f = 0.46$), melt temperature at fabrication process is between 670 and 760°C, pressure on the melt is about 2 to 4 MPa, process time is less than 60 s, $\sigma_{co}^* = 570$ MPa. Experimental data by PORTNOI et al. (1981).

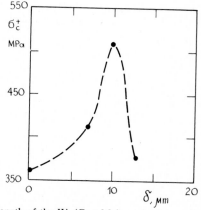

FIG. 32. Dependence of strength of the W–(Cu + Mn) composite on the thickness of the interface zone arising during annealing at temperature 850°C, $V_f = 0.148$. Experimental data by UMAKOSHI et al. (1974).

parameters on the compressive strength of a composite can be treated very simply (see Section 2.1.3).

Systematic data on the influence of the interface strength on other mechanical properties of composites are absent at present.

3.4.3.2. The additive contribution of a new phase. If a method of measuring the effective characteristics of the interface layers were found, then it would be easy to analyze its contribution to the composite strength. Such a method should certainly be based on measuring characteristics of composites. However, simple procedures of a type described in Section 2.1 applied to a three-component composite can lead to wrong results. For example, microcracking of a brittle phase at the interface can cause an expansion of the process zone at the tip of a macrocrack accompanied by plastic flow of the fibre and the matrix in the vicinity of circumferential microcracks. An increase in the effective fracture toughness of a composite according to the scheme described in Section 2.1.2.2 can be a result of this expansion.

An attempt to take into account a contribution of the brittle phase at the interface has been made by FRIEDRICH et al. (1974). The authors assume the strength of the interface layer of small enough thickness h to be higher than that of a part of the fibre replaced by the layer, so $\partial \sigma_c^+/\partial h > 0$ as $h \to 0$ (see Fig. 33). But when the thickness h increases, the strength of the layer decreases proportionally to $h^{-1/\beta}$, where β is the Weibull's parameter for a material of the brittle layer. Hence the layer's contribution to the composite strength varies as

$$\Delta \sigma^+ = \sigma_f^+ \left[\frac{\bar{\sigma}}{\sigma_f^+} \left(\frac{\bar{h}}{h} \right)^{1/\beta} - 1 \right],$$

i.e., the strength goes down, starting with a particular value of h. Therefore the dependence of σ_0^+/σ_{c0}^+ on h (σ_{c0}^+ is the strength of a composite at $h = 0$) should have a maximum as shown in Fig. 33.

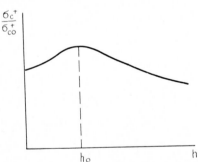

FIG. 33. Schematic dependence of the composite strength on the thickness of a brittle interface layer predicted by a model by FRIEDRICH et al. (1974).

3.4.3.3. Change in effective fibre characteristics. The influence of the physical–chemical interaction on the fibre strength can be observed in two ways.

Firstly the recrystallization and other structural changes of a fibre material can occur. This was revealed soon after carbon fibres had started to be studied. JACKSON and MARJORAM (1968) discovered the recrystallization of carbon and graphite fibres at a temperature of 1000°C was stimulated by the presence of nickel. It led to a drastic drop of the fibre strength (JACKSON, 1969). However, BARCLAY and BONFIELD (1971) believed that this effect had been caused not by nickel but elements soluted in nickel. They deposited pure nickel from the vapour and conducted the experiments in a good enough vacuum and did not observe recrystallization of nickel. But from the technological point of view the first result remains important.

A similar situation was then observed in other fibre–matrix systems. The tungsten fibres were investigated most widely. Diffusion of elements of a matrix to the tungsten fibre, its recrystallization and degradiation were studied in detail.

The second way is connected with the influence of a brittle interfacial zone containing chemical compounds of the fibre and the matrix elements on the fibre strength. A circumferential crack at a ductile fibre can cause a local increase of the fibre yield stress if the interface between the brittle layer and the fibre is strong enough (OCHIAI and MURAKAMI, 1976). The same situation in the case of a brittle fibre can obviously be unfavourable (USTINOV, 1979). Note that the dependence of strength of boron fibres coated with a layer of silicon carbide containing defects leading to microcracking on the coating thickness (see Fig. 34) illustrates in a convincing way the influence of a brittle interface on the fibre strength.

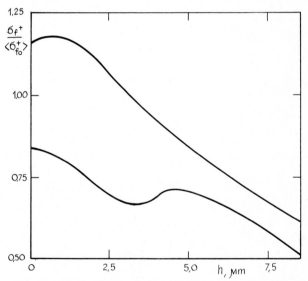

FIG. 34. A strength scatter band of boron fibres with silicon carbide coating depends upon the coating thickness according to USTINOV (1979), $\langle \sigma_{f0}^* \rangle = 2.98$ GPa.

Both the first (denoted as I in Table 3) and the second (denoted as II in Table 3) way lead to a decrease of characteristic fibre strength, usually measured by testing fibre specimens of a particular length. The schematic dependence of the fibre strength on time of exposure in contact with a matrix material are shown in Fig. 35. There are various mechanisms of fibre strength decrease and they have been studied in more detail. We are not going to

TABLE 3. Stability of some fibres in metal matrices.

Fibre	Matrix (coating)	t_c, h	T (°C)	Type of the curve (Fig. 35)	Type of inter-action	Sources[a]
C	Aℓ	24	580	A	I	1
C–HT	Aℓ	100	475	A	II	2
C–HM	Aℓ	100	550	A	II	2
C	Ni	1	600		I	3
C	Ni	5	600–800		I	4
C	Ni	<1	900	A	I	4
C–II	Ni	1	900	A	I	2
C	Ni	24	1000	A	I	1
C–I	Ni	1	1230	A	I	2
C	Ni	1	>1270			5
C	Ni–Cr	24	500	A	I	1
C	Co	24	700	A	I	1
C	Cu	24	800	B		1
SiC	Aℓ	24	700			1
SiC	Aℓ–3% Mg	10	580			6
B/SiC	Ti	0.5	870	B	II	7
B/SiC	Ti–6 Aℓ–4 V	0.5	850	B	II	7
B	Aℓ–3% Mg	100	400		II	6
		>10	500			
		1	540			
		0.1	580			
B		1000	230	B	II	6
		100	370	B	II	
	Aℓ-alloy of 6061-type	~0	540			
B		2	505			7
		0.1	560			7
B	Ti	1200	630	B	II	7
		0.15	870	B	II	
B	Ti–6Aℓ–4V	1500	540	B	II	7
		4.3	760	B	II	
B	Ni	24	400	B		1
Aℓ$_2$O$_3$	Ni	<1	1000			8
Aℓ$_2$O$_3$	80 Ni–20 Cr	<1	1000			8
Aℓ$_2$O$_3$	Ni–Cr–Fe	<16	1000			8
Aℓ$_2$O$_3$	TiC	0	1420			8
Aℓ$_2$O$_3$	HfC	>0	1320			8
Aℓ$_2$O$_3$	W	0	1320			8
		<16	1000			
Mo	Ni	<100	1100			9
W	Ni	<1	1100			9

[a] *References*: 1. JACKSON (1969); 2. BAKER and BONFIELD (1978); 3. WATANABE (1979); 4. DUDAREV et al. (1979); 5. WARREN et al. (1978); 6. MAXIMOVICH et al. (1979); 7. METCALF and KLEIN (1974); 8. TRESSLER (1974); 9. PORTNOI et al. (1979a).

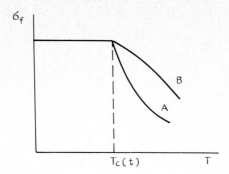

FIG. 35. Typical dependence of fibre strength on time of exposure with a metal matrix.

discuss the details, but just give a general table (Table 3), where characteristic parameters shown in Fig. 35 are collected.

To conclude we note the following:

(i) Optimal fabrication parameters can often be determined by two tendencies in a composite structure evolution. A fibre–matrix interaction leads to an increase of the interface strength but at the same time it gives birth to a brittle layer with defects or causes fibre recrystallization, i.e., leads to a decrease of the fibre strength. SUTTON and FEINGOLD (1966) applied such a consideration to sapphire–nickel composites and presented the results schematically as shown in Fig. 36. Certainly such a scheme can be considered as a first approximation to a procedure of choosing optimal fabrication parameters. However, it has not been developed quantitatively. An obvious reason for this is that an optimal composite structure is determined by a type of loading under

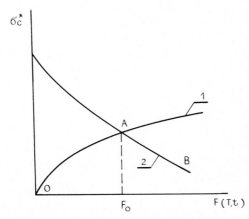

FIG. 36. Schematic diagram to explain the dependence of composite strength upon time–temperature regime of a fabrication process. Curve 1 accounts for the influence of the regime on the interface bond only (both the fibre and the matrix remain in the initial state) and curve 2 determines the influence of the regime on the fibre strength. Curve OAB is the strength–regime dependence. The idea of such a representation is due to SUTTON and FEINGOLD (1966).

service conditions (tension or compression, creep, fatigue and so on) and technological conditions (productivity rate, environments and so on).

(ii) The influence of the fibre on the matrix has been studied in less detail (because such influence is of less importance).

(iii) When choosing fabrication parameters Table 3 should be considered only for the very initial orientation.

3.4.4. Diffusion barriers

A fibre–matrix combination desirable from the point of view of mechanical properties of the components is often characterized by a tendency to unwanted physical or chemical interaction. A natural way to restrict this interaction is to introduce an interface layer to serve as a diffusion barrier which also should provide a necesary value of the interface strength. This is often absolutely necessary in the case of heat-resistant composites, when the interaction rate can be relatively high even under service conditions.

In principle, the problem of choosing and forming diffusion barriers can be solved on the basis of the research results described above in Sections 3.4.1 and 3.4.2 and in more detail in the references given there. Certainly the influence of interface layers on the composite strength (Section 3.4.3) should also be taken into account. We will just give two examples to illustrate possible solutions to the problem.

The difficulties that arose in the case of sapphire–nickel composites are well known (see Table 3). A great number of the possible barriers has been studied including refractory metals, carbides and so on. Quite a good result has been obtained using a combined fibre coating containing Y_2O_3–W–Ni (TRESSLER, 1974).

Many publications can be found on diffusion barriers in tungsten–nickel composites. CORNIE et al. (1978) have conducted a most detailed study of the thermal stability of the system tungsten–a barrier–nickel. Among the following substances—HfO_2, Y_2O_3, $A\ell_2O_3$, TiC, ZiN, HfC—they have chosen hafnium carbide as the appropriate candidate. The HfC-layer provides the stability of a composite at temperature 1175° for at least 2000 h. However, the authors conclude that their study has raised as many questions as answers. In particular, it is not clear how matrix impurities affect the barrier stability, how barrier stoichiometry affects the barrier stability, and what is the efficiency of the barriers during thermocycling.

Other examples of diffusion barriers for nickel matrix composites are given by PORTNOI et al. (1979a) who also give recommendations on matrix alloying to decrease its interaction with fibres.

4. Principles of technological processes

The requirements for a fabrication process, as well as some general guidelines to organize the technology were discussed in two preceding sections. Here we shall give a brief summary of the main technological schemes, emphasizing, not the engineering details (which can be extremely varied), but rather on techniques to satisfy the requirements mentioned above. It should also be noted that a proper fabrication process should provide a possibility to obtain a structural component which needs only minor forming and fastening operations.

4.1. Fabrication of composite precursors

Many technological schemes include manufacturing semi-fabricated products as a first stage. It is similar to the well-known prepregs in the fibre reinforced plastics technology. Preliminary bonding of fibres to a matrix appears to be convenient for further fabrication of structural components. A most widely used method of manufacturing semi-fabricated products at present is plasma spraying.

4.1.1. Plasma sprayed tapes

A jet of low temperature plasma had been used for deposition of metals and various coating on a solid surface long before the necessity of obtaining composite semi-fabricated products arose. Hence the suggestion to use such a process to deposit a layer of a matrix material on a set of fibres by KREIDER (1970) was a very natural step. In KREIDER's disclosure a technical scheme was also present and it was to form a basis for the present fabrication method.

The process is usually conducted in the following way. Fibres are wound onto a cylindrical mandrel with a fixed pitch. Then drops of the the molten matrix material are carried by a low temperature plasma jet to the mandrel surface, coating the fibres with a matrix layer. Then a composite layer with a weak matrix is cut along a generatrix of the cylinder and a precursor sheet or tape removed from the mandrel. Such a product is to be densified and sintered in a process of producing a particular structural component.

Besides the factors important in assessing any fabrication method (output, energy consumption and so on), in this particular method we have to be interested in the following factors: properties of the matrix obtained, the influence of the process on the mechanical properties of the fibres, the quality of fibre packing, and the possibility of effective processing of the semi-fabricated material. We shall consider some of these factors, looking mainly at the process of making aluminium–matrix composites.

Porosity of the plasma sprayed tape is inevitable; in some cases it can be useful—oxide layer at the fibre–matrix interface is being broken and an interface bond can then be formed as a result of the large displacement at the interface during densification of the matrix (PREWO, 1975). The porosity depends on spraying parameters, such as the electric power, the powder size and so on. The dependence of the porosity on the powder size has a minimum (KARPINOS et al., 1979).

During spraying the chemical composition of a matrix can change due to oxidation if the process is conducted in air or in atmosphere of an impure inert gas. RYCALIN et al. (1977) measured oxygen content in aluminium–magnesium alloys after spraying and found that it changed from the initial value equal to 0.06–0.09 per cent to 0.7–0.8 per cent after spraying in air, to 0.27–0.31 per cent after spraying with a local protection by argon gas, about 0.25 per cent after spraying in a box with an argon atmosphere, and about 0.22 per cent after spraying in the same box with a zirconium getter.

The authors also studied the formation of a structure of the aluminium–magnesium alloys after spraying and found temperatures between 560 and 580°C to be optimal for hot pressing. The alloys hot pressed in this temperature interval have the highest tensile strength, although such temperatures are too high for the boron fibres to preserve their strength (see Table 3), so the optimal temperature interval would appear to be lower. Decreasing the oxygen content in a matrix permits a reduction in hot pressing temperature (corresponding to a maximum value of the matrix tensile strength) by about 30–40°C.

Melting and subsequent rapid cooling of a matrix alloy during plasma spraying can fundamentally change the structure of the future matrix, when compared with the structure of a nominal alloy (which is usually taken from a series of wrought alloys). A possible deterioration of mechanical properties of the alloy can lead to a corresponding deviation of composite properties from those predicted on the basis of nominal alloy characteristics. That is why the investigation of the alloy properties after its spraying and various treatments is of importance.

ALIPOVA et al. (1979) carried out such a study of the aluminium–zinc–magnesium alloy which could be strengthened by heat-treatment. They directed their study towards fabrication of boron–aluminium composites by hot isostatic pressing. The specimens were obtained by depositing the matrix layer-by-layer in air, the density of the sprayed material being from about 10 to 15 per cent less than that of the nominal alloy. The oxygen content in the original alloy was 0.01 per cent, in sprayed specimens about 0.11 per cent by weight. The strength of these specimens was about 50 MPa. Then the specimens were placed into a vacuum container and hot pressed in a high temperature autoclave. The authors gave a table of the pressing parameters, namely temperature T, pressure q and time t, as well as strength σ_m^+ of final materials. They noted that the ultimate elongation of the material does not depend on pressing parameters and was equal to about 2.5 per cent. They also noted that a set of fabrication parameters previously found to be optimal (see Table 4) does not give matrix material with the highest strength.

Let us attempt to analyze the results obtained by ALIPOVA et al. (1979) in the following way. We shall assume that there exists the equivalent time expressed as

$$\vartheta = t \left(\frac{\sigma}{\sigma_*}\right)^n \exp\left\{-\frac{T_*}{T}\right\} \tag{3.12}$$

where n, σ_* and T_* are constants. It is obvious that such an assumption does not contradict common sense, and further, a set of the constants can be obtained which would supply a best correlation between σ_m^+ and ϑ. Becasue of the shortage of experimental data we shall not perform a standard procedure to search for such constants, and will just choose the following values: $n = 2$, $\sigma_* = 40$ MPa, $T_* = 1000$ K. The result obtained is shown in Fig. 37.

The experimental points in the σ_m^+–ϑ plane form two continuous lines, each with quite a small scatter. Obviously it is impossible to combine two groups of points into one by a better choice of the constants. So one must assume that lower values of the strength in three tests are the result of factors which have not been studied in the experiment. Then the upper line is the important one. It should be noted that the dependence of the strength on equivalent time ϑ is nearly exponential.

The ultimate matrix elongation can be made higher by annealing in vacuum and a standard heat treatment after annealing leads to the strength increase and the ultimate strain decrease.

During plasma spraying of a matrix, the fibre is subjected to the impacts of molten drops and the short-time action of liquid metal. A possible degradation of fibres as a result of action of the plasma jet alone was studied by SHOR-SHOROV et al. (1978) who showed that a combination of conditions could be

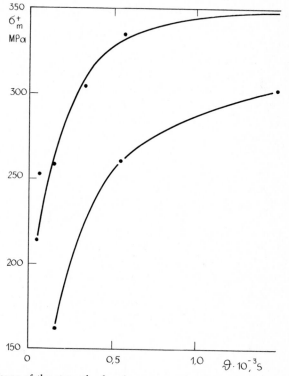

FIG. 37. Dependence of the strength of a plasma sprayed Aℓ–Zn–Mg-matrix alloy on equivalent time of hot pressing. Experimental data by ALIPOVA et al. (1979).

found which did not degrade boron and silicon carbide fibres. An optimal set of plasma spraying parameters should be chosen on the basis of specific experiments.

4.1.2. Infiltration to obtain wires and tapes

Pulling a fibre bundle through a bath of a molten metal (see Fig. 38) can be used to produce a composite precursor cable with a large enough fibre volume fraction. The time of contact between the fibre and the molten matrix, as well as the coating necessary to form a diffusion barrier or to promote wetting, can be estimated prior to the experiments (see Sections 3.2 and 3.4). Expected mechanical properties can also be estimated and analyzed (Section 2.1).

We shall mention a technical scheme of producing boron–aluminium cables with a matrix of commercially pure aluminium (SEMENOV et al., 1978) as an example. The rate of pulling is about 1 m/s. The composite precursor emerges after infiltration with a melt which begins to crystallize at a temperature about 90°C below the melting point. Crystallization starts at the fibre surface and gives a matrix with properties better than after normal casting.

A similar scheme is used by GIGERENZER et al. (1975) to produce a cable containing seven filaments of silicon carbide in an aluminium matrix (the 6061-alloy). The fibres are in contact with the melt for less than 1 s, and the fibre volume fraction can be varied by changing the melt temperature, pulling rate and the distance between fibres.

The third example is the production of the maraging steel–aluminium tapes (FRIEDRICH et al., 1975). A plane array of the steel cables is pulled through the melt, the pulling rate being equal to about 1.5 m/s and after the bath the tape is slightly compressed. The width of the tape is about 30 mm, and the fibre volume fraction is up to 60 per cent.

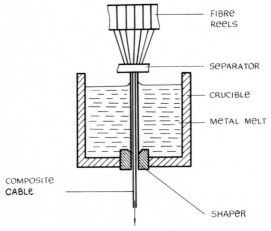

FIG. 38. A scheme of producing a composite cable by pulling a fibre bundle through a matrix melt.

Certainly the most effective use of liquid infiltration to form composite cables is in the case of carbon fibres. The fabrication scheme includes two main stages (GIGERENZER et al., 1978; MEYERER et al., 1978). At first fibres are coated by a mixture of titanium and boron to promote wetting. The procedure is based on the reduction of titanium tetrachloride and boron tetrachloride by zinc vapour and yields thin (10–20 nm) and uniform coating of each fibre in a bundle. Then the fibre bundle is pulled through the melt of an aluminium alloy and a compact composite wire is obtained. The strength of the cables at $V_f = 0.35$–0.40 is about 1.0 to 1.3 GPa. Photomicrographs of the cross-section reveal rather inefficient fibre packing.

An interesting method of prebonding fibres and a matrix, which can be used for making precursors as well as structural elements, was suggested by DIVECHA (1974). A thin cable of matrix material is wound onto a reinforcing fibre. Then such wound fibres are packed and an almost ideal fibre distribution is expected to be obtained.

4.2. Hot pressing

This is a variety of fabrication methods of processing semifabricated materials to produce structural components. All these methods are mainly based on using static pressure in procedures of non-continuous nature.

4.2.1. Processing parameters

We have already discussed the factors which determine the choice of processing parameters. But at present this procedure has been carried out mainly in an empirical fashion. Now it is obvious that there should exist an optimal set of temperature T, pressure q and pressing time t. The dependency of a mechanical property (for example tensile strength) on fabrication parameters is of the form shown in Fig. 39. An optimal combination T_0, q_0, t_0 is usually found in a particular study. There is a possibility that a few of the combinations of T, q and t give rise to a maximum for a mechanical property. It is obvious that different combinations can correspond to maximum values of different properties.

Table 4 gives the processing parameters published in the literature,

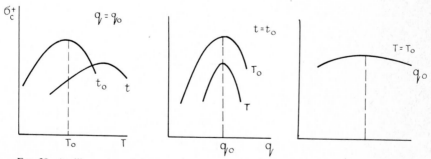

FIG. 39. An illustration of the choice of an optical set of the hot pressing parameters.

TABLE 4. Examples of hot pressing parameters for some composites.

Fibre	Matrix	Type of precursor	Atmosphere	T (°C)	q_0 (MPa)	t, h	Property of optimize	Sources
Stainless steel	Aℓ	foil	vacuum	510	30	4		PINNEL, LAWLEY (1970)
	Aℓ	foil	vacuum	525	35	0.5	σ_c^-	MILEIKO, KHVOSTUNKOV (1970)
B	Aℓ	plasma sprayed		560		0.33	σ_c^+	RYKALIN et al. (1977)
B	Aℓ	foil	vacuum	500–530	30	0.25	σ_c^+	SHORSHOROV et al. (1978)
B	AD33	plasma sprayed	vacuum	550	35	1		BREINAN, KREIDER (1970)
B	AD33	plasma sprayed	air	550	69	0.17	σ_c^+	PREWO (1975)
B	D16	foil	vacuum	485	25	1.5	σ_c^+	MILEIKO et al. (1973)
B	Aℓ	plasma sprayed		500	40	0.33	σ_m^+	ALIPOVA et al. (1979)
B	SAP	foil	vacuum	620–690	40–60	0.25–0.5		KARPINOS et al. (1977)
B/SiC	Ti	foil	vacuum	780–920		0.5		KARPINOS et al. (1977)
B/SiC	Ti-alloy	foil + powder	vacuum	870	85			METCALF (1974)
B	Mg	foil	air	580	50	0.33	σ_c^+	SWETLOV et al. (1975)

which are optimal or close to optimal according to the authors. These values can serve as a first approximation only because the really optimal values can be influenced by factors which are not shown in Table 4—for example, parameters of fabrication of precursor elements, the quality of preparation of semifabricated materials to processing, the fibre volume fraction, temperature gradient and heating method and so on.

4.2.2. Techniques

The techniques of hot pressing, except those under the isostatic condition, are simple. However the necessity of repeated pressure application, which is a specific feature of hot pressing, usually means that the process is time-consuming.

To obtain components of a simple shape is possible using a rigid die (see Fig. 40). The temperature should be about 500°C for aluminium, about 800°C for titanium, and about 1000 to 1200°C for nickel matrices. So corresponding materials have to be chosen for the dies with built-in heaters. A chamber should provide the necessary displacement of a die, and a vacuum or protective atmosphere. The choice of a press is determined by the necessary pressure (see Table 4), which is usually equal to about 10 to 100 MPa.

Using plasma sprayed precursors it is possible to conduct hot pressing in air. Large relative displacements of the surfaces coated with oxide layers lead to fracturing of such layers and provide conditions for bonding. PREWO (1975) has shown that the process parameters in this case can be chosen such as to decrease pressing time down to 10 minutes.

Using dynamic hot pressing studied in detail by KARPINOS et al. (1972) also leads to a decrease of the total process time.

One other possibility to decrease the total process time and to obtain large elements is to use the step-wise pressing described by PORTNOI et al. (1979b).

Fabrication of tubes and shells can be done if a 'soft' die is used (see Fig. 41).

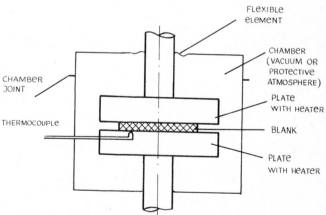

FIG. 40. A sketch of a chamber for hot pressing a composite.

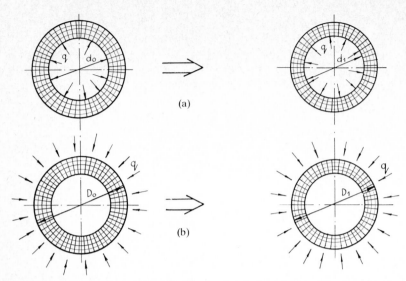

FIG. 41. A scheme for making a composite tube in an autoclave. (a) Densification of a blank with the outer rigid wall; (b) The same at the inner rigid wall.

Hence the isostatic pressure is wanted. But simple liquid autoclaves are of no use because pressing temperatures are too high (see Table 4). Therefore more expensive high pressure gas autoclaves should be used. WEISINGER (1967) was certainly the first to suggest applying gaseous pressure to densify and sinter boron–aluminium tubes according to the scheme shown in Fig. 41(a). Precursor boron–aluminium tape is rolled on a thin-walled steel mandrel. Then it is inserted into a thick-walled outer steel tube. The steel tubes are welded together at the ends to give a vacuum-tight assembly. Evaluation can be done via a special tube welded to the thick-walled steel tube. After diffusion bonding in an autoclave, the outer tube is machined to almost the same thickness as the inner tube and then both steel tubes are etched in nitric acid.

This method has many advantages as well as disadvantages. A scheme suggested by HEARN et al. (1978) permits decreasing the volume of gas and producing composite tubes without using an autoclave. Nearly the same final result, i.e., decreasing the product (gas volume times pressure) can be reached using low-pressure bonding of plasma sprayed braze tapes or specially prepared tapes which produce eutectics at the tape interface (CHRISTIAN, 1975).

4.3. Explosive welding

Explosive welding, discovered by LAVRENTJEV about 35 years ago (see DERIBAS, 1980), is an efficient way to bond together two similar or dissimilar materials. While the possibility of such welding had not been predicted, many theoretical treatments followed the discovery. The investigators were looking for optimal process parameters as well for possible applications. DERIBAS (1980) described the state-of-the-art in detail.

As a rule the impact of two plates is accompanied by a stationary periodic process leading to a characteristic undulating waved interface (see Fig. 42). All the important events take place in the vicinity of point A, which runs behind the detonation front (see Fig. 43). Certainly the conditions for forming the cumulative jet occur periodically here.

In the first attempts to produce metal–metal composites by explosive welding (see, for example, CROSSLAND et al., 1971; JARVIS and SLATE, 1968) the direction of detonation front propagation was chosen to coincide with the fibre direction. In this case some new features were observed and they were associated with the flow of the matrix layer around a more rigid fibre (SLATE, 1975). Many experimental data show that the choice of technological parameters in this case is not a difficult task. The situation is easily reached when the mechanical behaviour of a composite obtained by explosive welding corresponds to that of a composite with an ideal interface bond. Moreover, the very short time of the process excludes the formation of brittle interface layers as well as annealing the reinforcing wires. Another advantage of explosive welding is the possibility of making sheets and plates of large sizes with a high productivity rate.

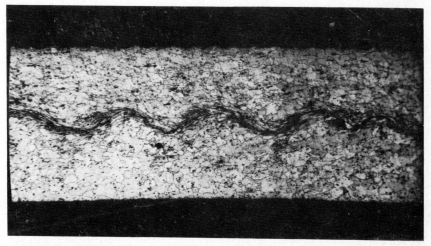

FIG. 42. Wave formed at the interface obtained by explosive welding.

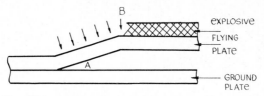

FIG. 43. A scheme of explosive welding of two plates.

When making composite tubes and shells with non-axial fibre directions it is not clear a priori what will be the result of the interaction of wave-forming processes at the matrix–matrix interface with a set of the reinforcing fibres lying, for example, along the impact front. A corresponding experiment is described by KONDAKOV and MILEIKO (1977) and MILEIKO et al. (1974). An assembly before detonation is shown in Fig. 44, and an example of a structure can be seen in Fig. 45.

An interesting situation arises when the wave formation is completely excluded—if, for example, the reinforcing element is a wire mesh (BHALLA and

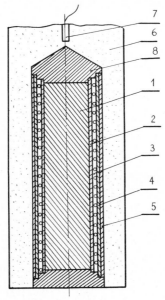

FIG. 44. An assembly to fabricate a composite tube with circumferencial reinforcement. 1—mandrel; 2 and 4—matrix layers; 3—wire reinforcement layer; 5—protective shell; 6—explosive; 7—detonator; 8—cone (KONDAKOV and MILEIKO, 1977).

FIG. 45. The steel–aluminium composite obtained by explosive welding (KASPEROVICH and KONDAKOV).

WILLIAMS, 1977). In this case the bonding mechanism appears to differ from that described above. In going through the mesh, the matrix surfaces are cleaned and then the physical contact arises and the bond forms.

4.4. *Liquid infiltration*

Wetting conditions and fibre stability in a molten matrix, possibilities of fibre coating to promote wetting and fibre protection as well as possibilities of matrix modification to provide conditions for conducting liquid infiltration were analyzed in Section 3.2. This analysis obviously leads to the conclusion that it is possible to bring a liquid matrix into a composite. Such fabrication methods can be expedient for at least two reasons. Firstly, filaments of a small diameter (carbon, silicon carbide, polycrystalline sapphire, various whiskers) can hardly be introduced into a solid matrix. Secondly, liquid infiltration methods are supposed to be simple.

Certainly both these reasons led to the development of the fabrication method for graphite fibre–aluminium matrix composites, based on making a composite wire using liquid infiltration (see Section 4.1.2). So possible schemes of infiltration method have already been described and it merely remains to give some more examples.

The early paper by MORRIS (1971) contains a scheme of infiltration under pressure by pressing a container with aluminium as matrix material and unidirectionally oriented carbon fibres. The pressure is to be applied after melting of the matrix.

A scheme of the vacuum infiltration of a bundle of fibres wetted by a molten matrix was also suggested at the very beginning of composite technology (DEAN, 1967). Here a driving force is the sum of capillary and atmospheric pressure which appears to be enough to lift a liquid column to a necessary height if fibre volume fraction is large enough. PORTNOI et al. (1979a) estimated the time for filling capillary channels by a molten matrix under such conditions, as well as the time for cooling and crystallization of the matrix. The second time naturally appears to be essentially longer, and limits the possibility of making composites with interactive components.

PORTNOI et al. (1979b) presented a good review of fabrication schemes based on the liquid infiltration. They are the result of an intensive activity of inventors. We shall not try to give our own review of this activity as it would be outside the projected plan of this chapter. We should just note that, in choosing a liquid infiltration method, one has to bear in mind the following points.

(i) A uniform fibre packing. The importance of this was discussed in Section 2.1. The effectiveness of the solution of this problem has been demonstrated in experiments conducted in a space laboratory (TAKAHASHI, 1978).

(ii) A choice of a matrix with satisfactory mechanical properties.

(iii) Providing the means to control the time of contact between the fibre and the molten matrix.

4.5. Rolling and drawing

These traditional methods of metal processing are attractive because they are very usual. A possibility to use them to produce composite sheets and profiles has been shown in a number of publications.

4.5.1. Rolling

The possibility of producing metal–metal composites by rolling is well established. Perhaps the only essential feature of the process is a necessity to conduct it in vacuum or with a protective atmosphere if foil is used for a matrix. It can be achieved either using a vacuum-tight container or a vacuum rolling machine. In this way the molybdenum–nickel specimens were obtained by KOPECKY et al. (1970). The tungsten–nichrome composites were produced by SEVERDENKO et al. (1974) who used evacuated containers. In the latter case the temperature was 1100–1200°C and the total reduction was about 30 to 35 per cent.

To use rolling for making brittle fibre composites one needs to choose processing parameters more carefully. GUSEV et al. (1980) were able to get a boron–aluminium composite by rolling plasma sprayed monotapes, the matrix being an $A\ell$–Zn–Mg alloy. A composite of good quality was obtained after 5 to 6 passes with total reduction of about 50 per cent, at a temperature of 400–450°C.

DOBLE and TOTH (1975) have reported to have obtained a boron–aluminium composite by rolling a package of the matrix foils and the filament mat, kept together by a polystyrene binder. The package is placed in between two stainless steel cover plates of about 8 mm thickness. The assembly is preheated either in an argon atmosphere or in air. (In the latter case it has to be wrapped with a protective foil.) The most important parameters of the process are temperature and rolling pressure. When temperature increases, the pressure necessary to form a bond goes down and at high values of the pressure fibres tend to fracture. The optimal temperature for the 6061-alloy is 565°C, which is slightly higher than that for static hot pressing. Rolling speed does not seem to influence the composite strength. The same method has also been used for producing precursor tapes. The largest sizes of the tape obtained by the authors are 1200 × 15 mm.

It should be noted that rolling composite sheets can be used to provide the matrix and the fibre with the required mechanical properties (GETTEN and EBERT, 1969; KOPJOV et al., 1980).

4.5.2. Drawing and extrusion

Drawing and extrusion are widely used for manufacturing metal rods, and profiles can be successfully used for combining composite cables together, as well as for making whisker-reinforced metal. In the latter case a process is to be carried out to achieve a unidirectional fibre orientation.

An example of drawing in an evacuated container (about 0.1 torr) has been

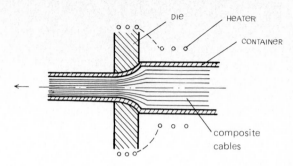

FIG. 46. Drawing in an evacuated container.

given by GIGERENZER et al. (1978) (see Fig. 46). For obtaining graphite–aluminium composites using precursor wires, the temperature of the process is about 500°C, the speed about 5 cm/s and the material of the container is inconel. When the diameter of the container is 20 mm, its wall thickness is about 0.9 mm. To get a composite with a high enough strength, the value of the reduction should be between 5 and 15 per cent. It has been noted that a minor modification to the fabrication scheme allows tubes with various profiles to be made.

4.6. Powder metallurgy methods

Powder metallurgy methods can bring to the composite technology all their advantages and at least two more. Firstly, they lead to decreasing technological temperatures and therefore give more possibilities to control the fibre–matrix interface properties. Secondly, the extrusion of a mixture of matrix powder and fibres can provide orientation of a whisker-type fibre.

Some elements of powder metallurgy can be seen in solid state fabrication methods considered above, the formation of a matrix from a plasma sprayed precursor being the example.

DIVECHA et al. (1981) have recently demonstrated an example of application of a pure powder metallurgy method to the fabrication of the SiC–aluminium composite. Their fibres are the whiskers of β–SiC obtained from rice hulls by a cheap (pyrolitic) process. The composite fabrication scheme includes blending of a mixture of whiskers and a matrix powder (possibly with binder additions), cold compacting, hot pressing of a billet, extrusion or rolling. Fibre packing appears to be quite uniform and whisker clusters are not observed on photomicrographs.

Note that the classical physics of sintering can only be applied to an analysis of sintering of a matrix with fibres in the same way as solid state mechanics applies to the analysis of the mechanical behaviour of non-homogeneous bodies. At present only preliminary attempts to do this are known (for example, TUL'CHINSKY, 1974).

5. Conclusion

It does not seem to be necessary to conclude this chapter with a brief summary, because the whole chapter had to be rather an expanded summary on most items. We had to omit many important items of composite technology, in particular the bonding of composite elements to metal elements, processing of composites to make structural components, and others. Nevertheless, after this chapter has been written, the author has become even more convinced of the usefulness of making composite technology a research field incorporating mechanics, physics and chemistry as its main components.

Acknowledgement

The author thanks Mrs A. Kucack, Mrs L. Kapustina, Mrs G. Peregudova and Miss T. Chernova for their help in preparing the manuscript. Many thanks to Anthony Kelly who made the author's English more readable for the English speaking reader.

References

AHMAD, I. and J.M. BARRANKO (1973), *Met. Trans.* **4**, 793.
ALIPOVA, A.A., V.V. IVANOV, V.G. IVANOV, V.V. KUCHKIN and A.A. KHVOSTUNKOV (1979), *Fizika i Chimia Obrabotki Materialov* **6**, 116 (in Russian).
ALLEN, B.C. and W.D. KINGERY (1959), *Trans. AIME* **215**, 30.
ARCHANGELSKA, I.N. and S.T. MILEIKO (1976), *J. Mater. Sci.* **11**, 356.
BAKER, S.J. and W. BONFIELD (1978), *J. Mater. Sci.* **13**, 1329.
BARCLAY, R.B. and W. BONFIELD (1971), *J. Mater. Sci.* **6**, 1076.
BHALLA, A.K. and J.D. WILLIAMS (1977), *J. Mater. Sci.* **12**, 522.
BREINAN, E.M. and K.G. KREIDER (1970), *Met. Trans.* **1**, 93.
CHAMPION, A.R., W.H. KRUEGER, H.S. HARTMAN and A.K. DHINGRA (1978), Fibre FP reinforced metal matrix composites, in: B. NOTON, R. SIGNORELLI, K. STREET and L. PHILLIPS, eds., *Proc. 1978 Internat. Conf. Composite Materials* (Met. Soc. AIME, New York).
CHRISTIAN, J.L. (1975), Fabrication—method and evaluation—of boron aluminium composites, in: E. SCALA, E. ANDERSON, I. TOTH and B.R. NOTON, eds., *Proc. 1975 Internat. Conf. Composite Materials* (Met. Soc. AIME, New York).
COOPER, G.A. and A. KELLY (1967), *J. Mech. Phys. Solids* **15**, 279.
CORNIE, J.A., J.J. SHREURS and R.W. PALMQUIST (1978), A kinetic and microstructural study of oxide, carbide, and nitride diffusion barriers in HSTW (tungsten) reinforced MAR-M-200 composites, in: B. NOTON, R. SIGNORELLI, K. STREET and L. PHILLIPS, eds., *Proc. 1978 Internat. Conf. Composite Materials* (Met. Soc. AIME, New York).
CROSSLAND, B. and A.S. BAHRANI (1971), *Welding and Metal Fabrication*, p. 20.
DEAN, A.V. (1967), *J. Inst. Met.* **95**, 79.
DERIBAS, A.A. (1980), *Physics of Explosive Strengthening and Welding* (Nauka, Novosibirsk) (in Russian).
DIVECHA, A.P. (1974), US Patent N 3 **828**, 417.
DIVECHA, A.P., S.G. FISHMAN and S.D. KARMAKAR (1981), *J. Met.* **33**, 12.
DOBLE, G.S. and I.J. TOTH (1975), Roll diffusion bonding of boron aluminium composites, in: E.

SCALA, E. ANDERSON, I. TOTH and B.R. NOTON, eds., *Proc. 1975 Internat. Conf. on Composite Materials* Vol. 2 (Met. Soc. AIME, New York).

DUDAREV, E.E., L.A. TOROVEC, V.E. OVCHARENKO, G.P. BACACK and V.F. TREGUBOV (1978), *Fizika i Chimia Obrabotki Materialov* **6**, 108 (in Russian).

DUNDURS, J. and M. COMNINOU (1979), The interface crack in retrospect and prospect, in: G.C. SIH and V.P. TAMUZS, eds., *Proc. 1st USA–USSR Symp. on Composite Materials* (Sijthoff and Noordhoff, Alphen a/d Rijn).

FRENKEL, YA.I. (1945), *J. Physics USSR* **9**, 385.

FRIEDRICH, E., I.M. KOPIEV, YU.E. BUSALOV, G.YU. WEISS and S. WIELHELM (1975), *Fizika i Chimia Obrabotki Materialov* **6**, 115.

FRIEDRICH, E., W. POMPE and I.M. KOPJEV (1974), *J. Mater. Sci.* **9**, 1911.

GARMONG, G. and L.A. SHEPARD (1971), *Met. Trans.* **4**, 863.

GEGUSIN, J.E. (1967), *Sintering Physics* (Nauka, Moscow) (in Russian).

GETTEN, J.R. and L.J. EBERT (1969), *ASM Trans.* **62**, 869.

GIGERENZER, H., R.T. PEPPER and W.L. LACHMAN (1978), Hot drawing of fiber (filament) reinforced metal–matrix composites, in: B. NOTON, R. SIGNORELLI, K. STREET and L. PHILLIPS, eds., *Proc. 1978 Internat. Conf. on Composite Materials* (Met. Soc. AIMS, New York).

GODDARD, D.M. (1978), *J. Mater. Sci.* **13**, 1841.

GODDARD, D.M. and E.G. KENDALL (1977), *Composites* **8**, 103.

GORBATKINA, YU.A. and T.N. KHAZANOVICH (1981), *The 5th USSR Congress on Theoretical and Applied Mechanics*, Abstracts of papers (Nauka, Alma-Ata) (in Russian).

GUSEV, E.D., V.V. IVANOV, V.V. KUCHKIN and T.T. TARAKANOVA (1980), *Fizika i Chimia Obrabotki Materialov* **1**, 137 (in Russian).

HARROD, D.L. and R.T. BEGLEY (1966), Some general aspects of interfaces in composites, in: *Proc. 10th Nat. SAMPE Symp.* (SAMPE, San-Diego).

HEARN, D., Y. FAVRY and A.R. BUNSELL (1978), Fabrication de tubes en material composite bore—aluminium, Presented at *Tournees Nat. sur les Composites*-1, Paris.

HERRING, N.W. and D.R. TENNEY (1973), *Met. Trans.* **4**, 437.

HILL, R. (1948), *Proc. Roy. Soc.* **A193**, 281.

HITCHCOCK, S.J., N.T. CAROLL and M.G. NICHOLAS (1981), *J. Mater. Sci.* **16**, 714.

HOOVER, W.R. and R.E. ALLRED (1974), *J. Compos. Mater.* **8**, 55.

JACKSON, P.W. (1969), *Met. Engrg. Quart.* **12**, 22.

JACKSON, P.W., D.M. BRADDICK and P.J. WALKER (1972), *Fibre Sci. Technol.* **5**, 219.

JACKSON, P.W. and J.R. MARJORAM (1968), *Nature* **218**, 83.

JARVIS, C.V. and P.M.B. SLATE (1968), *Nature* **220**, 782.

KALNIN, I. (1977), US Patent 4.056, 874.

KARAKOZOV, E.S. (1976), *Solid State Binding of Metals* (Metallurgia, Moscow) (in Russian).

KARPINOS, D.M., V.P. MORZE, V.G. ZILBERBERG, V.KH. KADYROV and V.P. DOROCHOVICH (1979), *Poroshkowaja Metallurgia* **7**, 57 (in Russian).

KARPINOS, D.M., L.I. TUCHINSKY and L.R. VISHNJAKOV (1977), *Modern Composite Materials* (Vysha Shkola, Kiev) (in Russian).

KELLY, A. (1970), *Proc. Roy. Soc.* **A319**, 95.

KELLY, A. (1973), *Strong Solids* (Clarendon Press, Oxford).

KELLY, A. and J. LILHOLT (1969), *Phil. Mag.* **20**, 311.

KELLY, A. and K.N. STREET (1972), *Proc. Roy. Soc.* **A328**, 283.

KENDALL, E.G. (1974), Graphite fibres—aluminium composites, in: L.J. BROUTMAN and R.H. KROCK, eds., *Composite Materials*, Vol. 4; K.G. KREIDER, ed., *Metal Matrix Composites* (Academic Press, New York–London) Ch. 7.

KIIKO, V.M. (1981), *Mechanica Compositnich Materialov* **4**, 615.

KIM, W.H., M.J. KOCZAK and A. LAWLY (1978), Interface reaction and characterization in $B/A\ell$ Composites, in: B. NOTON, R. SIGNORELLI, K. STREET and L. PHILLIPS, eds., *Proc. 1978 Internat. Conf. on Composite Materials* (Met. Soc. AIME, New York).

KOLESNICHENKO, G.A. (1972), On wetting of covalent crystals of high melting point by metal melts in: V.N. EREMENKO and YU.V. NAIDICH, eds., *Wettability and Surface Properties of Melts and Solids* (Naukova Dumka, Kiev) (in Russian).

KONDAKOV, S.F. and S.T. MILEIKO (1974), *Mashinovedenie* **3**, 73 (in Russian).

KONDAKOV, S.F. and S.T. MILEIKO (1977), *Mechanica Polymerov* (Polymer Mechanics) **7**, 90 (in Russian).

KOPECKY, CH.V., V.L. ORZHECHOVSKY and A.M. MARKOV (1970), *Fizika i Chimia Obrabotki Materialov* **1**, 70 (in Russian).

KOPJOV, I.M., YU.E. BUSALOV, R. KRUMPHOLD and H. PAUL (1980), *Fizika i Chimia Obrabotki Materialov* **5**, 109 (in Russian).

KOSTIKOV, V.I., V.S. DERGUNOVA, V.S. KILIN and YU.M. PETROV (1979), Carbon fibres as a reinforcing material for producing high temperature composites, in: E.M. SOKOLOVSKAYA, ed., *Composite Materials, Proc. of the Soviet–Japanese Symp. Composite Materials* (Moscow Univ. Press, Moscow).

KOSTIKOV, V.I. and A.N. VARENKOV (1981), *Interaction of Carbon with Molten Metals* (Metallurgia, Moscow) (in Russian).

KREIDER, K.G. (1970), Patent of France, Publication N 2, **030**, 043.

KURKJIAN, C.R. and W.D. KINGERY (1956), *J. Phys. Chem.* **60**, 961.

KUZMIN, A.M., A.N. KOZLOV, I.M. KOPJOV, YU.N. PETROV and L.M. TERENTJEVA (1975), *Fizika i Chimia Obrabotki Materialov* **5**, 101 (in Russian).

MAXIMOVICH, G.G., A.V. PHILLIPOVSKY and YE.M. LUTY (1979), *Fiziko-chimicheskaja Mechanica Materialov* **15**(1) 49 (in Russian).

MANEVICH, L.I. (1981), in: *Abstracts of Papers Presented at 5th USSR Conf. on Composite Materials* Moscow (in Russian).

METCALFE, A.G. (1974a), Physical chemistry of the interfaces, in: L.J. BROUTMAN and R.H. KROCK, eds., *Composite Materials* Vol. 1: A.G. METCALFE, ed., *Interfaces in Metal Matrix Composites* (Academic Press, New York–London) Ch. 3.

METCALFE, A.G. (1974b), Fibre reinforced titanium alloys, in: L.J. BROUTMAN and R.H. KROCK, eds., *Composite Materials* Vol. 4; K.G. Kreider, ed., *Metallic Matrix Composites* (Academic Press, New York–London) Ch. 6.

METCALFE, A.G. and M.J. KLEIN (1974), Effect of the interface on tensile properties of composites, in: L.J. BROUTMAN and R.H. KROCK, eds., *Composite Materials* Vol. 1; A.G. METCALFE, ed., *Interfaces in Metal Matrix Composites* (Academic Press, New York–London) Ch. 4.

MEYERER, W., D. KIZER, S. PAPROCKI and H. PAUL (1978), Versatility of graphite–aluminium composites, in: B. NOTON, R. SIGNORELLI, K. STREET and L. PHILLIPS, eds., *Proc. 1978 Conf. on Composite Materials* (Met. Soc. AIME, New York).

MILEIKO, S.T. (1969), *J. Mater. Sci.* **4**, 974.

MILEIKO, S.T. (1970), *J. Mater. Sci.* **5**, 254.

MILEIKO, S.T. (1971), *Problemi Prochnosti* **7**, 3 (in Russian).

MILEIKO, S.T. (1975), Creep and creep-rapture of a continuous brittle fibre composite, in: V.V. NOVOZHILOV, ed., *Mechanics of Deformable Solids and Structures* (Mashinostroenie, Moscow) (in Russian).

MILEIKO, S.T. (1979), *Mechanica Compositnikh Materialov* **2**, 276 (in Russian).

MILEIKO, S.T. (1982), Mechanics of metal–matrix composites, in: I.F. OBRASTSOV, ed., *Mechanics of Composite Media* (Mir, Moscow) in press.

MILEIKO, S.T. and V.M. ANISHENKOV (1980), *Mechanica Compositnich Materialov* **3**, 409 (in Russian).

MILEIKO, S.T. and V.I. KAZMIN (1979), *Mechanica Compositnich Materialov* **4**, 723.

MILEIKO, S.T. and A.A. KHVOSTUNKOV (1971), *Zhurnal Pricladnoi Mechaniki i Technicheskoi Fiziki* **4**, 155 (in Russian).

MILEIKO, S.T., S.F. KONDAKOV and V.B. KASPEROVICH (1974), *Problemi Prochnosti* **1**, 32 (in Russian).

MILEIKO, S.T. and YU.N. RABOTNOV (1980), *Advances in Mechanics* **3**, 3 (in Russian).

MILEIKO, S.T., N.M. SOROKIN and A.M. ZIRLIN (1973), *Mechanika Polymerov* **5**, 840 (in Russian).

MILEIKO, S.T., N.M. SOROKIN and A.M. ZIRLIN (1976), *Mechanica Polymerov* 6, 1010.

MILEIKO, S.T., N.M. SOROKIN, I.N. ARCHANGELSKA and A.M. ZIRLIN (1980), Patent of France, Publication N 2 **416**, 270.

MILEIKO, S.T. and F.KH. SULEIMANOV (1981), *Mechanica Compositnich Materialov* **3**, 421 (in Russian).

MIMURA, K., O. OKUNO and I. MIURA (1974), *J. Japan Inst. Met.* **38**, 757 (in Japanese).

MORRIS, A.W.H. (1971), The fabrication and evaluation of carbon-fibre—reinforced aluminium composites, Presented at *Internat. Conf. Carbon Fibres, Composites and Applications.*

NAIDICH, YU.V. (1972), *Solid-Metal Melts Interfaces* (Naukova Dumka, Kiev) (in Russian).

NEUMAN, P. and P. HAASEN (1971), *Phil. Mag.* **23**, 285.

NICHOLAS, M.G. and B.A. MORTIMER (1971), The wetting of carbon by liquid metals and alloys, Paper presented at *Internat. Conf. on Carbon Fibres, Composites and Applications.*

NICHOLAS, M.G., T.M. VALENTINE and M.J. WAITE (1981), *J. Mater. Sci.* **15**, 2197.

NIKITIN, L.V. and A.N. TUMANOV (1981), Analysis of microfracture in a composite, to be published in *Proc. of the 2nd USSR–USA Symp. on Fracture of Composites.*

OCHIAI, SH. and Y. MURAKAMI (1976), *Met. Sci.* **11**, 401.

OCHIAI, SH. and Y. MURAKAMI (1980), *J. Mater. Sci.* **15**, 1790.

PANASUK, A.D. (1981), *Poroshkovaja Metallurgia* **10**, 75 (in Russian).

PEPPER, R.T. and T.A. ZACK (1978), US Patent N 4 072, 516.

PINES, B.YA. (1964), *Zhurnal Technicheskoi Fiziki* **16**, 137 (in Russian).

PINES, B.YA. (1956), *Zhurnal Technicheskoi Fiziki* **26**, 2086.

PINNEL, M.A. and A. LAWLEY (1970), *Met. Trans.* **1**, 1337.

PORTNOI, K.I., B.N. BABICH and I.L. SWETLOV (1979a), *Nickel Based Composite Materials* (Metallurgia, Moscow) (in Russian).

PORTNOI, K.I., S.E. SALIBEKOV, I.L. SWETLOV and V.M. CHUBAROV (1979b), *Structure and Properties of Composite Materials* (Mashinostroenie, Moscow) (in Russian).

PORTNOI, K.I., N.I. TIMOFEEVA, A.A. ZABOLOTZKY, V.N. SAKOVICH, B.F. TREFILOV, M.KH. ZEVINSKA and N.N. POLJAK (1981), *Poroshkovaja Metallurgia* **2**, 45 (in Russian).

PRAGER, W. and P.G. HODGE (1951), *Theory of Perfects, Plastic Solids* (Wiley, New York).

PREWO, K.M. (1975), The fabrication of boron fibre reinforced aluminium matrix composites, in: E. SCALA, E. ANDERSON, I.J. TOTH, B.A. NOTON, eds., *Proc. 1975 Internat. Conf. on Composite Materials* Vol. 2 (Met. Soc. AIME, New York).

RABOTNOV, YU.N. (1966), *Creep of Structural Elements* (Nauka, Moscow) (in Russian).

RABOTNOV, YU.N. and S.T. MILEIKO (1970), *Short Time Creep* (Nauka, Moscow) (in Russian).

RABOTNOV, YU.N. and S.T. MILEIKO (1983), Mechanics of composite failure, in: A. KELLY and YU.N. RABOTNOV, eds., *Handbook of Composites* Vol. 3 (North-Holland, Amsterdam) in press.

RASHID, M. and C. WIRKUS (1972), *Ceram. Bull.* **51**, 836.

RYCALIN, N.N., M.KH. SHORSHOROV, V.V. KUDINOV, S.A. MAKAROV, L.V. KATINOVA, YU.A. GALKIN and YU.A. SHESTERIN (1977), *Fizika i Chimia Obrabotki Materialov* **6**, 102 (in Russian).

SAMSONOV, G.V., A.D. PANASUK and M.S. BOROVIKOVA (1972), Wettability of borides of IVa–VIa subgroups metals by liquid metals, in: V.N. EREMENKO and YU.V. NAIDICH, eds., *Wettability and Surface Properties of Melts and Solids* (Naukova Dumka, Kiev) (in Russian).

SERA, R.V. (1974), US Patent N 3 796, 587.

SEMENOV, B.I., E.I. KHANIN and YU.B. MARKEVICH (1978), *Fizika i Chimia Obrabotki Materialov* **1**, 138 (in Russian).

SEVERDENKO, V.P., A.S. MATUSEVICH and A.E. PISKAREV (1974), *Poroshkovaja Metallurgia* **6**, 51 (in Russian).

SHIOIRI, J. (1977), Tokyo University, Private communication.

SHORSHOROV, M.KH., V.A. KOLESNICHENKO, R.S. YUSUPOV and L.M. USTINOV (1978), *Fizika i Chimia Obrabotki Materialov* **4**, 117 (in Russian).

SHORSHOROV, M.KH., V.V. KUDINOV, V.I. ANTIPOV and L.V. KATINOVA (1974), *Fizika i Chimia Obrabotki Materialov* **5**, 59 (in Russian).

SLATE, P.M.B. (1975), Explosive fabrication of composite materials, in: E. SCALA, E. ANDERSON, I.J. TOTH and B.A. NOTON, eds., *Proc. 1975 Internat. Conf. on Composite Materials* Vol. 2 (Met. Soc. AIME, New York).

SOKOLOVSKAYA, E.M. and L.S. GUSEY (1978), *Physical-Chemical Aspects of Composite Materials* (Moscow Univ. Press, Moscow) (in Russian).

STANKING, R. and M. NICHOLAS (1978), *J. Mater. Sci.* **13**, 1509.

SUTTON, W.H. and E. FEINGOLD (1966), Role of interfacially active metals in the apparent adherence of nickel to sapphire, in: W.W. KRIEGEL and H. PALMOUR, eds., *Materials Science Research* Vol. 3 (Plenum Press, New York) Ch. 31.

SWETLOV, I.L. and V.F. STROGANOVA (1975), *Fizika i Chimia Obrabotki Materialov* **6**, 125 (in Russian).

TAKAHASHI, S. (1978), *AIAA J.* **16**, 452.

TRESSLER, R.E. (1974), Interfaces in oxide fibres reinforced metals, in: L.J. BROUTMAN and R.H. KROCK, eds., *Composite Materials* Vol. 1: A.G. METCALFE, ed., *Interfaces in Metal Matrix Composites* (Academic Press, New York).

TOTH, I.J., W.D. BRENTNALL and G.D. MENKE (1972), *J. Metals* **24**, 9.

TUCHINSKY, L.I. (1974), *Poroshkovaja Metallurgia* **12**, 35 (in Russian).

UMAKOSHI, Y., K. NAKOI and T. YAMANE (1974), *Met. Trans.* **5**, 1250.

USHIZKY, M.U. (1978), *J. Pricladnoi Mechaniki i Technicheskoi Fiziki* **4**, 183 (in Russian).

USTINOV, L.M. (1979), *Fizika i Chimia Obrabotki Materialov* **5**, 82 (in Russian).

WARREN, R., C.H. ANDERSON and M. CARLSON (1978), *J. Mater. Sci.* **13**, 178.

WATANABE, O. (1979), The compatibility and mechanical properties of the carbon fibre–metal composite materials, in: E.M. SOKOLOVSKAYA, ed., *Composite Materials, Proc. of the Soviet–Japanese Symp. Composite Materials* (Moscow Univ. Press, Moscow).

WEISINGER (1967), US Patent N 3 **788**, 926.

WRIGHT, P.K. and L.J. EBERT (1972), *Met. Trans.* **3**, 1645.

YAJIMA, S., K. OKAMURA, J. TANAKA and T. HAYASE (1981), *J. Mater. Sci.* **16**, 3033.

YUE, A.S., T.T. YANG and T.S. LIN (1975), On a model of calculating bond strength, in: E. SCALA, E. ANDERSON, I.J. TOTH and B.A. NOTON, eds., *Proc. 1975 Internat. Conf. Composite Materials* Vol. 2 (Met. Soc. AIME, New York).

Chapter VI

Directionally Solidified Composites for Application at High Temperature

M. Rabinovitch, J.F. Stohr, T. Khan and H. Bibring

Office National d'Etudes et de Recherches Aérospatiales (ONERA)
92320 Châtillon
France

Contents

HANDBOOK OF COMPOSITES, VOL. 4 – Fabrication of Composites
Edited by A. Kelly and S.T. Mileiko
© 1983, Elsevier Science Publishers B.V.

1. Introduction

Composites obtained by directional solidification (D.S. composites) occupy a specific place among metallic matrix fibre reinforced materials. These composites are obtained in a single operation by the simultaneous crystallisation of the matrix and the reinforcing phase during the directional freezing of a metallic melt having a eutectic or near-eutectic composition.

At the solid–liquid interface, the three phases present (the two solids and the liquid) are in quasi-equilibrium. For this reason, and provided that the regular microstructure obtained during solidification is not subsequently modified during cooling by solid-state transformations (e.g., eutectoid decomposition), the microstructure is stable up to the melting point. One can thus envisage the use of such materials up to temperatures of the order of 1100°C where diffusion would be expected to destroy the structure of any non-equilibrium mixture such as those found in more conventional 'fabricated' composites.

Nonetheless, directionally solidified composites are not totally exempt from problems of structural stability, which can arise, for example, during thermal cycling or during prolonged exposure to high operating temperatures.

In the case of 'fabricated' composites, the choice of the reinforcing fibre, its volume fraction and the type of matrix may be selected, within limits, according to the application in view. With the D.S. composite, one is, in addition, constrained by the limitations of the phase diagrams of the systems available. We shall therefore review the forms of the phase diagrams of the most promising candidate D.S. systems, and shall show how the controlled addition of certain elements allows us to adapt properties of the chosen system to the application in view.

In general, an ingot or cast blank of D.S. composite exhibits a columnar grain structure. At the microstructural level, the reinforcing phase is present in the form of monocrystalline fibres or lamellae which are nearly continuous, and are, in fact, whisker crystals. These are much finer than the reinforcing fibres found in most fabricated composites, being typically of the order of one micron in diameter, separated by a few microns.

The reinforcing phase may be brittle, 'semi ductile', with a limited capacity for plastic deformation, or ductile. The matrix is normally ductile, and may be further strengthened by precipitation and/or solid–solution hardening.

Finally, the simultaneous crystallisation of the fibres and the matrix in the direction of the thermal gradient results in a definite crystallographic relationship both between the lattices of fibre and matrix, and also in the orientation of the interface.

All of the above-mentioned factors, i.e., morphology and size of the grains, morphology, type and volume-fraction of the reinforcing phase, the structure of the fibre–matrix interface and the mechanism and degree of hardening of the matrix, will influence the mechanical properties of the composite. With this in mind, we shall give detailed consideration to the mechanical properties of the two families of composites which have been extensively studied because of their potential utility as blade materials for the gas-turbine engine.

The first of these, often designated as γ/γ'–metal carbide (γ/γ'–MC), has been studied at ONERA and covers a range of materials in which the matrix is a nickel-base superalloy, reinforced by a low volume-fraction (6%) of brittle fibres of TaC or NbC. Fig. 1 shows a typical microstructure of such a material.

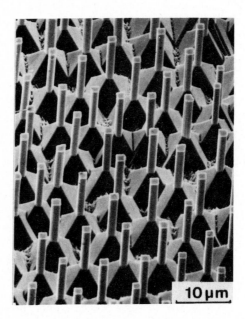

FIG. 1. Ni–Cr–TaC composite. Scanning electron microscope (S.E.M.) image obtained after selective etching of the matrix.

The second family, γ/γ'–δ, studied at United Aircraft Research Laboratory (UARL), is based on the same matrix, but reinforced by a high volume-fraction (44%) of 'semi-ductile' lamellae of the intermetallic Ni_3Nb (see Fig. 2).

Other types of composites will be discussed briefly. Finally, the various problems of fabrication and manufacture will be examined in detail. These are important, because the mechanical properties of D.S. composites depend in a large measure upon the regularity of the microstructure, and are strongly affected by the presence of solidification defects. In other words, the quality of the composite depends upon the quality of the manufacturing process.

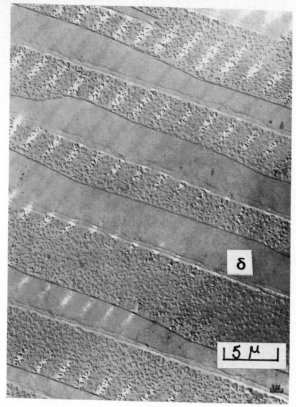

FIG. 2. Microstructure of the lamellar $\gamma/\gamma'-\delta$ composite (Ni/Ni$_3$Aℓ: Ni$_3$Nb) (LEMCKEY, 1973).

2. Preparation of composites by directional solidification

2.1. Composites obtained by the solidification of binary eutectics

Composites with a potential for application at high temperatures are based upon multi-component alloys of the refractory metals, but it was the pioneering work on the directional solidification of binary eutectics which opened the way to their use. Much work on these systems was published at the beginning of the sixties. Without, therefore, embarking upon an exhaustive study of this question (which has already been extensively reviewed by, inter alii, CHALMERS (1964), CHADWICK (1965) and FLEMINGS (1974)), it is nonetheless essential to devote a few pages to the particular problems posed by the preparation of these composites.

2.1.1. Growth of binary eutectics

Consider the phase diagram of the system AB having an eutectic E sketched in Fig. 3. The eutectic alloy, with concentration C_E has the lowest solidification

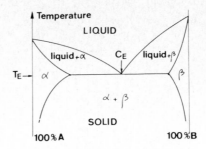

FIG. 3. Binary eutectic diagram.

temperature of any of the alloys AB. It has a zero solidification range, and the α and β solid solutions crystallise simultaneously from the liquid C_E at the melting point. Application of the phase rule to the equilibrium $L \rightleftharpoons \alpha + \beta$ at constant pressure shows that the eutectic reaction is invariant. The composition of the liquid, and the temperature of the reaction are fixed, and the proportions of phases α and β formed during solidification are given by the lever rule. These fractions may, however, vary during subsequent cooling if the solubility of A in β or B in α changes in a different way with change in temperature.

Turning now to the growth of the eutectic, one observes that the fibres or lamellae of α and β grow perpendicular to the solidification front, as was shown by HUNT and JACKSON (1966) and DAIGNE and GIRARD (1973). Fig. 4 shows an example of this behaviour. At the solid–liquid interface, the liquid is enriched in A in front of a lamella of β, and, conversely, in B in front of the α

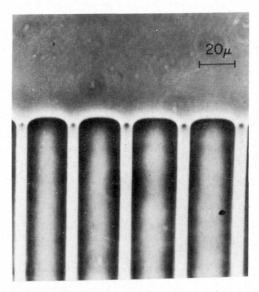

FIG. 4. Direct observation of the solid–liquid interface in the eutectic system carbon tetrabromide: hexachloroethane (HUNT and JACKSON, 1966).

phase. Concentration gradients which thus appear in the liquid immediately ahead of the growth front cause a lateral redistribution of elements A and B by short-range diffusion. This redistribution governs the growth of the lamellae of α and β. Numerous studies, notably those of CHADWICK (1963) and LEMKEY et al. (1965) have shown that the interlamellar spacing λ varies with the speed of advance of the front V according to a law of the type $\lambda^2 V = $ constant.

As the speed of the front increases, the time available for the diffusion of B towards β and A towards α decreases, with the consequence that the inter-lamellar spacing λ decreases also.

The advance of the front depends upon the undercooling of the solid–liquid interface, which is thus below the equilibrium temperature T_E shown on the phase diagram. It is recognised that this undercooling ΔT is, for the case of a eutectic, the sum of two terms, the first being purely kinetic in origin, and the other being related to the work done in creating new interfaces. The first is small and may be ignored, being, for example, $10^{-2}\,°C$ for tin solidified at 10^{-2} cm/sec as was shown by KRAMER and TILLER (1965).

The second term is of the order of 1°C, according to the measurements of HUNT and CHILTON (1963) and MORE and ELLIOT (1967). The undercooling follows a relation of the type $\Delta T = BV^{0.5}$; in the case of the lead–tin eutectic, ΔT is 0.5°C for a freezing rate of 1 cm/min.

Non-equilibrium structures are only found at high cooling rates of the order of $10^6\,°C$/sec. In the majority of solidification situations found in metallurgy where the cooling rates are a few degrees per second, the solid–liquid interface is effectively at thermodynamic equilibrium. Under these conditions, the equilibrium phase diagram may be safely used, at least to a first approximation. It should, however, be noted that this is no longer valid where one of the phases has a strong tendency towards facetted growth at the solid–liquid interface (HUNT and HURLE, 1968).

A number of authors has sought to combine these observations into a general theory of the growth of eutectics; notable among these are the works of TILLER (1958), CHALMERS (1964), CHADWICK (1965) and in particular JACKSON and HUNT (1966).

Without further pursuing the study of eutectic growth, one can already identify the essential condition necessary to obtain a regular structure. Because the two phases of the eutectic grow perpendicular to the solidification front, a composite reinforced by fibres or lamellae running parallel with the growth direction will only be obtained if the solid–liquid interface is plane. In what follows, we shall study the stability conditions for such a planar front, and the characteristic defects which are the results of particular instabilities.

2.1.2. Summary of conditions for unidirectional solidification

The structure of an alloy ingot reflects the state of supercooling of the liquid during solidification. This supercooling controls not only the nucleation, but also the growth and the microstructure of the grains which form the ingot.

In this discussion, supercooling should be understood to include not only the thermal undercooling discussed above, but also the constitutional supercooling arising from the rejection of solute elements at the solidification front of the alloy.

Unidirectional solidification is a procedure which allows us to control the supercooling and hence the macroscopic and microscopic structure of a given ingot.

The technique consists in general of arranging and controlling the heat flow in a vertical mould which is heated in the upper region and cooled below. Such an arrangement tends to limit the problems of natural convection. The solidification front is arranged to lie in the middle of an insulated zone which lies between the heat source and sink (see Fig. 5). In this way, in the region of the front, there is no radial heat flux, and the isothermal surfaces are plane. The temperature at the front increases in the upward direction, and thus the thermal gradient in the liquid avoids all tendency towards thermal supercooling and any nucleation ahead of the front. Under these conditions, the grains already existing in the solid portion continue to grow in a direction normal to the solidification front.

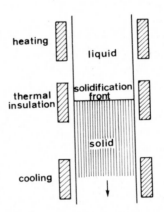

FIG. 5. Sketch of a directional solidification installation.

In this way an ingot of 'columnar grains' is obtained, with no grain boundaries normal to the growth direction, except in the starting zone at the bottom of the ingot. In this region, a large number of nuclei begins to grow, but those which are favourably oriented with respect to the thermal gradient tend to grow faster, and eventually succeed in filling the whole cross-section of the ingot.

In fact, it is found that all the columnar grains have a common crystallographic orientation parallel with the growth direction. In the F.C.C. alloys, this direction is $\langle 100 \rangle$.

The interior structure of each grain depends upon the fine-scale morphology of the solid–liquid interface. Thus, for example, the rejection of solute elements ahead of the solidification front can cause constitutional supercooling, and lead to the development of cellular or dendritic growth.

2.1.3. *Planar front solidification of monophase binary alloys*

In this section, we take the simple model of RUTTER and CHALMERS (1953) and TILLER et al. (1953). During the unidirectional solidification of an alloy AB rich in A, the first solid to be crystallised appears at temperature T_L°C (see Fig. 6). This solid has a concentration $k_0 C_0$, where k_0 is the partition coefficient of the solute between solid and liquid ($k_0 = C_{sol}/C_{liq}$). The solid thus formed is less rich in solute than the liquid ($k_0 < 1$) and its growth is accompanied by the rejection of solute into the surrounding liquid.

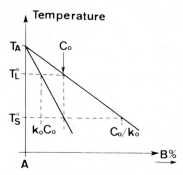

FIG. 6. Equilibrium diagram.

The concentration of solute in the solid will increase as the concentration of solute in the liquid increases, until a steady-state condition is reached when the composition of the solid approaches C_0; the composition of the liquid in equilibrium with this solid is then C_0/k_0. The composition of the liquid changes from C_0/k_0 at the interface to C_0 in the bulk of the melt. To examine this variation, we make the following assumptions:

(a) The solidus and liquidus are considered to be straight lines. Under these conditions, the partition coefficient k_0 is constant.

(b) Diffusion is the only mechanism for the redistribution of the excess solute ahead of the front; there is no convection in the liquid.

(c) Elements A and B are very pure, and the effect of trace impurities can be neglected.

(d) It is considered that the solid–liquid interface is in thermodynamic equilibrium, and that the liquid column is sufficiently long for steady-state conditions to be obtained.

(e) The solidification is carried out with a planar front, and with constant thermal gradient G and advance rate of the front V.

Under these conditions, in the steady state, the equation for the diffusion of solute in the front (and taking the origin at the front itself) may be written as

$$D \frac{\mathrm{d}^2 C}{\mathrm{d}x^2} + V \frac{\mathrm{d}C}{\mathrm{d}x} = 0$$

where D is the diffusion coefficient of the solute in the liquid.

With the boundary conditions defined by Fig. 7,

$$x = 0, \quad C_L = C_0/k_0 ; \qquad x = \infty, \quad C_L = C_0 ,$$

we obtain the solution

$$C_L = C_0 \left(1 + \frac{1 - k_0}{k_0} \exp\left\{-\frac{V}{D} x\right\}\right).$$

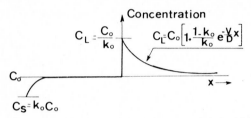

FIG. 7. Solute concentration profile ahead of the solid–liquid interface.

Knowing the composition of the liquid at any point, one can refer to the equilibrium diagram and obtain the corresponding liquidus temperature T_L. Thus:

$$T_L = T_0 - mC_L$$

where T_0 is the melting point of pure A and m the slope of the liquidus line. We thus obtain

$$T_L = T_0 - mC_0 \left(1 + \frac{1 - k_0}{k_0} \exp\left\{-\frac{V}{D} x\right\}\right).$$

One can thus show, on the same diagram, the variation of the liquidus temperature T_L ahead of the front, and also the actual temperature T_{eff} in the bath under the given experimental conditions,

$$T_{\text{eff}} = T_f + Gx ,$$

where T_f is the temperature of the front and G the thermal gradient in the liquid at the interface.

It can now be seen, in Fig. 8, that for a temperature gradient of G_2, all the liquid between 0 and X_M, is below the liquidus temperature. Under such conditions the planar solidification front becomes unstable, and proturberances may grow as far ahead of the front as X_M. The consequences of such behaviour upon the shape of the front on segregation and the microstructure of the alloy will be considered later.

In Fig. 8 it can be seen that this 'constitutional supercooling' can be avoided

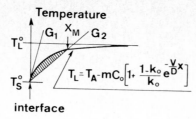

FIG. 8. Liquidus temperature in the liquid near the interface and the actual temperature gradient.

by increasing the temperature gradient so that it is at least as great as the initial slope of the curve $T_L(x)$. This may be expressed as

$$G \geqslant \frac{dT_L(x)}{dx} \quad \text{or} \quad G \geqslant -mC_0 \left(\frac{1-k_0}{k_0}\right) \frac{V}{D}$$

or, separating the process variables from those peculiar to the alloy itself,

$$\frac{G}{V} \geqslant -m \left(\frac{1-k_0}{k_0}\right) \frac{C_0}{D}.$$

This criterion for the stability of the front can be cast in another form, if we note that

$$mC_0 \left(\frac{1-k_0}{k_0}\right) = -m \left(\frac{C_0}{k_0} - C_0\right) = \Delta T$$

where ΔT is the solidification range for the alloy. From the above, we obtain

$$G/V \geqslant \Delta T/D.$$

This form for the stability criterion of the front allows one to make a good estimate of the critical ratio G/V directly from the solidification range of the alloy shown on the equilibrium diagram, and hence to determine the conditions necessary for successful fabrication.

2.1.4. Planar front solidification of two-phase binary alloys

Arguments closely similar to the above may be applied to a two-phase binary alloy. We consider the unidirectional solidification of a liquid column of concentration C_0 of the alloy AB whose phase diagram is shown in Fig. 9. As the column moves through the furnace, a single phase solid of composition k_0C_0 will be thrown down at temperature T_{L_0}. As before, the liquid ahead of the front will be enriched in B, and the concentration in the solid will rise from kC_0 to C_α as the liquid concentration rises from C_0 to C_E.

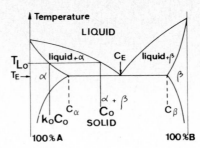

FIG. 9. Binary eutectic phase diagram.

At this point, the steady-state conditions are established. The solid which crystallises out will be two-phase (α, β) with an average composition C_0, and these conditions will be maintained until the solidification is almost complete.

As the last sections of liquid crystallise, their concentration in B will rise progressively, followed by corresponding rises in the solid material, until the final parts of the ingot have the eutectic composition.

We can now solve the diffusion equation for the solute ahead of the front, applying the same assumptions are in the case of the monophase alloy. The boundary conditions are:

$$x = 0, \quad C_L = C_E ; \qquad x = \infty, \quad C_L = C_0 .$$

The solution is then

$$C_L = C_0 + (C_E - C_0) \exp\left\{-\frac{V}{D} x\right\} .$$

Similarly, one can deduce the liquidus temperature T_L corresponding to each value of C_L, and obtain by comparison with the actual temperature the criterion for the stability of the front,

$$\frac{G}{V} \geqslant \frac{m}{D} (C_E - C_0) .$$

This simple criterion has given good agreement with observation in many cases. A more complex theory, however, has been developed by MULLINS and SEKERKA (1963) who studied the stability of a perturbation in the front in terms of thermal effects and surface energies.

2.1.5. *Influence of constitutional supercooling on the microstructure*

2.1.5.1. Monophase binary alloys. In the case of monophase binary alloys containing impurities the liquidus temperature may be modified locally, as a result of the segregation of impurities (particularly in the region of grain boundaries). This impurity question will be considered below; in this discussion we restrict our attention to alloys where the effect of impurities may be ignored.

In this case, if the ratio G/V is sufficient to maintain a planar front, the cast solid is homogeneous.

If, for the same alloy, the ratio G/V falls to allow some supercooling, but over a limited range only, typically $100 \mu m$ or so, the growth becomes cellular.

Fig. 10 shows schematically the phenomenon of cellular growth. Fig. 10(a) shows the equilibrium diagram of the alloy being considered. In Fig. 10(b) we show the curve $C_L = f(x)$, giving the change in concentration of solute in the liquid ahead of the front. This arrangement allows us to obtain point by point the curve of $T_L = f(x)$, giving the temperature of the liquidus for any value of C_L. This is shown in Fig. 10(c), together with the curve of the experimental temperature $T_{eff} = f(x)$. The layer of liquid in the supercooled state has thickness OX_M.

Fig. 10(d) shows the form of the front. Fig. 11 shows that cellular growth is associated with a transverse segregation of element B. The solid at the centre of a cell has a concentration C_c lower than C_0, and the average concentration in a section of the solid bar will be C_0.

Fig. 11 shows the morphology of a cellular growth front in a Pb–Sb alloy. If the thickness of the supercooled layer increases, the growth morphology will become progressively more dendritic, passing through a transitional state where a dendrite forms at the centre of each cell.

If the process proceeds further, one may even observe the formation of 'parasite grains' by nucleation within the melt.

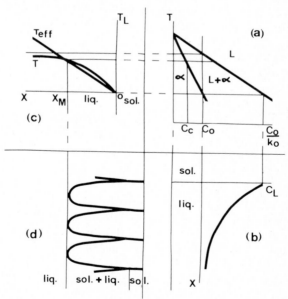

FIG. 10. Sketch of cellular growth.

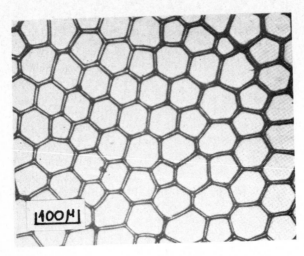

FIG. 11. Transverse section of an ingot of Pb–Sn alloy showing cellular structure (MORRIS and WINEGARD, 1969).

2.1.5.2. Two-phase alloys. In the case of a two-phase alloy, and provided that the ratio G/V is sufficiently large to guarantee the stability of a planar front, it is possible to obtain a two-phase solid of regular structure even in a non-eutectic alloy. This was demonstrated experimentally by MOLLARD and FLEM-MING (1967) in the lead–tin system. Working with a strong temperature gradient (400°C/cm) they obtained growth on a planar front with an alloy of 12 atomic percent Pb, whereas the eutectic composition is at 26 atomic percent Pb. The composite which is obtained at the eutectic composition is lamellar, but this becomes fibrous as the lead content is decreased.

Such a procedure should, in principle, enable one to choose a priori the desired volume fraction of reinforcement. In practice, however, problems arise due to the necessity of maintaining the steep temperature gradient needed to keep the solidification front planar.

In the special case of high melting point composites, the high temperature gradient and the large superheat of the melt that this entails, very often leads to problems with the ceramic mould materials, and thus to significantly reducing the available possibilities.

The influence of constitutional supercooling on the structure of a two-phase alloy is shown schematically in Fig. 12, arranged in the same way as Fig. 10. If, as in the case of the single-phase alloy, the impurity concentration is low, the appearance of constitutional supercooling will be recognised by the growth of dendrites of the majority phase without passing through an intermediate cellular stage.

In fact, the growth of these dendrites ahead of the front absorbs the majority elements and thereby reduces the concentration of the interdendritic melt to a level compatible with the ratio G/V in such a way that the interdendritic solid

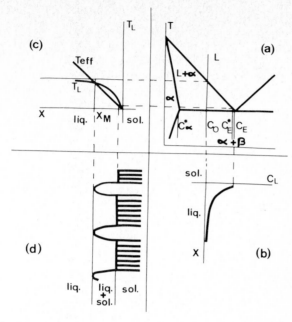

FIG. 12. Sketch of the dendritic growth of a binary alloy of composition near to the eutectic.

FIG. 13. Direct observation of the solid–liquid interface in an off-eutectic mixture of carbon tetrabromide and hexachloroethane.

grows with a planar front and a regular structure. Fig. 13 shows a direct observation of this phenomenon, obtained by JACKSON and HUNT (1966).

This type of growth is accompanied by a transverse segregation:
- the centre of the dendrites has composition C_c,
- the surface of the dendrites has composition C_α,
- the interdendritic solid has average composition C_E^*, lying between C_0 and C_E such that the criterion for supercooling is reached for the ratio G/V. If f_D is the volume fraction of dendrites and $\bar{C}_0$ the mean concentration in a dendrite, the value of C_E^* may be estimated from the rule of mixtures:

$$C_0 = f_D \cdot \bar{C}_D + (1 - f_D)C_E^* .$$

In the case where the impurity content of the alloy may no longer be ignored, impurities may diffuse over considerable distances ahead of the front and give rise to supercooling. This phenomenon may lead to the appearance of cellular growth, including the case of eutectic alloys. Fig. 14 shows cellular growth in a eutectic caused by the presence of impurities.

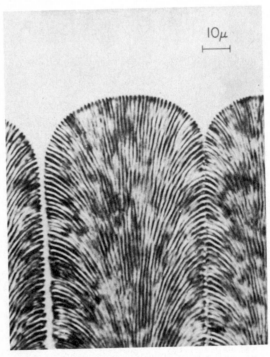

FIG. 14. Direct observation of a cellular interface in the camphor–succinonitrile eutectic containing an impurity.

2.1.6. Influence of non-steady phenomena

The above discussion presupposed steady-state conditions of solidification. In practice non-steady phenomena may appear, either from thermal or mechanical sources, or from convection current in the liquid.

Thermal or mechanical changes cause a change in the speed of movement of the front and are responsible for the appearance of 'banding' in the ingot. The structure of the bands varies according to the composition of the alloy and the severity of the perturbation.

In the case of a eutectic alloy without impurities, a variation in the speed of the front gives rise only to a change in the spacing of the lamellae according to the relation $\lambda^2 V = $ constant.

In the case of an off-eutectic alloy, an increase in the speed of the front can give rise to the appearance of a cellular or dendritic structure. Further, MOLLARD and FLEMMINGS (1967) have studied the influence of speed on the composition of the cast solid. They find that for a two-phase alloy, with concentrration C_0 lower than C_E, an accleration of the front causes a temporary increase of B and thus an increase in the concentration of the reinforcement.

Consequently, a decrease can cause a reduction in the concentration of B in the solid, which can drop as low as $k_0 C_0$ and at this point cause the formation of a band of single-phase material. It should be noted that such a band will weaken the ingot over its entire cross-section and thereby preclude its use for any practical application. The production of reliable material clearly depends upon the use of regulating systems of good quality both for thermal control and for the mechanical displacement.

The convection current may appear in the liquid alloy if there is insufficient heat input at the top of the melt column to compensate for the heat loss. It is therefore important to avoid these currents so that they do not perturb the heat exchange and solute diffusion in the vicinity of the solidification front.

Convection currents can give rise to transport of solute elements away from the enriched zone near the front. The concentration of the liquid far from the front will then cease to be constant and equal to C_0, but will rise steadily during solidification.

This question was studied by PFANN (1958) and BURTON et al. (1953). Fig. 15 summarises the results of these studies.

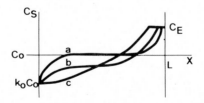

FIG. 15. Changes in the composition of an ingot solidified (curve a) without and (curve c) with convection stirring.

Curve a shows the variation of the concentration of a long bar solidified without convection. As expected, it shows a plateau of constant composition C_0. Curve c shows the case where the melt is continuously and thoroughly stirred, while curve b represents an intermediate case where the solute-rich zone ahead of the front is only partly affected by the stirring of the melt.

2.2. *Composites obtained by the solidification of ternary alloys*

The study of the directional solidification of multiconstituent alloys is immediately hindered by the lack of data concerning the ternary-phase diagrams. The available stock of complete or even partial ternary diagrams is very limited, and it is practically non-existent for quaternary- or higher-order alloys. For this reason, we shall here limit our study to the ternary alloys, following the work of HAUSER et al. (1977). In what follows, we first review the types of phase diagrams likely to be able to provide composites, and then we shall deal with the problem of solidifying the ternary mono- or two-phase alloys.

2.2.1. *Monovariant solidification*

In the ternary phase diagram ABC shown in Fig. 16 the binary system AB has a binary eutectic e, and the systems AC and BC form continuous solid solutions. A continuous region of (single-phase) solid solution $\alpha(\beta)$ surrounds a two-phase region $\alpha + \beta$. The monovariant trough starts from the binary eutectic e and eventually disappears at e_1 into the liquidus surface.

Any liquid e_i corresponding to a point in the eutectic trough will give rise to a two-phase solid $(a_i\,b_i)$. The composition of phases a_i and b_i in equilibrium with the liquid e_i is obtained using the tie triangle $e_i\,a_i\,b_i$. The vertical section shown in Fig. 17 reveals the existence of a three-phase region $(L + \alpha + \beta)$ and hence a

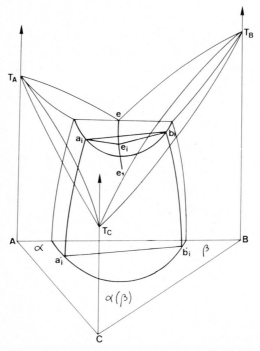

FIG. 16. Ternary equilibrium diagram with a monovariant trough $e\,e_1$ (after PRINCE, 1966).

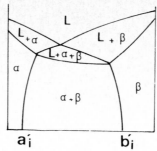

FIG. 17. Section ($a_i'\, b_i'$) of the diagram in Fig. 16.

solidification range for any alloy e_i. Of all the possible alloys ABC, the alloys e_i have the minimum solidification range.

Finally, the application of the phase rule at constant pressure at the equilibrium point $e_i \rightleftarrows \alpha + \beta$ shows that the reaction is monovariant.

Monovariant troughs may also be found in other types of ternary diagrams.

Among these is the ternary system ABC containing two binary eutectics, in AB and BC, while AC forms a continuous solid solution. Fig. 18 shows such a diagram, in which the monovariant trough can be seen starting at the eutectic e_{AB} and running to the eutectic e_{BC}.

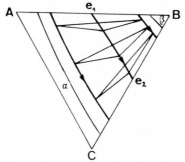

FIG. 18. Monovariant trough in a ternary diagram with two binary eutectics e_1 and e_2: projection onto the composition plane (after PRINCE, 1966).

Similarly in the case of the ternary shown in Fig. 19, where AB, BC and AC each have a eutectic. Here, three monovariant troughs start at eutectics e_{AB}, e_{BC} and e_{AC} to meet at the ternary invariant E.

When a ternary diagram has one or more binary or ternary intermetallic phases having congruent melting points, a pseudo-binary section may be found between one of the elements A, B or C and the intermetallic phase. In this particular case, the tie lines joining a liquid in the section to the solids in equilibrium with it are also to be found in the plane of the section. Furthermore it is quite common to find a pseudo-binary eutectic whose solidification range is zero (see Fig. 20).

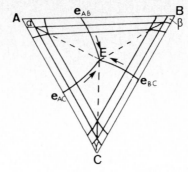

FIG. 19. Monovariant troughs in a ternary diagram containing three binary eutectics: projection onto the composition plane (after PRINCE, 1966).

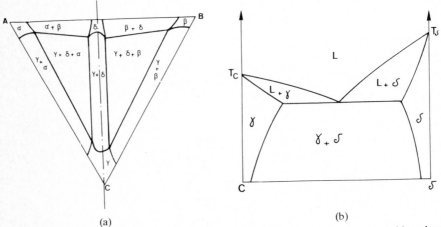

(a)

(b)

FIG. 20. Ternary diagram having a pseudo-binary section. (a) Projection on the composition plane; (b) Sketch of the pseudo-binary section.

Such a configuration is particularly well suited to the fabrication of a multi-constituent composite. In practice, one starts from a pseudo-binary eutectic, and by adding a fourth element, one studies the solidification of a range of alloys along the trough leading away from the pseudo-binary eutectic.

2.2.2. Planar front solidification of ternary alloys under steady-state conditions

The question has been addressed by COATES et al. (1968) and RINALDI et al. (1972) following an approach similar to that taken for the binary alloys.

The model considered is one-dimensional since it considers the front to be planar. The speed of advance of the front V and the temperature gradient G at the interface are constant. The redistribution of the excess solute occurs exclusively by diffusion, and it is supposed that the length of the liquid column is sufficiently long for steady-state conditions to be established. Under these conditions, one can calculate the solute concentration ahead of the front, and hence a stability criterion for the maintenance of the planar front.

Consider therefore the ternary diagram in Fig. 21 and a liquid alloy of composition (C_{0m}, C_{0n}) which under steady-state conditions leads to the formation of a solid of average composition (C_{Sm}, C_{Sn}) equal to that of the liquid. The diffusion equations for the solute m and n in the liquid, taking the origin of z at the front, are

$$D_{mm}\frac{\mathrm{d}^2 C_m}{\mathrm{d}z^2} + D_{mn}\frac{\mathrm{d}^2 C_n}{\mathrm{d}z^2} + V\frac{\mathrm{d}C_m}{\mathrm{d}z} = 0 ,$$

$$D_{nm}\frac{\mathrm{d}^2 C_m}{\mathrm{d}z^2} + D_{nn}\frac{\mathrm{d}^2 C_n}{\mathrm{d}z^2} + V\frac{\mathrm{d}C_n}{\mathrm{d}z} = 0$$

where C_m and C_n are the concentrations of the solutes m and n in the liquid ahead of the front, and D_{mm}, D_{mn}, D_{nm} and D_{nn} are the diffusion coefficients of the solute elements in the liquid.

The boundary conditions are

$$z = 0, \quad \begin{cases} C_m = C_{Lm}^*, \\ C_n = C_{Ln}^*; \end{cases} \quad\quad z = \infty, \quad \begin{cases} C_m = C_{0m}, \\ C_n = C_{0n}. \end{cases}$$

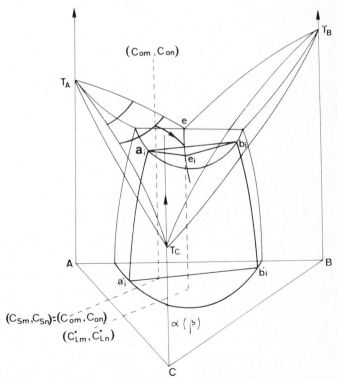

Fig. 21. Monovariant solidification.

The composition of the liquid (C^*_{Lm}, C^*_{Ln}) in equilibrium with the solid (C_{0m}, C_{0n}) is determined on the liquidus by the tie line corresponding to this solid in the case of a single-phase alloy, or by the tie line triangle in the case of a two-phase solid.

To obtain more convenient solutions to the equations, we make, in addition, the assumption that the diffusion of the elements m and n are independent, which amounts to stating that $D_{mn} = D_{nm} = 0$.

Under these conditions the curves for the concentration of solute ahead of the front are

$$\frac{C_m - C_{0m}}{C^*_{Lm} - C_{0m}} = \exp\left\{-\frac{V}{D_{mm}} z\right\}, \qquad \frac{C_n - C_{0n}}{C^*_{Ln} - C_{0n}} = \exp\left\{-\frac{V}{D_{nn}} z\right\}.$$

As in the case of the binary alloys, the stability criterion for the planar front is written as

$$G \geqslant \frac{dT_L}{dz}\bigg|_{z=0}$$

where G is the effective temperature gradient in the liquid and $T_L(z)$ the temperature of the liquidus at any given point.

Finally, we obtain the simplified expression

$$\frac{G}{V} \geqslant -\left(\frac{p(C^*_{Lm} - C_{0m})}{D_{mm}} + \frac{s(C^*_{Ln} - C_{0n})}{D_{nn}}\right)$$

where $p = (\partial T_L/\partial C_m)_{C^*_{Lm}}$ and $s = (\partial T_L/\partial C_n)_{C^*_{Ln}}$ are the slopes of the tangents to the liquidus surface parallel to the planes $C_m = 0$ and $C_n = 0$.

It may be noted in this case that the partition of each of the solutes is additive.

Furthermore, the quantities $p(C^*_{Lm} - C_{0n})$ and $s(C^*_{Ln} - C_{0n})$ are homogeneous with respect to change in temperature. Thus, if we assume that the liquidus surface is locally planar, and that $D_{mm} \approx D_{nn}$, the criterion becomes $G/V > \Delta T/D$ where ΔT is the solidification range of the alloy.

This admittedly questionable simplification to the supercooling criterion has nonetheless the merit of indicating from first principles that the stability of the planar front will be increasingly easy to maintain as the solidification range becomes smaller. This condition is best met along monovariant troughs of gentle slope, in the neighbourhood of binary or pseudo-binary eutectics.

The predictions of the theory have been confirmed by the experimental work of RINALDI (1972) and DUNN et al. (1976) on the Aℓ–Cu–Ni system for alloys rich in aluminium.

2.2.3. Influence of constitutional supercooling on the microstructure

For the most part, the influence of supercooling results in similar types of microstructure as in the binary alloys. Rather than go into exhaustive detail, we

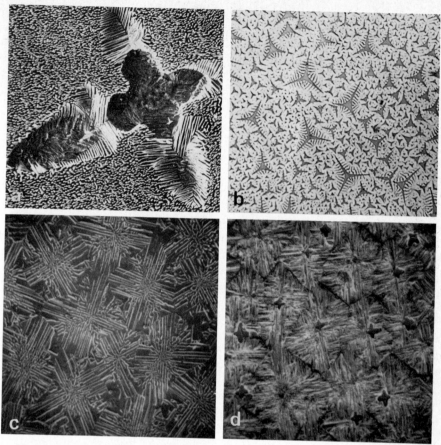

FIG. 22. Influence of constitutional supercooling on the structure of complex alloys. (a) Cobalt dendrites in a hypoeutectic alloy Co–NbC; (b) Dendrites of niobium carbide NbC in a hypereutectic Co–NbC alloy; (c) and (d) Cellular and cellular–dendritic structures in a multicomponent alloy of the COTAC 74 type.

shall limit ourselves to the illustrations of Fig. 22. Micrographs (a) and (b) show the growth of dendrites of cobalt and niobium carbide in hypo- and hyper-eutectics in the neighbourhood of the pseudo-binary eutectic Co–NbC. Micrographs (c) and (d) show cellular and transitional cellular–dendritic structures in two multi-constituent alloys.

The analysis of the structure of these imperfections allows us to determine the way in which the composition of the alloy should be changed in order to obtain a composite of regular structure.

2.3. *Multi-constituent refractory composites*

The needs of the engine-builders of the aerospace industry for turbine blade materials which outperform the present superalloys have led several teams, particularly in France and the United States, to develop multi-constituent refractory composites. The properties demanded in such materials are numerous: they must have a good resistance to creep, good performance in fatigue and also a satisfactory resistance to oxidation and corrosion.

The development of multi-constituent composites meeting these demands has required certain choices to be made, based upon prior experience gained in the metallurgy of the superalloys.

As regards the matrix, one is naturally drawn towards nickel and cobalt, which allow substantial additions of chromium, effective solid–solution hardening, and, in the case of nickel, precipitation hardening by the coherent $\gamma'(Ni_3A\ell)$ phase.

The lack of information concerning complex phase diagrams, and the number of additions necessary have led, via a semi-empirical approach, to the selection of a number of binary or pseudo-binary eutectics which allow such additions to be made without modifying the nature of the phases formed upon solidification and without significantly increasing the solidification range. The concentrations of these additional elements have been progressively optimised by studying their influence upon the manufacturing conditions, upon the structural stability and the physical and mechanical properties of the material.

2.3.1. The γ/γ'–MC family

The study of composites containing γ–MC and γ/γ'–MC began at ONERA in 1967; here we shall limit our discussion to the main stages in the development of materials which are well adapted to industrial applications.

The starting point of this work was the identification by BIBRING et al. (1971) of the pseudo-binary eutectics between cobalt, nickel or iron and the mono-carbides of the transition metals TaC, NbC and TiC.

The possibility of making additions of chromium, tungsten, aluminium, etc. to these simple eutectics allowed the development of the composites named COTAC 3 and NITAC 5, whose compositions are given in Table 1 (BIBRING et al., 1971; BIBRING et al., 1972). The tensile behaviour and resistance to creep of these alloys were found to be remarkable, particularly for COTAC 3, but the material had to be abandoned because of its instability in thermal cycling.

The next stage was the development of composites which were unaffected by thermal cycling, which demanded a strongly hardened matrix. For this reason, the search turned towards nickel-base matrices, where one can make use of precipitation hardening by γ'. The replacement of tantalum carbide by nio-

TABLE 1. Composition of COTAC 3 and NITAC 5 (Wt %).

Composite	Ni	Co	Cr	W	Aℓ	Ta	C
COTAC 3	10	base	20	–	–	12.8	0.78
NITAC 5	base	20	10	10	3	9	0.4

bium carbide allowed an increase in the volume fraction of reinforcing fibres and a decrease in its density. This stage of development ended with the development of a γ/γ'–NbC composite named COTAC 74 (BIBRING et al., 1977).

Finally, it was found necessary to further harden the matrix of COTAC 74 to correct a relative weakness of its creep behaviour at intermediate temperatures. The increase in volume-fraction of the $\gamma'(Ni_3A\ell)$ phase required an adjustment of the concentration of the other elements to maintain structural stability. Finally, we arrive at the composition COTAC 744 whose industrial application is today being seriously considered by French turbine engine designers (KHAN, 1978).

In parallel with this work, the development of methods for the precision casting of shapes has been completed (RABINOVITCH and HAUSER, 1978).

2.3.2. Structure and morphology of γ–MC or γ/γ'–MC composites

The composites in the γ–MC and γ/γ'–MC family all have the same structure and morphology. They are all based upon an austenitic matrix reinforced by fibres of the monocarbides of metals in group IV or V, generally tantalum or niobium. The volume-fraction of fibres is low, being 6–10% depending on the particular alloy. The composite is a columnar-grained polycrystal. The grains are composed of two phases, the matrix being of nickel- or cobalt-base solid solution, and the fibres of the carbide. Within each grain, the lattices of fibre and matrix (both F.C.C.) are parallel, and in epitaxy: $\langle 001 \rangle$ m$/\!/\langle 001 \rangle$ carb. In all composites whose matrix contains more than 2% chromium the growth direction is $\langle 001 \rangle$. The fibres have straight prismatic form, whose cross-section is a bevelled square. The large faces are [110] planes and the bevelled corners are [100] planes (see Fig. 23). The two lattices are not, however, coherent, as

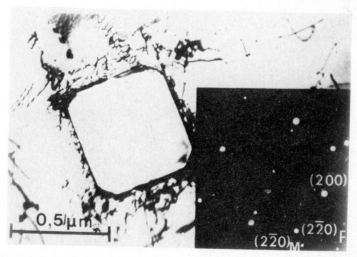

FIG. 23. Transmission electron micrograph showing the cross-section of a tantalum carbide fibre in a Ni–Cr–TaC composite.

their parameters differ by 25%. The fibres, which are perfect monocrystals of TaC or NbC according to the alloy, have an aspect ratio L/D (ratio of length to diameter) which is very high, in excess of 10 000.

The [110] and [100] planes which form the surface of the fibre are not, however, perfectly smooth, but have steps (see Fig. 24). A transmission electron-microscope study of the fibre–matrix interface has shown the existence of linear arrays of defects, whose analysis has shown that the true contact interface is, in reality,

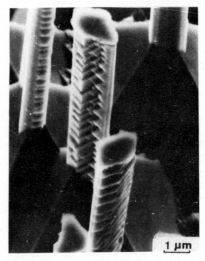

FIG. 24. Scanning electron micrograph showing facets on the [110] and [100] faces of the fibres in the Ni–Cr–TaC composite.

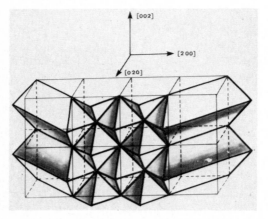

FIG. 25. Sketch of the fibre–matrix interface, showing that the contact surfaces are the close-packed [111] planes.

formed of [111] planes (STOHR, 1978). One can thus reasonably propose that the interface has the following structure (see Fig. 25): The [110] planes are formed of dihedra whose faces are [111] planes, whereas the [100] planes are formed from a series of pyramids whose faces are also [111].

2.3.3. The $\gamma/\gamma'-\delta$ family

The team at Pratt & Whitney has developed refractory composites from the Ni–Aℓ–Nb ternary system. An isothermal section of its phase diagram is shown schematically in Fig. 26. The system has two binary eutectics e_1(Ni–Ni$_3$Aℓ) and e_2(Ni–Ni$_3$Nb), and a pseudo-binary eutectic e_3(Ni$_3$Aℓ–Ni$_3$Nb). Monovariant troughs start at these three eutectics, and converge at the ternary eutectic E_T(Ni–Ni$_3$Aℓ–Ni$_3$Nb) (LEMKEY, 1973).

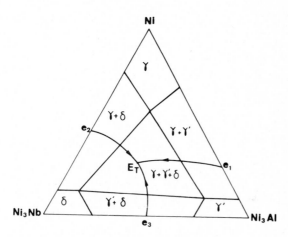

FIG. 26. The Ni–Aℓ–Nb equilibrium diagram.

By choosing compositions in the e_2E_T trough, it is possible to obtain D.S. composites in which the matrix is nickel hardened by γ' precipitates, reinforced by lamellae of the δ-phase (Ni$_3$Nb), whence the denomination $\gamma/\gamma'-\delta$.

In addition, the influence of chromium additions has been studied and from this work an optimal composition has been developed: Ni–19.7 Nb–6.0 Cr–2.5 Aℓ. The melting point of this alloy is 1242°C, the solidification range 18°C, the density 8.6 gm/cc and the volume fraction of lamellae of the δ phase 40%.

Fig. 2 shows the microstructure of this alloy. The crystallographic relations between the F.C.C. matrix and the orthorhombic δ phase are

$$\langle 1\bar{1}0 \rangle_\gamma /\!/\langle 100 \rangle_\delta /\!/\text{direction of growth}.$$

The interfacial planes between the phases are [111]$_\gamma$ and [010]$_\delta$.

This material has been studied and fully characterised: Pratt & Whitney are now actively engaged in a development programme for turbine blades and engine tests (SHEFFLER et al., 1976; SALKELD et al., 1978; CETEL et al., 1978).

2.3.4. Other types of refractory composites

2.3.4.1. NITAC 13. Simultaneously with the above work, an alloy of the γ/γ'–MC type has been developed from the NITAC group by the General Electric Company. This material is reinforced by fibres of the mixed carbide (Ta, V)C, and has the nominal composition Ni: 3.3 Co, 4.4 Cr, 5.8 Aℓ, 8.1 Ta, 0.54 C, 3.1 W, 6.2 Re, 5.6 V. It has very high performance in creep (WOODFORD, 1977) and is being benched-tested in engines (BRUCH et al., 1978).

2.3.4.2. γ/γ'–α composites. These materials have a γ/γ'-matrix, reinforced by ductile molybdenum fibres, and have attractive mechanical properties (LEMKEY, 1976; PEARSON et al., 1977).

2.3.4.3. Composites reinforced by chromium carbide fibres. In this category two families are found, composites of the type (Co, Cr)–Cr_7C_3 (THOMPSON and LEMKEY, 1970; SAHM et al., 1978), and composites based on the system γ/γ'–Cr_3C_2 (MILES et al., 1976; BULLOCK et al., 1978).

2.4. Solidification techniques

2.4.1. High temperature gradient

The directional solidification of composite materials containing a number of elements necessitates a gradient of temperature at the solidification front greater than 100°C/cm. Such temperature gradients impose such high temperatures on liquid metals that these cannot be withstood by normal ceramic mould materials. The problem then consists of finding a system by which one can obtain sufficiently high temperature gradient in order to obtain a good composite while maintaining a maximum temperature in the liquid compatible with the chemical and mechanical properties of the moulds. In the case of composites γ/γ'–MC having a melt temperature of 1350°C the maximum admissible for the mould is 1650–1700°C. It is then a question of obtaining a gradient of 140 to 150°C/cm without superheating the liquid by more than 300°C.

In order to show in a clear fashion parameters which one has at one's disposal to effect such a compromise, we shall study the solidification of a rod of small cross-section within which one can neglect radial heat flows.

Consider the solidification apparatus represented schematically in Fig. 27. The source of heat consists of an isothermal wall at temperature T_H, and the heat sink is provided by a wall at temperature T_C. The height of the thermally isolated zone is $2h$. The apparatus functions in a steady-state regime; the mould is moved with respect to the sources at velocity V and the temperature T_H is controlled in such a manner that the solidification front is situated in the middle of the isolated zone, at the temperature T_M. It is assumed that the temperature gradient at the surface of the rod is linear.

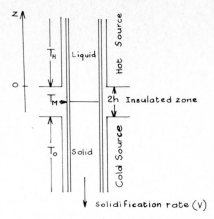

FIG. 27. Sketch of a directional solidification apparatus.

$$\lambda \frac{\mathrm{d}T}{\mathrm{d}z} = Hp(T_H - T)$$

where
- λ (w × cm^{-1} °C^{-1}) is the thermal conduction of the liquid metal,
- H (w × cm^{-2} °C^{-1}) is the coefficient of heat transfer between the hot source and the rod,
- p (cm) is the perimeter of the rod.

Under conditions of equilibrium, the thermal flux for a small element of the rod with respect to the hot source can be written as

$$\lambda s \frac{\mathrm{d}^2T}{\mathrm{d}z^2} + \rho Vcs \frac{\mathrm{d}T}{\mathrm{d}z} + Hp(T_H - T) = 0$$

where
- s (cm^2) is the cross-section of the rod,
- ρ (g × cm^{-3}) is the density of the liquid,
- V (cm × sec^{-1}) is the velocity of movement of the mould,
- c (w × g^{-1} °C^{-1}) is the specific heat of the liquid.

The integration of this function taking $z = 0$ at a distance h above the solidification front, $T_M = 0$, and with the boundary conditions at $z = 0$, $T = Gh$ where G is the gradient at the front and at $z = \infty$, $T = T_H$ leads to the solution

$$T_H - T = (T_H - Gh)\,e^{-\alpha z}$$

where

$$\alpha = \frac{\rho Vc}{2\lambda} + \left[\left(\frac{\rho Vc}{2\lambda}\right)^2 + \frac{Hp}{\lambda s}\right]^{1/2}.$$

The velocity of solidification of eutectics is generally small, of the order of 3×10^{-4} cm sec^{-1}. The term

$$\frac{\rho V c}{2\lambda} \ll 1$$

and one can take

$$\alpha = \left(\frac{Hp}{\lambda s}\right)^{1/2}.$$

We may write

$$\frac{dT}{dz}\bigg|_{z=0} = G, \quad G = \left(\frac{Hp}{\lambda s}\right)^{1/2} (T_H - Gh)$$

where

$$T_H = G\left[\left(\frac{H}{\lambda}\frac{p}{s}\right)^{-1/2} + h\right].$$

This simple relation shows well the role of the principal parameters. In particular the geometry of the rod comes in through the ratio p/s. In the case of a rod of circular cross-section, diameter D, $p/s = 4/D$ and T_H will increase as $D^{1/2}$. In order to obtain T_H within the conditions, it will be necessary to impose a smaller temperature gradient or to alter the construction of the furnace in a manner so as to increase the value of the transfer coefficient H.

Regarding the choice of value of h, it is usually necessary to use small values of h in such a way as to lead to an increase of G without increase of T_H, but in this case control of the value of T_H becomes critical and a small departure from the optimal value of T_H leads to a notable curving of the solidification front. It is for this reason that for rods of several cm^2 in cross-section it is most convenient to adopt for H a value approximately equal to half the smallest transverse dimension of the rod. Thus for a cylinder of diameter D, one takes $h = \frac{1}{2}D$, for a rod of rectangular section of width 1 and of thickness e, one takes $h = \frac{1}{2}e$.

In fact, even in the case of rods of several cm^2 of cross-section, only a computer calculation treating the problem in two dimensions for the case of a rod having an axis of revolution or in three dimensions in the case of a rod of irregular cross-section becomes necessary in order to control the procedure. When the melting temperature of alloys is high ($T_M > 1000°C$), the loss of heat by radiation becomes important and the flux conditions at the walls of the rod are no longer linear. Under these conditions one must carry through calculation in two or three dimensions with non-linear boundary conditions. Such an investigation must be considered in the case of a furnace to be used in an industrial pilot plant.

In practical cases for the fabrication in small quantity of rods, a simplified model is quite sufficient using the simple procedure just described and adjusting the parameters by trial and error during experiments.

2.4.2. Directional solidification of rods of large p/s ratio

The actual conditions under which one can solidify rods of an alloy of the type γ/γ'–MC ($T_M = 1350°C$, $\lambda = 0.5 \, w \times cm^{-1}°C^{-1}$) of diameter $D = 0.8 \, cm$ ($p/s = 5 \, cm^{-1}$) and plates of rectangular cross-section $6.5 \times 1.2 \, cm$ ($p/s = 1.97 \, cm^{-1}$) will now be given.

The values of the temperature gradients measured at the solidification front are respectively $240°C \, cm^{-1}$ and $150°C \, cm^{-1}$. In these two cases, the construction of the furnaces is identical: the source of heat is a graphite resistor heated by induction. The condition T_H constant is effectively obtained by choosing the separation of the turns of the induction heater. The heat sink is a wall of copper cooled by water circulation. Graphite felt is used in order to obtain insulated zone. The values of h are respectively $0.4 \, cm$ and $0.6 \, cm$.

Thermal conditions are complex since one must take into account the heat exchange due to radiation on the one hand, the transverse conduction through a layer of argon of $0.1 \, cm$ thickness between the resistor and the mould as well as the conduction through the $0.1 \, cm$ thick mould wall on the other hand.

Under these conditions the coefficient of transport H varies with the temperature. Nonetheless the valuation of $T_H = G(1/\alpha + h)$ for the case of rods of diameter $0.8 \, cm$ and for the case of plates of cross section $6.5 \times 1.2 \, cm$, taking $H = 0.1 \, w \times cm^{-2}°C^{-1}$ give for the two cases $(T_H + T_M) = 1680°C$, a value in good agreement with the experimental results.

2.4.3. Directional solidification of pieces of complex shape

The study of the directional solidification of rods of constant cross-section has shown in influence of the geometric factors on the heat exchanges. In the case of a specimen of complex form, such as the blade of the turbine, this factor changes along the length of the piece. The largest variation occurs at the vicinity of the airfoil-fir tree root junction.

The problem then consists of finding a procedure which permits one to maintain steady-state conditions of solidification (planar front, G and V constant) in a situation where the thermal transport parameters can vary rapidly along the length of the piece.

We can examine the procedures put into effect at ONERA (RABINOVITCH and HAUSER, 1978).

The solidification apparatus is shown in Fig. 28. It consists of a graphite container with a column of liquid metal of constant cross-section. The metal liquid chosen here is tin. The upper part of the piece of graphite is heated by induction; the lower part is cooled by radiation. Between these two parts the thermally isolated zone is situated. Such a system enables one to establish within the column of tin a region of stable thermal conditions, planar isotherms and a high temperature gradient at the level of the isolated zone.

The mould containing the eutectic alloy is lowered into the column of liquid

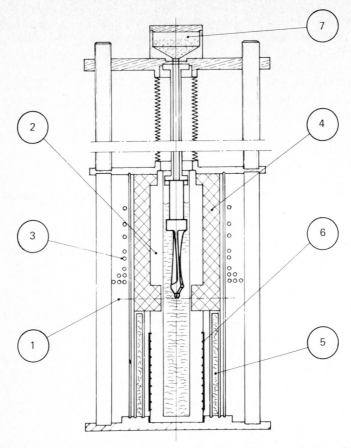

FIG. 28. Directional solidification of cast-to-size turbine blade. Schematic diagram of the apparatus. 1. Location of the solidification front; 2. Liquid tin column and graphite container; 3. Induction heating coil; 4. Thermal insulation; 5. Cold source; 6. Auxiliary heating device; 7. Prealloyed powder feeder.

tin at a constant velocity and solidification takes place at the 1350°C isotherm within the isolated zone.

It is thus possible to solidify a specimen of changing form under quasi-static conditions comparable to those obtained in the case of a rod of constant cross-section provided that

(1) the thermal conductivities of the alloy and of the tin are of the same order of magnitude,

(2) the thickness of the mould wall is made as small as possible in order to render the thermal barrier presented by the ceramic wall least effective.

The conditions of operation of the apparatus are:

- diameter of column of tin: 4 cm,
- height of the isolated zone: $2h = 2$ cm,
- maximum temperature: $T_H = 1670°C$,

- thermal gradient: $G = 150°C \times cm^{-1}$,
- velocity of solidification: $V = 3.3 \ 10^{-4} \ cm \times sec^{-1}$.

The empty mould is placed in the upper part of the column of tin. The eutectic alloy is introduced into the mould in the form of a pre-alloyed powder of which the rate of delivery is controlled in such a manner that the melting of the powder and the filling of the mould take place gradually. When thermal equilibrium is reached, solidification is started and at the end of the operation an auxiliary heater keeps the tin liquid when the main heating system is switched off.

Experience has demonstrated the validity of the procedure. Even with large variations in cross-section of the specimens, conditions of solidification vary very little and the fibrous microstructure of the composite in these critical zones is well maintained.

2.4.4. Ceramic moulds

As one can see from the above, the conditions of fabrication of complex composite materials lead one to utilise moulds under quite unusual conditions. In fact the mould must be able to resist chemical and mechanical damage while in contact with a bath of metal at 1700°C for some tens of hours. In the case of the alloys γ/γ'–MC, the carbon content is high and it is not possible any more to use materials containing silica in the fabrication of the moulds.

At 1700°C silica is reduced by carbon from the alloy and the presence of silica, even as little as 1% in the wall of a mould, produces decarburisation of the alloy and at the same time pollution of the alloy with silicon.

It has been shown that pure alumina of low porosity (5–10%) is the only material capable of providing a solution to the problem of the chemical content of the mould.

The mechanical properties required of the mould show two contradictory aspects. On the one hand, the existence of high thermal gradients in the ceramic wall induces thermal stresses which increase as the thickness of the wall increases. In order to avoid cracking of the mould, it is necessary to limit the thickness to between 0.1 and 0.2 cm. Such a small thickness is also favourable to rapid heat exchange.

On the other hand, resistance to creep of alumina at 1700°C is small and a thin mould is likely to deform under the hydrostatic pressure of a liquid metal.

Which solution is adopted depends on the problem in hand. For example, for the fabrication of a small quantity of rods of constant cross section, the solution adopted was to make thin moulds (0.1 cm thickness) by plasma spraying of alumina. To ensure the mechanical strength of this mould it was surrounded with a sheath of carbon–carbon composite or of graphite of small thickness in order to reduce the creep deformation of the mould.

In the case when shaped moulds must be employed, one uses the lost wax method of making them as in a high quality foundry. It is possible to obtain a thin mould which is also high density by plasma spraying of alumina on to a form (RABINOVITCH and HAUSER, 1978) or to make a shell mould by quenching a piece of wax into a slurry of ceramic (GRESKOVITCH et al., 1977) followed by thermal treatment of the shell thus obtained.

3. Mechanical properties of directionally solidified composites

3.1. Tensile behaviour of composite materials

The tensile test is a simple and rapid test to characterise a material. It allows one simultaneously to examine the mode and mechanism of deformation. The study of the tensile properties of fabricated composites has been addressed by many workers, most of whom consider the composite as a simple association of two phases which obey the law of mixtures (KELLY and DAVIES, 1965),

$$P_c = P_f V_f + P_m(1 - V_f) .$$

This simple relationship has the value of having worked well in a large number of cases, but does, nonetheless, ignore any interaction between the two phases. By contrast with the resin matrix composites, where the interaction is usually small, the matrix in a metal–matrix composite is often significantly hardened by the presence of the fibres. The effect is most marked in composites with a fine structure and a good fibre–matrix bond. As these attributes are commonly found in D.S. composites, the law of mixtures is seldom found to be obeyed in such materials.

Such is the case with γ–MC and γ/γ'–MC composites, where the fibres are perfect, dislocation-free monocrystals. If one applies the law of mixtures to such a composite, it is found that the tensile stress carried by the fibres at failure (at a strain in the composite of about 1.5%), is well below their tensile strength ($\sim \mu/10 \approx 25\,000$ MPa). It is thus necessary to make a detailed study of the tensile behaviour of such a material in order to determine the contribution of each phase to the failure mechanism.

3.1.1. Tensile behaviour of γ–MC and γ/γ'–MC composites at low temperature

The shape of the tensile stress–strain curve for such composites changes markedly with temperature. In this discussion we shall consider only those composites in which the matrix is a solid solution (Cobalt-base composite, COTAC 3).

The tensile curve of cobalt-base composites at temperatures up to 600°C is sketched in Fig. 29. Such curves show four distinct stages: In stage 1 both fibres and matrix deform elastically; in stage 2 the fibres are elastic and the matrix plastic; the end of this stage is indicated by a yield-drop; stage 3 is characterised by plastic deformation at constant stress, and stage 4 by substantial work hardening. Above 600°C (or 800°C, depending upon the matrix), stage 4 disappears. At even higher temperatures (at or above 1000°C) stage 3 also

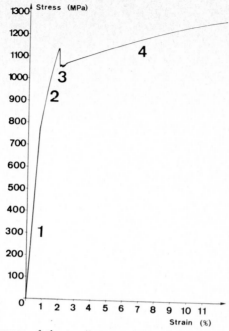

FIG. 29. Typical appearance of the tensile stress–strain curve for a γ–MC composite at low temperature.

disappears and rupture occurs immediately after the yield-drop. To understand the low-temperature tensile behaviour, and its change with temperature, microscopic examinations have been made on flat specimens. At stresses below the yield-drop fibre failure is never observed. The yield-drop corresponds to the appearance of a band of heterogeneous deformation inclined at 45° to the specimen axis, similar to the Lüders bands seen in steels. Detailed micrographic examination shows that within the deformation band the fibres are broken into short segments of three to five fibre diameters long. Each break is associated with a shear plane in the matrix (see Fig. 30) (BIBRING et al., 1972). The plateau (stage 3) corresponds to the propagation of the band and the failure of the fibres throughout the length of the test section. At the end of stage 3 the total extension is high, of the order of 6%. In stage 4 the composite behaves as an alloy work-hardened both by the fibre debris and by the marked strain-hardening capacity of the matrix.

3.1.2. Failure mechanism of the fibres in γ–MC composites

At the yield-drop the stress carried by the fibres, estimated from the law of mixtures, is of the order of 10 000 MPa. This stress is far lower than their theoretical yield stress ($\sim$25 000 MPa). To explain the failure of the fibres, the existence of stress-concentrations must be envisaged. As each fibre failure is associated with a shear plane, it is reasonable to suppose that these are the

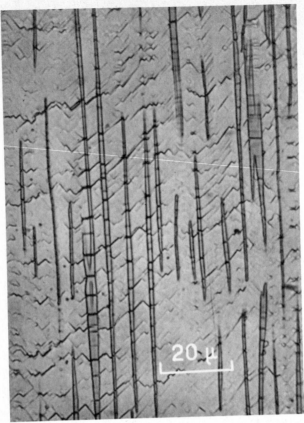

FIG. 30. Slip lines associated with the failure of the fibres in the deformation band (optical micrograph).

origin of the stress concentrations. One possible mechanism would be the formation of dislocation pile-ups against the fibre. To investigate this possibility, transmission electron microscope studies have been undertaken upon previously-deformed massive samples. These studies have added nothing to the information obtained in optical examinations; within the deformation band, the work-hardening of the matrix is so great that the dislocations cannot be resolved, while outside this region the dislocation structures formed in stage 2 are not sufficiently stable to be observed. Such structures change during the thinning of the foil, on the one hand because of surface effects, and, more significantly, on the other hand because of relaxation during the unloading of the specimen. The dislocation structures formed in the earliest stages of the deformation have only been able to be observed in in-situ test carried out in an 1 MeV instrument (STOHR et al., 1975).

During the deformation of these micro-test specimens, pile-ups containing more than a hundred dislocations form against the fibres (see Fig. 31). Quan-

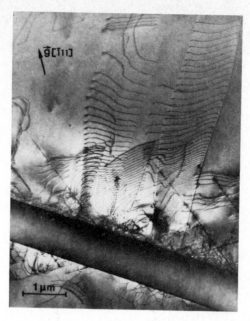

FIG. 31. Dislocation pile-ups against a TaC fibres (in-situ test in 1 MeV electron microscope).

titative analysis, based on the theory of dislocations, allows the stress applied to the micro-specimen to be estimated, as well as the stress at the head of the pile-up. The shear stress at this point on the [111] planes reaches 25 000 MPa. This is of the same order of magnitude as the theoretical elastic limit of tantalum carbide ($\mu/10$). In the presence of such high local stress concentrations, it becomes possible to nucleate a crack in the monocrystalline carbide fibre. This crack can then propagate under the influence of the applied stress and leads to the failure of the fibre.

These observations taken together have fully confirmed the theory of fibre failure. The deformation mechanism of γ–MC composites at room temperature is the following: In stage 2, dislocation sources on the [111] planes are activated, and pile-ups form against the fibres. As the test proceeds, the numbers of dislocations increase, and the stress at the head of the pile-up rises until it reaches the theoretical yield stress of the carbide. The fibre breaks, and the matrix is subjected to a higher stress in this region. Pile-ups then form against the neighbouring fibres which then break in their turn. The deformation then propagates from neighbour to neighbour, leading to the formation of a heterogeneous deformation band. This corresponds to the yield drop. During stage 3 the deformation band extends over the whole test length of the test section, and in stage 4, the material behaves much as a dispersion-hardened alloy.

3.1.3. *Influence of the temperature on the behaviour of the different phases*

The hardening effect of small sections of fibres has been demonstrated by the following experiments: specimens of COTAC 3 composites were deformed in tension to the end of stage 3, so as to break the fibres into short sections. After annealing at 1000°C these same specimens were again loaded in tension at room temperature (see Fig. 32). The failure stress of these specimens is very close to the stress found in the composite at the lower yield-point. These experiments allow us to explain the propagation of the deformation throughout the whole test section of the specimen during stage 3. In this case, the matrix which is hardened both by the short lengths of fibre and its own work-hardening is sufficiently strong to carry the applied load even when the fibres are broken. Furthermore, the stress–strain curve of the matrix containing broken fibres (Fig. 32, curve b) shows a rate of work-hardening which is greater than that of the matrix alone (Fig. 32, curve c).

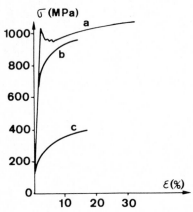

FIG. 32. Tensile curves at room temperature. Curve a—COTAC 3 composites; Curve b—The same, but with the fibres previously broken by straining to the end of stage 3, followed by annealing at 1000°C; Curve c—D.S. alloy having the same composition as the *matrix alone* in COTAC 3.

This type of behaviour is seen at temperatures up to 600°C. Thus, in the temperature range 20–600°C, the reinforcement of the material is essentially due to the hardening of the matrix by the presence of the fibres, and not because of a true fibre reinforcement by load transfer. In contrast, above 600°C, the matrix loses its work-hardening capacity; in fact, tensile tests carried out at 1000°C on specimens of COTAC 3 where fibres had been previously broken into short segments, show that the fibre stress in this case scarcely reaches $\frac{1}{3}$ of the stress carried by the composite at the lower yield-point (see Fig. 33). because of a true fibre reinforcement by load transfer. In contrast, above the matrix when hardened by the broken fibre segments, the failure of the fibres in any cross-section results in the failure of the composite. Above 650°C, the work-hardening of the matrix loses importance and use is then made of reinforcement by load transfer. Thus, at 1000°C the load is almost completely

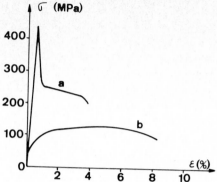

FIG. 33. Tensile curves at 1000°C of (curve a) COTAC 3 and (curve b) the same material with the fibres broken.

carried by the fibres, and the material behaves very much as a 'fabricated' composite.

The tensile behaviour described above for COTAC 3 is observed in all composites of the metal–MC carbide type. However, the temperature at which stages 2 and 3 gradually disappear, depends essentially upon the work-hardening capacity of the matrix. In the nickel-base composites, where the matrix is hardened by precipitation of the coherent $\gamma'(Ni_3A\ell)$ phase, the transition temperature is of the order 850°C instead of 650°C in the cobalt-base materials.

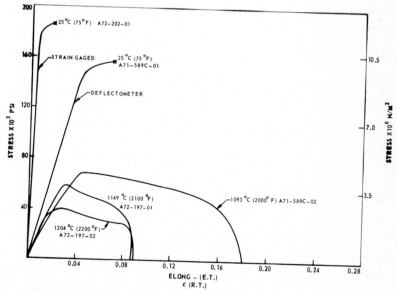

FIG. 34. Tensile curve at various temperatures for $\gamma/\gamma'-\delta$ composites.

3.1.4. Tensile behaviour of $\gamma/\gamma'-\delta$ composites

In this case, we are dealing with a composite reinforced with a high volume fraction (44%) of lamellae of $\delta-(Ni_3Nb)$, which have a certain ductility. In tension at room temperature the failure strain is about 2.5%. Optical and T.E.M. studies have identified two mechanisms which contribute to the deformation of the lamellae of the δ phase: twinning on [112] and slip on [100] planes. Numerous cleavage fractures on the twin planes are also observed (LEMKEY, 1973).

Above 815°C macroscopic deformation bands are observed, at 45° to the specimen axis, and confined to the necking zone. [112] twins are always observed in the δ phase. Fig. 34 shows the tensile curves of this material between 25 and 1293°C.

3.1.5. Comparison of behaviour of different composites

Values of UTS between room temperature and 1070°C for various D.S. composites are given in Tables 2 and 3.

Table 2 compares the tensile strength of COTAC 3 (cobalt-base), COTAC 74 and COTAC 744 (nickel-base) composites, while Table 3 compares the best ONERA composite, COTAC 744, with material developed in the USA. It will be noticed that in the case of COTAC 3 (cobalt-base) the loss of strength between room temperature and 800°C is 44% whilst it is only 22% for COTAC 744 whose nickel matrix has a much greater capacity for work-hardening. Furthermore, the strength of COTAC 744 at high temperatures is always distinctly superior to that of COTAC 74, because of the

TABLE 2. Tensile properties of composites COTAC 3, COTAC 74 and COTAC 744.

Temperature (°C)	Rupture stress (MPa)		
	COTAC 3	COTAC 74	COTAC 744
20	1020	1550	1505
800	540	1030	1170
1000	400	415	570
1070	350	287	406

TABLE 3. Tensile properties of composites COTAC 744, NITAC 13 and $\gamma/\gamma'-\delta$.

Temperature (°C)	Rupture stress (MPa)		
	COTAC 744	NITAC 13	$\gamma/\gamma'-\delta$
20	1505	1140	1240
800	1170	1130	970
1000	570	800	750
1070	406	600	600

high volume-fraction of γ' precipitates in the former material. In Table 2, we note that the American composites are stronger than COTAC 744 in the range 1000–1070°C. We shall see below that this fact does not lead to a superior performance by the American materials NITAC 13 and $\gamma/\gamma'-\delta$ in creep in this same temperature range, however.

3.2. *Creep of directionally solidified composites*

Directionally solidified composites for high temperature use have been principally developed for the fabrication of the rotating turbine blades, and vanes of the gas-turbine engine. For the blading, the major performance criterion is in creep, because of the centrifugal loading of the rotating parts (disk and blades). For this reason, work has been mainly concentrated upon the improvement of the creep performance. At present gas-turbine engines operate with turbine inlet temperatures running up to 1350°C. It should be noted, however, that the temperature of the metal in the high pressure (H.P.) blades does not exceed 980°C thanks to the various cooling methods used. The stresses to which the H.P. blades are subjected do not generally exceed 100 to 120 MPa at 980°C. In this case, the low pressure (L.P.) blading, which is not necessarily cooled, reaches very similar temperatures, but at distinctly lower stress levels. Fig. 35 shows the way in which the design office uses the creep data (e.g., 1% extension in 2000 hours) to determine the operating conditions of the blade (stress and temperature) as a function of the desired life. It should be noted that the critical point on the blade, that is to say, the point where the

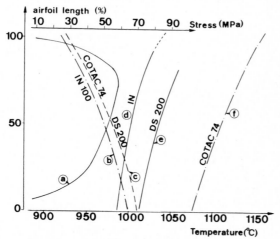

FIG. 35. Variation of stresses and temperature in a L.P. turbine blade from a modern gas turbine. Curve a—Variation of temperature along the blade; Curves b and c—Variation of stresses along the blade for IN 100, DS 200 and COTAC 74 alloys; Curves d and e—Allowable temperatures for a deformation of 1% in 2000 hours for IN 100 and DS 200; Curve f—Allowable temperature for a half-life of 2000 hours for COTAC 74.

curve of allowable temperature for the materials reaches its maximum, is found around 985°C. If one wished to increase the temperature of the blade by 40–50°C, the super alloy IN 100 could no longer be used in the motor, and one would then have to envisage the use of a stronger material such as MAR M 200 in directionally solidified form (DS 200).

Below, we consider the metallurgical procedure adopted to develop materials responding to the demands of the designer, beginning with the cobalt-base alloys.

3.2.1. *Creep of cobalt-base composites*

The creep curves of COTAC 3 composites show the well-known primary, secondary and tertiary stages. Primary creep is generally small at all temperatures. The second stage is very long, with a very low strain rate, of the order of 10^{-6}/hour. Tertiary creep, which coincides with the failure of the fibres on a given cross-section, occurs at a strain of about 1.5%. The tertiary creep regime normally lasts a few tens of hours. This rapid failure may be easily understood in the light of what is known from the short-term tensile testing: in the range 800–1100°C the work-hardening of the matrix is low, and failure of the fibres on any section results in the failure of the composite.

The COTAC 3 material (see Fig. 36) shows a distinct gain in creep performance over the superalloys above 900°C. On the other hand, at lower temperatures (750–850°C) it fares distinctly less well, and the potential temperature gain is thereby reduced. In effect, the engine designers generally demand that in the blade root region, the material should be able to withstand a stress of the order of 400 to 500 MPa for several hundred hours. In the case of COTAC 3, the thousand-hour rupture stress is only 320 MPa (or 350 MPa for a life of 100 h).

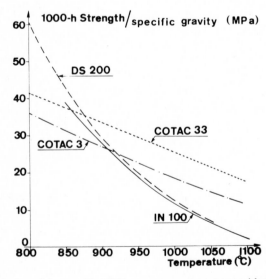

FIG. 36. Variation of the specific 1000 hour creep-rupture stress with temperature.

For use in the gas turbine, such a weakness in the root would require a redesign of the blade, and this could become troublesome from the aerodynamic point of view. For all of the above reasons it became necessary to improve the mid-range creep characteristics. The matrix of the COTAC 3 composite, which is a cobalt-base superalloy, cannot be strengthened by precipitates analogous to the γ'(Ni$_3$Aℓ) used in the nickel-base materials. Even if an intermetallic of the type of Co$_3$Aℓ is found in the cobalt-base alloys, it has neither the stability nor the effectiveness necessary to sufficiently strengthen the matrix. As a result, improvements in the creep resistance of cobalt-base superalloys are normally obtained by the precipitation of carbides MC, M$_{23}$C$_6$, . . . , etc.

In the COTAC 3 composite, the solubility of tantalum carbide in the matrix varies with the temperature (see Fig. 37). This variation has allowed us to develop a fine and homogeneous precipitation of TaC in the matrix of the composite (TROTTIER et al., 1974) by use of an appropriate heat treatment. This comprises a solution treatment at 1300°C for two hours, which causes a uniform decrease in fibre cross-section by about 10% without altering their morphology. This is followed by aging for 24 hours between 700 and 900°C. The microstructure of the material is then as shown in Fig. 38. The size of the precipitates is less than 1000 Å, and they remain stable even under long exposure to temperatures up to 1100°C.

The hardening of the matrix which accompanies this precipitation results in an increase of the yield-point of the composite. For example, at 800°C, the yield stress rises from 540 MPa for the untreated COTAC 3 to 600 MPa for the treated material (COTAC 33). Fig. 36 shows the specific creep resistance of COTAC 3 and 33 compared with that of the superalloys IN 100 and DS 200.

It will be noticed that, at high temperatures, the performance of COTAC 33 is some

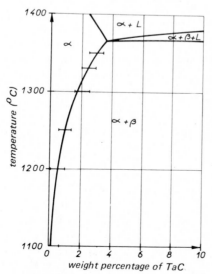

FIG. 37. Variation of the solubility of TaC with temperature in the alloy Co–10Ni–20Cr–TaC.

FIG. 38. Morphology of the carbide precipitates in COTAC 33.

80°C better than COTAC 3. It is important to emphasise here that the creep resistance of COTAC 33 at 800°C is very comparable with that of the nickel-base superalloys. For this reason, COTAC 33 became a very interesting material for blade applications. However, in spite of its excellent creep performance, particularly at high temperature, its poor resistance to thermal cycling above 1000°C led us to abandon it in favour of the nickel-base composites.

3.2.2. Creep of nickel-base composites

In view of the fact that the main factor governing the thermal-cycling resistance of these materials is the plastic deformation of the matrix, one was naturally drawn to the nickel-base alloys where the matrix can be substantially hardened by precipitation of the coherent $\gamma'(\text{Ni}_3\text{A}\ell)$ phase, and by solid solution hardening through the addition of W and Mo. In the early stages of this work, the composition COTAC 74 was thoroughly characterised by Bibring (BIBRING, 1977).

The creep curves of COTAC 74 are distinctly different from those of the cobalt-base alloys. Up to about 1000–1020°C COTAC 74 has a much greater creep extension (7–10%) than the COTAC 3 (1.5–2%). Nevertheless, at the end of stage 2, which corresponds as usual to the failure of the fibres, the strain is still only about 1.5%. It follows that the ductility of COTAC 74 arises from the hardening of the matrix due to the precipitation of the γ' phase, a hardening which allows the composite to carry the applied load even after failure of the fibres. A comparison of the creep characteristics of COTAC 74 and those of the best present superalloys IN 100 and DS 200 (see Fig. 39) shows that the improvement offered by COTAC 74 only becomes apparent for lives in excess of 1000 hours.

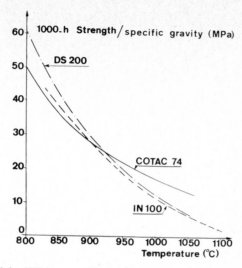

FIG. 39. Variation of the 1000 hour specific creep-rupture stress as a function of temperature.

TABLE 4. Creep stress (MPa) at 900°C for a
rupture life of 150 h and 1000 h.

Material	150 h	1000 h
IN 100	290 MPa	210 MPa
COTAC 74	260 MPa	240 MPa

Furthermore, by contrast with what is observed in the nickel-base super-alloys, a minor change in the applied stress causes a large change in the life of COTAC 74. Thus, at 900°C, Table 4 shows that, in the case IN 100, one must reduce the stress from 290 to 210 MPa, i.e., 80 MPa to increase the life from 150 to 1000 hours, whereas for COTAC 74 it is sufficient to reduce the stress by only 20 MPa (260 to 240 MPa). On the other hand, however, a slight increase in the creep stress will result in a substantial reduction in the life of the composite. This special behaviour of the COTAC composites in response to small changes in the applied stress is clearly shown in a representation of the creep results in a $\log \sigma = F(\log t)$ diagram. In such a graph (see Fig. 40), the experimental points are found on straight lines whose slope is very small. In the same diagram, the slopes obtained at high temperature for the superalloys (for lifes less than 1000 h) would be greater.

As regards COTAC 74, the marked effect of stress on the creep life limits the practical value of the composites to applications which demand very long lives at moderate stress levels (see Table 5), which is the case of the L.P. turbine blades in present day civil engines with high dilution (bypass type). For such applications, COTAC 74 can give a very large improvement in stress and temperature (cf. Fig. 35). The weak point of COTAC 74 then appears to be its creep

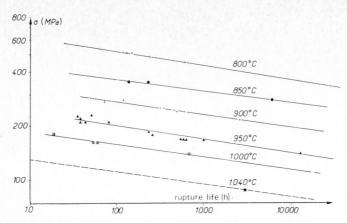

FIG. 40. Representation of the creep data on the $\log \sigma - \log t_R$ diagram.

TABLE 5. Long-term creep lifes of COTAC 74.

Temperature (°C)	Stress (MPa)	Creep lifes (hours)		
		COTAC 74	IN 100	DS 200
900	220	2600	~800	~2400
950	140	13000	~1100	~2500
1040	90	3028	~160	~250
1080	60	> 4000	~160	~250

resistance at high stress levels and medium temperatures. It thus became important to develop daughter composites with a creep resistance equivalent to that of the superalloy DS 200 under these conditions. To reach this objective, we tried to harden the matrix of COTAC 74 in a similar manner to that found in DS 200. This was achieved by simultaneously hardening in solid solution (either by increasing the tungsten content or by adding molybdenum) and by precipitating the γ' phase by increasing the aluminium content. These changes in the composition resulted finally in the alloy COTAC 744, whose composition and properties are shown in Table 6.

TABLE 6. Characteristics of COTAC 744.

Composition. Ni, 10 Co, 4 Cr, 10 W, 2 Mo, 6 Aℓ, 3.8 Nb, 0.42 C

Properties

Solidus temperature (±5°C)	γ' solvus temperature (±10°C)	γ' volume fraction (%)
1340	1200	~58

The specific creep properties for a 1000 hour life for COTAC 744, IN 100 and DS 200 are compared in Fig. 41. As regards the creep resistance under high stress (leading to failure in 300 hours) COTAC 744 is always superior to IN 100 but falls below the performance of DS 200 at temperatures above 850°C.

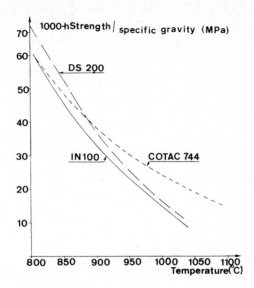

FIG. 41. Variation of the 1000 hour specific creep-rupture stress as a function of temperature.

In summary, therefore, the creep performance of COTAC 744 is practically equivalent to that of the superalloys up to 850°C, and is distinctly superior at temperatures above. Thus, for example, Table 7 compares the potential operating temperature of the ONERA composites with those of IN 100, DS 200 and the American composites NITAC 13 and $\gamma/\gamma'-\delta$. These temperatures are for a life of 1000 hours and a stress of 150 MPa.

TABLE 7. Potential use temperature for a 1000 hour rupture life at a stress of 150 MPa.

COTAC 74	COTAC 744	NITAC 13	$\gamma/\gamma'-\delta$	DS 100	IN 100
1000°C	1045°C	1005°C	1000°C	965°C	945°C

3.2.3. Prediction of long-term creep-rupture properties

The moving blades of aero gas turbines are designed to operate for thousands of hours. At present, certain motors for civil applications are required to operate for 10–20 000 hours; in view of these requirements it would be useful to know the long-term creep behaviour of these materials.

Unfortunately, it is practically impossible, because of the time required, to carry out long-term tests at a range of temperatures and stresses. So as to facilitate the work of the designers, it is important to be able to determine the very long term creep behaviour (10 000 h) by extrapolation from tests over relatively short times. LARSON and MILLER (1952) proposed a plot which allows the reduction onto a single curve ($\sigma = f(\text{LMP})$) of all creep data. The Larson–Miller Parameter (LMP) $= T(C + \log t_r)$, where T is the absolute temperature, t_r the failure time and C a constant, takes into account the joint effects of time and temperature. For the superalloys, the constant has typical values around $C = 20$.

The Larson–Miller curve allows us to make predictions based upon a few short-time tests of the life expectancy of the material over a wide range of stresses and temperatures. This type of relation is also very useful to compare the creep performance of a range of superalloys.

In the particular case of the COTAC composites the extrapolation of the short-term creep data via the Larson–Miller relationship leads to an under-estimation of the performance of the material at long times. As Fig. 42 shows, for COTAC 74 the 100 hour data (high stresses) and the 1000 hour data (moderate stresses) cannot both be reduced onto the same curve.

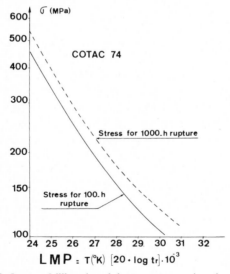

FIG. 42. Larson–Miller plot of the creep properties of COTAC 74.

By contrast, the creep performance of the COTAC composites can be des-cribed by a law of the type $\sigma = At_r^M$ (KHAN, 1978) where A is a constant and M the time exponent. Such a relation between the stress and the time to failure has become particularly valuable for the predictions of the long-time per-formance. The value of this relationship lies in the fact that the time exponent appears to be essentially independent of the temperature. The value of the exponent M varies from one alloy to another: as an example, it has a value 0.066 for COTAC 74 (see Fig. 40) and 0.1 for COTAC 744. The good agreement between the lines and the experimental points allows us, on the one hand to extrapolate to lifes of 10 000 hours from data obtained in the 100–1000 hour range, and on the other hand, bears witness to a good structural stability of the material in creep. Any instability would have made itself known by an increase in the absolute value of the slopes with temperature.

We now turn to the creep criteria which are used by the design office. In the case of the conventional superalloys, the life allowed for a moving blade is equal to the time for an overall deformation of 1%.

For the aligned eutectics of the COTAC type a criterion based upon a 1% deformation is not valid, above all if one wishes to use the potential creep performance to the full. For example, in the case of COTAC 744 up to about 1000°C, a deformation of 1% is obtained at only $\frac{1}{10}$ of the failure time, whereas it requires $\frac{6}{10}$ of the life time to reach 1.5%. At higher temperatures (around 1070°C), failure occurs at 1.5–2% strain. For these reasons, it is reasonable to take as a failure criterion half the total rupture time for the composite blade in question. Fig. 43 shows the specific stress–temperature curves which give either a 1% strain for the superalloys or the half life, 10 000 hours, for COTAC 74 and COTAC 744 composites. These curves show that the use of D.S. eutectics can provide a significant gain in temperature and stress. They also show the improvements in long-term creep loading which can be obtained with COTAC 744.

To summarise, then, the D.S. composites in the COTAC 74 family have a creep performance which is sufficiently good to allow a gain in temperature of the order of 80°C over the superalloys. Furthermore, the majority of the other characteristics of the COTAC group are also superior to those of the con-ventional superalloys. For this reason, the practical application of these alloys will not demand any major redesign of the turbine blade.

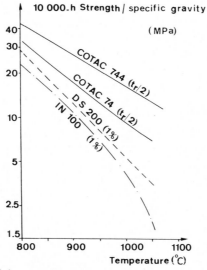

FIG. 43. Comparison of the potential performance in creep of COTAC composites and superalloys.

Finally, a comparison between the D.S. composites studied in the USA (Pratt & Whitney, General Electric) and the ONERA materials of the COTAC 744 type show that at moderate temperatures the American composites are distinctly superior in creep performance to COTAC 744; on the other hand, above 940°C, COTAC 744 becomes superior to NITAC 13 and $\gamma/\gamma'-\delta$ (see Fig. 44).

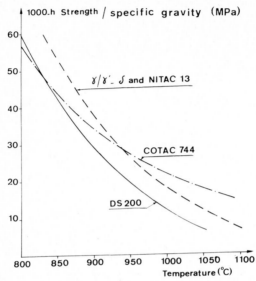

FIG. 44. Variation of the 1000 hour specific creep-rupture stress as a function of temperature.

3.3. Other properties

As well as a good performance in creep and thermal cycling, a turbine blade material must possess certain other mechanical properties. Most important are impact resistance, creep resistance in shear, and resistance to thermal and mechanical fatigue.

3.3.1. Impact bending

Tests in impact bending have been carried out on smooth or notched specimens of COTAC 74 and IN 100 at various temperatures. To allow for the anisotropy of COTAC 74, the specimens were prepared either parallel or perpendicular to the fibres. The values of fracture energy obtained on smooth and notched specimens are compared for the two alloys in Table 8.

In general, the results obtained on smooth or notched specimens of COTAC 74, (specimens cut parallel to the fibres), are superior to those of the superalloy IN 100. In contrast, however, when the notch is parallel to the fibres, COTAC 74 has a lower performance than IN 100 at temperatures up to 700°C, and becomes equal to 1000°C.

TABLE 8. Fracture energy of COTAC 74 and IN 100 (J/cm²).

Temperature (°C)	Longitudinal unnotched specimens		Longitudinal notched specimens[a]		Transverse notched specimens[b] COTAC 74
	COTAC 74	IN 100	COTAC 74	IN 100	
20	77–110	57–120	25–31	14–28	6–7
700	37.5–40	43	21–23	12	8.3–8.7
1000	48–50	25	25–26	11	13–14

[a] Notch ⊥ to fibres.
[b] Notch ∥ to fibres.

3.3.2. Shear creep

The connection between the disk and the moving blades is often made through a 'fir-tree' root, whose teeth are loaded in shear. Comparative tests have been carried out on COTAC 74 and IN 100 under particularly severe conditions (700°C at 400 MPa) so as to detect any potential weakness in the composite under these conditions. The rupture time for IN 100 was very low (about 1 hour) whereas for COTAC 74 it was found to be about 100 hours. In view of the large difference in these two lifes, it appears that the use of the fir-tree root for COTAC 74 should not give rise to any undue fears.

3.3.3. Thermal fatigue

The object of these tests should be to study the formation and propagation of cracks under conditions as close as possible to those experienced by the turbine blade. In this type of test, thin edged wedge specimens are heated by a flame in 60 seconds to 1100°C and then cooled by a forced air jet in 20 seconds to about 100°C. The tests were carried out under identical conditions on COTAC 74, on IN 100 (nickel base) and on one of the alloys commonly used for the fixed blading (nozzle

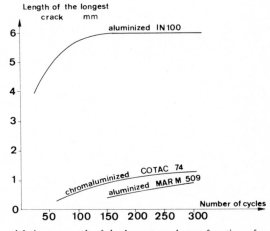

FIG. 45. Thermal fatigue: growth of the largest crack as a function of number of cycles.

guides vanes), MAR M 509, which is cobalt-base. The three materials were previously given an appropriate protective coating. Cracks appeared on the thin edge of the specimens, and Fig. 45 (p. 345) shows the length of the longest crack as a function of cycle number for each material. It is seen that COTAC 74 is distinctly better than IN 100, and very close to MAR M 509, an alloy well known for its very good resistance to thermal fatigue.

3.3.4. *Mechanical fatigue*

In general, the fatigue resistance in low- and high-cycle fatigue loading of fibrous D.S. composites of the COTAC type is superior to the conventional superalloys. In low-cycle fatigue, tests have been carried out in tension–tension at 650°C on notched specimens (stress concentration factor $K_t = 2.4$); Table 9 compares the endurance limits at 10^4 cycles for COTAC 74 and IN 100. These results show the distinct superiority of COTAC 74, with nearly 40% increase in allowable stress.

Regarding high-cycle fatigue, the endurance limit at 10^7 cycles for transverse specimens (perpendicular to the fibres) is equal to 75% of the value obtained on material loaded parallel with the fibre direction (see Table 10).

TABLE 9. Low cycle fatigue of notched COTAC 74 and IN 100 at 650°C ($K_t = 2.4$).

Materials	Temperature (°C)	Number of cycles	Rupture stress (MPa)
IN 100	650	10^4	600
COTAC 74	650	10^4	830

TABLE 10. High cycle fatigue characteristics of COTAC 74.

Temperature (°C)	Hour glass specimens	Test mode		Stress for 10^7 cycles (MPa)
		Type	Frequency (Hz)	
20	$\parallel$ to fibres	Rotating	49	±400[a]
20	$\perp$ to fibres	Bending	49	±310[a]
700	$\parallel$ to fibres	Tension–tension	87	20–680[a]
800	$\parallel$ to fibres	Tension–tension	87	20–520[a]

[a] Unbroken specimens.

4. Thermal stability

The problem of thermal stability in turbine blade materials is a many-sided problem in view of the wide range of temperatures and conditions encountered in the aero gas turbine.

In brief, the turbine blade is subjected to the following conditions during its life:
- Large variations in temperature during take-off or landing.
- Short periods of high temperature exposure (980°C) when maximum power is required, e.g., at take-off.
- Long periods at moderate temperatures (800–900°C) under cruise conditions.

Taken together, these conditions may give rise to significant structural changes in the alloys chosen for the blading.

In the case of composite materials whose temperature potential is around 1100°C, i.e., 100°C greater than that of the superalloys, the problem of thermal stability is particularly acute, for any degradation of the reinforcing phase will be reflected in a heavy reduction in the mechanical properties. In the case of COTAC composites, the thermal stability question can be considered under three heads:

(1) Thermal cycling.

(2) Coarsening or spheriodisation of the fibres which can occur when the composite is held at high temperature (1100°C).

(3) Finally, the possibility of solid-state reactions in the medium-temperature range (800–1000°C) which may alter the phases which were produced at high temperature during manufacture.

4.1. Thermal cycling

In γ–MC composites, where the two phases are very different—metallic matrix and carbide fibres—large or rapid variations in temperature, for example from 20 to 1070°C may cause a degradation of the material. This may lead to loss of the mechanical properties.

In the case of COTAC 3, changing the temperature from 20 to 1070°C (test temperature) twice per hour results in a reduction of the cyclic creep life by a factor 10 with respect to the isothermal creep life. This loss of properties, which could become catastrophic, is due to a degradation of the fibres during thermal cycling. The nature of this degradation is very different depending upon whether the maximum test temperature is above or below 1000°C (see Fig. 46).

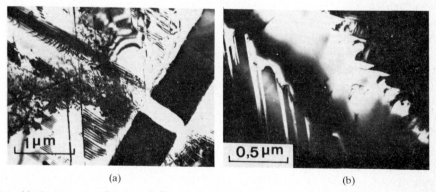

(a)   (b)

FIG. 46. Appearance of COTAC 3 fibres after thermal cycling. (a) At 1000°C; (b) At 1070°C (transmission electron micrographs).

At temperatures below 1000°C the fibres are simply snapped without their shape being changed; in contrast, above 1000°C there is a shape change, as may be seen after a few hundred cycles between 20 and 1070°C. These two modes of degradation give rise to a purely mechanical effect at 1000°C, and a combined mechanical and chemical effect at 1070°C. The two phenomena to be taken into account under thermal cycling are:
- the very high internal stresses arising from the differential thermal expansion of the two phases $(\alpha_m - \alpha_f) = 10^{-5} \,°C^{-1}$ in the case of the COTAC family),
- the variation in solubility of the carbide with temperature.

We now consider these two mechanisms separately.

4.1.1. Thermal cycling at temperatures below 1000°C: Mechanical effects

To understand the mechanism which leads to the failure of the fibres, it is important to know the internal state of stress of the composite and its variation with temperature.

During the manufacture of the composite, at the solidification temperature, the two phases are without internal stress. During cooling, because the average thermal expansion coefficient in the matrix is much greater than that of the fibres, the matrix will shrink much more. Thus, at room temperature, the fibres will be in compression and the matrix in tension. It is possible to estimate the internal stresses to which the fibres and matrix will be subject using elasticity theory (EL GAMMAL, 1972).

The following simplified assumptions were adopted.

In the composite, the fibres are regularly arranged in a hexagonal array: thus, one can subdivide the composite into a series of hexagonal prisms of matrix, each with a fibre at the centre. In a further simplification, both fibre and prism of matrix are made of circular cross-section.

One can then calculate the stresses in fibre and matrix if both are taken as elastic. In a composite with 10 % of fibres, with a temperature interval of 700°C, the axial stress in the fibre is around 2200 MPa, while it reaches 300 MPa in the matrix. The shear stress in the neighbourhood of the fibres is about 900 MPa. While these values are drawn from a highly simplified calculation, they nonetheless show that the shear stress in the matrix is very probably in excess of the elastic limit. As a consequence, the differential expansion of the two phases will cause, with each cycle, a plastic deformation of the matrix in the region of the fibres. In fact, if $\tau_E(T)$ is the elastic limit in shear of the matrix which is supposed perfectly elastic-plastic, the stress cycle to which the matrix is subject can be represented on a temperature–shear stress diagram such as that shown in Fig. 47. In this figure, we show the elastic limit τ_E of the matrix as a function of the temperature and the shear stress τ_m calculated from elasticity theory. At the point A, the true shear stress in the matrix is equal to its elastic limit. During heating, fibres and matrix expand, and the shear stress in the matrix follows the line AB (elastic) and then BC (plastic deformation in compression). During cooling, the shear stress follows CD (elastic) and then DA (plastic deformation in tension). For the matrix, therefore, each cycle is

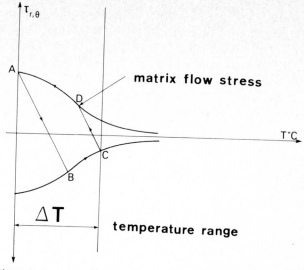

FIG. 47. Shear stress–temperature diagram showing the plastic deformation introduced into the matrix during each cycle.

equivalent to a compression–tension fatigue cycle. Instead of showing the variation in matrix stress caused by the change in temperature, one could equally well show it as a function of deformation. Nonetheless, only a calculation taking into account plasticity can give the real deformation of the matrix as a function of temperature. However, dilatometric curves can give us some ideas, as witnessed by the data of Fig. 48 obtained on COTAC 3. The cycle shown in the figure is equivalent, for the composite, to the stress–temperature

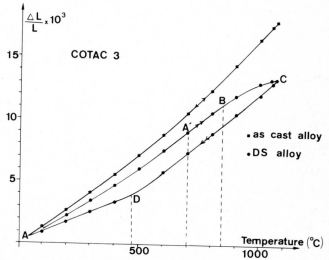

FIG. 48. Dilatometer curve from COTAC 3 showing evidence for plastic deformation in the matrix caused by internal stresses.

cycle described above for the matrix. During heating the composite expands almost linearly to point B. The temperature at this point, T_B, is the same for the cycles $\Delta l(T)$ of the composite and for the matrix $\tau_m(T)$. At point B the shear stress in the matrix reaches its elastic limit, which then deforms in compression. As a result, from this point on, the measured rate of expansion of the composite decreases. During cooling, the fibres and matrix contract, and at D, the matrix yields plastically in tension. It is thus possible to derive the stress cycle followed by the matrix as a function of its plastic deformation ε_m^P by the use of these two diagrams (see Fig. 49). Hence, during each cycle, the matrix undergoes mechanical work whose value can be represented, to a first approximation, by the shaded area in the diagram $(\tau_m, \varepsilon_m^P)$.

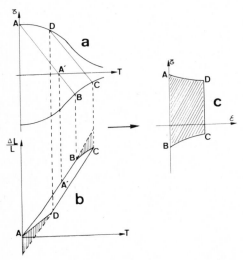

FIG. 49. Various representations of plastic deformation of the matrix during thermal cycling of the composite. (a) Shear stress vs. temperature diagram; (b) Dilatometric curve $(\Delta L/L)(T)$; (c) Shear stress vs. strain diagram deduced from (a) and (b).

On the microscopic scale this work gives rise to an increased density of dislocations, especially in the neighbourhood of the fibres. This effect can be observed by making thermal cycling experiments in the H.V.E.M. To facilitate the experimental work, the observations were made on a nickel-base alloy reinforced by fibres of tantalum carbide. After only three cycles between 20 and 1000°C, one can see dislocations near the fibres (see Fig. 50). Thermal cycling will therefore cause a significant plastic deformation in the matrix by the accumulation of the contributions from each cycle. The increase in dislocation density in the neighbourhood of the fibres can then give rise to stress concentrations which lead, in some cases, to fracture of the fibres.

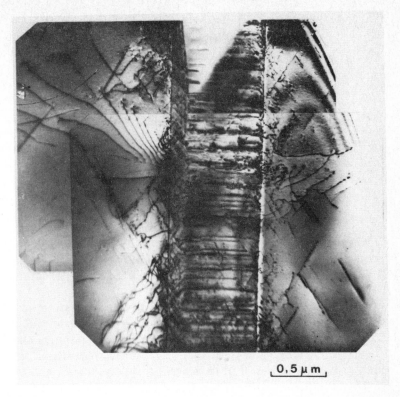

FIG. 50. Transmission electron micrograph showing the dislocation structure in the neighbourhood of the fibre after three cycles between 20 and 1000°C.

4.1.2. *Thermal cycling at temperatures above 1000°C*

In COTAC 3, when the maximum temperature of the cycle exceeds 1000°C, the fibres develop a characteristic saw-edge profile (cf. Fig. 46(b)).

This shape results from the diffusion of the metal and carbon atoms which form the fibre into the matrix. The diffusion is naturally favoured by the increased solubility of the fibre components with increasing temperature.

The solubility has been determined in a certain number of composites with the help of the electron microprobe analyser (see Table 11).

The measure of solubility is obtained by measuring the content in the matrix of the transition element which forms the carbide fibre; this measure is made upon the as-solidified composite and after a solution-treatment of two hours at a temperature 30°C below the melting point. In COTAC 3, which is very sensitive to thermal cycling, the variation in the solubility of the fibre between 20 and 1300°C is very large. On the other hand, in (Co, Cr, Ni)–HfC and COTAC 74, which are practically unaffected by thermal cycling, the variation is much smaller. One would thus be tempted to attribute the damage observed in COTAC 3 to this large variation in fibre solubility with temperature. In fact, the situation is not so simple, because, if this same COTAC 3 is cycled between 700

TABLE 11. Comparative solubility of the carbide fibers in various heat-treated composites.

Composite	Solution temperature (°C)	Transition metal in the matrix of the composite (Wt %)	
		Non-treated	Heat-treated
COTAC 3	1300	1.76 ± 0.06	3.12 ± 0.15
Co, Cr, Ni–HfC	1330	0.19 ± 0.02	0.46 ± 0.02
COTAC 74	1300	0.75 ± 0.07	1.02 ± 0.06
Ni, Cr–TaC	1300	13.2 ± 0.2	13.0 ± 0.25

and 1070°C, no degradation of the fibre is observed even after 10 000 cycles, in spite of the fact that the change in solubility of the fibres between 700 and 1070°C is very similar to that between 20 and 1070°C. Thus, the change in solubility of the fibres does not alone allow us to explain the degradation of the fibres above 1000°C.

This is confirmed by the fact that, in the NiCr–TaC composite where no change in fibre solubility at all has been detected, the carbide fibres are nonetheless degraded by thermal cycling. A more detailed examination of the form of the damaged fibres shows that the outgrowths which form on the fibres always develop along [111] planes of the matrix, as if these planes were in some way 'short-circuit' diffusion paths. These planes are, in fact, slip planes in the matrix and will therefore contain dislocations introduced as a consequence of the thermal cycling (cf. Fig. 50). This high dislocation density in the neighbourhood of the fibres will both aid the diffusion of the fibre elements into the matrix, and also increase the local solubility of the carbide. All of the above, therefore, helps to explain, at least qualitatively, the formation of outgrowths on the fibres when the maximum cycle temperature exceeds 1000°C.

In conclusion, even if changes in solubility of the carbide with temperature play a role in the degradation of the fibres under high temperature thermal cycling (>1000°C), the fibre damage is most directly attributable to the plastic deformation of the matrix.

4.1.3. Development of composites unaffected by thermal cycling

To obtain composites which are almost unaffected by thermal cycling, one must harden the matrix so as to limit its plastic deformation. Clearly, an increase in the elastic limit of the matrix would reduce the size of the hysteresis loop, and hence the plastic deformation accumulated in the matrix during each cycle.

In the case of COTAC 3 composites, where the matrix is cobalt-base, the matrix could be hardened by a fine, homogeneous precipitation of TaC which would improve the creep performance throughout the whole temperature range (700–1070°C). One could therefore also hope that such a treatment would lead to improvements in thermal cycling. In fact, when tests are carried out between 20 and 1000°C under an applied stress of 160 MPa, the number of cycles to failure rises from 1000 for COTAC 3 to 5000 for COTAC 33. Nonetheless, as soon as the upper temperature reaches about 1070°C, the latter alloy is no better than the earlier material.

This was a further reason for us to abandon the cobalt-based alloys in favour of nickel, where the matrix can be hardened in a much more effective manner by the precipitation of the coherent $\gamma'(Ni_3A\ell)$ phase.

In composites of the COTAC 74 type, the hardening of the matrix, obtained both in solid solution by the addition of 10% tungsten, and above all by the high volume fraction of γ', results in a satisfactory behaviour in thermal cycling at temperatures up to 1100°C.

In spite of this, however, the damaging effects of thermal cycling have not been completely suppressed, but only driven upwards to higher temperatures. In fact, a slight loss in both tensile and creep properties in COTAC 74 still occurs after 5000 cycles between 20 and 1070°C. Detailed metallographic examination of these samples shows a very slight degradation of the fibres.

In order to more exactly approximate to aero engine operating conditions, tests have also been made in creep under thermal cycling. Thus, in the case of a test carried out at 1070°C under 120 MPa, the life in thermal-cyclic-creep (500 hours or 1000 cycles of $\frac{1}{2}$ hour) is slightly less than under isothermal conditions (600 h). Taken together, these results clearly show that COTAC 74 has a satisfactory thermal stability over the whole temperature range foreseen for its use.

Nonetheless, if the maximum temperature is raised, degradation of the fibres reappears: after a hundred cycles of three minutes between 20 and 1250°C, the fibres are badly damaged (see Fig. 51).

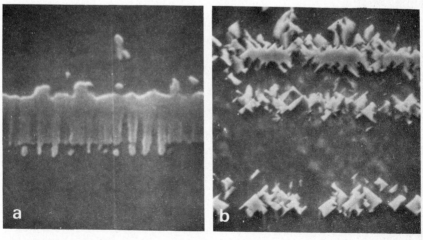

FIG. 51. Degradation of the fibres in COTAC 74 cycled between 20 and 1250°C. (a) 100 cycles; (b) 1000 cycles.

In COTAC 744, where the matrix is even more strongly hardened, no damage is seen after several thousand cycles between 20 and 1150°C. In thermal-cycling creep between 250 and 1070°C, the life of COTAC 744 is double that of COTAC 74.

In conclusion, the family of γ/γ'–NbC composites is practically insensitive to thermal cycling throughout the temperature range foreseen for their application.

4.2. High temperature thermal stability

In a two-phase (α, β) system, while the proportion of the α and β phases present at any given temperature is determined from the phase diagram, a further evolution is nonetheless possible. The energy of the system may be further reduced if the total area of interface between the two phases is reduced: this is the phenomenon of coarsening. The effect, well known in chemistry, has also become important in metallurgy, because high performance metal alloys are often obtained by the dispersion in them of a second phase.

An alternative development may be in the breakup of the fibres. In the case where the interfacial energy is isotropic, the equilibrium form of the second phase is a sphere; if perturbations in shape occur along the length of a fibre, it can break down into globular particles. This is known as spheroidisation.

For D.S. composites which are aimed at long time high temperature service, questions of coarsening and spheroidisation are important, for any decrease in the number of fibres, or tendency to fibre breakup, will lead to loss of properties. It is thus particularly important to know the kinetics of coarsening or spheroidisation of the reinforcing phase so as to determine whether a composite can be used or not.

4.2.1. Driving force for coarsening

The driving force for coarsening of dispersed particles in a two-phase system derives from the variation in the chemical potential with size and shape of the precipitates. The formalism for solid phases (precipitates, or reinforcing phase in the case of composites) has been derived from that established previously for liquids. We consider two cases: first, where the interfacial energy is isotropic (case of spherical particles or circular cylindrical fibres), and second, the case where the reinforcement is facetted.

4.2.1.1. Isotropic interfacial energy. The equilibrium form in this case is a sphere, and the Gibbs–Thompson relation gives the variation of the chemical potential,

$$\mu_B(r) = \mu_B(\infty) + 2\gamma\Omega_B/r,$$

where Ω_B is the atomic volume of solute B and γ the interfacial energy. This equation shows that the chemical potential of the solute B is larger in the neighbourhood of small particles than of larger ones. Atoms of B will therefore diffuse from small particles to large, and the latter will grow at the expense of the former.

4.2.1.2. Anisotropic interfacial energy. If δN atoms of solute B pass from an infinitely large precipitate to one of size λ, without change of shape, the interfacial energy of the two-phase system may be written as

$$\delta N[\mu_B(\lambda) - \mu_B(\infty)] = \bar{\gamma} \frac{\partial S}{\partial V} \Omega_B \delta N$$

where $\partial S/\partial V$ is the change in interfacial area resulting from a unit change of volume and $\bar{\gamma}$ the average value of the interfacial energy. If the precipitate conserves its equilibrium form, the above equation can be written as

$$\mu_B(\lambda) - \mu_B(\infty) = \bar{\gamma} \frac{\partial S}{\partial V} \Omega_B .$$

The kinetics of coarsening or spheroidisation are clearly governed by the slowest process: this can be either the bulk diffusion of an atom of precipitate from one particle to another, the surface diffusion of an atom along the surface of the particle, or alternatively the surface reaction which allows the release or acquisition of an atom on the particle surface. Such are, in general, the rate controlling diffusion mechanisms which govern the kinetics, and are those which have been the most widely studied. Before addressing the question of the kinetics of coarsening, it is worth spending some time in the examination of the shape changes which can occur in certain composites: essentially a question of the spheroidisation of cylindrical fibres or lamellae.

4.2.2. Change of morphology: Spheroidisation
Different authors have recently shown that slight changes in the cylindrical shape of the fibres can lead to their breakdown into little spheres, the sphere being the thermodynamically stable shape if the interfacial energy is isotropic. This spheroidisation has been observed in certain composites such as Fe–FeB or Fe–FeS, and it is, in general, relatively rapid (MARICH and JAFFREY, 1971).
    In lamellar composites of the $A\ell$–$CuA\ell_2$ type, Ho and WEATHERLEY (1975) have similarly observed a spheroidisation of the lamellae linked to the presence of grain boundaries in grains perpendicular to the lamellae, which aid the spheroidisation by the mechanism of intergranular and interfacial diffusion.

4.2.3. Kinetics of coarsening
Two mechanisms can influence the coarsening of D.S. composites, these are on the one hand the migration of growth defects, and on the other, Ostwald ripening, governed by diffusion.

4.2.3.1. Migration of defects. In the case of lamellar eutectics, an intercalated lamella will dissolve at its end, where the surface is convex, and the atoms will deposit on the concave surfaces of the neighbouring lamellae (see Fig. 52). The lamella C will thus disappear as A and B grow. In the case of fibrous eutectics (NiAℓCr and NiAℓMo, for example) coarsening by the migration of defects has also been observed (WALTER and KLINE, 1973).

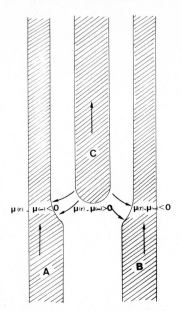

$\mu_{(r)}$ $\mu_{(\infty)} < 0$ $\mu_{(r)} - \mu_{(\infty)} > 0$ $\mu_{(r)} - \mu_{(\infty)} < 0$

FIG. 52. The coarsening mechanism for lamellae.

4.2.3.2. Coarsening by Ostwald ripening. For coarsening which is governed by diffusion, ARDELL (1972) has adapted to the two-dimensional case of composites the model of WAGNER (1961) and LIFSHITZ and SLYOZOV (1961) which was developed for the coalescence of precipitates. We summarise here the major arguments:

ARDELL supposed that the Gibbs–Thomson equation was still valid for cylinders, and that, in the neighbourhood of a curved surface, the variation in concentration was given by

$$C_{(r)} - K^* = \frac{\alpha \gamma K^* \Omega}{rRT}$$

where K^* is the equilibrium concentration far from the interface, α a geometric factor, R the gas constant and T the absolute temperature.

ARDELL showed, in solving the diffusion equations in cylindrical geometry, that the radius of the fibres varies with time according to the law,

$$r^3 - r_0^3 = At,$$

where A is a constant which depends only on the temperature,

$$A = \frac{6\gamma S\Omega^2 D}{RT} f(\Phi)$$

where D is the diffusion coefficients of the element whose diffusion controls the kinetics, and $f(\Phi)$ is a function of the fibre volume fraction.

The application of this model to the γ–MC family of composites allows their limiting useful temperature to be determined.

4.2.4. *Application to γ–MC composites*

In the metal–carbide composites (COTAC 3 or COTAC 74) the carbide fibres coarsen rapidly near the melting point. The kinetics of the coarsening, and an estimation of the damage which could result to the reinforcing phase have been studied for the composite COTAC 74.

The coarsening of the fibres results both in a reduction in the number of fibres per unit cross-sectional area (ρ) and a broadening of their size range, as is shown in Fig. 53 (STOHR, 1978).

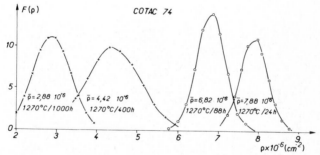

FIG. 53. Influence of coarsening on the size distribution $F(\rho)$ and number of fibres per unit cross-sectional area ρ.

In the case of the COTAC composites, because there is no three-dimensional coarsening of the fibres, the product $\rho \cdot \bar{r}^2$, where $\bar{r}$ is the average radius of the fibre, is constant, and ARDELL's equation can be re-written as

$$\Delta(\rho^{-3/2}) = \rho^{-3/2} - \rho_0^{3/2} = [A'(T)f(\Phi)]t.$$

In Fig. 54 the variation in the number of fibres per unit cross-sectional area to the power $-\frac{3}{2}$ is shown as a function of the holding time at different temperatures.

In the temperature range 1180–1270°C the experimental points fall onto straight lines passing through the origin, in good agreement with ARDELL's model. At 1295°C, however, there is an evident breakdown. At this temperature, grain boundaries play a preferential role in the coarsening (see Fig.

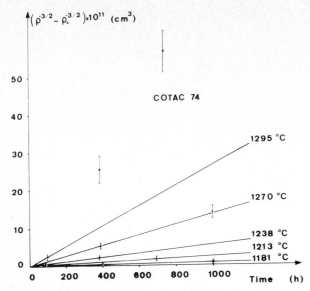

FIG. 54. Variation of the number of fibres per unit cross-sectional area with holding time.

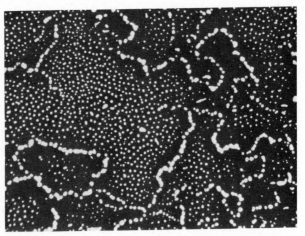

FIG. 55. Influence of the grain boundaries on fibre coarsening at 1300°C.

55) with the consequence that the spacial distribution of the fibres can no longer be considered uniform at temperatures of 1295°C for holding times greater than 100 hours.

The activation energy for the coarsening of the fibres has been determined by replotting the slopes of the lines $\Delta(\rho^{-3/2})$ as a function of temperature against the inverse of the absolute temperature. The value of the activation energy was found to be 560 kJ/mol. This high value may be explained as follows.

If we re-examine ARDELL's relation for the kinetics of coarsening of the fibres, it may be re-written as

$$\bar{r}^3 - \bar{r}_0^3 = \left(\frac{6\gamma S\Omega^2 D}{RT} f(\varPhi) \right) t$$

where S is the solubility of NbC at the test temperature, D the diffusion coefficient of the rate-determining species and T the absolute temperature. If we suppose that the solubility of NbC varies with temperature according to the thermodynamics of dilute solutions, we have

$$S = S_0 \exp\{-Q_s/RT\}$$

and the activation energy for the coarsening of the fibres may be written as the sum of two terms,

$$Q_c = Q_s(\text{NbC}) + Q_d(\text{Nb}),$$

where Q_s is the energy of solution of NbC and Q_d the activation energy for the rate-determining species (here niobium). The heat of solution of NbC can be written as

$$Q_s(\text{NbC}) = -\Delta H(\text{NbC}) + Q_s(\text{Nb}) + Q_s(\text{C})$$

where $\Delta H(\text{NbC})$ is the enthalpy of formation of NbC. In the temperature range 1100–1300°C the value of this term is 221 kJ/mol. The heats of solution of niobium and carbon in the matrix have been presumed equal to those of niobium and carbon in pure nickel: these heats have been obtained from the solubility curves of niobium and carbon in nickel, viz. $Q_s(\text{Nb}) = 48$ kJ/mol and $Q_s(\text{C}) = 43$ kJ/mol. The activation energy for the diffusion of niobium in the matrix of COTAC is 207 kJ/mol. The coarsening energy of NbC fibres resulting from the solution of the carbide and the diffusion of niobium has a value (560 ± 20 kJ/mol), very close to the measured values (MARONI et al., 1978).

In COTAC 74, and in general in the whole COTAC family, the kinetics of coarsening of the fibres is governed by the dissolution of the fibres and the diffusion of the metallic atom which they contain.

It now becomes possible to estimate the damage to the reinforcing phase which could result from this coarsening. Thus, at 1100°C, a reduction of 10% in the number of fibres per unit cross-sectional area would require 18 000 hours holding time, about 50 times greater than the expected exposure time of a turbine blade to its maximum temperature (about 300 hours).

In addition, micrographic observations show clearly that the coarsening of the fibres by Ostwald ripening cannot produce a structural degradation in COTAC composites in the temperature range forseen for their use.

4.2.5. Influence of a transverse thermal gradient

A final source of instability to the reinforcing phase at high temperature could arise from thermal gradients perpendicular to the fibres, which are of the order of 100°C/mm and are caused by the internal cooling of the blade. The important effects of a transverse thermal gradient have been demonstrated by JONES (1974) in the case of the Pb–Ag and $A\ell$–$A\ell_3Ni$ metallic eutectics. MACLEAN (1975) has also shown evidence of a thermal transport of the fibres of carbide in the (CoCr)–Cr_7C_3 system.

According to JONES (1974) the accelerated coarsening of the fibres in the presence of a transversal thermal gradient results in a thermal transport of the fibres through the interfacial diffusion of the elements of the matrix: the fibres should be seen to move sideways with speed

$$V = \frac{D_i \delta Q^* G}{2RT^2 r}$$

where D_i is the coefficient of interfacial diffusion, δ the 'thickness' of the interface, G the gradient, r the fibre radius, R the gas constant, T the absolute temperature and Q^* the diffusion activation energy of the diffusing element. According to JONES, the thermal transport of the fibres could lead to collisions between fibres, and to a decrease in the inter-fibre spacing which would in turn lead to an increased rate of coarsening.

In COTAC 74, the speed of migration of the fibres due to the interfacial diffusion of nickel should be of the order of 2×10^{-3} nm/s at 1100°C under a temperature gradient of 100°C/mm. The speed of displacement of the fibres due to volume diffusion is of the same order of magnitude. In the same conditions as given above, i.e., 1100°C, and $\nabla T = 100$°C/mm, it would require more than 1500 hours before the fibres move more than one inter-fibre distance in the hottest region of the specimen. Since a turbine blade only spends a few hundred hours of its total life in these conditions, the decrease in the interfibre spacing will be much too small to cause an acceleration over the coarsening rate observed under isothermal conditions.

Another model of fibre coarsening under thermal gradients has recently been proposed by MACLEAN (1975); this model appears more satisfactory because it does not require collisions between fibres to account for their accelerated coalescence under the influence of a transverse thermal gradient. In this model, the fibres move parallel or anti-parallel to the thermal gradient with a speed V_0 whilst their average radius $\bar{r}$ increases as follows:

$$\bar{r} = \bar{r}_0(1 + \mu \alpha V_0 t)$$

where μ is a geometric factor near unity, and $\alpha = Q^*G/RT^2$ where Q^* is the diffusion activation energy of the element whose diffusion controls the kinetics of coarsening—here niobium—and G the thermal gradient. In COTAC 74 the ratio of the speeds of coarsening due to the thermal gradient on the one hand and to Ostwald ripening on the other is of the order of 0.01 at 1100°C for a thermal gradient normal to the fibres of 100°C/mm. It follows that the dominating mechanism will be that of Ostwald. This has been fully confirmed by tests under thermal gradient carried out on COTAC 74: the maximum temperature was 1120°C, the gradient 100°C/mm, and the time 250 hours. No effect at all could be detected due to the thermal gradient under these conditions (STOHR et al., 1976).

We can thus conclude that under the conditions normally encountered in aero engines, COTAC composites will be perfectly stable at high temperature, and, furthermore, offer a potential gain of about 100°C over the superalloys.

4.3. Stability at moderate temperatures

4.3.1. Problems associated with the formation of chromium carbide

In composites of COTAC 74 (see Table 12) the niobium carbide fibres are stable at all temperatures; in COTAC 741, however, which contains 5% of aluminium instead of 4% as in COTAC 74, the NbC fibres transform into particles of chromium carbide of the type $M'_{23}C_6$ after holding at 900°C for 1000 hours (see Fig. 56). This transformation, which occurs in the temperature range 700–1000°C, and is not found at temperatures above 1100°C , is analogous to the phenomenon observed in the nickel base superalloys.

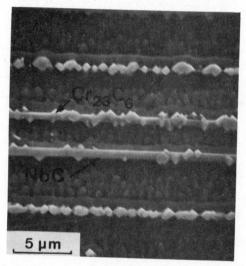

FIG. 56. Appearance of niobium carbide fibres after holding for 1000 hours at 900°C: Scanning electron micrograph.

TABLE 12. Composition (Wt %) of COTAC 74 and COTAC 741.

Composite \ Element	Ni	Co	Cr	W	Aℓ	Nb	C
COTAC 74	50.55	20	10	10	4	4.9	0.55
COTAC 741	59.6	10	10	10	5	4.9	0.5

A systematic study carried out on the γ/γ'–NbC composites derived from COTAC 74 has shown that the stability of the NbC fibres basically depends upon the concentration of aluminium and chromium in the alloy. For example, if the concentration of chromium is increased, or the concentration of aluminium in COTAC 74 (which is initially stable), the fibres of niobium carbide transform into chromium carbide $M'_{23}C_6$. To a lesser extent, tungsten and molybdenum play a similar role in the stability of MC fibres.

The similarity in the influence of chromium and aluminium in the appearance of the MC $\rightarrow$ $M'_{23}C_6$ transformation suggests that the instability of the NbC fibres in certain γ/γ'–NbC composites depends only upon the composition of the γ phase. The γ' phase would then only be indirectly involved: its precipitation would result in changes in the composition of the γ phase between 1300 and 800°C, for example. If this hypothesis is correct, a three-phase zone γ, NbC, $M'_{23}C_6$ must exist at moderate temperatures in the γ, Nb, C, Cr phase diagram.

4.3.1.1. Determination of the three-phase zone γ–NbC–$M_{23}C_6$. In order to prove the existence of this zone, calculations have been made starting in the nickel-rich part of the quaternary Ni, Cr, Nb, C system, with the object of establishing limits to the three-phase zone at moderate temperatures (ANSARA, to appear).

The calculation was made from thermodynamic data obtained from the four ternary diagrams: (Ni, Cr, C), (Ni, Cr, Nb), (Ni, Nb, C) and (C, Cr, Nb). The boundaries between the two-phase regions (γ, NbC) and the three-phase region (γ, NbC, $Cr_{23}C_6$) was calculated by minimising the Gibbs' free energy of systems containing either two or three phases in equilibrium. Fig. 57 shows the projection of the nickel-rich end of the Ni, Cr, C, Nb diagram onto the Ni, Cr, Nb plane at 1000°K. The two-phase (γ, NbC) region is separated from the three-phase region (γ, NbC, $Cr_{27}C_6$) by the surface (bcd).

In the absence of sufficiently precise thermodynamic data at temperatures other than 1000°K, the boundary between the two- and three-phase regions was determined experimentally as a function of temperature and chromium content using the compositions in the simple Ni, Cr, NbC system.

The following method was used to determine if an alloy of given composition lay in the two- or three-phase area: A fraction of the niobium carbide fibres was put into solution by a treatment at 1300°C. The dissolved carbon was then precipitated by annealing between 700 and 1200°C. If the carbide which precipitated at a given temperature was niobium carbide, the alloy was in the two-phase region of the diagram (γ + NbC). On the other hand, if chromium carbide was precipitated, the composition lay in the three-phase region γ + NbC fibres in certain γ/γ'–NbC composites depends only upon the com-microdiffraction, is shown in Table 13 as a function of the annealing temperature for the two quarternary composites Ni–Cr–NbC whose chromium contents are, respectively, 20 and 25%.

It is clear that the composite with 20% chrome lies in the two-phase area of

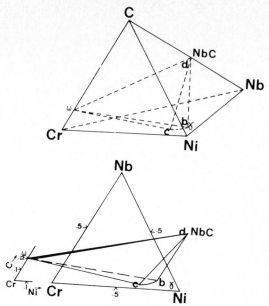

FIG. 57. The quaternary Ni, Cr, Nb, C diagram at 1000°K.

TABLE 13. Nature of the precipitated carbide at various heat-treatment temperature in two quarternary composites.

Composite	Temperature	700	800	900	1000	1100	1200
Ni–20 Cr NbC		NbC	NbC	NbC	NbC	NbC	No carbide precipitation
Ni–25 Cr NbC		$Cr_{23}C_6$	$Cr_{23}C_6$	$Cr_{23}C_6$	NbC $+Cr_{23}C_6$	NbC	No carbide precipitation

the diagram, whilst that which contains 25% is in the three-phase region. In the temperature range 700–900°C, one can consider that the concentration of chrome which corresponds to the limit between the two- and three-phase regions is 22%. This experimental value for the critical chromium concentration is in good agreement with that calculated by ANSARA (to appear). The limit between the regions of stability of chromium carbide and niobium carbide has been further confirmed by the above-mentioned composites by soaking treatments of 1000 hours in the same temperature range.

At this point, we can now conclude that the NbC–$Cr_{23}C_6$ transformation which is observed in the simple Ni, Cr–NbC system is indeed due to the existence of a three-phase region, γ, NbC, $Cr_{23}C_6$ which is only found at intermediate temperatures.

4.3.1.2. Stability of γ/γ'–*NbC composites.* Regarding the γ/γ'–NbC composites, the instability of the niobium carbide fibres arises from the interaction between the chromium-rich γ phase (enriched because of the precipitation of the γ' phase) and the niobium carbide. In the complex γ/γ'–NbC composites, the niobium carbide fibres will be stable if the chromium content of the γ phase remains below a critical value C_0. This value will, of course, be different from the value of 22% cited above for the quaternary Ni, Cr, Nb, C system. In the three-phase zone of the quaternary Ni, Cr, Nb, C system, the three phases in equilibrium are the (Ni, Cr, Nb) solid solution, niobium carbide NbC and chromium carbide $Cr_{23}C_6$. By contrast, in the γ/γ'–NbC system, the analogous phases are much more complex; the γ solid solution, for example, contains a large number of elements such as Ni, Co, Cr, W, Mo, Aℓ, etc. and the MC and $M_{23}C_6$ carbides may contain molybdenum or tungsten, as is shown in Table 14 for COTAC 741.

TABLE 14. Composition of MC and $M_{23}C_6$ carbide particles in COTAC 741.

MC		$M_{23}C_6$	
Element	Composition (%)	Element	Composition (%)
Nb	80.6	Cr	75.6
W	17.1	W	15.2
Cr	2.3	Ni	5.9
		Co	3.2
		Nb	traces

The composition of the different phases present in γ/γ'–NbC composites, discussed above, shows that the determination of the boundary between the two- and three-phase regions would require the determination of complex phase diagrams containing six or seven elements, a task which is at present impossible. The determination of the boundary has been restricted to systems with five constituents, such as Ni, Cr, W, NbC and Ni, Cr, Mo, NbC, taking as the variable parameter the concentration of chromium in the alloy.

The experimental procedure described above, of partial re-dissolution of niobium carbide, followed by annealing, was again used, and Table 15 shows the results obtained for the five-component alloys, Ni, Cr, W, NbC containing 10 weight percent of tungsten. In these alloys, the critical chromium concentration, above which no chromium carbide precipitation occurs, is 16%. In these alloys containing 4 weight percent of molybdenum, the critical value is 18% chrome.

Because of uncertainties in the exact value of the critical chromium concentration, it was fixed at 15 weight percent for all alloys, provided that the total concentration of tungsten plus molybdenum lay below four atomic percent, e.g.

TABLE 15. Nature of the precipitated carbides in (Ni, Cr, Mo–NbC) and (Ni, Cr, W–NbC) as a function of chromium content and temperature.

Cr (%) \ Temperature (°C)	800	900	1000	1100	1200
13.5	NbC	NbC	NbC	NbC	No carbide precipitation
17	Nbc +$M_{23}C_6$	NbC +$M_{23}C_6$	NbC	NbC	id.
20	$M_{23}C_6$	$M_{23}C_6$	$M_{23}C_6$	$M_{23}C_6$	id.
23	$M_{23}C_6$	$M_{23}C_6$	$M_{23}C_6$	$M_{23}C_6$	id.

$10^w/_o$ W + $2^w/_o$ Mo. Such a stability criterion may be defined for any γ/γ'–MC alloy, by using the simple and rapid procedure outlined above.

To determine if a given γ/γ'–NbC composite will be stable or not, it is sufficient to calculate the composition of the γ phase, starting from the nominal composition of the alloy. We have used a method similar to that used to predict the σ phase. It thus becomes possible to represent the region of stability of COTAC composites on a chromium–aluminium diagram (see Fig. 58). Those composites whose composition lies below the full line will be completely stable throughout the whole temperature range, as is shown by the micrograph of a specimen which had been held at 900°C for 1000 h. For composites falling above the dotted line, the MC–$M_{23}C_6$ transformation will occur, as witnessed by the micrograph, also from a specimen which was held for 1000 hours at 900°C.

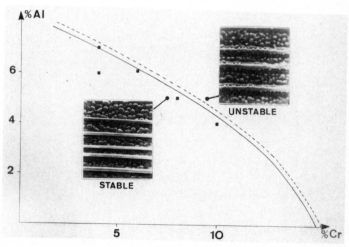

FIG. 58. Cr–Aℓ diagram showing the region of stability for γ/γ'–NbC composites.

4.3.1.3. Compositions of stable alloys. Starting from the basic alloy COTAC 74, three compositions were selected which have both a γ'-loaded matrix and good thermal stability at medium temperatures (see Table 16).

TABLE 16. Composition of new composites.

COTAC 741/8	Ni, 10 Co, 8 Cr, 10 W, 5 Aℓ, 4.5 Nb, 0.5 C
COTAC 742	Ni, 10 Co, 6 Cr, 10 W, 6 Aℓ, 4.2 Nb, 0.47 C
COTAC 744	Ni, 10 Co, 4 Cr, 10 W, 6 Aℓ, 3.8 Nb, 0.46 C, 2 Mo

The thermal stability of these new variants has been thoroughly studied in order both to verify their stability in the intermediate temperature range and also to test the validity of the stability criterion which was used. Considering COTAC 744, the chromium content of the γ phase (10% by weight) is such as to give complete stability at intermediate temperatures: in fact soaking tests of 1000 hours in the range 800–1000°C have not revealed the formation of the $M'_{23}C_6$ phase.

The chromium and aluminium contents of the grades COTAC 741/8 and COTAC 742 are such that they lie very close to the stability boundary of Fig. 59. These alloys

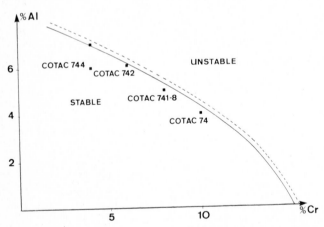

FIG. 59. Cr–Aℓ diagram: Points show the positions of the various γ/γ'–NbC grades.

were therefore exposed to very long soaking treatments, up to 5000 hours, in the temperature range 850–1000°C. Both grades were found to be stable up to 5000 hours. Nonetheless, if the chromium content exceeds 8% in COTAC 741/8 or 6% in COTAC 742, the NbC fibres transform into chromium carbide.

4.3.2. Problems associated with the formation of the σ phase

In superalloys which contain a high concentration of chromium, molybdenum, tungsten manganese, vanadium, etc. we observe the formation of the

σ phase after holding for long periods in the range 700–1000°C. This phase, which appears as platelets or needles in the grain boundaries, tends to embrittle the alloys concerned (Ross, 1963). The dominant parameter governing the appearance of the σ phase is the composition of the γ phase material.

The σ phase has a complex tetragonal lattice, and its chemical composition may vary widely. The similarity of conditions under which this phase occurs in different nickel- or cobalt-base alloys has led several teams of workers to attempt to predict its appearance by supposing it to have a particular electronic composition. Experimentally, it has been shown that the σ phase does not normally appear unless the number of electron vacancies in the γ phase is less than about 2.5. The average number of vacancies in the γ phase can be calculated from the relationship,

$$N_v = \sum_i a_i (N_v)_i ,$$

where $(N_v)_i$ is the number of electron vacancies of element i, and a_i its atomic concentration in the γ phase.

Table 17 shows the values for the number of electron vacancies for different elements, obtained by use of Pauling's model (PAULING, 1938).

TABLE 17. Electron hole vacancy for various elements.

Element	N_v	Element	N_v	Element	N_v
Ti	6.66	Zn	6.66	Hf	6.66
V	5.66	Nb	5.66	Ta	5.66
Cr	4.66	Mo	4.66	W	4.66
Mn	3.66	–		Re	3.66
Fe	2.66	–		–	
Co	1.71	–		–	
Ni	0.66	–		–	

The values given in the table should be used with caution. Thus, if the concentrations of molybdenum or tungsten are high, the σ phase may appear even if the number of electron vacancies is less than 2.5.

In the case of the COTAC 74 family, the number of electron vacancies remains below 1.7 for all grades. The σ phase should therefore not appear, and it is not found experimentally. In COTAC 74, no precipitation of the σ phase was detected after 13 000 hours creep at 900°C. In a similar way, in the derivatives of COTAC 74, COTAC 742 and above all COTAC 744, no precipitation of the σ phase was detected after holding for 5000 hours in the temperature range 800–1000°C.

In conclusion, the grades derived from COTAC 74 will be completely stable throughout the whole temperature range, whether it be a question of the formation of chromium carbide or of the topologically compact structures, such as σ, Γ or Laves' phases. Nonetheless, it should be borne in mind that the stability of the niobium carbide fibres was only obtained in the grades containing γ' by progressively lowering the chromium content to 10% in COTAC 74, then to 6% in COTAC 42 and finally to 4% in COTAC 744, which contains the highest concentration of aluminium.

This reduction in the chromium content could lead to a reduction in the resistance to oxidation, and above all, the corrosion, in these new compositions.

5. Resistance to oxidation and corrosion of directionally solidified composites

D.S. composites developed for application at high temperature should possess an intrinsic resistance to oxidation and corrosion at least sufficient to avoid their rapid destruction following a failure of their protective coating. In a turbine the blades are subject to high temperature oxidation ($> 950°C$) and to corrosion at temperatures below 950°C.

The behaviour of D.S. composites in oxidation or corrosion can be very different from that of the superalloys because of the different natures of the two phases and the lack of overall isotropy. This may be most clearly seen in the metal–carbide composites where the γ/γ'-matrix is little different in composition from that of the normal superalloys.

The object of the oxidation and corrosion testing was to compare the composites with the superalloys. As a result, emphasis was placed upon cyclic testing conditions, which are most representative of the service conditions foreseen.

5.1. Oxidation

The oxidation behaviour of the metal–carbide composites, e.g., COTAC 74 and COTAC 744, is strongly anisotropic. In a cyclic oxidation test at 1050°C with cooling to room temperature once per hour, the weight loss in specimens taken perpendicular to the fibre direction is approximately twice that in specimens taken parallel (see Fig. 60). When the testing temperature is increased, this disparity decreases. The anisotropic oxidation is the consequence of the fibres being more susceptible to oxidation than the matrix: in the perpendicular orientation oxygen attack occurs preferentially down the fibres.

It is worth noting that, in the case of a turbine blade, the fibres are essentially parallel with the surface of the component, and therefore the specimens used for comparison with the superalloys were taken parallel with the fibre direction.

The elements which are in the main responsible to oxidation resistance are chromium and aluminium. In the medium-temperature range (900 to 1000°C), chromium is the most important, and the oxidation resistance decreases in passing from COTAC 74 (10% Cr) to COTAC 744 (4% Cr). The oxidation resistance of all grades of COTAC 74 is nonetheless superior to that of the superalloy IN 100.

At temperatures above 1000°C the oxidation resistance is principally

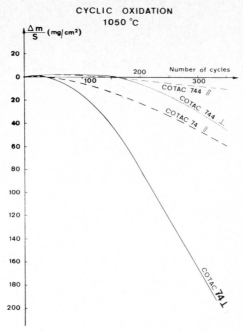

FIG. 60. Weight loss ($\Delta m/S$) cyclic oxidation at 1050°C depending upon whether the specimen surface was parallel or perpendicular to this fibre direction.

achieved by the formation of a protective layer of alumina, and the oxidation resistance increases from COTAC 74 (4% Aℓ) to COTAC 744 (6% Aℓ). The resistance of COTAC 74 at high temperature (~ 1100°C) is equal to that of IN 100, while that of COTAC 744 is distinctly better.

5.2. Corrosion

In a gas turbine, the environment surrounding the blades is formed from the combustion products. Because of the presence of sulphur in the kerosene and sodium chloride in the air, the corroding medium in the gases is essentially sodium sulphate. This material deposits on those parts of the blades whose temperature lies between 800 and 950°C, causing an increased degradation through corrosion.

Experiments have been carried out in the corrosion test cell, using kerosene containing 0.15% sulphur and a salt content in the air of 1 p.p.m.

As was observed in the oxidation tests, the material behaves anisotropically. Corrosion occurs preferentially along the carbide fibres and leads to a greater loss of properties in the direction parallel with the fibres (see Fig. 61).

The corrosion resistance depends principally upon the chromium content of the material. Thus the corrosion resistance decreases from COTAC 74 (10% Cr) to COTAC 744 (4% Cr). Nonetheless, the corrosion resistance of COTAC 744 is very much better than that of IN 100.

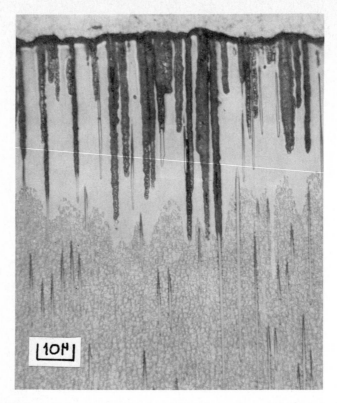

FIG. 61. Optical micrograph showing preferential corrosion attack down the fibre.

We conclude that, in the composites containing a high concentration of $\gamma'(744)$, the high aluminium content confers a good resistance to oxidation at high temperatures. The resistance to corrosion of the composites is distinctly superior to that of a superalloy of the IN 100 type, in spite of this low chromium content.

References

ANSARA, I., to appear.

ARDELL, A.J. (1972), *Acta Met.* **1**, 61.

BIBRING, H., J.P. TROTTIER, M. RABINOVITCH and G. SEIBEL (1971), *Mem. Sci. Rev. Met.* **68**(1), 23.

BIBRING, H., M. RABINOVITCH and T. KHAN (1972), *C.R. Acad. Sci.* **275** C, 1475.

BIBRING, H., G. SEIBEL and M. RABINOVITCH (1972), *Mem. Sci. Rev. Met.* **69**(5), 341.

BIBRING, H., T. KHAN, M. RABINOVITCH and J.F. STOHR (1976), *3rd Internat. Symp. on Superalloys: Metallurgy and Manufacture* (Claitor's Publishing Division, Louisiana).

BIBRING, H. (1977), in: *A.G.A.R.D. Proc. Conf. on High Temperature Problems in Gas Turbine Engines* Vol. 229.

BRUCH, C.A., R.W. HARRISSON, M.F.X. GIGLIOTTI, M.F. HENRY, R.C. HAUBERT and C.H. GRAY (1978), *Conf. on In Situ Composites* Vol. 3 (Ginn Custom Publishing, Lexington, MA).

BULLOCK, E., P.N. QUESTED and M. MCLEAN (1978), *Conf. on In Situ Composite* Vol. 3 (Ginn Custom Publishing, Lexington, MA).

BURTON, J.A., R.C. PRIMM and W.P. SLICHTER (1953), *J. Chem. Phys.* **21**, 1987.

CETEL, A.D., M. GELL and J.W. GLATZ (1978), *Conf. In Situ Composites* Vol. 3 (Ginn Custom Publishing, Lexington, MA).

CHADWICK, G.A. (1963), *J.I.M.* **91**.

CHADWICK, G.A. (1965), Eutectic alloy solidification, in: B. CHALMERS, ed., *Progress in Materials Science* Vol. 12 (Pergamon Press, Oxford).

CHALMERS, B. (1964), *Principles of Solidification* (Wiley, New York).

COATES, D.E., S.V. SUBRAMANIAN and G.R. PURDY (1968), *Met. Trans.* **242**, 800.

DAIGNE B. and F. GIRARD (1973), *Mem. Sci. Rev. Met.* **4**, 335.

DUNN, E.M., R.A. WASSON, K.P. YOUNG and M.C. FLEMINGS (1976), *Conf. on In Situ Composites* Vol. 2 (Xerox Publishing, Lexington, MA).

EL GAMMAL, M. (1972), Tech. Rept. ONERA.

FLEMINGS, M.C. (1974), *Solidification Processing* (McGraw-Hill, New York).

GRESKOVITCH, C., M.F.X. GIGLIOTTI and P. SVEC (1977), G.E. Tech. Info Series Rept. 77 CRD 038.

HAUSER, J.M., M. RABINOVITCH and T. KHAN (1977), *20th Coll. de Metallurgie — I.N.S.T.N. Saclay* (Commissariat à l'Energie Atomique, Saclay).

HUNT, J.D. and K.A. JACKSON (1966), *Trans. AIME* **236**, 843.

HUNT, J.D. and J.P. CHILTON (1963–64), *J. Inst. Metals* **92**, 21.

HUNT, J.D. and D.T.J. HURLE (1968), *Trans. AIME* **242**, 1043.

HUNT, J.D. and K.A. JACKSON (1967), *Trans. AIME* **239**, 844.

HO, E. and G.C. WEATHERLY (1975), *Acta Met.* **23**, 1451.

JACKSON, K.A. and J.D. HUNT (1966), *Trans. AIME* **236**, 1129.

JACKSON, K.A. (1972), *Solidification* (A.S.M., Metals Park, OH).

JONES, D.R.H. (1974), *Met. Sci.* **8**, 37.

KELLY, A. and G.J. DAVIES (1965), *Met. Rev.* **10**(37).

KHAN, T. (1978), *Conf. on In Situ Composites* Vol. 3 (Ginn Custom Publishing, Lexington, MA).

KHAN, T. (1978), *Rech. Aérosp.* **1**, 49.

KRAMER, J.J. and W.A. TILLER (1965), *J. Chem. Phys.* **42**, 257.

LARSON, F.R. and J.M. MILLER (1952), *Trans. AIME* **74**, 765.

LEMKEY, F.D. (1973), *NASA Contr. Rept.* 2278.

LEMKEY, F.D. (1976), *3rd Internat. Symp. on Superalloys: Metallurgy and Manufacture* (Claitor's Publishing Division, Louisiana).

LEMKEY, F.D., R.W. HERTZBERG and J.A. FORD (1965), *Trans. AIME* **233**, 334.

LIFSHITZ, T.P. and V.V. SLYOZOV (1961), *J. Phys. Chem. Solids*, **19**.

MACLEAN, M. (1975), *Scripta Met.* **9**, 439.

MARICH, S. and D. JAFFREY (1971), *Met. Trans.* **2**, 2680.

MARONI, E., J.F. STOHR and M. AUCOUTURIER (1978), *Conf. on In Situ Composites* Vol. 3 (Ginn Custom Publishing, Lexington, MA).

MILES, D.E., E. BULLOCK and M. MACLEAN (1976), *Conf. on In Situ Composites* (Xerox Publishing, Lexington, MA).

MOLLARD, F.R. and M.C. FLEMING (1967), *Trans. AIME* **239**, 1534.

MOLLARD, F.R. and M.C. FLEMING (1967), *Trans. AIME* **239**, 1526.

MOORE, A. and R. ELLIOT (1967), *Proc. Conf. on Solidification of Metals* (Iron and Steel Inst., London).

MORRIS, L.R. and W.C. WINEGARD (1969), The development of cells during the solidification of a dilute Pb–Sb alloy, *J. Crystal Growth* **5**, 361.

MULLINS, W.W. and R.F. SEKERKA (1963), *J. Appl. Phys.* **34**, 323.

PAULING, L. (1938), *Phys. Rev.* **54**, 899.

PEARSON, D.D. and F.D. LEMKEY (1979), in: *Solidification and Casting of Metals* (Metals Soc., London).

PFANN, W.G. (1958), *Zone Melting* (Wiley, New York).

PRINCE, A. (1966), *Alloy Phase Equilibria* (Elsevier Publishing Company, Amsterdam–New York).
RABINOVITCH, M. and J.M. HAUSER (1978), *Conf. on In Situ Composites* Vol. 3 (Ginn Custom Publishing, Lexington, MA).
RINALDI, M.D., R.M. SHARP and M.C. FLEMING (1972), *Met. Trans.* **3**, 3133.
ROSS, E.W. (1963), AIME Ann. Meeting, Dallas.
RUTTER, J.W. and B. CHALMERS (1953), *Acta Met.* **1**, 428.
SAHM, P.R. and F. SCHUBERT (1978), *Conf. on In Situ Composites* Vol. 3 (Ginn Custom Publishing, Lexington, MA).
SALKELD, R.W., A.F. GIAMEY and K.E. TAYLOR (1978), *Conf. on In Situ Composites* Vol. 3 (Ginn Custom Publishing, Lexington, MA).
SHEFFLER, K.D., R.H. BARKALOW, J.J. JACKSON and A. YEN (1976), NASA Conf. Rept. 135 000.
STOHR, J.F. (1978), *Rech. Aerosp.* **1**, 13.
STOHR, J.F. and R. VALLE (1975), *Phil. Mag.* **32**(1), 43.
STOHR, J.F., J.M. HAUSER, T. KHAN, M. RABINOVITCH and H. BIBRING (1976), *Scripta Met.* **10**, 729.
THOMPSON, E.R. and F.D. LEMKEY (1970), *Met. Trans.* **1**, 2799.
TILLER, W.A. (1958), *Liquid Metals and Solidification* (A.S.M., Cleveland).
TILLER, W.A., K.A. JACKSON, J.W. RUTTER and B. CHALMERS (1953), *Acta Met.* **1**, 428.
TROTTIER, J.P., T. KHAN, J.F. STOHR, M. RABINOVITCH and H. BIBRING (1974), *Cobalt* **3**, 54.
WAGNER, C. (1961), *Z. Electrochem.* **65**, 581.
WALTER, J.L. and H.E. CLINE (1973), *Met. Trans* **4**, 33.
WOODFORD, D.A. (1977), *Met. Trans.* **8A**, 639.

Fibre Reinforced Ceramics

D.C. Phillips

Materials Development Division
Atomic Energy Research Establishment Harwell
Oxfordshire
United Kingdom

Contents

HANDBOOK OF COMPOSITES, VOL. 4 – Fabrication of Composites
Edited by A. KELLY and S.T. MILEIKO
© 1983, Elsevier Science Publishers B.V.

List of Symbols

T – Temperature
α – Thermal expansion coefficient
α_f – Thermal expansion coefficient of fibre
α_r – Radial thermal expansion coefficient of fibre
α_a – Axial thermal expansion coefficient of fibre
σ_c – Tensile stress in a composite
$(\sigma_c)_u$ – Tensile strength of a composite
$(\sigma_c)_B$ – Stress at which matrix microcracking occurs
E_f – Young's modulus of fibre
E_m – Young's modulus of matrix
V_f – Fibre volume fraction or percentage
V_m – Matrix volume fraction or percentage
P – Matrix porosity expressed as a fraction of matrix volume
τ – Shear strength
γ – Fracture energy

1. Introduction

The fibre reinforced ceramics which are going to be discussed in this chapter are composite materials which consist of a ceramic, glass–ceramic or inorganic glass matrix reinforced with strong stiff fibres. Considerable work has been carried out over the last twenty years in the development of such materials and recent technological advances make this an opportune time for a review of their science and technology.

The main aim in the development of fibre reinforced ceramics has been the achievement of a composite material which has the advantages of a ceramic combined with greatly increased toughness. Ceramics possess many attractive attributes as engineering materials. Their most important properties are their retention of strength at high temperatures coupled with chemical inertness. Secondary advantages are low density compared with metals, stiffness, hardness and their electrical insulating properties. Their main disadvantage is their low toughness or resistance to crack propagation. A consequence of this low toughness is that although high strengths can be achieved under ideal laboratory conditions, these strengths are easily lost by surface damage. Ceramics tend to be susceptible to cracking by thermal or mechanical shock, and intolerant to damage during service or to defects introduced during fabrication, such as pores or weak grain boundaries. Improvements in toughness will reduce all of these faults and can lead to improved strength and damage tolerance, and equally importantly to reduced variability in strength.

Low toughness is a consequence of the very limited ductility of ceramics. The retention of this low ductility to high temperatures is the principal reason why the mechanical properties of ceramics are retained, or even improved, at the high temperatures at which most metals creep or flow too readily. Consequently attempts to improve the toughness of ceramics by increasing their ductility at low temperatures will almost certainly reduce the upper temperatures at which they can be employed usefully. Fibre reinforcement causes toughening by other mechanisms than plastic flow and will not result, in the same way, in reductions in upper operating temperatures. Indeed, for glass it could lead to an increase in upper operating temperature by reducing the effect of decreasing glass viscosity.

Experimental work has demonstrated adequately that fibre reinforced ceramics can be tough materials whose toughness and strength are retained to high temperatures, provided chemical reaction between the two phases or with the environment does not occur. Chemical compatibility of the fibres and matrix with one another and with the environment is of crucial importance if these materials are to be employed as high temperature, tough systems. Tough, high temperature resistant, fibre reinforced ceramics could have a role in high temperature gas turbines and other applications where their lower density compared with refractory metals would be an advantage. Problems still exist, however, in maintaining their properties in the presence of oxidising atmospheres and this is currently the main obstacle to their wider use, and where further development work is necessary.

There are, however, other potentially important applications where extreme temperature capability is not required. For example, fibre reinforced glasses and glass–ceramics have been shown to have mechanical properties comparable with high performance fibre reinforced polymers, and could be used at intermediate temperatures higher than attainable with polymeric systems. They are more stable to ionising radiations than fibre reinforced plastics and this could make them more attractive for nuclear and space environments. They are less susceptible to hygrothermal effects and therefore could be attractive for aircraft or marine applications. Their hardness could make them more erosion resistant than fibre reinforced plastics and therefore attractive as radome materials, while again hardness combined with a tailorable toughness could make them useful in armour.

2. Historical development

The idea of a fibre reinforced ceramic is not new. Since early times fibres have been added to brittle materials to improve their properties. For example, grass and animal hair were mixed with clays during the manufacture of pottery jars and vessels, principally to improve their strength prior to firing. It has been suggested that even carbon fibre reinforced ceramics may have an ancient history, goat hair added to clay being converted to a form of carbon fibre during firing. Such early carbon fibre reinforced ceramic artefacts were reputed to have much better impact properties than unreinforced ceramics, being less prone to break when dropped. It seems that appropriate choice of reinforcing phase was important even in that early technology as goat hair was considered superior to that of other animals.

Most of the development of fibre reinforced ceramics has occurred very recently and over the last twenty years the level of interest has fluctuated widely as various technical problems were encountered, served as obstacles, and were then solved.

The early work has been discussed by TINKLEPAUGH (1965), KROCHMAL (1967), BOWEN (1968) and BUSALOV and KOP'EV (1970). Randomly orientated, short, refractory metal fibres were incorporated into a range of ceramics. Generally this led to an increase in toughness and sometimes to an increase in strength, but there were significant disadvantages in that the increases in strength were only modest, the densities were increased over monolithic ceramics, and the matrix failed to protect the fibres from rapid degradation in oxidising environments at high temperatures. By the middle to the late nineteen sixties it was generally concluded that the potential of this type of material was very limited because of its lack of environmental compatibility, highish density and only moderate strength increase.

Some work was also carried out on the in-situ growth of whiskers in a ceramic and on the growth of oriented ceramic eutectics. For example, LAMBE et al. (1969) precipitated calcium zirconate whiskers in magnesia but were able to obtain only low volume fractions of whiskers which failed to toughen the magnesia significantly. ROWCLIFFE et al. (1969) obtained oriented lamellae or rod-like structures in aluminium titanate–titania and alumina–aluminium titanate. These structures appeared to deflect cracks but any increases in toughness appeared to be small and the materials were weak. In general whisker reinforcement and in-situ growth of oriented structures did not produce significant toughening, both because the toughening effect of fibres decreases with decreasing fibre diameter and coherent interfaces fail to deflect cracks.

The development of light-weight stiff carbon (including graphite) fibres of potentially low cost led to an upsurge of interest and much work was carried out, principally in the U.K. but also in the U.S.A., on the development of carbon fibre reinforced glasses and ceramics (CRIVELLI-VISCONTI and COOPER, 1969; SAMBELL, 1970; BOWEN et al., 1972). Techniques were developed for the manufacture of preferentially (partially) aligned short fibre composites and these gave encouraging improvements in strength and toughness (SAMBELL et al., 1972a). A major advance was the development in the late sixties and early seventies of techniques for the manufacture of continuous carbon fibre reinforced glasses and glass–ceramics of high quality (SAMBELL et al., 1972b; LEVITT, 1973; SAMBELL et al., 1974). These demonstrated high strengths and toughnesses, comparable with the fibre reinforced plastics of the time. Further, they could be laminated in the same way as polymeric matrix composites. Much research was carried out at that time into understanding the effects of fabrication parameters and into the significance and understanding of their mechanical properties (PHILLIPS et al., 1972; PHILLIPS, 1972; PHILLIPS, 1974). Interest waned, however, due to their poor oxidation resistance. It had been hoped that the ceramic or glass matrix would protect the carbon fibres, but this protection was not achieved and losses of strength occurred on operating at about 400°C in air (SAMBELL et al., 1974). Although this was better than fibre reinforced plastics the improvement was marginal and there was insufficient commercial interest for continued development. Attempts to improve oxidation resistance by coating the surface of the composite with ceramic glazes were unsuccessful.

By the mid seventies the technology existed for the fabrication of moderate-to-high performance carbon fibre reinforced glasses and ceramics for use in inert atmospheres at temperatures up to at least 1000°C, but for use in oxidising environments there was a need for the development either of a technique for protecting the fibres or for new low-cost fibres which would be resistant to oxygen. Work up to this time has been reviewed by DONALD and McMILLAN (1976). Some research had also been carried out on silicon carbide reinforced glass–ceramic (AVESTON, 1971). These silicon carbide fibres were manufactured by deposition onto a tungsten substrate and although the system was oxidation resistant, it suffered the disadvantages of a relatively dense, high-cost fibre which lost its strength at about 800°C.

Research continued into the development of continuous ceramic fibres which would combine low density and oxidation resistance with good mechanical properties and, hopefully, be relatively cheap. For example, there was much work on the deposition of silicon carbide onto carbon fibre substrates leading eventually to the present production of such fibres in pilot plant quantities (reported by PREWO and BRENNAN, 1980); continuous silicon carbide yarn has been synthesised from an organo-metallic polymer (YAJIMA et al., 1976); alumina and aluminium borosilicate fibres have become more readily available.

Recently there has been another upsurge in interest in fibre reinforced ceramics containing these newer fibres, using and extending the fabrication technology developed in the late sixties (BACON et al., 1978; PREWO and BRENNAN, 1980). The properties of these new materials, which have been produced both as uni-directional and laminated composites, are sufficiently encouraging to warrant further study and development. Their properties are interesting but pose some questions, particularly with respect to the extent that the strength of the fibre can be developed in the composite.

Now is an opportune time to review the subject, to consider our understanding of the fabrication parameters and performance of the composites, and to attempt to assess both the short-term and longer-term potential applications of these materials. In this chapter, because of their superiority in performance, we shall be concerned primarily with composites made from continuous fibres, but for comparison the work on short fibre composites will also be described briefly.

3. Fibre and matrix compatibility

Some of the fibres which have been used for the reinforcement of ceramics are listed in Table 1 with their properties and those of some typical ceramics and glasses. Additional data are provided in the review by DONALD and McMILLAN (1976). In order that a particular fibre and matrix will combine to form a successful composite it is necessary that three types of compatibility be satisfied. These are the thermal expansion mismatch between fibres and matrix, the relative elastic moduli of the two components and chemical compatibility of the two components with each other and their external environment under the temperatures of fabrication and use.

Because of the limited ductility of ceramics it is necessary to minimise the

TABLE 1. Comparison of the properties of some fibre and matrix materials.

Fibres	Young's modulus (GPa)	Tensile strength (MPa)	Thermal expansion coeff. α ($10^{-6}\,°C^{-1}$)	Density e ($g \times cm^{-3}$)	Fibre diameter d (μm)
High modulus carbon[a]	360	2400	radial 8 axial 0	1.8	8
Silicon carbide yarn from organo-metallic polymer[b]	220	2060	4.8	2.7	10
Silicon carbide deposited on carbon filament[b]	415	3450	4.8	3.2	140
α-alumina[c]	385	>1400	8.5	3.9	20
Alumina–borosilicate[d]	152	1550		2.5	11
Boron	420	3000	8.2	3.4	100
Tungsten	340–410	2900–3800	4.8	19.3	
Niobium	83–124	500–1030	8.1	8.6	
Stainless steel	150–210	2050–2550		7.7–8.0	
Matrix ceramics					
Borosilicate glass	60	100	3.5	2.3	
Soda–lime glass	60	100	8.9	2.5	
Lithium aluminosilicate glass–ceramic	100	100–150	1.5	2.0	
Magnesium aluminosilicate glass–ceramic	120	110–170	2.5–5.5	2.6–2.8	
Mullite	143	83	5.3		
MgO	210–300	97–130	13.8	3.6	
Si_3N_4	310	410	2.25–2.87	3.2	
$A\ell_2O_3$	360–400	250–300	8.5	3.9–4.0	
SiC	400–440	310	4.8	3.2	

[a] Courtaulds Grafil HT.

[b] PREWO and BRENNAN (1980).

[c] Dupont Fiber FP.

[d] 3M brand ceramic fibres (EVERITT, 1977).

stresses which occur between fibres and matrix. The most successful routes for the fabrication of fibre reinforced ceramics have all required high temperatures and the thermal mismatch stresses on cooling can cause cracking of the matrix or fibres unless the thermal expansion coefficients are appropriately matched. A number of simple one-dimensional theoretical formulae have been developed to predict the matrix stresses due to thermal mismatch (Aveston, 1971; Phillips et al., 1972; Cooper and Sillwood, 1972; Donald and McMillan, 1976). For aligned fibre systems the accuracy of these formulae is reasonable but for random short fibre systems the complexity of stressing due to the random orientation of the fibres and the effects at fibre ends allow only very qualitative comparisons. When assessing these thermal effects it is necessary to consider the axial and radial effects separately. On cooling from the fabrication temperature to ambient a temperature is reached at which stresses can be relieved no further and elastic stresses are set up. If the further decrease in temperature is T, then the thermal mismatch stresses and strains will depend on $T \Delta\alpha$ where $\Delta\alpha$ is the difference in thermal expansion coefficient between fibres and matrix ($\Delta\alpha = \alpha_f - \alpha_m$) over the temperature of interest. The expansion coefficients of fibres are not, in general, equal in radial and axial directions, particularly those of carbon fibres, and the radial ($\Delta\alpha_r$) and axial ($\Delta\alpha_a$) differences may not be the same. If $\Delta\alpha_a$ is positive, the matrix is placed in compression and this could lead to a significant increase in the stress at which the matrix cracks on loading. The fibres however are placed in tension and this might lead to a significant decrease in their apparent strengths in the composite. If $\Delta\alpha_a$ is negative the matrix is placed in tension and if $T \Delta\alpha_a$ is sufficiently high, the matrix cracks, the fibres being set in compression. Radially, if $\Delta\alpha_r$ is positive, the fibre shrinks away from the matrix and this can significantly reduce the bond strength of the fibre–matrix interface and in turn the way the axial stresses are set up. Evidence will be shown that the shear strength, toughness and mode of fracture can be very significantly altered, and in the extreme the fibre–matrix bond can be virtually completely destroyed. If $\Delta\alpha_r$ is negative, the fibre–matrix bond can be increased and tensile stresses can be set up in the matrix tangentially to the fibres leading again to matrix cracking.

Matrix cracking due to axial thermal expansion mismatch is more detrimental to strength in a short fibre composite than in a continuous fibre composite, and in addition it occurs more readily. Microcracks in an aligned continuous fibre system transverse the material with a regular spacing but the fibres hold the blocks of matrix together and the composite can display reasonable strengths. In a random short fibre system the cracks occur in all directions and the composite is very weak. Table 2 illustrates how cracking developed in a range of carbon fibre reinforced glass and ceramic composites as the parameter $T \Delta\alpha$ varied. In the soda–lime glass system considerable cracking occurred in the short fibre composite, similar to that shown in Fig. 1 for the carbon fibre reinforced magnesia. When the fibres were continuous and aligned, matrix cracking was confined to very localised regions best observed by the inter-

TABLE 2.[a] A comparison of damage due to thermal expansion mismatch in some carbon fibre reinforced systems (SAMBELL et al., 1972a).

Matrix	α_m ($°C^{-1}$) $\times 10^6$	T_c (°C)	E_m (GNm^{-2})	$(\sigma_m)_u$ (MNm^{-2})	σ_a (MNm^{-2})	σ_r (MNm^{-2})	ϕ_a	ϕ_r	Damage
MgO	13.6	1200	300	200	4900	2020	25	10	Severe cracking
Aℓ$_2$O$_3$ (80% dense)	8.3	1400	230	300	2660	97	9	0.3	Severe cracking
Soda–lime glass	8.9	480	60	100	260	26	2.6	0.3	Localised cracks
Borosilicate glass	3.5	520	60	100	110	−140	1.1	−1.4	Uncracked
Glass–ceramic	1.5	1000	100	100	150	−650	1.5	−6.5	Uncracked

[a] Type I carbon fibres $\alpha_a \sim 0$, $\alpha_r \sim 8 \times 10^{-6} \, °C^{-1}$. T_c = the temperature below which little stress relaxation can occur; $(\sigma_m)_u$ = the matrix α_m = the matrix thermal expansion coefficient; strength; σ_a, σ_r = the calculated stresses in the matrix axially and radially; ϕ_a, ϕ_r = thermal expansion mismatch parameters $(\alpha_m - \alpha) \Delta T E_m / \sigma_m$.

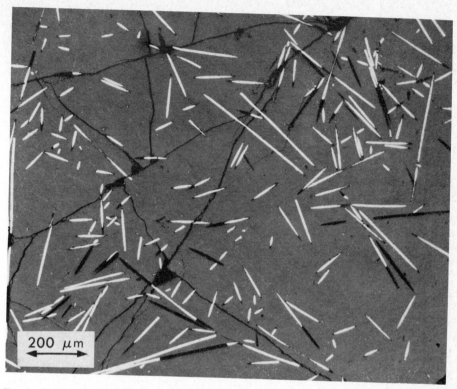

200 μm

FIG. 1. Cracking due to thermal expansion mismatch in magnesia containing short, random carbon fibres (SAMBELL et al., 1972a). The carbon fibres appear white because of their higher reflectivity than the matrix after polishing.

ference bands seen in reflected light microscopy shown in Fig. 2. Increased cracking in short fibre systems is to be expected due to the increased stresses in the matrix at fibre ends and to the lesser effect of the matrix crack suppression mechanism to be described later.

The relative elastic moduli of fibres and matrix are particularly important in determining the extent to which cracking of the matrix can be prevented. Generally fibres are stronger than matrix materials and the aim in a composite is to enable the fibres to bear a greater proportion of the load than the matrix. This sharing of load depends on their relative moduli. Potential ceramic matrices differ very significantly from polymer matrices in two important ways. Their moduli are much closer to those of the fibres, and their failure strains tend to be lower than those of the fibres. On the simple isostrain model of the law of mixtures, the tensile stress (σ_c) in a unidirectional composite strained to ε_c where $\varepsilon_c = \varepsilon_f = \varepsilon_m$, the fibre and matrix strains respectively, is

$$\sigma_c = (\sigma_f)_u V_f + \sigma_m V_m \qquad (3.1)$$

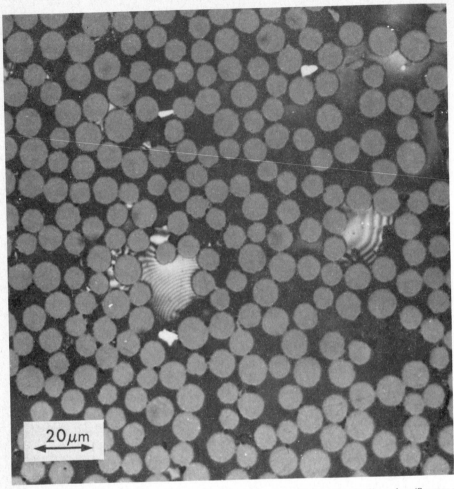

FIG. 2. Localised cracking perpendicular to fibres in carbon fibre reinforced soda–lime glass (SAMBELL et al., 1972b).

and the ratio of stress carried by fibres and matrix respectively is $E_f V_f / E_m V_m$. As the stress in the composite is increased, a strain may be reached at which the matrix will crack. From (3.1) it can be shown that the stress $(\sigma_c)_B$ at which this occurs is

$$(\sigma_c)_B = (\sigma_m)_u [1 + V_f (E_f / E_m - 1)] \tag{3.2}$$

where $(\sigma_m)_u$ is the unreinforced strength of the matrix (BOWEN, 1968; PHILLIPS et al., 1972). This simple model has been shown to be inadequate under certain circumstances and these will be discussed later. However, to a first approximation (3.1) and (3.2) may be regarded as two different failure criteria representing an upper and lower bound to composite strength, (3.1) reducing to $(\sigma_c)_u = (\sigma_f)_u V_f$. The importance of the ratio E_f / E_m is well illustrated by considering the reinforcement of a ceramic and a glass by carbon fibres. Table 3 shows typical strength and moduli of these materials and the calculated strengths for 60 °/$_o$ composites ignoring the thermal mismatch effect. Cracking occurs at much lower stresses in the high modulus ceramic than in the glass, the stress at which the ceramic cracks being hardly increased at all in this example. For this reason, low modulus glasses and ceramics offer some advantages over high modulus matrices.

TABLE 3. Theoretical matrix cracking stresses $(\sigma_c)_B$ for a borosilicate glass and magnesia reinforced with 60 °/$_o$ of high modulus carbon fibres ($E_f = 360$ GPa). Thermal expansion mismatch stresses have been neglected.

	E_m	$(\sigma_m)_u$	$(\sigma_c)_B$
Borosilicate glass	60 GPa	100 MPa	400 MPa
Magnesia	250 GPa	120 MPa	124 MPa

The fact that matrix cracking may occur at lower stresses than ultimate is not necessarily a reason for discarding a system, as strength may be retained to $(\sigma_c)_u$ and toughness is retained. The main disadvantage of matrix cracking is that it provides an easy path for environmental attack of the fibres and fibre–matrix interface.

Chemical compatibility is a more complicated subject and it is necessary to consider the thermodynamics of potential fibre and matrix systems. KROCHMAL (1967) has reviewed thoroughly many potential systems and reference should be made to his report or other compilations (e.g., LEVIN et al., 1964) for more detailed comments. While such compilations are useful for making an initial selection of possible compatible systems, they can serve only as a very rough guide. Fibres themselves may permanently lose their strengths if heated above certain critical temperatures. For example boron and silicon carbide fibres formed by deposition on tungsten lose strength permanently when heated above ~1000°C (KROCHMAL, 1967; AVESTON, 1971).

4. Fabrication techniques

A wide range of fabrication routes exists for the manufacture of monolithic ceramic, glass and glass–ceramic materials (see, for example, WANG, 1976). Because of the refractoriness of crystalline ceramics it is normally necessary to use a powder route employing high temperatures for consolidation. Glasses soften and melt at lower temperatures and techniques employing the molten phase are used to produce monolithic glass components. Glass–ceramics are materials which can be processed initially in the glassy state at relatively low temperatures and then devitrified by heat-treatment to produce a fine-grained polycrystalline ceramic. Table 4 summarises the main types of process which

TABLE 4. Basic fabrication processes for ceramics (after SAMBELL, 1970).

Method	Typical materials fabricated	Starting materials	Forming process	Advantages	Disadvantages
Plastic forming	Complex clay-based ceramics	Fine powder + plasticiser + binder	Naturally plasticised clays are moulded or extruded to shape and fired	Traditional, cheap, simple process. Suitable for complicated shapes	High firing temperatures required. Close dimensional tolerances are not possible because of large volume shrinkages. Considerable porosity usually
Slip-casting	Complex clay-based ceramics and similar systems	Fine powder + liquid carrier	Thin slip poured into a porous mould and fired at high temperatures	As above	As above
Cold-pressing and sintering	Engineering ceramics such as oxides, nitrides, etc.	Fine powder + binder	Compacted at high pressure in a metal die at room temperature. Removed and fired at high temperature	Unfired green compact is often tough enough to be machined to shape. Lower porosities than plastic forming and slip casting	High firing temperatures Moderate shrinkage Suitable only for relatively simple shapes
Vitreous phase sintering	As above	As above	As above but with the inclusion of a glassy phase which has low viscosity at temperature	As above but reduced porosity	As above but the glassy phase can reduce the high temperature capability of the artefact
Hot-pressing	Suitable, with modification, for virtually all ceramics but because of cost mainly used for high performance materials	Fine powder	Simultaneous application of moderate temperature and pressure in refractory dies, e.g., graphite, silicon carbide, alumina, refractory metal alloys	One stage process, binders not required, close dimensional tolerances and near-theoretical densities possible	Only simple and relatively small shapes possible, usually a slow one-off process, low length-to-diameter ratio, high-cost product

Process	Materials	Description	Advantages	Limitations
Isostatic hot-pressing	As above. Used for complex shapes	Artefact cast or plastically formed to appropriate complex shape. Pressure applied at temperature through a medium capable of transmitting hydrostatic forces such as a liquid metal or inert powder	Can produce complex shapes	Dimensional tolerances not as good as for hot-pressing, high-cost product
Activated hot-pressing, reaction hot-pressing, pressure calcintering		Modifications of hot-pressing in which additions are made to the powder to promote chemical reactions, lubricate grain boundaries, or in other ways reduce the necessary temperatures and pressures. Used for specialised systems (SAMBELL, 1970)		
Reaction bonding	Si_3N_4, SiC	For Si_3N_4, silicon powder. For SiC, silicon carbide powder and graphite	Silicon or silicon carbide and graphite powder with a binder are cold-pressed to shape, heated in an atmosphere of N_2 (for Si_3N_4) or Si vapour (SiC) to produce a reaction-bonded body of Si_3N_4 or SiC	High dimensional tolerances and low porosities. Complex shapes possible because forming is carried out at low temperatures	Limited to a small number of materials. Porosities not as low as hot-pressed material
Glassification and devitrification	Glasses and glass–ceramics	Silica based powders	Mixed silicates heated to high temperature to produce liquid glass which is then formed to shape at temperature. Glass-ceramics are produced from the glass by carefully controlled heat-treatment leading to devitrification and a fine-grained ceramic	Glass-making is a traditional, cheap technology. Capable of complex shapes. Glass-ceramics have the advantages of glasses but with improved strengths and higher temperature capability

TABLE 5. Fabrication processes which have been used for the manufacture of fibre reinforced ceramics.

Process	Composites (fibre–matrix)	Reference	Comments
Hot-pressing	W-glass, Ni-glass	EINMAHL (1966)	Fibres and matrix powder mixed together (see Section 5) and hot-pressed. Produces low porosity composites, with uncracked matrices provided thermal expansion coefficients are matched. Continuous, aligned composites can have very high strengths
	Mo-thoria	BASKIN et al. (1959, 1960)	
	Mo–alumina, W–ceramic	TINKLEPAUGH (1965)	
	Stainless steel–alumina	BORTZ and BLUM (1968)	
	C-glass	SAMBELL et al. (1972a, b, 1974)	
		PREWO and BACON (1978)	
	C-glass–ceramic	SAMBELL et al. (1972a, b, 1974)	
		LEVITT (1973)	
	C-MgO	SAMBELL et al. (1972a, b)	
	C-Aℓ$_2$O$_3$	SAMBELL et al. (1972a, b)	
	ZrO$_2$–MgO	SAMBELL et al. (1972a)	
	ZrO$_2$–ZrO$_2$	GRAVES et al. (1970)	
	SiC-glass	PREWO and BRENNAN (1980)	
	Aℓ$_2$O$_3$-glass	BACON et al. (1978)	
Cold-pressing and sintering	C-glass	SAMBELL et al. (1974)	Fibres and matrix mixed, cold-pressed and sintered. Disappointing results because the large shrinkage of the matrix during firing produces cracked composites
	Metal fibre–ceramic	TINKLEPAUGH (1965)	

Method	Material	Reference	Description
Devitrification	C-glass–ceramic	SAMBELL et al. (1974)	Carbon fibres and glass powder hot-pressed at relatively low temperatures to give a carbon reinforced glass. Further high temperature heat-treatment used to devitrify the glass to a glass–ceramic. Disappointingly low strengths, probably because of volume changes during devitrification
Reaction-bonding	Reinforced Si_3N_4	Reported in SAMBELL (1970)	Fibres incorporated into flame-sprayed silicon which was subsequently reaction-sintered in nitrogen
Slip-casting	Ceramic fibre-fused silica	CORBETT et al. (1965, 1966)	Ceramic fibres incorporated into slips of finely divided fused silica and fired. Increased porosity resulting from the presence of fibres usually resulted in degradation of properties
Plasma-spraying	Mo-Al_2O_3, W-Al_2O_3	Moss et al. (1972)	Al_2O_3 powder plasma-sprayed. Unlikely to be commerically viable
Glass-casting	Al_2O_3-glass	KLIMAN (1962)	Suitable only for short random fibres. In general, possible reaction between fibres and glass at the temperature of glass casting
In-situ growth	Calcium–zirconate–MgO Eutectic structures	LAMBE et al. (1969) ROWCLIFFE et al. (1969)	Whisker reinforcement generally has been disappointing because significant toughening is not obtained with small diameter fibres. In general coherent interfaces also contribute little toughening

are used to manufacture these materials in their unreinforced forms (after
SAMBELL, 1970). Extensions of many of these techniques have been tried with
varying success for fibre reinforced ceramics (the term fibre reinforced ceramic
will be used generally to include reinforced glasses and glass–ceramics unless it
is necessary to differentiate between these classes of materials).

The manufacture of a fibre reinforced ceramic in general involves two stages.
Incorporation of fibres into the unconsolidated ceramic matrix material, followed by consolidation of the matrix. Sometimes these two stages can be
combined in one process. The initial fibre-incorporation stage can be done in
different ways depending on the final form required of the reinforcement. For
random short fibres a simple mixing of fibres and powdered matrix material
may suffice. Techniques exist for aligning, at least partially, short fibres. For
example, by mixing the fibres and powder in a liquid to form a slurry and then
causing the slurry to flow, the shear stresses tend to align the fibres. Continuous
fibres may be impregnated by passing the fibre through a slurry of powdered
matrix material and binder to produce a pre-preg sheet. The objective of the
consolidation stage is to produce a low porosity matrix while damaging the
fibres as little as possible. Porosity in the matrix causes low matrix strength,
resulting in poor mechanical properties of the composite, particularly in stress
states where the matrix plays a significant role, such as in flexure, shear and
compression. For low porosity to be achieved during consolidation, any organic
binder must be removed and a small matrix powder particle size relative to the
fibre diameter is important. Fibre damage can be caused by oxidation or other
chemical reactions and by indentation by the hard matrix particles if pressure is
applied at too low a temperature.

SAMBELL (1970) has produced a useful review of the techniques which have
been, or might be, used to manufacture fibre reinforced ceramics. Table 5 lists
techniques which have been used to manufacture fibre reinforced ceramics,
together with some of the systems manufactured. In this table they are divided
into a number of classes and a brief description of the techniques and their
relative advantages and disadvantages are listed.

The most widely used technique producing the highest performance composites is hot-pressing. This has been used to produce reinforced crystalline
ceramics, glass–ceramics and glasses. Although hot-pressing has the disadvantage of being a relatively expensive technique, the superiority of the
material produced in this way has outweighed the possible advantages of other
routes.

The hot-pressing route has enabled the production of fibre reinforced
materials with reproducible properties which have been studied in some depth,
and the effects of fabrication parameters on the properties of hot-pressed
materials have been more quantitatively studied than those produced by other
routes. Although in the future other fabrication techniques for fibre reinforced
ceramics may be developed which provide significant cost or property advantages over hot-pressed material, it is instructive to consider the effects of
micro-defects introduced during manufacture by hot-pressing. These micro-

defects such as matrix porosity and fibre damage can occur in other routes but little or no quantitative data exists for these other techniques.

Since hot-pressing requires the initial incorporation of fibres into the matrix powder, techniques available for incorporation will first be described and then the effects of fabrication parameters discussed.

5. Incorporation and alignment of fibres

In the manufacture of a composite it is desirable to control the volume fraction, alignment and dispersion of the fibres. For components which are to be mechanically stressed alignment of the fibres is important both to achieve maximum reinforcement in the required direction and also because higher packing densities of fibres can be achieved in aligned systems.

Randomly orientated short fibre composites can be produced readily by mixing fibres and matrix powder together, as, for example, by SAMBELL et al. (1972a). Mixing can be accomplished in many ways, for example, by just shaking the fibres and powder together or by blending with a high speed blender or mixing machine. Uniform dispersion of the fibre into the matrix becomes more difficult as the fibre length or volume fraction is increased and clustering of fibres into bundles can occur. During the mixing process the fibres can be broken into shorter lengths leading to a distribution of fibre length in the composite. On hot-pressing a random mixture there tends to be an alignment of fibres into the plane perpendicular to the hot-pressing direction, but within this plane the fibres are randomly oriented.

Alignment of short fibres has been carried out by a number of different techniques generally involving hydrodynamic alignment of the fibres through shear during flow in some carrying medium. The addition of the matrix powder into the flowing medium can cause the intimate co-deposition of fibres and powder into either an unbound, aligned mat or felt, or the addition of an organic binder can result in a more easily handleable pre-preg sheet. For example, PARRATT (1969) developed an alginate process in which fibres or whiskers were mixed with a few percent of ammonium alginate in water to produce a viscous mass which could then be extruded through an orifice into an acid precipitating bath. This is illustrated in Fig. 3. By this means fairly well aligned pre-preg sheet could be produced continuously. The process has been used to produce metal matrix composites by the addition of metal powder to the mix before extrusion and could be used for ceramics. BAGG et al. (1969) have dispersed fibres in glycerine and by allowing the mix to flow through a narrow slit nozzle in a head which traversed backwards and forwards over a table produced well-aligned short fibre pre-preg (Fig. 4). SAMBELL et al. (1974) have used a doctor blade technique in which the fibres and matrix powder in an organic binding medium were carried on a paper sheet through a comb to align the fibres. Moderate to good alignment may be achieved by techniques of this sort and the volume fraction of fibres in the composite can be controlled

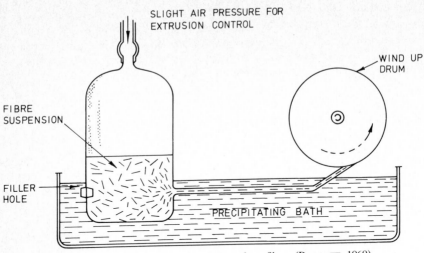

FIG. 3. The alginate process for aligning short fibres (PARRATT, 1969).

through the quantities of fibre and matrix powder in the initial mix.

Electrophoretic processes have also been used to produce aligned short fibre composites, and to impregnate tows of fibres. This process depends on the fact that solid particles suspended in a fluid tend to pick up a surface charge. If a d.c. electric field is generated by immersing a pair of electrodes in the liquid the

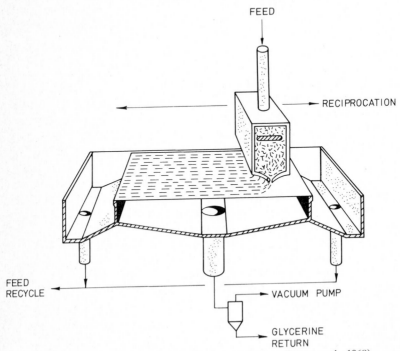

FIG. 4. The glycerine process for aligning short fibres (BAGG et al., 1969).

suspended particles will be transported through the liquid to the anode or the cathode. KIRKPATRICK et al. (1969) used a continuous electrophoretic process to deposit aligned whiskers on to metal films which were then stacked and hot-pressed. SAMBELL has suggested that the process could be adapted for ceramic matrices by the co-deposition of fibres and powder onto a conducting substrate which could then be stripped away before the consolidation stage. Electrophoretic deposition has a further advantage in that if a bundle of (conducting) fibres is used as one electrode, then mutual repulsion should occur between the fibres. This has been used with carbon fibres to open up the tow thus allowing the depositing ceramic particles to penetrate to the innermost fibres (SAMBELL, 1970; SAMBELL et al., 1974). Because deposited particles produce a shielding effect there is a self-regulating mechanism that leads to uniform thickness of coating. This process has been used to produce continuous fibre composites and could be adapted to a continuous process.

The most favoured route for the continuous production of preimpregnated aligned continuous fibre is illustrated in Fig. 5 (SAMBELL et al., 1974). This technique has been used to manufacture successfully composites of carbon

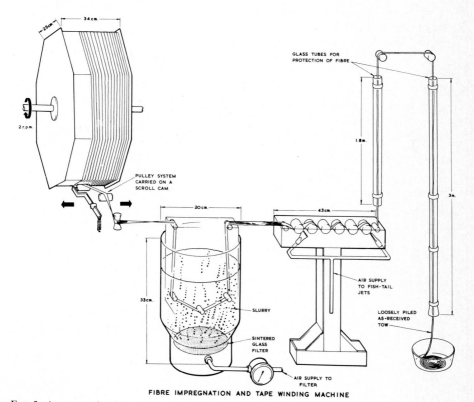

FIBRE IMPREGNATION AND TAPE WINDING MACHINE

FIG. 5. Apparatus for the continuous impregnation of fibre and the production of pre-impregnated tape (SAMBELL et al., 1974).

(SAMBELL et al., 1974; PREWO and BACON, 1978), alumina (BACON et al., 1978) and silicon carbide (PREWO and BRENNAN, 1980) fibres in glass, glass–ceramic and ceramic matrices. Basically, fibre in single filament or tow form, is passed through a slurry of powdered matrix material suspended in an organic solvent containing an organic binder. Powder adheres to the fibre which is then wound on to a drum where the solvent evaporates off. This figure illustrates one way in which the process has been used for carbon fibres. These fibres are supplied as tows consisting typically of 10 000 filaments (of 8 μm diameter) which may contain a twist. The tow is pulled vertically for some distance to enable kinks and twists to fall out, and then under and over a series of PTFE rollers where it is subjected to two jets of air, the function of which is to fan the fibres out over the width of the rollers and spread the tow into a tape. The jets are arranged in a counter-current so that there is no net force tending to contract the tow. The tape is passed through a tank and saturated by total immersion in a slurry of organic binder, matrix powder and solvent, which is agitated by air fed through a sintered glass disc which forms the bottom of the tank. The impregnated tape is reduced to a controlled width, typically 1 cm, and wound onto a drum while still wet. In this case each turn bonds with its neighbour and with the underlying layer. After drying out the solvent, the selfsupporting sheet can be cut to appropriate sizes for stacking and hot-pressing. Close control over the ultimate volume fraction of fibre in the composite can be achieved by controlling the quantities of powder and binder in the slurry, and fibre volume fractions from 20% to in excess of 60% have been achieved. For example, Fig. 6 shows how volume fraction of fibre in the composite varied with ceramic powder content in the slurry for a particular system. Unidirectional composites produced by hot-pressing stacked layers of such pre-preg showed very homogeneous fibre distributions without ceramic-rich layers at the interply regions.

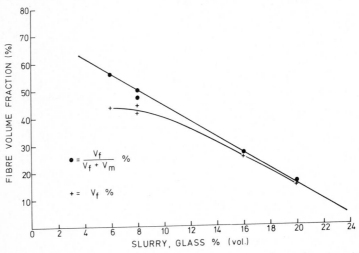

FIG. 6. The dependence of the volume fraction of fibres in the composite on the composition of the slurry for a carbon fibre reinforced glass–ceramic.

6. Hot-pressing

The properties of a composite produced by hot-pressing depend critically on fabrication parameters such as the size-distribution of the powder and the temperatures and pressures which are employed. These control the damage introduced into the fibres and the porosity of the matrix, and can affect the fibre–matrix bond strength. The optimum fabrication parameters have to be evaluated experimentally by trial and error for any given system, but useful guidance is provided by previous work.

6.1. Powder preparation

The size-distribution and shape of the matrix powder particles might be expected to affect the properties of a composite. A high proportion of powder of smaller size than the fibre diameter is necessary to ensure a homogeneous distribution of the fibres in the matrix, and angular particles might be expected to cause greater indentation damage in the fibre surface.

The effect of powder size-distribution has been well illustrated by work on hot-pressed carbon fibre reinforced glass–ceramics (SAMBELL et al., 1974). Composites were produced from powder some of which had been reduced from frit by ball-milling and some by electrohydraulic crushing (EHC). In each case only powder which passed through a 53 μm mesh was employed but the ball-milled material gave consistently better composites than the EHC material. For example a single tow was used to make up two different batches of pre-preg tape with two slurries, one containing ball-milled powder and the other EHC powder. Regular tow strength measurements were carried out and showed that the carbon fibre strength did not vary significantly. Several composites were hot-pressed under identical conditions. One contained all ball-milled powder, another all EHC powder and a third was made of alternate layers of the two. Table 6 shows the results. The EHC powder produced much worse composites than the ball-milled powder while the composite consisting of alternate layers fell between the two. Microscopic examination showed that the effect in this case could not be explained by particle shape, but particle size analyses carried out by β-back

TABLE 6. Effect of powder preparation on the properties of a composite containing high modulus carbon fibres in a glass–ceramic matrix (SAMBELL et al., 1974).

Powder type	V_f (%)	V_m (%)	V_{op} [a] (%)	V_{cp} [a] (%)	Mean σ [b] (MPa)	Mean τ [b] (MPa)
Ball-milled	23.8	71.9	1.4	2.8	432	49.5
EHC	21.5	74.2	1.6	2.6	226	40.0
Alternate layers	23.3	72.6	1.1	3.0	350	39.0

[a] V_{op} and V_{cp} are the volume per cents of open and closed porosities respectively.
[b] σ and τ are flexural and shear strengths.

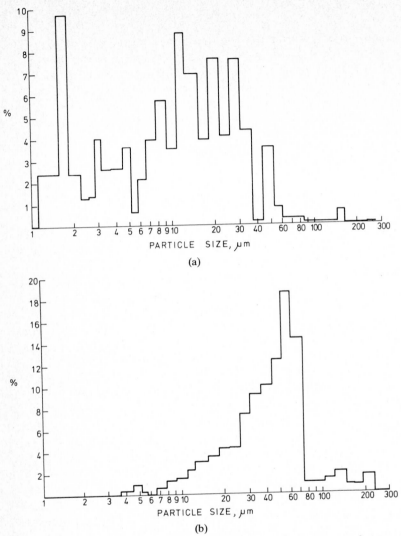

FIG. 7. Histograms of particle size analyses of the powders used in the manufacture of the composites described in Table 7. (a) Ball-milled, glass–ceramic powder; (b) Electrohydraulically crushed, glass–ceramic powder (SAMBELL et al., 1974).

scattering gave the results shown in Fig. 7(a) and (b), showing that the EHC material contained a higher percentage of larger powder particles than the ball-milled material.

6.2. Hot-pressing dies

When hot-pressing a ceramic or fibre reinforced ceramic, problems which need to be considered include: the strength of the die material at the temperatures of pressing, thermal expansion mismatch between die materials and compact, and chemical compatibility between die, compact and environment at the temperatures of pressing. In general it is desirable to employ a die material which has a lower expansion coefficient than the compact so that on cooling the die or compact are not unduly stressed. If this is not possible and thermal mismatch stresses are set up, the die or compact can be fractured or, at least, removal of the compact from the die can be difficult. This latter effect can be circumvented either by removing the compact while still hot, or by employing a dismantlable die.

A range of die materials has been used for producing fibre reinforced ceramics including refractory metal alloys, stainless steel, silicon carbide, alumina and graphite. Refractory metal alloys have the advantage of strength but are expensive and require protection from oxidation. Alumina is resistant to wear and oxidation and is inert to most other ceramics with the notable exception of silica-based materials but has the disadvantages of high cost, difficulty of machining, and a relatively high coefficient of thermal expansion of $\sim 9 \times 10^{-6}\,°C^{-1}$. Silicon carbide has similar properties to alumina but its lower expansion coefficient of $4.5 \times 10^{-6}\,°C^{-1}$ is an advantage.

For carbon fibre reinforced glasses and ceramics graphite has proved to be a very successful die material, used either by itself or bolstered by a metal. Graphite has a number of very significant advantages. It is relatively cheap, available in large stock sizes and is easily machined to shape; chemical compatibility with glasses and ceramics is very good; it has a relatively low coefficient of thermal expansion of $5 \times 10^{-6}°C^{-1}$; its strength increases with increasing temperature. In practice, as a die, it can be protected easily from oxidation either by surrounding it with a thermal insulating blanket which maintains the CO_2 atmosphere which is formed close to the die, or else by surrounding it with an inert gas. Significant oxidation of the carbon fibres in the composites produced with graphite dies does not occur during hot-pressing, presumably because of the CO_2 atmosphere formed by the limited oxidation of the die material. Metal and alumina dies have proved troublesome due to seizure of the parts caused by extrusion of glass between die barrels and plunger, or to chemical reaction. The components of graphite dies could be separated under even these conditions.

Surface preparation of the die is important. Glass and glass–ceramic materials separate easily from mould surfaces provided they are first coated with colloidal graphite. Metals react differently with different ceramics and glasses. For example molybdenum foil, without any surface coating, separated readily from hot-pressed borosilicate glass leaving a good, glaze-like surface finish but lithia aluminosilicate glass–ceramics bonded strongly to molybdenum under the same conditions. The surface of the finished compact can also be modified by lining the faces of the mould with thin sheets of appropriate material. In this way glass composites have been glazed with thin layers (0.1 mm) of glass and glass–ceramic, and other ceramic or metal surface coatings might be deposited by this method.

Two types of die which have been used are illustrated in Figs. 8 and 9. The graphite die in Fig. 8 has been used for hot-pressing small bars ($50 \times 12.5 \times$

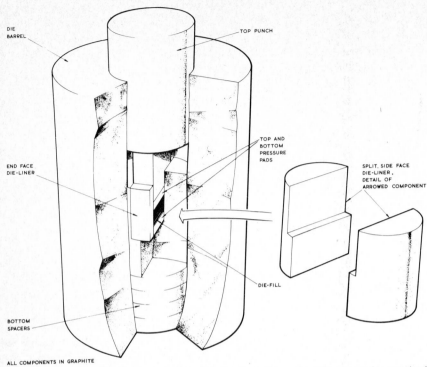

DIE
BARREL

TOP PUNCH

TOP AND
BOTTOM
PRESSURE
PADS

END FACE
DIE-LINER

SPLIT, SIDE FACE
DIE-LINER,
DETAIL OF
ARROWED COMPONENT

DIE-FILL

BOTTOM
SPACERS

ALL COMPONENTS IN GRAPHITE

FIG. 8. A small graphite die used for hot-pressing small specimens for laboratory characterisation
(SAMBELL et al., 1974).

6.2 mm) of carbon fibre reinforced ceramics for laboratory characterisation.
The die punch is assembled from several parts both for ease of machining, and
to enable the die assembly to be dismantled if thermal expansion mismatch
causes contraction of the die onto the compact, thus avoiding the need for hot
ejection. Much larger components have been produced using die assemblies of
the type shown in Fig. 9. The die illustrated was used to produce plates
$216 \times 100 \times 15$ mm but more complicated components including hoops and
aerofoil-type sections as shown in Fig. 10 have been produced in a similar way.
The illustrated die consisted of a set of graphite liners in the shape of an open
box, $230 \times 115 \times 150$ mm deep, bolstered successively with alumina and stainless steel plates, the latter being held in position by stainless steel bolts set in a
water-cooled, mild steel yoke. The die-fill was heated by conduction from
graphite punches which were heated by a pair of induction coils in series. The
spaces between the punches and coils were lagged with thermal insulating
blanket to protect the coils, minimise heat loss and reduce oxidation. With this
arrangement plates could be hot-pressed successfully at 1200°C and 21 MPa.
For higher temperatures lower pressures were necessary to prevent deformation of the stainless steel.

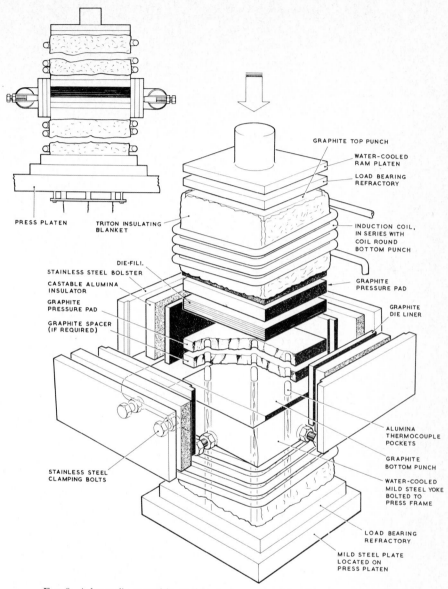

GRAPHITE TOP PUNCH

WATER-COOLED
RAM PLATEN

LOAD BEARING
REFRACTORY

INDUCTION COIL,
IN SERIES WITH
COIL ROUND
BOTTOM PUNCH

GRAPHITE
PRESSURE PAD

GRAPHITE
DIE LINER

ALUMINA
THERMOCOUPLE
POCKETS

GRAPHITE
BOTTOM PUNCH

WATER-COOLED
MILD STEEL YOKE
BOLTED TO
PRESS FRAME

LOAD BEARING
REFRACTORY

MILD STEEL PLATE
LOCATED ON
PRESS PLATEN

PRESS PLATEN

TRITON INSULATING
BLANKET

DIE-FILL
STAINLESS STEEL BOLSTER
CASTABLE ALUMINA
INSULATOR
GRAPHITE
PRESSURE PAD
GRAPHITE SPACER
(IF REQUIRED)

STAINLESS STEEL
CLAMPING BOLTS

FIG. 9. A large die assembly used for hot-pressing plates (SAMBELL et al., 1974).

6.3. *Hot-pressing parameters*

The parameters involved in hot-pressing are temperature, pressure and heating
rate. For any new system it is necessary to determine the optimum hot-pressing
cycle by trial and error. A detailed study of the effects of these parameters has
been made for a small number of systems, notably the carbon fibre reinforced

FIG. 10. Some examples of carbon fibre reinforced glass and glass–ceramic components.

glasses and glass–ceramics, and these indicate the importance of these parameters (PHILLIPS et al., 1972; SAMBELL et al., 1974). These parameters can affect the composite properties through the damage introduced into fibres, the porosity produced in the matrix, and the bond between fibres and matrix.

For example, Figs. 11 and 12 show the effects of temperature and pressure on the flexural and shear strengths of borosilicate glass composites containing 40 v/o of fibre and Table 7 shows the porosities. These composites were produced from pre-preg tape containing organic binder by the process illustrated in Fig. 5, and it was necessary to drive off the binder before consolidating the matrix. Each compact was therefore slowly heated to 550°C under a pressure of 0.35 MPa, the minimum necessary to hold the die system together, until the binder was removed. At 550°C, when the glass had softened, pressure was increased to 1.05 MPa to gently compact fibres and glass and prevent the glass powder settling to the bottom of the die. At 800°C the pressure was increased to either 6.3 or 10.5 MPa and maintained at that level to a maximum temperature of 1000, 1200 or 1400°C, and then on cooling to 500°C when it was removed. When the pressure was not maintained on cooling to 500°C, large amounts of porosity were obtained presumably due to the escape of gases from solution in the matrix. Figs. 11 and 12 show that the 6.3 MPa, 1200 and 1400°C pressing conditions gave significantly higher flexural strengths, and that temperature also had an effect on shear strength.

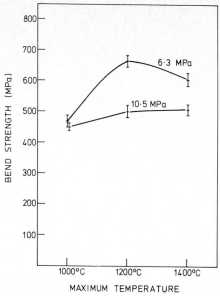

FIG. 11. The effect of hot-pressing temperature and pressure on the flexural strength of carbon fibre reinforced borosilicate glass (SAMBELL et al., 1974).

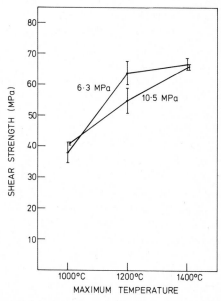

FIG. 12. The effect of hot-pressing temperature and pressure on the interlaminar shear strength of carbon fibre reinforced borosilicate glass (SAMBELL et al., 1974).

TABLE 7. The porosity of carbon fibre reinforced borosilicate glass after different hot-pressing conditions (SAMBELL et al., 1974).

Maximum pressure (MPa)	Maximum temperature (°C)	Volume of fibres (%)	Volume of glass (%)	Open porosity (%)	Closed porosity (%)	Total porosity (%)
6.3	1000	40.4	56.5	3.2 ± 0.2	0 ± 0.3	3.2 ± 0.5
	1200	40.5	57.6	0.9 ± 0.2	0.9 ± 0.3	1.8 ± 0.5
	1400	38.8	59.4	1.1 ± 0.2	0.7 ± 0.3	1.8 ± 0.5
10.5	1000	37.8	55.4	6.1 ± 0.2	0.6 ± 0.3	6.7 ± 0.5
	1200	40.8	57.2	1.0 ± 0.2	0.9 ± 0.3	1.9 ± 0.5
	1400	37.8	61.1	0.8 ± 0.2	0.3 ± 0.3	1.1 ± 0.5

The specimens produced at 1000°C had lower flexural and shear strengths than those produced at higher temperatures because of their higher porosities as shown in Table 7. The table shows that the decrease in strength was due primarily to open porosity rather than closed porosity. Porosity can decrease the strength of a composite in two ways. Firstly its presence reduces the strength of the matrix, and secondly it reduces the area of contact between fibre and matrix and hence the efficiency of reinforcement. This latter effect is aggravated because open porosity tends to occur preferentially at the fibre–matrix interface. Open porosity cannot be completely eliminated in carbon fibre reinforced borosilicate glass composites because the fibres contract from the matrix on cooling. Calculations showed that this shrinkage was 0.01 μm representing a limiting open porosity of 1% in a 40% composite. Later it will be shown that varying the ratios of coefficient of thermal expansion, and hence open porosity, affects shear strength significantly. Above 1000°C this table shows that the porosities were not significantly different at 1200 and 1400°C and that porosity therefore was not responsible for differences in strength. The 6.3 MPa specimens were stronger than the 10.5 MPa specimens because of damage to the fibres. This was verified by dissolving the matrix of the different composites and examining the fibres. These tended to occur in shorter lengths in weak composites than in strong ones. The 1200 and 1400°C materials had higher shear strengths than the 1000°C specimens again because of their low porosities. Shear strength increased with temperature after the porosity had been minimised and it was speculated that this might be due to an increased bonding. The fibre–matrix bond in these materials was thought to be merely a mechanical keying. If there is an increase in mechanical keying with increased temperature or pressure, the effect of temperature would be more marked than that of pressure because of the strong dependence of glass viscosity on temperature. The fact that the shear strengths increased with temperature but did not vary greatly with pressure tended to support this view.

During hot-pressing or other heat-treatments the bond between fibres and matrix can also be affected by the maximum temperatures used, either through chemical reaction between fibres and matrix or due to a change of phase of the fibres or matrix leading to different thermal expansion mismatch effects.

Changes in the chemical bond between fibres and matrix have been shown
for example in zirconia fibre reinforced magnesia (SAMBELL et al., 1972a).
Zirconia and magnesia react significantly at 1600°C. In composites produced at
temperatures less than this the fibre–matrix interface had little strength, as
evidenced by pluck-out of the fibres during mechanical polishing for optical
microscopy. After heat-treatment at 1600°C the fibre–matrix interface was
significantly altered and mechanical grinding damage no longer occurred.
Modification of the interface by chemical reaction in this way has to be
carefully controlled as evidenced by the fact that heat-treatment at 1700°C
caused complete destruction of the fibres and distribution of the zirconia to the
grain boundaries.

Changes in composite properties due to alteration of the nature of the
fibre–matrix bond through phase changes in the matrix have been well illustrated in silicon carbide fibre reinforced cordierite (AVESTON, 1971). Fig. 13
shows how the room temperature strength of SiC reinforced cordierite, hot-
pressed at 900°C, varied with heat-treatment temperature. On heat-treatment
below 1000°C the matrix consisted of μ-cordierite which has a higher thermal
expansion coefficient than SiC fibre, so that in the composite the matrix is in
tension and the fibre under radial and axial compression. These composites
were relatively weak and brittle, with planar fracture surfaces due to the strong
bond between fibre and matrix. After treating above 1000°C the μ-cordierite is
changed into high cordierite which has a lower expansion coefficient than the
fibres. The matrix is thus in compression parallel to the fibres and radially the
fibre contracts from the matrix producing a weaker interface. These composites
were much stronger and tougher displaying fracture surfaces with protruding
fibres.

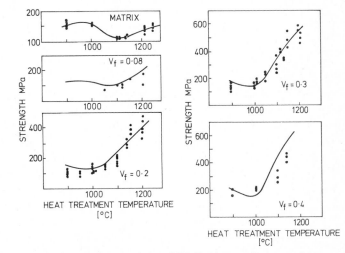

FIG. 13. The strength of composites consisting of SiC fibre in a cordierite matrix, pressed at 900°C and
then modified by heating for one hour at the indicated temperature. Also shown is the strength of
unreinforced matrix material (after AVESTON, 1971).

7. The properties of fibre reinforced ceramics

7.1. Strength

Fibre composites differ from monolithic ceramics in that account must be taken of the anisotropy of elasticity and strength of the composite in designing a structure. In the engineering design of a ceramic component it is normally sufficient to assume that the tensile or compressive properties do not vary with the direction of applied stress but this is certainly not true for fibre composites. For unidirectional fibre composites the strength and stiffness in the direction of the fibres are considerably greater than perpendicular to the fibres or on shear planes parallel to the fibres. In the design of high performance fibre composite structures much use is made of laminate theory (see, for example, JONES, 1975). The composite is recognised as being made up of unidirectional plies which are stacked together in an appropriate sequence to provide a laminate with the required properties. The elastic and strength parameters in the principal symmetry directions within a ply need to be known and from these can be predicted the behaviour of the laminate. Testing techniques for the measurement of the components of the elastic and strength tensors have been developed and used for high performance composites. In the development of fibre reinforced ceramics this degree of sophistication has not been appropriate. Strength measurements have generally been confined to determining the tensile properties through flexural measurements of the sort normally used for isotropic, monolithic ceramics, because of the ease with which these measurements can be carried out. In addition for unidirectional fibre reinforced ceramics the shear strength on planes parallel to fibres has been measured, usually by a short beam interlaminar shear strength test.

7.2. Tensile and flexural strength

The factors which control the strength of a fibre reinforced ceramic are the length and orientation of the fibres, the strengths and elastic moduli of the fibres and matrix, the relative thermal expansion coefficients and hence mismatch strains, matrix porosity and fibre damage. These effects tend to be more significant in fibre reinforced ceramics than fibre reinforced polymers because of the more brittle nature of the ceramic matrix.

Composites containing short random fibres tend to be weak, frequently being weaker than the unreinforced matrix as shown in Fig. 14 for a range of carbon fibre reinforced glasses and ceramics. This weakness is a result of the stress concentrating effect of the misaligned fibres and the thermal expansion mismatch stresses. At the extreme the expansion mismatch stresses may be sufficient to cause extensive cracking in the matrix as shown for carbon fibre reinforced magnesia in Fig. 1, but even where such cracking does not occur strengths tend to be low. Improvements in strengths over that of the matrix have been obtained in a few systems but these have been obtained either by

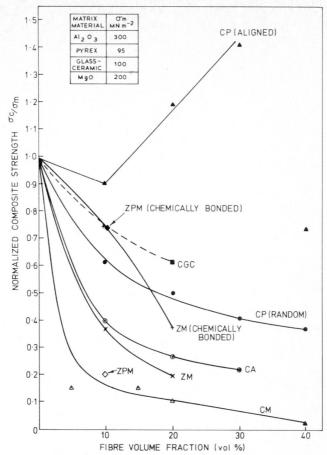

FIG. 14. The flexural strengths of some composites consisting of short carbon fibres in ceramic and glass matrices (SAMBELL et al., 1972a). The matrices are: CP, borosilicate glass (Pyrex); CGC, lithium aluminosilicate glass–ceramic; CA, alumina; CM, magnesia. ZPM and ZM are magnesia reinforced with zirconia powder and zirconia fibres respectively.

avoiding thermal mismatch stresses by using a low temperature fabrication technique or by using a matrix which was originally very weak. Accurate calculation of the thermal mismatch stresses in a random discontinuous composite is not possible. The exact functional relationship of the thermal stresses to $TE_m \Delta\alpha$ depends upon the interaction of stresses from all the fibres and thus depends upon the geometry of the fibre array, fibre dimensions, elasticity of fibres and matrix, fibre–matrix bonding and on local stress concentrations. Such a theory does not exist but some feel for the way in which strength varies with mismatch stress is provided by considering the mismatch parameter

$$\phi = (\alpha_m - \alpha_f)TE_m/\sigma_m$$

(SAMBELL et al., 1972a). For the carbon fibre composites in Fig. 14 the axial

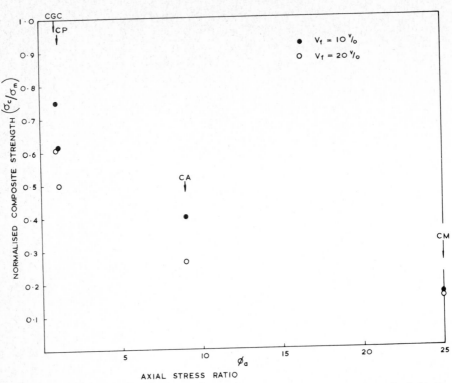

FIG. 15. Normalised composite strength (σ_c/σ_m) plotted as a function of axial stress ratio ϕ_a for the random, short fibre, composites shown in Fig. 14. Composite strengths increase as the thermal expansion mismatch parameter decreases (SAMBELL et al., 1972a).

stress ratios were all positive giving rise to tensile stresses in the matrix. Fig. 15 shows the normalised composite strength plotted against the axial stress parameter ϕ_a. Strength clearly increases as ϕ_a decreases. DONALD and McMILLAN (1976) have considered the range of strengths quoted for discontinuous fibre reinforced ceramics and, by comparing experimental data with the theoretical values derived from (3.1) and (3.2), have shown that the best strengths are obtained when $\Delta\alpha$ approaches zero. They argue that when $\Delta\alpha$ is increasingly positive, the mismatch stresses increase and weaken the matrix while as $\Delta\alpha$ becomes increasingly negative, the bond between fibre and matrix decreases resulting in reduction in reinforcement efficiency.

Alignment of short fibres can result in improvements in strength in the alignment direction as shown for carbon fibre reinforced borosilicate glass in Fig. 14. The aligned fibre composites in this case were not particularly well aligned as shown in Fig. 16 but even so significant increases in strength were obtained.

Use of continuous aligned fibres results in very much greater increases in strength due to the increased efficiency in reinforcement and a minimising of the stress concentration effect due to misaligned fibres. A further advantage is that

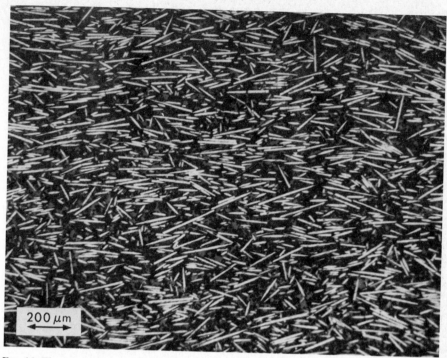

FIG. 16. The microstructure of an 'aligned' short fibre system, the 30 volume percent carbon fibre reinforced borosilicate glass whose strength is shown in Fig. 14. Alignment was obtained by the doctor blade process and significant increases of strength were obtained in the alignment direction (SAMBELL et al., 1972a). The carbon fibres appear white because of their higher reflectivity than the matrix after polishing.

higher volume fractions of fibre may be achieved in the composite, although at high volume fractions problems arise due to the practical difficulty of removing porosity during fabrication. Thus, in Fig. 17 there is an increase in strength on increasing V_f up to 55 $^v/_o$ but a dramatic fall-off at higher V_f (PHILLIPS et al., 1972). This decrease is directly attributable to increased porosity, porosity values for the composite being shown in Table 8. The effect of porosity on strength is shown in Fig. 18. Here the results of measurements on composites containing different volume fractions of fibre have been normalised by calculating the apparent strength of the fibre in the composite through the law of mixtures. The apparent strength of the fibre is plotted against the percentage of porosity in the matrix.

Thermal mismatch cracking is also suppressed on going from short random fibres to continuous aligned. For example, as described in Section 3, short random carbon fibres in soda–lime glass produced extensive cracking while for aligned continuous fibres the cracking was reduced to small localised regions.

The effect of fibre misorientation is shown in Fig. 19 where the strengths of unidirectional continuous fibre composites are plotted against the angle of the applied maximum stress (PHILLIPS et al., 1972). The theoretical curves have been

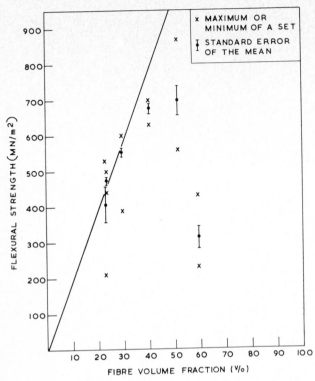

FIG. 17. Flexural strength as a function of volume fraction of fibre in aligned, continuous fibre, carbon fibre reinforced borosilicate glass (PHILLIPS et al., 1972).

plotted from the tensile (0°), transverse tensile (90°) and shear strengths using the maximum stress theory. For these composites the transverse tensile strengths were extremely low. These was no evidence of a chemical bond between the fibres and matrix and the experimental transverse strengths were close to the theoretical values calculated assuming no bond between fibre and matrix. This low transverse strength of fibre reinforced ceramics is clearly a severe disadvantage. Combining plies with different orientations however gives laminates with more isotropic properties, and the strengths and moduli of these can be calculated, at least approximately, from single ply data by laminate theory.

A number of different systems of aligned fibres in ceramic matrices has been manufactured and studied in recent years. As described earlier the strength of aligned fibre composites may be approximated theoretically by (3.1) and (3.2). Both of these are approximations, (3.1) because it does not take into account the statistical nature of fibre failure and the fibre length–strength effect, and (3.2) because there is an enhancement in matrix strength due to the incorporation of fibres for reasons to be explained later. Also they do not take into account thermal mismatch stresses in matrix and fibres.

It is worthwhile, however, to relate the strengths of real systems to these

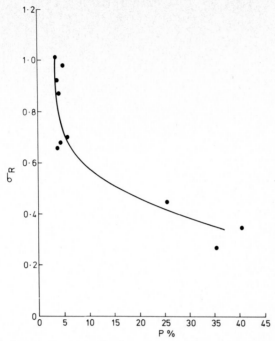

FIG. 18. The variation of experimental strength, σ_R, expressed as a fraction of the theoretical strength calculated from the law of mixtures, with the total porosity, expressed as a percentage of the matrix volume, for a continuous, carbon fibre reinforced borosilicate glass (SAMBELL et al., 1974).

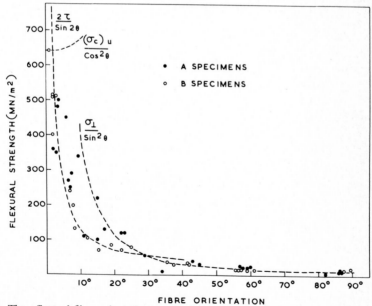

FIG. 19. The effect of fibre orientation on the flexural strength of a unidirectional, continuous, carbon fibre reinforced borosilicate glass (PHILLIPS et al., 1972).

TABLE 8. Microstructure and strength as a function of fibre volume fraction for carbon fibre reinforced borosilicate glass (PHILLIPS et al., 1972).

Volume of fibres (%)	Volume of glass (%)	Open porosity (%)	Closed porosity (%)	Porosity as a percentage of matrix (%)	Mean flexural strength (MPa)	Mean maximum shear stress (MPa)	Mean bendover stress (MPa)
22.9	74.2	2.2	0.7	3.8 ± 0.6	410	47	159
23.3	73.9	1.6	1.2	3.7 ± 0.6	480	44	136
29.3	67.1	2.5	1.1	5.1 ± 0.7	560	51	454
40.2	57.3	1.0	1.5	4.2 ± 0.8	680	63	338
51.4	45.7	2.8	0	5.8 ± 1.0	700	71	191
59.5	26.2	14.4	0	35.4 ± 1.5	310	18	247

TABLE 9. The experimentally achieved strengths (σ_{exp}) of some unidirectional, continuous fibre reinforced ceramics, compared with theoretical strengths derived from Eq. (3.1) (σ_{TU}) and Eq. (3.2) (σ_{TL}).

Reference	Fibre	Matrix	V_f (%)	σ_{exp} (MPa)	σ_{TU} (MPa)	σ_{TL} (MPa)
PHILLIPS, SAMBELL and BOWEN (unpublished)	Carbon	lithia–alumino–silicate glass–ceramic	15	397	281	160
			25	476	468	200
			42	544	787	260
LEVITT (1973)	Carbon	lithia–alumino–silicate glass–ceramic	36	890	760	240
PREWO and BRENNAN (1980)	SiC monofilament	borosilicate glass	35	650	1210	310
			65	830	2240	490
PREWO and BRENNAN (1980)	SiC yarn	borosilicate glass	40	290	824	210
BACON et al. (1978)	$A\ell_2O_3$	S glass	30	280	520	220
BACON et al. (1978)	$A\ell_2O_3$	SiO_2	37	187	640	250

formulae to determine to what extent experimental systems are achieving their potential strengths. Table 9 shows published composite strength data for continuous fibre systems and the strengths calculated from (3.1) and (3.2). Materials described in the table include carbon fibres in glass and glass–ceramic (Phillips et al., 1972; Sambell et al., 1974), alumina fibres in glass and fused silica (Bacon et al., 1978), and two types of silicon carbide fibre in glass (Prewo and Brennan, 1980). The carbon fibre reinforced glass and glass–ceramic composites approach or surpass the strengths predicted by the mixtures law at fibre volume fractions low enough for porosity to be kept low. The alumina and silicon carbide composites however display rather low strengths, considerably less than that predicted by the mixtures law and even lower than the matrix cracking value for alumina reinforced silica. The reason for this low achievement compared with their potential strengths is not presently known with certainty. On cooling from the fabrication temperature the fibres in the alumina and SiC fibre composites are set into tension but the magnitude of the mismatch stress is much less than necessary to explain the shortfall in strength. Other possible reasons include fibre damage, matrix porosity, and for the alumina fibre system a chemical bond between fibres and matrix which is too strong. The alumina and SiC fibre systems are potentially of importance because they possess much better environmental stability at high temperature than the carbon fibre composites. Improvement in their strengths is an important goal.

7.3. Shear strength

Shear strengths of unidirectional ceramic composites are affected by the bond between fibres and matrix, and by porosity in the matrix. The effects of porosity have been demonstrated earlier. The nature of the bond between fibres and matrix in ceramic matrix composites is not well understood. In carbon fibre reinforced glasses and glass–ceramics it is believed that there is little or no chemical bonding as the transverse strengths of these composites are low and there is little evidence of matrix material adhering to fibres after fracture. The bond is thought to be almost entirely due to mechanical keying of the glass into irregularities in the fibre surface. This keying is shown by the replication of the fibre surface in the glass interface in the fracture surface shown in Fig. 20. Table 10 shows shear strength data obtained from four composites made with two different types of carbon fibre in borosilicate glass and lithia–aluminosili-cate glass–ceramic matrices. The borosilicate glass composites had shear strengths almost double those of the glass–ceramic composites. This is due to the differing relative radial shrinkage of the fibres from the glass during cooling after fabrication. Using the data for the stress relaxation temperatures and expansion coefficients shown in this table, the calculated radial contractions of fibre from matrix are 2.4×10^{-8} m for the glass–ceramic and 0.9×10^{-8} m for the glass. The fibres thus contracted further from the glass–ceramic matrix than from the glass matrix. This shrinkage reduces the mechanical keying and thus the fibre–matrix bond.

TABLE 10. Mechanical properties and microstructural details of unidirectional composites made from two different types of carbon fibre in borosilicate glass[a] and lithia–alumino–silicate glass–ceramic[b] (PHILLIPS, 1974).

Material	Microstructural details (%)				Mean strengths (MPa)		Mean work of fracture (kJ/m²)
	Volume of fibre V_f	Volume of matrix V_m	Volume of open porosity V_{op}	Volume of closed porosity V_{cp}	Bend strength	Interlaminar shear strength	
Borosilicate glass and high modulus fibre	45.9	52.4	0.6	1.1	459	59	3.1
Borosilicate glass and high strength fibre	48.9	46.0	0.3	4.9	575	71	3.6
Glass–ceramic and high modulus fibre	49.5	41.6	7.3	1.6	558	32	4.5
Glass–ceramic and high strength fibre	45.7	40.7	4.7	9.0	574	26	10.3

[a] Borosilicate glass: $\Delta T = 500°C$, $\alpha = 3.5 \times 10^{-6} °C^{-1}$.
[b] Glass–ceramic: $\Delta T = 1000°C$, $\alpha = 2.0 \times 10^{-6} °C^{-1}$.

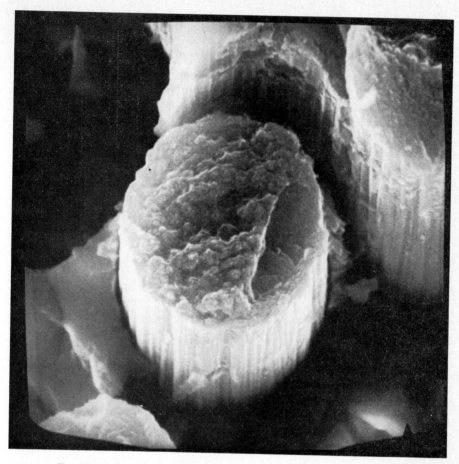

FIG. 20. Replication of a carbon fibre surface in a borosilicate glass matrix.

7.4. Young's modulus

The Young's modulus of fibre reinforced ceramics is given to a good approximation by the mixtures law,

$$E_c = E_f V_f + E_m V_m \tag{7.1}$$

where E_f and E_m are Young's moduli of fibre and matrix respectively. Fig. 21 shows data for a carbon fibre reinforced glass. This shows that Young's modulus increased linearly with fibre volume fraction between 23 and 51 $^v/_o$ but deviated from linearity at 59 $^v/_o$. On the same diagram is plotted the theoretical variation based on (7.1) and assuming a constant matrix modulus of 64 GPa and a measured fibre modulus of 353 GPa. In general, the matrix modulus differs from that of fully dense borosilicate glass because of porosity, and a correction

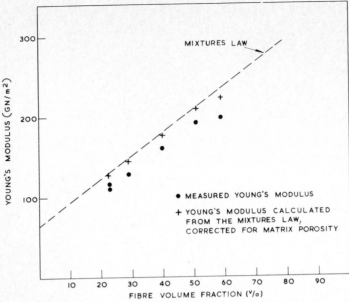

FIG. 21. Young's modulus of a unidirectional carbon fibre reinforced borosilicate glass as a function of fibre volume fraction (PHILLIPS et al., 1972).

has to be made to the theoretical value. This was carried out by assuming a relationship between matrix modulus and porosity due to MACKENZIE (1950),

$$E_m = (E_m)_0(1 - 1.9P + 0.9P^2) \tag{7.2}$$

where $(E_m)_0$ is the modulus in the absence of porosity and E_m is the modulus in the presence of a total porosity P. This correction has little effect between 23 and 51 % but is significant at 59 % and suggests that the deviation from linearity of the 59 % specimen is due to matrix porosity. Between 23 and 51 % the variation of modulus is parallel to the prediction of the mixtures law but smaller than the theoretical value by about 20 GN/m². REYNOLDS (1970) has shown that the Young's modulus of a carbon fibre decreases rapidly with angle to the fibre axis. Between 0° and 15° it decreases by an order of magnitude and it is probable that the modulus of the carbon fibre reinforced glass was decreased below its theoretical value by the lack of perfect alignment among the fibres.

7.5. Matrix microcracking

The microcracking which can occur in the matrix prior to total failure has already been described. This microcracking can occur both because of the thermal mismatch stresses which are induced during fabrication and because of the lower strain to failure of the matrix.

Typical microcracking which occurred during the bending of a carbon fibre reinforced glass is shown in Fig. 22. Eq. (3.2) which has been used to calculate the stress at which matrix microcracking occurs on loading generally underestimates the matrix cracking stress. Experimentally it is observed that matrix microcracking is suppressed. For example Fig. 23 shows the variation of the stress in the matrix at which microcracking occurred for a unidirectionally reinforced carbon fibre reinforced glass (PHILLIPS et al., 1972). In this figure the stress in the matrix at cracking due to the applied load has been calculated from (3.2) and the thermal mismatch stress due to differential contraction of

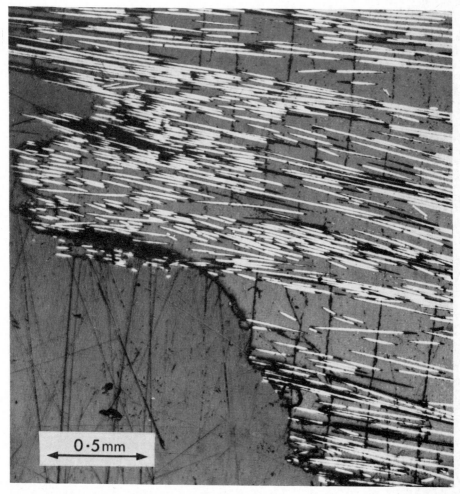

FIG. 22. Microcracking developed in a carbon fibre reinforced glass on loading (PHILLIPS et al., 1972). The carbon fibres appear white because of their higher reflectivity than the matrix after polishing.

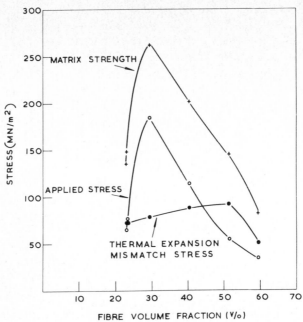

FIG. 23. The matrix stresses at the onset of matrix cracking in a carbon fibre reinforced glass (PHILLIPS et al., 1972).

fibres and matrix has been calculated from

$$\sigma_m = \frac{(\alpha_m - \alpha_f)E_f V_f \Delta T}{1 + V_f(E_f/E_m - 1)} . \tag{7.3}$$

These have been combined to give the matrix strength at which cracking occurred. The strength of hot-pressed borosilicate glass in the absence of fibres is typically about 100 MPa while the derived matrix strength increased from 140 MPa at 23% to a maximum of 260 MPa at 30% and then decreased with increasing fibre concentration to a minimum of 80 MPa. In unreinforced ceramics the growth of a single flaw and its propagation as a crack can result in total failure. The presence of fibres clearly influences the growth and propagation of cracks because cracks are observed to be localised during the early stages of matrix cracking (see Fig. 24). In addition, however, the increased strength of the matrix demonstrates that the presence of fibres inhibits the initiation of cracks.

AVESTON, COOPER and KELLY (1971) have developed a theory (in short: ACK theory) which explains the suppression of cracking in fibre composites. This theory has been applied to a wide range of materials including fibre reinforced ceramics and high performance polymer matrix composites (COOPER and SILLWOOD, 1972; COOPER, 1971). It is believed that the increase in matrix strength in Fig. 23 is due to the ACK mechanism and that the maximum in

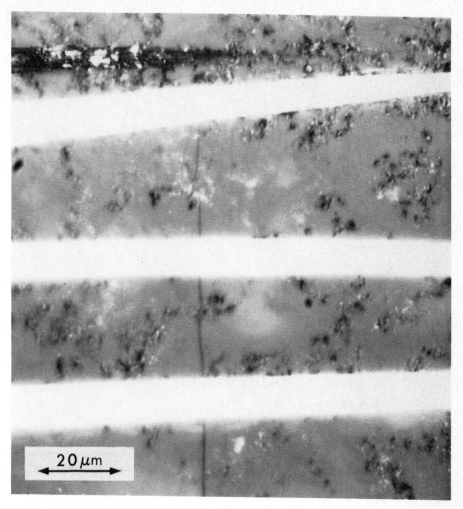

FIG. 24. The onset of matrix carcking in a carbon fibre reinforced glass, showing the localisation of the crack which ends at the top fibre in the photograph (PHILLIPS et al., 1972).

strength at 30% of fibres is due to a competing mechanism weakening the matrix. The most likely reason for this weakening effect is increasing matrix porosity.

The ACK theory essentially considers the energy requirement for the production of matrix cracking. The formation of a matrix crack is governed by a number of energy terms. These include: the work done by the applied stress in extending the composite, ΔW; the release of strain energy by relaxation of the matrix on either side of the crack, ΔU_m; the resulting increase in strain energy of the fibres in this region, ΔU_f; the frictional energy loss due to relative motion of fibres and matrix near the crack U_s; and the fracture surface energy of the matrix, $2\gamma_m V_m$. Summing these terms, the necessary condition for spontaneous cracking is

$$\Delta W + \Delta U_m \geqslant \Delta U_f + U_s + 2\gamma_m V_m \tag{7.4}$$

and it can then be shown that the expected failure strain of the matrix of a composite of high failure strain fibres in a low failure strain matrix is

$$\varepsilon_m = \left(\frac{12\gamma_m \tau E_f V_f^2}{E_c E_m^2 V_m r}\right)^{1/3} \tag{7.5}$$

where τ is the shear strength of the interface. Hence by increasing the fibre volume fraction or decreasing the fibre radius the failure strain may be increased. COOPER (1971) has applied this theory to data obtained by PHILLIPS, SAMBELL and BOWEN (1972) for carbon fibre reinforced glass and obtained reasonable agreement between predicted and measured values of matrix cracking stress.

The theory has been extended to polymer matrix laminates in which adjacent plies may have very different failure strains and in which cracking may occur during fabrication by thermal mismatch or during loading at strains lower than the ultimate. It has been shown experimentally that matrix cracking may be inhibited by using thinner plies and good agreement with the ACK theory has been demonstrated (PARVIZI et al., 1978; BAILEY et al., 1979). Although this has not been demonstrated experimentally for ceramic matrix laminates, the use of thin plies for suppressing matrix cracking is a potentially, important fabrication technique.

7.6. Toughness

The main aim in the development of fibre reinforced ceramics has been to achieve a tough ceramic. The toughness of a material is essentially its resistance to the growth of cracks. In practice it is measured in a number of different ways such as by impact testing, work of fracture, and linear elastic fracture mechanics techniques. The relationships between these for composites has been discussed elsewhere (DAVIDGE and PHILLIPS, 1972; PHILLIPS and TETELMAN, 1972; PHILLIPS and HARRIS, 1977).

In the development of fibre reinforced ceramics toughness has usually been measured by the work of fracture technique. The composite is fractured in three- or four-point bending in a hard testing machine to give a controlled fracture where all the work done is absorbed in processes occurring in the fracture process zone (TATTERSALL and TAPPIN, 1966; NAKAYAMA, 1965). Works of fracture measured in this way are very much higher for fibre reinforced ceramics than for unreinforced ceramics and glasses. Typically the work of fracture of unreinforced glasses and ceramics range from a few Jm^{-2} to a few hundred Jm^{-2}. Fibre reinforcement can increase these to about $10^4 \, Jm^{-2}$ (AVESTON, 1971; PHILLIPS, 1972). For non-fibrous materials the work of fracture corresponds roughly to a half the linear elastic fracture mechanics G_{IC} value and the reasons for high toughness values of fibre reinforced ceramics, and their implications for engineering design, have been the subject of much research.

Detailed studies of the toughness of fibre reinforced ceramics have been made by AVESTON (1971) and PHILLIPS (1972, 1974a, b). The toughness of fibre composites can arise from a number of different mechanisms such as fibre–matrix debonding, post-debond frictional work, fibre pull-out, delamination and multiple fracture. Different mechanisms dominate in different composites. For example, in unidirectional carbon fibre reinforced polymers pull-out appears to be the dominant mechanism while in glass fibre composites significant contributions are made by debonding and post-debond frictional work as well as pull-out (TETELMAN and PHILLIPS, 1972; HARRIS, MORLEY and PHILLIPS, 1976). Similarly in fibre reinforced ceramics presumably different mechanisms will dominate in different composites. In carbon fibre reinforced glass PHILLIPS (1972) has shown that the predominant toughening mechanism in a work of fracture test is pull-out, the work done against friction as the fibres pull out of the matrix behind the advancing crack, as the two surfaces separate. Fig. 25 shows carbon fibres bridging the two halves of a fracturing specimen as the halves separate and Fig. 26 shows carbon fibres protruding from the fracture surface of a higher volume fraction composite. PHILLIPS (1972, 1974a,b) measured toughness both by the work of fracture technique and by linear elastic fracture mechanics (LEFM) techniques. The work of fracture technique involves directly measuring the absorbed energy during the whole of the fracture process, while LEFM techniques measure the load at which unstable fracture commences, and use this in an analysis to determine the initial energy corresponding to the crack propagating over a relatively small distance. In his early measurements (PHILLIPS, 1972) it appeared that the fracture energy (G_{IC}) derived from the LEFM analysis was very much less than that obtained during a work of fracture test, casting a measure of doubt on the usefulness of large works of fracture. However later, more precise measurements (PHILLIPS, 1974a, b) on both unidirectional material and 0°/90° laminates demonstrated much better agreement between LEFM and work of fracture values, as shown in Table 11. The discrepancy in the earlier measurements was attributed to cracks initially propagating in low energy directions parallel to fibres. It now seems clear that high works of fracture can imply large values of LEFM G_{IC} even where pull-out is the dominant mechanism, provided the characteristic pull-out length is relatively small so that the process zone at the crack tip is small compared with the crack length. The corresponding values of critical stress intensity factor K_{IC} which have been achieved experimentally for carbon fibre and silicon carbide fibre reinforced glasses range from $10 \, \text{MNm}^{-3/2}$ to $20 \, \text{MNm}^{-3/2}$ (PHILLIPS, 1974b; PREWO and BRENNAN, 1980; PREWO et al., 1979).

The toughness of fibre reinforced ceramics can be affected both by microstructural parameters such as the fibre–matrix bond and matrix porosity, and by testing parameters such as crack speed (PHILLIPS, 1974a).

Too strong a bond between fibres and matrix leads to a reduction in

FIG. 25. Carbon fibres bridging the two halves of a fracturing specimen of carbon fibre reinforced borosilicate glass.

TABLE 11. Critical stress intensity factors and fracture toughness values for carbon fibre reinforced glasses (PHILLIPS, 1974b).

Fibre	V_f (%)	K_{IC} (MNm$^{-3/2}$)		G_{IC} (kJm^{-2})		
		TC [a]	CN [b]	TC [a]	CN [b]	WF [c]
High modulus	38	9.2 ± 0.5	6.3 ± 0.1	3.0 ± 0.8	1.4 ± 0.2	1.8 ± 0.1
High strength	40	10.3 ± 0.5	7.1 ± 0.2	1.8 ± 0.4	0.9 ± 0.2	3.1 ± 0.1

[a] TC = tapered cantilever measurements.
[b] CN = centre-notched measurements.
[c] WF = works of fracture.

FIG. 26. Carbon fibres protruding from the fracture surface of a carbon fibre reinforced borosilicate glass (PHILLIPS, 1972).

toughness. Table 10 shows shear strength data for four different carbon fibre reinforced systems, and demonstrated that the shear strength was significantly affected by the bond which was developed due to the thermal expansion mismatch. Also shown in Table 10 are works of fracture of the four systems. Compared with the borosilicate glass composites the two glass–ceramic systems, for which differential radial contraction of fibres from the matrix was greatest, had greater works of fracture corresponding to their lower shear strengths. A similar effect of decreasing toughness on increasing bonding has been observed by AVESTON (1971) in silicon carbide reinforced cordierite. PHILLIPS (1974a) has related the work of fracture to the composite shear strength through the fibre–matrix interfacial shear strength. Work of fracture (γ_F) due to pull-out is related to fibre–matrix interface shear strength τ_i by

$$\gamma_F = \frac{V_f r \sigma_f^2}{12 \tau_i},$$

$$(7.6)$$

while the shear strength τ_c of a composite sometimes can be related to fibre–matrix shear strength by an expression of the form

$$\tau_c = x\tau_i + (1 - x)\tau_m \tag{7.7}$$

where τ_m is the shear strength of the matrix and

$$x = \frac{(V_f\pi)^{1/2}}{(V_f\pi)^{1/2} + 1 - 2(V_f/\pi)^{1/2}} \, . \tag{7.8}$$

Combining (7.6), (7.7) and (7.8) we have that

$$\tau_c = A \frac{V_f\sigma_f^2}{\gamma_F} + B\tau_m \, . \tag{7.9}$$

Fig. 27 shows a plot of τ_c vs. $V_f\sigma_f^2/\gamma_F$. Quite good agreement is obtained between the theoretically predicted line and the experimental data for these carbon fibre reinforced glass and glass–ceramic materials.

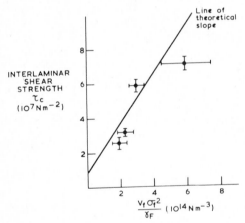

FIG. 27. A relationship between composite shear strength and work of fracture. A plot of τ_c vs. $V_f\sigma_f^2/\gamma_F$ for carbon fibre reinforced glass and glass–ceramic (PHILLIPS, 1974a).

Matrix porosity can alter the mode of failure of a composite, high porosity resulting in a weaker composite but one whose mode of failure can in some applications be advantageous by being more yielding. Fig. 28 shows the behaviour in three-point bending of a strong, low porosity, carbon fibre reinforced glass (51 % of fibre) and a weaker, high porosity, carbon fibre reinforced glass (60 % of fibre). The weaker material displays a more controlled failure, with a capacity for increasing load even after the onset of fracture.

The effect of testing rate, and crack velocity, on the work of fracture is shown in Fig. 29. The test specimens used in these measurements fractured in a controlled manner so that there was a direct correspondence between testing

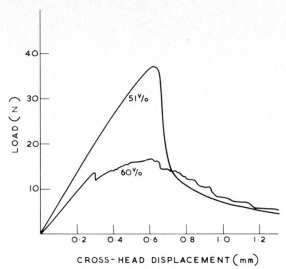

FIG. 28. Load vs. deflection in 3-point bending for 51 and 60 %/₀ for carbon fibre reinforced glass (PHILLIPS et al., 1972).

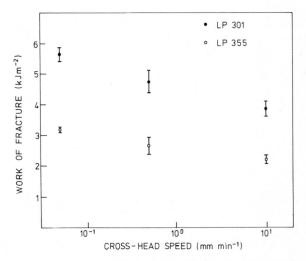

FIG. 29. The variation of work of fracture with testing rate for carbon fibre reinforced glass. LP 301 and 355 are two different materials.

machine cross-head speed and crack velocity. After low cross-head speeds the fracture surfaces exhibited relatively long fibre pull-out lengths corresponding to the higher works of fracture, while at high cross-head speeds the pull-out lengths were shorter. Fortunately this effect does not continue at higher testing rates but appears to reach a limiting value at the high rates of impact testing as shown in Table 12 (PHILLIPS, 1974a).

TABLE 12. Comparison of Charpy energy and low speed work of fracture of carbon fibre reinforced glass (PHILLIPS, 1974a).

Test	Fracture speed (m/sec)	Fracture energy (kJ/m)
Work of fracture	8.4×10^{-7}	3.1 ± 0.1
	8.4×10^{-6}	2.7 ± 0.3
	1.7×10^{-4}	2.2 ± 0.2
Charpy impact	2	2.5 ± 0.3

AVESTON (1971) has attempted to theoretically calculate the maximum fracture energies that might be obtained from fibre reinforced ceramics. In his calculations he considered delamination, fibre pull-out, debonding and multiple fracture. For silicon carbide reinforced cordierite he obtained maximum values somewhat in excess of $10^4 \, Jm^{-2}$ and it seems probable that this is about the maximum which could be obtained with a brittle fibre–brittle matrix system. Correspondingly, the maximum values of critical stress intensity factor which have been obtained are $\sim 20 \, MNm^{-3/2}$. These values of toughness are much greater than for unreinforced ceramics and glasses but much less than high temperature metallic alloys. On toughness alone fibre reinforced ceramics are therefore unlikely to compete with metals for high temperature applications. However, if they can be developed with superior oxidation resistance, then this combined with their low densities could make them more attractive than metals.

7.7. Fatigue

Very little is known about the fatigue of fibre reinforced ceramics. PHILLIPS et al. (1972) and LINGER (1974) have carried out measurements on carbon fibre reinforced glasses, primarily to determine what happens when material is cycled at stresses at which matrix microcracking occurs. LINGER cycled one set of specimens to 80% of the static flexural strength for 3×10^7 cycles, and another set to 57–93% of ultimate for 3×10^6 cycles. He then measured their remaining static strengths which all fell within the scatter band for the static strengths of unfatigued specimens, a result also obtained by PHILLIPS et al. (1972). Although there appears to be no significant loss of strength, changes do occur. The density and penetration of matrix cracking increased during cyclic loading and the fracture behaviour altered. Non-fatigued specimens failed in a controlled way with a high work of fracture, while the fatigued specimens failed in a more catastrophic way with a reduced work of fracture.

7.8. Thermal shock

Fibre reinforcement can improve the thermal shock resistance of ceramics by increasing their thermal conductivity and by increasing the resistance to crack growth. Tinklepaugh and his co-workers in their extensive studies of short metal fibre reinforced ceramics (TINKLEPAUGH, 1965; KROCHMAL, 1967) obtained data such as that shown in Fig. 30 for a molybdenum reinforced alumina. These were obtained by heating specimens to 1200°C and then placing them on a thick steel plate at room temperature. Unreinforced material was unable to survive this treatment but the fibre reinforced alumina, although cracked prior

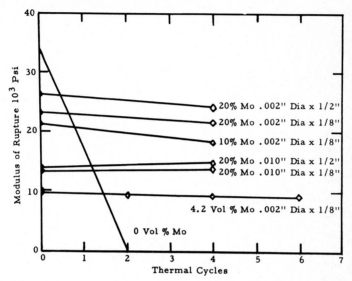

FIG. 30. Modulus of rupture as a function of thermal cycling for an alumina–molybdenum fibre system (TINKLEPAUGH, 1965).

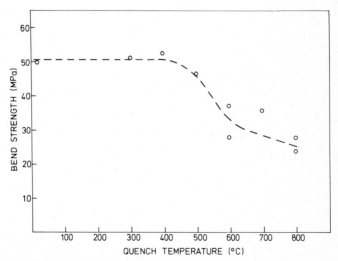

FIG. 31. The effect of quench temperature on the bend strength of $10\,^v/_o$ random, short carbon fibre reinforced borosilicate glass (SAMBELL et al., 1972a).

to thermal shocking, retained its strength. SAMBELL et al. (1972a) also obtained a much improved thermal shock resistance in carbon fibre reinforced borosilicate glass. On plunging into water the unreinforced glass cracked on cooling from about 300°C with large severe cracks and an almost complete loss of strength. Similar glass reinforced with $10\,^{v}/_{o}$ of short carbon fibres did not display any damage until quenched from temperatures as high as 500°C, and even after quenching from 600°C cracking was confined to a fine surface cracking. The loss of strength was much reduced too as shown in Fig. 31.

7.9. *The effects of temperature*

The main disadvantage of carbon fibre reinforced glasses and ceramics is their poor resistance to oxidation when heated in air. In an inert environment they can be used to high temperatures for prolonged periods without loss of strength, as shown for a carbon fibre reinforced lithia–alumino–silicate glass–ceramic in Fig. 32. The presence of oxygen, however, causes a drastic reduction in strengths on heating above approximately 350°C due to degradation of the fibres, as shown in Fig. 33. Protection of the fibres by applying a thin coating of an oxygen barrier to the fibres themselves, or by means of a glaze on the surface of the composite, have been unsuccessful. The use of a glaze is unlikely to be successful, even with further development, as oxygen appears to diffuse into the composite not just along the fibre–matrix interface, but also perpendicular to fibres through the glass (LAMBE and PHILLIPS, unpublished work).

Composites based on other fibres such as silicon carbide, alumina or alu-

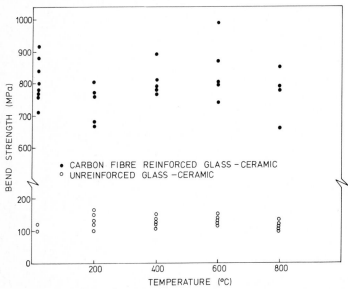

FIG. 32. The flexural strength of unreinforced and carbon fibre reinforced lithia–alumino–silicate glass–ceramic as a function of temperature in an inert environment (SAMBELL et al., 1974).

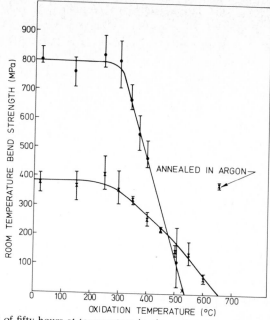

FIG. 33. The effect of fifty hours at temperature in air on the room temperature bend strength of carbon fibre reinforced glass–ceramic (SAMBELL et al., 1974).

minium borosilicate would be expected to be more resistant to oxidation, and preliminary results appear to show that this is so. BACON et al. (1978) have shown that the strength of an alumina reinforced silica is maintained to 1000°C, and PREWO and BRENNAN (1980) that silicon carbide reinforced glasses retained or improved strengths in excess of 600°C, above which temperature creep of the glass became excessive.

8. The future, applications and costs

The development work which has been carried out to date has shown that fibre reinforced ceramics offer the advantages of high strength and stiffness coupled with low density, moderate toughness and hardness. The major outstanding problem appears still to be the development of a system in which these advantages can be retained at high temperatures in oxidising atmospheres. The best properties have so far been achieved in carbon fibre reinforced ceramic systems in which degradation of the fibres occurs in air above about 400°C. Other systems such as silicon carbide or alumina fibre reinforced ceramics have better temperature capability but have not demonstrated the same efficiency of fibre reinforcement as the carbon fibre systems, apart from the silicon carbide fibre reinforced glass ceramic of AVESTON (1971). The development of a truly high temperature system does not yet appear to have been fulfilled and still remains a goal of future research.

The properties of a high temperature system which might potentially be achieved are very attractive. For example a ceramic matrix composite containing 50 % of available unidirectional silicon carbide fibres might, in principle, have a strength and stiffness of ~1.5 GPa and 200 GPa respectively with a density ~3 gm per cc and a toughness greater than $10 \, kJm^{-2}$. The consequent high values of specific strength and stiffness combined with the much enhanced notch toughness make such a material very attractive for applications requiring light weight such as in aerospace or high speed machinery.

Many potential applications exist for fibre reinforced ceramics. In the military sector these have been reviewed by HOVE and DAVIS (1977) and include radomes, re-entry space vehicle antenna windows and nosetips, armour, components of engines such as heat exchangers, gas turbine components, specialised incinerator linings, bearings and nozzles. Materials suggested by them include quartz filament reinforced silica, silicon carbide reinforced silicon, tantalum carbide reinforced graphite, boron carbide reinforced graphite, glasses reinforced with carbon, silicon carbide or alumina, and composites based on boron nitride.

Other applications outside military and aerospace include biomedical materials where the excellent corrosion resistance and biocompatibility of ceramics could be superior to metals, internal combustion engines where the search for light-weight materials for improved fuel consumption could provide an incentive for further development, and some specialised chemical plant.

The materials which might be suitable for these applications will not all require the same properties and would be based on different fibre and matrix systems. Assessment of their potential costs is not possible with any precision depending as it does on the future cost of the reinforcing fibres, capital cost of fabrication plant, processing costs and market size. A very approximate estimate may be obtained by considering the costs of fibres. The lowest price of moderate performance carbon fibres is currently of the order of £40 per kg but could be reduced to half that in tonnage quantities. The current costs of high temperature fibres are much higher and are unlikely to fall below £80 to £100 per kg in the foreseeable future. High temperature fibre reinforced ceramic systems are likely therefore to cost a minimum of several tens of pounds sterling per kg and possibly very much more than this. At these prices there clearly have to be very significant potential advantages before the materials can be seriously considered, but these prices are not very much greater than for existing high performance polymer matrix composites which are finding increasing engineering use.

The main thrust of further development needs to be in the areas of fabrication, both of fibres and of composite systems. This needs to be backed up by studies of the thermodynamic stability of different systems, basic fracture mechanisms, thermal and mechanical fatigue, and stress-corrosion effects. Knowledge of the fabrication and behaviour of fibre reinforced ceramics has advanced rapidly during the last ten years and these materials could make a modest penetration in sophisticated applications in the near future.

References

AVESTON, J. (1971), Strength and toughness in fibre reinforced ceramics, in: *Proc. Conf. on the Properties of Fibre Composites* (IPC Science and Technology Press Ltd., Guildford).

AVESTON, J., G.A. COOPER and A. KELLY (1971), Single and multiple fracture, in: *Proc. Conf. on the Properties of Fibre Composites* (IPC Science and Technology Press Ltd., Guildford).

BACON, J.F., K.M. PREWO and R.D. VELTRI (1978), Glass matrix composites 2—alumina reinforced glass, in: *Proc. 2nd Internat. Conf. on Composite Materials* (AIME, New York).

BAGG, G.E.G., M.E.N. EVANS and A.W.M. PRYDE (1969), *Composites* **1**, 97.

BAILEY, J.E., P.T. CURTIS and A. PARVIZI (1979), *Proc. Roy. Soc. London Ser. A* **366**, 599.

BASKIN, Y., C.A. ARENBURG and J.H. HANDWERK (1959), *Bull. Amer. Ceram. Soc.* **38**, 345.

BASKIN, Y., Y. HARADA and J.H. HANDWERK (1960), *J. Amer. Ceram. Soc.* **43**, 489.

BORTZ, S.A. and S.L. BLUM (1968), A metal-fibre reinforced ceramic system, in: P. POPPER, ed., *Special Ceramics 4, Proc. 4th Symp. on Special Ceramics* (BCRA, Stoke-on-Trent).

BOWEN, D.H. (1969), *Fibre Sci. Technol.* **1**, 85.

BOWEN, D.H., D.C. PHILLIPS, R.A.J. SAMBELL and A. BRIGGS (1972), Carbon fibre reinforced ceramics, in: *Mechanical Behaviour of Materials, Proc. 1971 Internat. Conf. on Mechanical Behaviour of Materials* (The Society of Materials Science, Japan).

BUSALOV, YU.E. and I.M. KOP'EV (1970), Metal-fibre reinforced ceramics, *Fizika i Khimiya Obrabotki Materialov* **1**, 57 (translated by J.A. BARTLETT, Information Branch, U.K.A.E.A.).

COOPER, G.A. (1971), *Rev. Phys. Technol.* **2**, 49.

COOPER, G.A. and J.M. SILLWOOD (1972), *J. Mater. Sci.* **7**, 325.

CORBETT, W.J., A.T. SALES and J.D. WALTON JR. (1965), Improving the properties of slip-cast fused silica by fibrous reinforcement, Tech. Rept. SC-DC-66-1130, Georgia Institute of Technology, Atlanta, Georgia.

CORBETT, W.J. and J.D. WALTON (1966), in: A. BOLTAX and J.H. HANDWERK, eds., *Proc. Conf. on Nuclear Applications of Nonfissionable Ceramics* (American Nuclear Soc. and American Ceramic Soc., Hinsdale, IL).

CRIVELLI-VISCONTI, I. and G.A. COOPER (1969), *Nature* **221**, 754.

DAVIDGE, R.W. and D.C. PHILLIPS (1972), *J. Mater. Sci.* **7**, 1308.

DONALD, I.W. and P.W. McMILLAN (1976), *J. Mater. Sci.* **11**, 949.

EINMAHL, G. (1966), M.S. Thesis, UCRL 16844, Univ. California, Berkeley.

EVERITT, G.F. (1977), Continuous filament ceramic fibre refractory insulation, in: *Proc. Elec./Electron Insulating Conf.*, pp. 236–240.

GRAVES, G.A., C.T. LYNCH and K.S. MAZDIYASNI (1970), *Bull. Amer. Ceram. Soc.* **49**, 797.

HARRIS, B., J. MORLEY and D.C. PHILLIPS (1975), *J. Mater. Sci.* **10**, 2050.

HOVE, J.E. and H. MAUZEE DAVIS (1977), Assessment of ceramic–matrix composite technology and potential DoD applications. Paper P-1307, Assession No. AD-A054 017, Institute for Defense Analyses, Arlington, Virginia.

JONES, R.M. (1975), *Mechanics of Composite Materials* (Pub. Scripta Book Co., Washington, D.C.).

KIRKPATRICK, M.E., K.P. STAUDHAMMER, J.L. REGER and A. TOY (1969), *J. Composite Mater.* **3**, 322.

KLIMAN, M.I. (1962), Transverse rupture strength of alumina fibre-ceramic composites, Tech. Rept. WAL TR 371/53, Watertown Arsenal Laboratories.

KROCHMAL, J.J. (1967), Fibre reinforced ceramics: A review and assessment of their potential, Tech. Rept. AFML-TR-62-207, Air Force Materials Laboratory, Wright–Patterson Air Force Base, OH.

LAMBE, K.A.D., N.J. MATTINGLEY and D.H. BOWEN (1969), *Fibre Sci. Technol.* **2**, 59.

LEVIN, E.M., C.R. ROBBINS and H.F. McMURDIE (1964), *Phase Diagrams for Ceramists* (American Ceramic Soc., Hinsdale, IL).

LEVITT, S.R. (1973), *J. Mater. Sci.* **8**, 793.

LINGER, K.R. (1974), Fatigue effects in carbon fibre reinforced glass, in: *Conf. Proc. Composites— Standards Testing and Design* (IPC Science and Technology Press Ltd., Guildford).

MACKENZIE, J.K. (1950), *Proc. Phys. Soc. Ser. B* **63**, 2.

MOSS, M., W.L. CYRUS and B.M. SCHUSTER (1972), *Bull. Amer. Ceram. Soc.* **51**, 107.

NAKAYAMA, J. (1965), *J. Amer. Ceram. Soc.* **48**, 583.

PARRATT, N.J. (1969), *Composites* **1**, 25.

PARVIZI, A., K.W. GARRETT and J.E. BAILEY (1978), *J. Mater. Sci.* **13**, 195.

PHILLIPS, D.C. (1972), *J. Mater. Sci.* **7**, 1175.

PHILLIPS, D.C. (1974a), *J. Mater. Sci.* **9**, 1847.

PHILLIPS, D.C. (1974b), *J. Composite Mater.* **8**, 130.

PHILLIPS, D.C. and B. HARRIS (1977), The strength, toughness and fatigue properties of polymer composites, in: M.O.W. RICHARDSON, ed., *Polymer Engineering Composites* (Applied Science, London) Chapter 2.

PHILLIPS, D.C. and A.S. TETELMAN (1972), *Composites* **4**, 216.

PHILLIPS, D.C., R.A.J. SAMBELL and D.H. BOWEN (1972), *J. Mater. Sci.* **7**, 1454.

PREWO, K.M. and J.F. BACON (1978), Glass matrix composites —1—graphite fibre reinforced glass, in: B. NOTON, R. SIGNORELLI, K. STREET and L.N. PHILLIPS, eds., *Proc. of the 2nd Internat. Conf. on Composite Materials* (AIME, New York).

PREWO, K.M., J.F. BACON and D.L. DICUS (1979), *SAMPE Quart.* **10**(4) 42.

PREWO, K.M. and J.J. BRENNAN (1980), *J. Mater. Sci.* **15**, 463.

REYNOLDS, W.N. (1970), Structure and mechanical properties of carbon fibres, in: *3rd Conf. on Industrial Carbons and Graphite* (Society of Chemical Industry, London).

ROWCLIFFE, D.J., W.J. WARREN, A.G. ELLIOT and W.S. ROTHWELL (1969), *J. Mater. Sci.* **4**, 902.

SAMBELL, R.A.J. (1970), *Composites* **1**, 276.

SAMBELL, R.A.J., D.H. BOWEN and D.C. PHILLIPS (1972a), *J. Mater. Sci.* **7**, 663.

SAMBELL, R.A.J., A. BRIGGS, D.C. PHILLIPS and D.H. BOWEN (1972b), *J. Mater. Sci.* **7**, 676.

SAMBELL, R.A.J., D.C. PHILLIPS and D.H. BOWEN (1974), The technology of carbon fibre reinforced glasses and ceramics, in: *Carbon fibres, their Place in Modern Technology, Proc. Internat. Conf.* (Plastic Institute, London).

TATTERSALL, H.G. and G. TAPPIN (1966), *J. Mater. Sci.* **1**, 296.

TINKLEPAUGH, J.R. (1965), Ceramic–metal fibre composite systems, in: J.J. BURKE, N.L. REED and V. WEISS, eds., *Strengthening Mechanisms, Metals and Ceramics, 12th Sagamore Army Materials Research Conf.* (Syracuse University Press).

WANG, F.F.Y. (1976), *Ceramic Fabrication Processes, Treatise on Materials Science and Technology,* Vol. IX, H. HERMAN, ed. (Academic Press, New York).

YAJIMA, SEISHI, K. OKAMURA, J. HAYASHI and M. OMORI (1976), *J. Amer. Ceram. Soc.* **59**, 324.

CHAPTER VIII

Fibre Reinforced Cements

D.J. Hannant

Department of Civil Engineering
University of Surrey
Guildford
Surrey GU2 5XH
United Kingdom

Contents

HANDBOOK OF COMPOSITES, VOL. 4 – Fabrication of Composites
Edited by A. KELLY and S.T. MILEIKO
© 1983, Elsevier Science Publishers B.V.

List of Symbols

A_f – cross-sectional area of fibre

A_g – (weight of aggregate greater than 5 mm)/(total weight of concrete) ratio

C – cost of fibre – per kg

E_c, E_f, E_m – modulus of elasticity of composite, fibre, matrix

N – number of fibres crossing unit area of composite

P_f – perimeter of fibre

SG_f – fibre specific gravity

V_f, V_m – volume fraction of fibre, matrix

V_f' – effective volume of fibre in the direction of stress calculated from V_f using efficiency factors

$V_{f(crit)}$ – volume of fibres which, after matrix cracking, will carry the load which the composite sustained before cracking

$V_{f(min)}$ – minimum fibre volume for strengthening in bending

W_f – weight of fibres per cent, compactable with normal size techniques

Y – cost of fibre per m^3 of composite

d – fibre diameter

l – fibre length

l_c – critical fibre length, i.e., twice the length of fibre embedment which would cause fibre failure in a pull-out test

l_c' – see Table 3

r – fibre radius

w – crack width for minimum crack spacing

w' – crack width for average crack spacing

x' – transfer length for stress for long fibres (also minimum crack spacing)

α – scaling factor $= E_m V_m / E_f V_f$

ε_c – strain in composite

ε_{mu} – ultimate strain of the matrix

σ_{comp} – compressive stress in composite

σ_{cu} – ultimate strength of composite

σ_f – stress in fibre

$\sigma_{f\ell}$ – flexural strength assuming elastic material and neutral axis at centre

σ_{fu} – failure stress of fully bonded fibres or pull-out stress of debonded fibre

σ_{mu} – ultimate strength of matrix

σ_t – tensile cracking strength of the composite
τ – average sliding friction bond strength
τ_d – bond strength after fibre slip
τ_s – bond strength before fibre slip

1. Introduction

Cement mortar and concrete differ from most other matrices used in fibre composites in that their essential characteristics are those of low tensile strength (generally less than 7 MPa) and very low strain to failure in tension (generally less than 0.05%) which result in materials which are rather brittle when subjected to impact loads.

The binder in the matrix is cement paste which is produced by mixing cement powder and water which harden after an exothermic reaction. The hardened material consists of complex chemical compounds which have a wide range of physical and chemical characteristics. These characteristics depend on a variety of factors including the manufacturing technique of the composite, for instance products autoclaved in steam at 180°C have a different micro-structure from those produced by wet curing at ambient temperatures. Also, low alkali or aluminous cements produce compounds which can significantly alter the performance characteristics of some of their fibre composites when compared with Portland cements. In the case of Portland cement, the hardened material consists mainly of hydrated calcium silicates with some calcium hydroxide crystals, an example of the type of structure being shown in Fig. 1.

The hardened cement is dimensionally unstable due to the complex effects of water movements at the molecular level within the crystal structure and cement paste is therefore rarely used by itself in construction but is filled with inert mineral fillers known as aggregates to increase its stability and also to reduce the cost.

The matrix is therefore generally a composite material in its own right and for the purposes of this text a cement mortar matrix will be defined as one containing particles less than 5 mm maximum size and a concrete matrix as containing particles less than 40 mm maximum size. In concrete, the filler may occupy 70% of the volume and hence the filler has a controlling influence on the properties of the matrix.

The volume of fibre which can be easily included in the matrix during the manufacture of the product is greatly affected by the volume and particle size of the filler and the weak aggregate–cement interfaces cannot be strengthened by fibre addition. Also, the rheological characteristics of the wet paste are not as favourable as resins to the inclusion of fibres. Volumes of fibre in excess of 10% of the composite volume are difficult to include in mortar and, in the case of concrete, 2% by volume of fibre is considered to be a high proportion which is generally insufficient to permit significant reinforcement of the matrix before

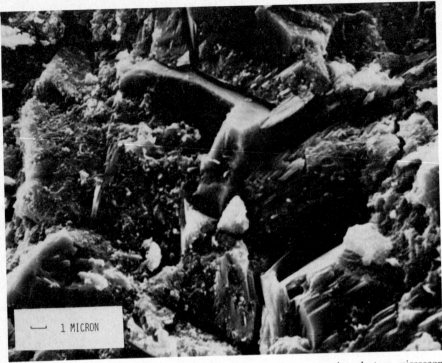

FIG. 1. Broken surface of hydrated Portland cement taken on a scanning electron microscope (published by permission of D.C. Montgomery, University of Wollongong).

cracking occurs, particularly where individual fibres may be spaced up to 40 mm apart by the aggregate particles.

The type of fibre composite described in this chapter is therefore one in which the fibre has little effect on the properties of the relatively stiff matrix until cracking has occurred at the micro-level. The fibres then become effective resulting in a composite with reduced stiffness and increased strain capacity providing increased toughness compared with the unreinforced matrix.

Although such a material may not at first sight appear to be very promising technically, the low energy costs of cement and concrete combined with the vast world bulk production of concrete, which is in excess of 4×10^9 tonnes/annum, render the potential of fibre reinforced cement based materials of considerable commercial significance.

The increasing emphasis on energy conservation has also stimulated interest in methods of replacing materials such as cast iron, glass reinforced plastics, and bituminous materials by the use of fibre cements and fibre concretes. This idea of replacing existing materials also extends into the asbestos-cement industry, where, due to problems relating to the supply of asbestos in the long-term and also because of possible health hazards, attention is being given to the replacement of asbestos with man-made fibres.

In order to be able to satisfy the performance requirements of these various interests, adequate material properties must be achieved in the fibre composite and the main objectives of the materials engineer in the attempting to modify the properties of cement or concrete are the following:

(a) to improve the tensile or flexural strength,

(b) to improve the impact strength,

(c) to control cracking and the mode of failure by means of post-cracking ductility,

(d) to change the rheology or flow characteristics of the material in the fresh state.

There is no doubt that these objectives can be achieved in the short term but a degree of caution is necessary for most fibre concretes and cements regarding the long-term performance, which, in the civil engineering sense may well be in excess of 50 years. This is likely to result in the bulk of applications being non-structural in that no great danger to life should result from component collapse in the event of unforseen deterioration with time.

Fibre reinforced cement mortars and fibre reinforced concretes have sufficiently different fabrication techniques and consequently very different properties in their products to warrant separate sections in this chapter. For instance, the production of fibre reinforced mortars has often relied on modifications of ideas developed for the glass reinforced plastics industry or from the asbestos-cement industry whereas the fabrication techniques for fibre concretes have been influenced to a large extent by the existing batching and mixing plant for concrete which has been developed by the industry during the past 100 years.

2. Materials

The basic technology and properties of unreinforced cement and concrete have been well documented in all the major languages, standard works in English speaking areas being those by NEVILLE (1978) and by LEA (1970).

The major factors which affect fabrication techniques are the rheological properties of the cement or concrete whereas the performance of the hardened material is controlled by the volume and physical properties of the fibres and the matrix and the strength of the bond between the two.

2.1. Rheological characteristics

2.1.1. Concrete
Concrete is a thixotropic material in which flow is encouraged by vibration with or without the application of external pressure. De-watering by vacuum is not

commonly used and therefore the mixing operation during which the short fibres are added has to take place with a matrix of the same workability which is required for the final compacting operation on site. Transportation to the point of compaction may take up to one hour after mixing and the composite material must then be made to flow into position and entrapped air must be removed by vibration possibly combined with pressure. The fibre concrete may also be pumped along a pipeline several hundred metres long.

Careful control of the mix design of the matrix is therefore required in order to ensure that the concrete will flow in all these situations without segregation and with minimum air voids in the final product. If a high fibre volume is required ($\sim$2%) this is generally made possible by a high total fines content, preferably with less than 30% by volume of particles exceeding 5 mm in dimension, a continuous aggregate grading to prevent segregation under pressure, and possibly a workability admixture.

The very nature of the compaction procedure, in which flow is involved, will cause fibre orientation and an anisotropic hardened composite and this may be used to advantage if the fibres can be orientated in the direction of greatest stress in the product. The rheological properties may be less important where sprayed fibre concrete is used as described in Section 5.1.

2.1.2. Cement paste and mortar

The physical characteristics required of fine grained cement matrices in the fresh state for fabrication into fibre composites are generally different from those for concretes. Vibration and flow are often not required in the production of thin sheet products in which an important requirement may be the filtration characteristics of the fresh composites in preventing removal of cement fines during vacuum de-watering. For instance it may be necessary to add a flocculating agent such as a polyacrylamide of anionic character.

Alternatively, where sprayed composites are manufactured without vacuum de-watering a suitable fluidity must be achieved to enable flow through the spray gun without an excessively high water/cement ratio which would result in a weak matrix. Water reducing admixtures assist in this process.

In cases where extrusion or pressing are involved, thickening agents such as polyethelene oxide are necessary in the matrix to prevent the flow of water under pressure gradients and blocking of the die.

2.2. Properties of the matrix in relation to the fibres

It is important to have an awareness of the properties of the hardened matrix in comparison with those of the reinforcing fibres because the philosophy behind the mechanism of reinforcement is very different from that of reinforced thermoplastics or of metal matrix composites. The fibre properties and volumes also have a major influence on the fabrication procedures and hence a brief description of the most relevant properties is given below.

Typical ranges for the properties of the relevant fibres are shown in Table 1 and properties for some of the matrices are shown in Table 2.

TABLE 1. Typical fibre properties (HANNANT, 1978, by permission of John Wiley Ltd.).

Fibre		Diameter (μm)	Length (mm)	Density (kg/m³ × 10³)	Young's modulus (GPa)	Poisson's ratio	Tensile strength (MPa)	Elongation at break[a] (%)	Typical volume in composite (%)
Asbestos	Chrysotile (white)	0.02–30	<40	2.55	164	0.3	200–1800 (fibre bundles)	2–3	10
	Crocidolite (blue)	0.1–20	–	3.37	196	–	3500	2–3	–
Carbon	Type 1 (High modulus)	8	10-continuous	1.90	380	0.35	1800	~0.5	2–12
	Type 2 (High strength)	9		1.90	230		2600	~1.0	
Cellulose		10–50	1–4	1.2	10	–	300–500	~15	10–20
Glass	E	8–10	10–50	2.54	72	0.25	3500	4.8	2–8
	Cem-Fil filament	12.5			80	0.22	2500	3.6	
	204 filament strand	110 × 650		2.7	70	–	1250	–	
Kevlar	PRD 49	10	6–65	1.45	133	0.32	2900	2.1	<2
	PRD 29	12		1.44	69	–	2900	4.0	
Nylon (Type 242)		>4	5–50	1.14	Rate dependent up to 4	0.40	750–900	13.5	0.1–6
Polypropylene	Monofilament	100–200	5–50	0.9	Rate dependent up to 5	–	400	18	0.1–6
	Fibrillated	500–4000	20–75	0.9	up to 8	0.29–0.46	400	8	0.2–1.2
Steel	High tensile	100–600	10–60	7.86	200		700–2000	3.5	0.5–2
	Stainless	10–330			160	0.28	2100	3	

[a] *Note*: 1% elongation = $10\,000 \times 10^{-6}$ strain.

TABLE 2. Typical properties of matrix (HANNANT, 1978, by permission of John Wiley Ltd.).

Matrix	Density (kg/m^3)	Young's modulus (GPa)	Tensile strength (MPa)	Strain at failure $\times 10^{-6}$
Ordinary Portland cement paste	2000 −2200	10–25	3–6	100–500
High alumina cement paste	2100 −2300	10–25	3–7	100–500
O.P.C. mortar	2200 −2300	25–35	2–6	50–150
O.P.C. concrete	2300 −2450	30–40	1–4	50–150

It is apparent from these tables that the elongations at break of all the fibres are two or three orders of magnitude greater than the strain at failure of the matrix and hence the matrix will crack long before the fibre strength is approached. This fact is the reason for the emphasis on post-cracking performance in the theoretical treatment.

On the other hand, the modulus of elasticity of the fibre is generally less than five times that of the matrix and this, combined with the low fibre volume fraction, means that the modulus of the composite is not greatly different from that of the matrix.

The fibre types in Table 1 can be divided into two main groups, those with moduli lower than the cement matrix, such as cellulose, nylon and polypropylene and those with higher moduli such as asbestos, glass, steel, carbon, and Kevlar, which is a form of aromatic polyamide introduced by DuPont. The last two are included for the sake of completeness but high cost would seem to rule them out for major engineering applications.

The low modulus organic fibres are generally subject to relatively high creep which means that if they are used to support permanent high stresses in a cracked composite, considerable elongations or deflections may occur over a period of time. They are therefore more likely to be used in situations where the matrix is expected to be uncracked, but where transitory overloads such as handling stresses, impacts or wind loads are significant.

Another problem with the low modulus fibres is that they generally have large values of Poisson's ratio and this, combined with their low moduli, means that if stretched along their axis, they contract sideways much more than the other fibres. This leads to a high lateral tensile stress at the fibre–matrix interface which is likely to cause a short aligned fibre to debond and pull out. Devices such as woven meshes or networks of fibrillated fibres may therefore be necessary to give efficient composites.

Even the high modulus short fibres may require mechanical bonding to avoid pull out unless the specific surface area is very large. Thus steel fibres are commonly produced with varying cross-sections or bent ends to provide anchorage and glass fibre bundles may be penetrated with cement hydration products to give an effective mechanical bond after a period of time.

Four types of matrices are shown in Table 2 and the two cement types shown are important mainly because they have different degrees of alkalinity which affect the durability of glass and steel fibres.

In order to avoid shrinkage and surface crazing problems in finished products it is advisable to use at least 50 per cent by volume of inert mineral filler, which may be aggregate or could include pulverized fuel ash, or limestone dust. However, if the inert filler consists of a large volume of coarse aggregate, the volume of fibres which can be included will be limited which will in turn limit the tensile strength and ductility of the composite.

Strength of the matrix is mainly affected by the free water/cement ratio and this parameter also has a lesser effect on the modulus so that the properties of the matrices shown in Table 2 can vary widely.

2.3. Main types of fibres

The main types of fibre which have been successfully included in commercial products are asbestos, cellulose, glass, polypropylene and steel and a brief description of the physical and geometrical characteristics which are of relevance to fabrication techniques is given below.

2.3.1. Asbestos fibres

Asbestos is a general name for several varieties of naturally occurring crystalline fibrous silicate minerals which possess a rather unique range of physical and chemical properties. The two main groups are the serpentines and the amphiboles and generally both types have developed as cross-fibre seams or veins in the host rocks. The width of the seams determines the fibre length which is commonly in the range 0.8–19 mm, and the fibres exist in extremely tight packed parallel formations. Certain types of asbestos occur in fibrous masses with randomly oriented blocks of fibres up to 25 mm in length and it is not uncommon to find fibres up to 100 mm long.

By far the most abundant mineral is chrysotile ($3MgO$–$2SiO_2$–$2H_2O$) or white asbestos and this is the sole member of the serpentine group.

Chrysotile constitutes more than 90 per cent of the world asbestos reserves and is used to a large extent in the manufacture of asbestos-cement. The fibre is white and silky with a minimum diameter of about $0.01\,\mu m$ and electron microscope work suggests that even the finest chrysotile fibres are hollow which may partly account for their affinity for cement. Also, the fibres well resist the severe mechanical attrition received during processing possibly due to their high strength and flexibility.

Very high values ($>3000\,MPa$) are quoted in the literature for the tensile strength of chrysotile but the strengths of the fibre bundles in the composite are much lower. For example, Klos (1975) considers 560–750 MPa to be a practical range whereas MAJUMDAR et al. (1977) have measured 300–1800 MPa for fibre bundles. Fibre strengths of this order are easily obtained with several fibres other than asbestos and this consideration is largely responsible for the current interest in substitutes for asbestos as cement reinforcement. The chemical resistance offered by chrysoltile asbestos, particularly to strongly alkaline conditions is considered to be excellent and this is reflected in the durability which asbestos-cement products generally enjoy. The fibres are, however, susceptible to strength loss at elevated temperatures and above 400°C their strengthening power is greatly reduced. As such high temperatures are not normally encountered in buildings, asbestos-cement products are considered to be reasonably safe in most applications in which they are currently used.

The strongest type of fibres in the amphibole group is crocidolite or blue asbestos ($Na_2O–Fe_2O_3–3FeO–8SiO_2–H_2O$) and the group also contains amosite, anthophylite, tremolite and actinolite. Crocidolite is considered to be the most dangerous form of asbestos from the point of view of hazard to health.

Asbestos fibres produced at the mines require further processing at manufacturing sites to make them suitable for many applications. This involves milling to split the coarser fibres into finer ones. A variety of methods is used, ranging from edgerunners and rod mills to hammer mills and attritors, the choice of particular methods depending on the application.

The hazards to health of asbestos fibres have been well documented but it is difficult to make a definitive statement regarding the health hazards of asbestos-cement. However, any operation which releases fibres or dust such as sawing or drilling will require precautions to be taken against inhalation or ingestion of the material.

2.3.2. Cellulose fibres

Cellulose fibres produced from wood pulp have been used for many years as additives in asbestos-cement products but their precise physical properties seem to vary widely depending on the authority quoted and this variability may also depend on the pulping techniques used. Fibre lengths between 1 and 4 mm have been quoted and the helical structure of the fibres gives a variable diameter between 10 and 50 microns. Thus a typical length/diameter ratio may be about 100. Tensile strengths between 50 and 600 MPa have been quoted with moduli between 10 GPa and more than the modulus of cement. It is probable that the cell wall of the hollow fibres has a higher modulus than the complete unit.

According to GORDON and JERONIMIDIS (1974), total longitudinal extensions in individual fibres of 15 to 20% are possible due to buckling of the walls into the lumen thus absorbing considerable energy, whereas the bulk fracture strain of timber may only be about 1%. However, according to PAGE et al. (1971) the stress–strain curves of fibres that are prevented from buckling are quite different from those of free fibres. Thus, while data obtained from isolated fibres are relevant to the behaviour of loosely bonded structures such as non-woven fabrics, they are not directly applicable to the behaviour of tightly bonded structures such as may exist in cement composites. Further details of mechanical properties have been given by Mark (1971).

One of the advantages of beaten or pulped cellulose fibres is that the ends tend to fray or fibrillate thus helping to mechanically anchor the fibre into the matrix in addition to the frictional interfacial stresses. It is also important that chemical bonding is not prevented by sugars and other chemicals present in the wood so that some chemical pre-treatment is necessary before their use in cement.

2.3.3. Glass fibres

Glass fibres in the form of single filaments and in the form of strands have been used for cement reinforcement. The major applications have resulted from the use of chopped strands and Biryukovich et al. (1964) were responsible for most of the early developments with strand using a matrix of low alkali cement or high alumina cement.

The Russian work stimulated the work of Majumdar and Nurse (1974) at the Building Research Establishment in the U.K. which was directed towards the development of a glass fibre which would resist attack by the highly alkaline ordinary Portland cements commonly used in Europe and America.

The identification of alkali-resistant glasses by the Building Research Establishment and their subsequent development and commercial production in the U.K. by Pilkington Brothers Ltd. has encouraged a world wide interest in glass reinforced cement.

The production of glass reinforced cement generally involves the use of continuous strands which are chopped just before coming in contact with the cement and these continuous fibres are produced mechanically be drawing filaments from the bottom of a heated platinum bushing or tank containing several hundred holes.

The glass fibres are collected in strands of about 200 filaments on a rotating drum and their final diameter depends on the speed of rotation of the drum, the viscosity of the melt and the size of the holes in the bushing.

The collection of 200 or so filaments is known as a strand, but, before reaching the drum, the strand is coated with a size which holds the filaments together in a lens shaped form as shown in Fig. 2.

The cross-sectional area (A_f) and perimeter (P_f) of the strand have been estimated to be 0.016 mm^2 and 2.64 mm respectively by Krenchel (1975a) and 0.027 mm^2 and 1.42 mm respectively by Ali et al. (1975). Oakley and Proctor

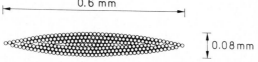

FIG. 2. Idealised view of a glass fibre strand containing 204 monofilaments each of a diameter of 10 μm (Krenchel, 1975, by permission of Construction Press Ltd.).

(1975) have shown, for strands of 204, 12.5 μm fibres about 0.65 mm wide by 0.11 mm thick, that $A_f = 0.074$ mm^2 and $P_f = 2.83$ mm. This means that the glass occupies 34 per cent of the strand volume.

Several strands may be lightly bonded to form a roving which may be wound as a 'cheese'. The roving can be later unwound from the inside of the cheese to be chopped to suitable lengths for use directly in cement or for the production of chopped strand mat.

Mechanically drawn glass fibres are commonly available in three groups: soda–lime–silica glass or A-glass, borosilicate glass or E-glass and Zirconia glass which is more resistant to attack by alkalis than A- or E-glass. Typical chemical compositions for these glasses have been given by LARNER et al. (1976).

The tensile strength of Zirconia glass filaments is about 2500 MPa with an elastic modulus of about 80 GPa.

Glass wool and mineral wool with fibre lengths up to 150 mm and diameter between 5 and 10 microns can also be used for special applications. These fibres are produced by blowing compressed air or steam at a stream of molten glass and similar but longer fibres may be manufactured by centrifuging molten glass. Combinations of monofilament glass and chopped strand glass may also prove beneficial for certain purposes.

2.3.4. Polypropylene fibres

The development of polypropylene in a new strong form, the isotatic configuration, and commerical production in the 1960's, offered the textile industry a potentially low-priced polymer capable of being converted into useful fibre. Polypropylene fibres then became available in two forms, monofilaments (or spinneret) fibres and film fibres.

Both these types of fibres have been used in concrete applications but the process of film extrusion produces cheaper fibres and is particularly suited to the processing of isotactic polypropylene. The extruder is fitted with a die to produce a tubular or flat film which is then slit into tapes and mono-axially stretched. The 'draw ratio' is a measure of the extension which is applied to the fibre during fabrication and draw ratios of about 8 are common for polypropylene film although higher ratios up to about 20 are possible. A molecular orientation results from the stretching and is the cause of the high tensile strength.

Having achieved the production of films with adequate properties, their use in concrete is made possible by fibrillation which is the generation of longitudinal splits and can be controlled by the use of carefully designed pin systems on rollers over which the stretched films are led.

The regular pattern of a pinned yarn is shown in Fig. 3. Fibrillated films twisted into the form of twine are supplied in spool form for cutting on site, or are chopped by the manufacturer, usually in staple lengths between 25 and 75 mm. Purchasing the fibres on spools and cutting to the required length in the precast works can lead to considerable savings. Spools are also easier to handle in transport and require less storage space.

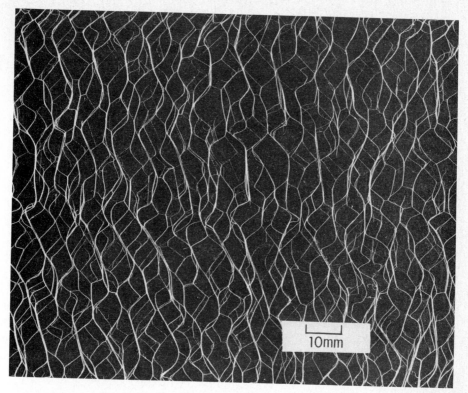

FIG. 3. Fibrillated polypropylene film split by pin rolling.

The types of fibre are characterised by the length in metres per kilogramme of twine, commonly used types being 700 m/kg and 1400 m/kg.

Tensile strengths may range between 200 and 500 MPa with moduli of elasticity between 2 and 8 GPa.

2.3.5. Steel fibres

The bond strength between the fibre and the concrete is one of the major factors which determines the properties of the hardened concrete. Manufacturers of fibre wire have attempted to improve the mechanical bond in a variety of ways and these have led to the different configurations shown in Fig. 4.

Wires with a circular cross-section are produced by normal wire-drawing techniques which are relatively expensive, but cheaper ways of fibre production have been developed. The production method utilising slit sheet results in rectangular section fibres which may be produced cheaply when supplies of scrap metal sheet are readily available and another economic technique is the 'melt extract' process in which fibres are produced directly from the molten steel by means of a spinning multi-edged extraction disc in contact with the surface of the molten metal. In this process the fibres are automatically thrown free of the disc giving the fibre shape as shown in Fig. 4(f) and cheap chromium

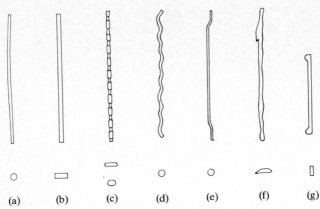

FIG. 4. Shapes of steel fibres. (a) Round; (b) Rectangular; (c) Indented (Duoform, National Standard Patent); (d) Crimped (G.K.N. and Johnson Nephew Ltd.); (e) Hooked ends (Dramix Z. Bekaert Ltd. Patent); (f) Melt extract process (Batelle Patent); (g) Enlarged ends (Australian Wire Industries Ltd. Patent) (HANNANT, 1978, by permission of John Wiley, Ltd.).

steel and stainless steel fibres can be produced from scrap materials in addition to carbon steel fibres.

Fibre diameters for drawn steel wires have commonly ranged between about 0.1 to 0.7 mm with lengths between 10 and 70 mm. Fibres of about 0.5 mm diameter have been used where there is a risk of corrosion but these stiffer fibres can cause local failure of the matrix due to lateral tensile stresses imposed on the concrete as the fibres pull out at random angles.

Wire strengths of between 700 and 2000 MPa are common but these strengths are rarely utilised in the composite due to pull out of the fibres.

2.4. Material costs

The commercial viability of fibre composites is critically dependent on the costs of the fibres which exercise a controlling influence on the cost of the product because the matrix is so cheap.

Thus, a commercial decision on whether an existing product can be profitably replaced with one made from fibre cement or fibre concrete may depend on using the cheapest suitable fibre in the lowest volume required to fulfil the strength and durability requirement.

Unfortunately it is not possible to give specific prices because costs of fibres are subject to relatively rapid changes depending on demand and energy costs in their production. However, the cost of fibre per cubic metre of composite (Y) can be calculated from (2.1) provided that the fibre cost per kg (C) is known:

$$Y = V_f \times C \times 10^3 \times SG_f \qquad (2.1)$$

where V_f is the volume fraction of fibre and SG_f the fibre specific gravity.

For example, a polypropylene fibre composite containing 6 per cent by volume of fibres costing 100 price units per kg would result in a cost of fibres in the composite of $0.06 \times 100 \times 10^3 \times 0.9$ price units/m³, i.e., 5.4×10^3 price units/m³.

Similarly 6 per cent by volume of glass fibres costing 150 price units per kg would cost $0.06 \times 150 \times 10^3 \times 2.7$ price units/m³, i.e., 24.3×10^3 price units/m³ would be the cost of the fibres alone.

It is apparent from these crude calculations that some careful commercial decisions may have to be taken regarding savings in labour or transport costs compared with the costs of equivalent products in timber, steel, aluminium, plastic or reinforced concrete before these new materials are used in large quantities.

3. Test techniques

Accurate testing is fundamental to the understanding of and to the improvement of the composites and it is also essential as a control on the fabrication techniques. Unfortunately, existing procedures for concrete testing are mainly based on compression tests and these are inappropriate for fibre composites which are mostly used in tension or in flexure. Also, bending or flexural tests for concrete or for asbestos-cement are based on the assumption of linear elastic behaviour up to fracture which is grossly inaccurate for the more recently developed fibre cements and concretes.

Extensive development work has gone into the area of testing and much of this has been summarised in the RILEM Symposium (1978).

Detailed specifications for test methods for fibre cement and fibre concrete products are currently being considered by the RILEM 19-FRC Technical Committee and for fibre concretes by the American Concrete Institute Committee 544 and these specifications may eventually be considered by International Standards Organisations for inclusion in the relevant national standards.

Another important area which has seen considerable activity during the past five years is that of the production of fibre cements specifically to provide alternatives to asbestos-cement products. The International Committee ISO/TC 77/SG4 for products in non-asbestos fibre reinforced cement is considering appropriate testing techniques to provide specifications for thin sheet cladding products and the Danish Standards Organisation has already produced a draft performance specification relating to fibre reinforced cement corrugated sheets.

Testing techniques for fibre reinforced cement mortars and for fibre reinforced concretes have been developed along two basically separate lines, i.e., techniques for thin sheet, fibre cement products, and techniques for bulk fibre concrete products. As a result, two different sets of testing procedures will probably be required depending on the type of material and its potential applications. In addition, further tests will be required for the control of production processes, probably based on performance specifications for the individual products.

4. Theoretical principles

The theoretical principles which govern the performance of fibres in brittle matrices are well known (AVESTON et al., 1971, 1974) and only the points of direct relevance to the fabrication of fibre cements and fibre concretes are stated below.

4.1. Critical fibre volume in tension

It has already been pointed out that it is unlikely to be very beneficial to include fibres in cement matrices to increase the cracking stress, and therefore, the merit, if there is any, of fibre inclusion must lie in the load carrying ability of the fibres after matrix cracking has occurred.

The cracked composite may carry a less or a greater load after cracking than the uncracked material and typical load–extension curves for tension tests are shown in Fig. 5(a) and (b).

In order that the material may follow the curve in Fig. 5(b) it is necessary for the critical fibre volume $V_{f(crit)}$ to be exceeded.

The critical fibre volume is defined as the volume of fibres which, after matrix cracking, will carry the load which the composite sustained before cracking.

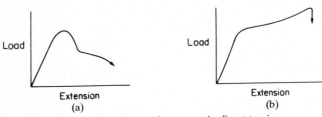

FIG. 5. Typical load–extension curves in direct tension.

This definition needs to be used with a little care because material which has less than the critical volume of fibre in tension as in Fig. 5(a) may have considerably greater than the critical fibre volume required for flexural strengthening (HANNANT, 1978).

However, it is common practice to assume that the above definition of critical fibre volume refers only to uniaxial tensile stresses and for the simplest case of continuous aligned fibres with frictional bond the critical volume can be expressed as a fraction of the elastic modulus of the composite (E_c), the ultimate strain of the matrix (ε_{mu}) and the fibre strength (σ_{fu}):

$$V_{f(crit)} = E_c \varepsilon_{mu} / \sigma_{fu} . \qquad (4.1)$$

Efficiency factors for fibre orientation and fibre length increase $V_{f(crit)}$ considerably from that calculated from (4.1) and most fabrication techniques for adding random, short, chopped fibres to concrete are unable to include sufficient fibre to reach this critical volume. However, many techniques are available which enable the inclusion of more than the critical volume of fibre in fine grained cement mortars.

4.2. Stress–strain curve, multiple cracking and ultimate strength in tension

If the critical fibre volume for strengthening has been reached, then it is possible to achieve multiple cracking of the matrix. This is a desirable situation because it changes a basically brittle material with a single fracture surface and low energy requirement to fracture, into a pseudo-ductile material which can absorb transient minor overloads and shocks with little visible damage. Therefore, the aim of the materials engineer is often to produce a large number of cracks at as close a spacing as possible so that the crack widths are very small (say <0.1 mm). These cracks are almost invisible to the naked eye in a rough concrete surface and the small width reduces the rate at which aggressive materials can penetrate the matrix when compared with commonly allowable widths in reinforced concrete of up to 0.3 mm.

High bond strength helps to give a close crack spacing but it is also essential that the fibres de-bond sufficiently local to the crack to give ductility which will absorb impacts.

The principles behind the calculation of the complete stress–strain curve, the crack spacing and the crack width for long aligned fibres for the simplified case where the bond between the fibres and matrix is purely frictional and the matrix has a well-defined single valued breaking stress has been given by AVESTON et al. (1975, 1976) and is published by permission of the Director of the National Physical Laboratory, Teddington, Middlesex (Crown Copyright reserved).

4.2.1. Long fibres with frictional bond

The idealised stress–strain curve for a fibre reinforced brittle matrix composite is shown in Fig. 6.

If the fibre diameter is not too small, the matrix will fail at its normal failure strain (ε_{mu}) and the subsequent behaviour will depend on whether the fibres can withstand the additional load without breaking, i.e., whether

$$\sigma_{fu} V_f > E_c \varepsilon_{mu} . \qquad (4.2)$$

If they can take this additional load, it will be transferred back into the matrix over a transfer length x' (see Fig. 7) and the matrix will eventually be

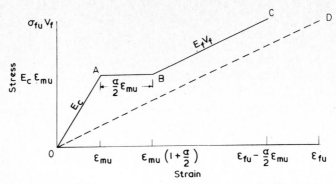

FIG. 6. Idealised stress–strain diagram for a brittle matrix composite (AVESTON et al., 1975, by permission of National Physical Laboratory, Crown Copyright reserved).

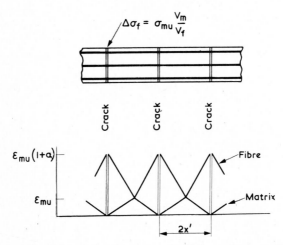

FIG. 7. Strain distribution after cracking of an aligned brittle matrix composite (AVESTON et al., 1975, by permission of National Physical Laboratory, Crown Copyright reserved).

broken down into a series of blocks of length between x' and $2x'$ with an average spacing of $1.364x'$.

We can calculate x' from a simple balance of the load $\sigma_{mu} V_m$ needed to break a unit area of the matrix and the load carried by N fibres of radius r across the same area after cracking. This load is transferred over a distance x' by the limiting maximum shear stress τ, i.e.,

$$N = V_f/\pi r^2 , \qquad 2\pi r N \tau x' = \sigma_{mu} V_m \qquad (4.3)$$

or

$$x' = \frac{V_m}{V_f} \frac{\sigma_{mu} r}{2\tau} . \qquad (4.4)$$

The stress distribution in the fibres and the matrix crack spacing $2x'$ will then be shown in Fig. 7.

The additional stress $(\Delta \sigma_f)$ on the fibres due to cracking of the matrix varies between $\sigma_{mu} V_m / V_f$ at the crack and zero at distance x' from the crack so that the average additional strain in the fibres, which is equal to the extension per unit length of composite at constant stress $E_c \varepsilon_{mu}$ is given by

$$\Delta \varepsilon_c = \tfrac{1}{2} \sigma_{mu} \frac{V_m}{V_f} \frac{1}{E_f},$$

i.e.,

$$\Delta \varepsilon_c = \frac{\varepsilon_{mu} E_m V_m}{2 E_f V_f} = \tfrac{1}{2} \alpha \varepsilon_{mu} \tag{4.5}$$

where

$$\alpha = E_m V_m / E_f V_f \tag{4.6}$$

and the crack width, w, bearing in mind that the matrix strain relaxes from ε_{mu} to $\tfrac{1}{2} \varepsilon_{mu}$, will be given by

$$w = 2x'(\tfrac{1}{2} \alpha \varepsilon_{mu} + \tfrac{1}{2} \varepsilon_{mu}) \quad \text{or} \quad w = \varepsilon_{mu}(1 + \alpha)x'. \tag{4.7}$$

At the completion of cracking the blocks of the matrix will all be less than the length, $2x'$, required to transfer their breaking load $\sigma_{mu} V_m$ and so further increase in load on the composite results in the fibres sliding relative to the matrix, and the tangent modulus becomes $E_f V_f$.

In this condition the load is supported entirely by the fibres and the ultimate strength, σ_{cu}, is given by

$$\sigma_{cu} = \sigma_{fu} V_f. \tag{4.8}$$

If the average crack spacing is used in the calculation instead of the minimum crack spacing, then the length of the multiple cracking region (AB in Fig. 6) will become 0.659α. For the case in which α is much greater than unity, the crack width w' at the end of the multiple cracking (B in Fig. 6) will be given by (4.9) for cracks at the average spacing:

$$w' \simeq \frac{0.9 \sigma_{mu}^2}{E_f \tau} \frac{A_f}{P_f} \left(\frac{V_m}{V_f}\right)^2. \tag{4.9}$$

Although the crack width at the end of multiple cracking in a given composite does not have a particular practical significance in itself, the smaller the crack width at this point, the smaller the crack widths will be thereafter at given strain or stress levels.

The relationship between the average crack width w' at the end of multiple cracking and fibre volume is shown in Fig. 8 from which it is apparent that, even using the most optimistic assumptions, large variations in crack width are likely to occur for small variations in fibre volume at less than 2% by volume of fibre and volumes above 5% are desirable if a uniformly and invisibly cracked composite is to be achieved. Hence the fabrication techniques must be adjusted to produce a uniformly dispersed fibre and preferably a high (5–10%) fibre volume if invisible cracking is required in the product.

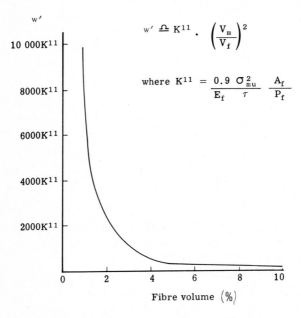

FIG. 8. Relation between average crack width w' at the end of multiple cracking and fibre volume V_f for the case in which α is much greater than unity.

4.2.2. *Stress–strain curves for real composites*

Real composites, with the exception of those containing continuous networks of polyolefin films, generally contain short fibres in two- or three-dimensional random orientation. Hence, the fibre volume fractions to be used in the theoretical treatment in Section 4.2.1 are less than the actual volume of fibre in the composites.

The efficiency factors of fibres in the post-cracking state are difficult to assess accurately but LAWS (1971) has provided some guide lines which are shown in Table 3. The efficiency factors allow for both fibre length and fibre orientation

TABLE 3.[a] Efficiency factors for post-cracking strength for restrained fibrous mat (LAWS, 1971, by permission of Institute of Physics).

Orientation	Efficiency factor	
	Continuous fibres	Short fibres
Aligned	1	$l/4l'_c$ $(l \leqslant 2l'_c)$ $1 - l'_c/l$ $(l \geqslant 2l'_c)$
Random 2-D	$\frac{3}{8}$	$\frac{9}{80}l/l'_c$ $(l \leqslant \frac{5}{3}l'_c)$ $\frac{3}{8}(1 - \frac{5}{6}l'_c/l)$ $(l \geqslant \frac{5}{3}l'_c)$
Random 3-D	$\frac{1}{5}$	$\frac{7}{100}l/l'_c$ $(l \leqslant \frac{10}{7}l'_c)$ $\frac{1}{5}(1 - \frac{5}{7}l'_c/l)$ $(l \geqslant \frac{10}{7}l'_c)$

[a] $l'_c = \frac{1}{2}l_c(2 - \tau_d/\tau_s)$ where τ_s is the static interfacial bond force and τ_d the sliding frictional bond force.

but their accuracy is limited by the lack of accurate experimental data for τ_s and τ_d which will in any case vary with time and ambient conditions.

By making allowance for some of these factors, typical stress–strain curves for real composites may be obtained and four such curves for different fibre types are shown in Fig. 9 in which V'_f is the effective volume of fibre in the direction of stress calculated from the total V_f using the appropriate efficiency factors.

The values for α shown in Fig. 9 have been calculated from (4.6), and, when substituted in (4.5), they give an indication of the possible range for real composites of the horizontal part ($\frac{1}{2}\alpha\varepsilon_{mu}$) of Fig. 6.

Fig. 9(b) shows that it is possible for glass reinforced cement to have a typical extension due to multiple cracking alone of about 9 times the matrix cracking strain whereas asbestos cement (Fig. 9(a)) can only extend by about 1 times the matrix cracking strain before the fibres take over completely. Likewise, a similar volume of polypropylene to that of asbestos could increase the strain due to multiple cracking alone by nearly 40 times the matrix cracking strain even under fairly rapid loading. This would probably be exceeded for extended loading periods because the modulus of polypropylene is time dependent.

The other features of the stress–strain curves are that for glass reinforced cement (Fig. 9(b)) the bond is good, allowing many cracks to form (cf. (4.4)), the matrix is fairly uniform but each matrix crack forms at a slightly higher stress and the fibres are able to take over completely, shortly before fibre failure or pull out occurs.

For asbestos-cement (Fig. 9(a)) the bond is very good resulting in many, very fine cracks with only short de-bonded regions and a much smaller increase in total strain due to cracking as predicted by (4.5). Because the slope of the post-cracking stress–strain curve is not greatly different from that of the matrix, the transition between the two can be relatively smooth. Thus, the change in

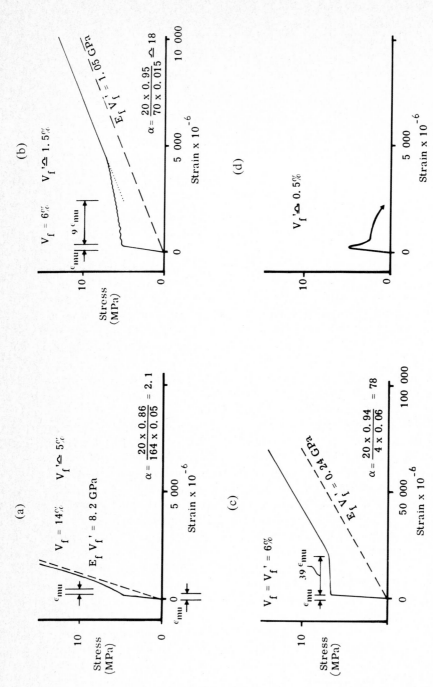

FIG. 9. Typical stress–strain curves for real composites. (a) Asbestos fibres in cement paste; (b) 2-D random glass fibres in cement paste; (c) Continuous aligned polypropylene fibres in cement paste; (d) Short, random, chopped steel or polypropylene fibres in concrete.

slope of the stress–strain curve is not always a good criterion by which the onset of multiple cracking may be judged. The lack of a horizontal portion in Fig. 9(a) is a serious deficiency in that it results in the material having limited capability to absorb shocks and accidental over strains. It is possible that the asbestos fibres have such a small diameter (0.02–20 μm) that matrix cracking is suppressed until successively higher strains are reached as suggested by AVES-TON et al. (1976) and final failure is by single fracture with some fibre pull-out.

The strain to failure for the polypropylene composite in Fig. 9(c) is an order of magnitude greater than for the other composites and hence the energy absorbing capability under overload conditions which depends on the area under the stress–strain curve is likely to be higher. However, the ultimate strains may never be reached in practice due to excessive deformation.

Fig. 9(d) is typical of concretes containing less than the critical volume of short chopped fibres, generally steel or polypropylene in random three-dimensional orientation. The fibres in these circumstances are generally sufficiently short to pull out, rather than break, when cracks occur in the matrix and the ultimate load on the composite is then controlled by the numbers of fibres across a crack, the length/diameter ratio of the fibres and the bond strength (HANNANT, 1978).

The stress in the fibre at pull out is often less than half the strength of the fibre and hence high bond strength, rather than high fibre strength may be the most important requirement for this type of composite.

4.3. Flexural strengthening

It is fortunate from the point of view of fabrication techniques that the critical fibre volume for flexural strengthening is much less than that for tensile strengthening. In fact, the mixing problems associated with chopped steel fibre and polypropylene fibre concrete are such that the tensile strength of the composite cannot usually be increased by fibre addition. However, quite useful strengthening of the composite can be achieved in flexure and, because many of the major applications of these materials are in flexural situations, an understanding of the principles of strengthening in flexure may be more important than an analysis of the direct stress situation.

4.3.1. Mechanism of flexural strengthening

The load which a beam of the fibre composite will carry in bending can be up to three times that which would be predicted using the ultimate tensile strength together with an analysis based on assumptions of linear elastic behaviour.

The main reason for this discrepancy is that the post-cracking stress–strain curve (cf. Fig. 6) on the tensile side of a fibre cement or fibre concrete beam is very different from that in compression and, as a result, conventional beam theory is inadequate.

The flexural strengthening mechanism is mainly due to this quasi-plastic behaviour of fibre composites in tension as a result of fibre pull-out or elastic extension of the fibres after matrix cracking, and the main principles have been described in detail by AVESTON (1974) and by HANNANT (1978).

The stress block for an elastic material in bending is shown in Fig. 10(a) and this is usually used to calculate the flexural strength ($\sigma_{f\ell}$) even although it is known to be grossly inaccurate for quasi-ductile fibre composites.

Fig. 10(b) shows a simplified stress block in bending for the type of tensile stress–strain curve OAB in Fig. 6. This is typical of a fibre concrete composite

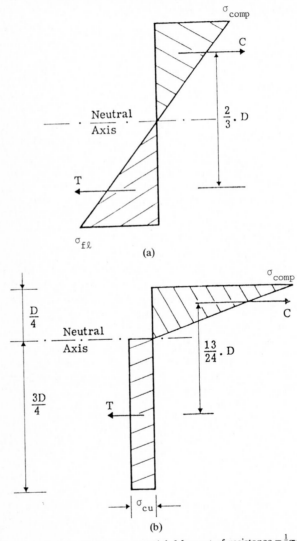

FIG. 10. Stress blocks in flexure. (a) Elastic material. Moment of resistance $= \frac{1}{6}\sigma_{f\ell}D^2$; (b) Elastic in compression. Plastic in tension. Moment of resistance $= \frac{13}{32}\sigma_{cu}D^2$ (HANNANT, 1975, by permission of Construction Press Ltd.).

after cracking, where the fibres are extending or are pulling out at constant load across a crack throughout the tensile section. The ultimate post-cracking tensile strength of the composite is σ_{cu} and σ_{comp} is the compressive stress on the outer face of the beam. Fig. 10(b) approximates to the stresses in steel–fibre concrete where the crack widths are small (<0.3 mm) compared with the fibre length and possibly to glass reinforced cement at early ages when the fibres are poorly bonded and extend before fracture or pull-out after fracture at roughly constant load.

A conservative estimate for the distance of the neutral axis from the compressive surface is $\frac{1}{4}D$ and using this assumption the moments of resistance of the two stress blocks can be compared:

$$\text{moment of resistance} = \tfrac{1}{6}\sigma_{f\ell}D^2 \quad \text{for Fig. 10(a)}, \tag{4.10}$$

$$\text{moment of resistance} = \tfrac{13}{32}\sigma_{cu}D^2 \quad \text{for Fig. 10(b)}. \tag{4.11}$$

In order that the two beams represented in Fig. 10 can carry the same load, their moments of resistance should be equal, i.e.,

$$\tfrac{1}{6}\sigma_{f\ell}D^2 = \tfrac{13}{32}\sigma_{cu}D^2 .$$

Therefore

$$\sigma_{f\ell} = 2.44\sigma_{cu} . \tag{4.12}$$

The limiting condition is when the neutral axis reaches the compressive surface of the beam while maintaining the maximum tensile strength (σ_{cu}) throughout the section. In this case

$$\tfrac{1}{2}\sigma_{cu}D^2 = \tfrac{1}{6}\sigma_{f\ell}D^2 , \quad \text{i.e.,} \quad \sigma_{f\ell} = 3\sigma_{cu} . \tag{4.13}$$

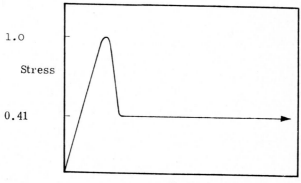

FIG. 11. Type of stress–strain curve in uniaxial tension which will cause no decrease in flexural load capacity after cracking (HANNANT, 1975, by permission of Construction Press Ltd.).

In practice this condition will rarely be achieved because compression failure will be initiated at the outer beam surface first.

This type of simplified analysis explains why the flexural strength for fibre cements and fibre concretes is often quoted to be between 2 and 3 times the tensile strength. It also follows that materials with fibre volumes of less than half the critical value for tensile strengthening as indicated by Fig. 11 can nevertheless at the same time contain the critical volume for flexural strengthening.

4.3.2. Effect of loss of ductility in tension on the flexural strength

The importance of the post-cracking tensile strain capacity in relation to the area of the tensile stress block in bending has already been demonstrated. A further result of this major factor is that changes in strain to failure in the composite can result in changes in the flexural strength even when the tensile strength remains constant. This is particularly relevant to glass reinforced

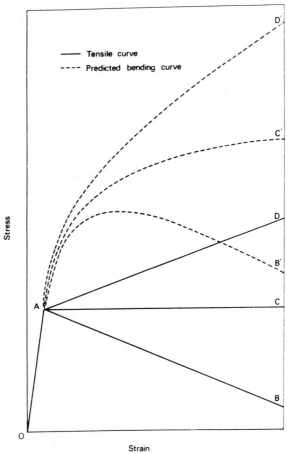

FIG. 12. Apparent bending (flexural strength) curves predicted for assumed direct tensile curves (LAWS and ALI, 1977, by permission of The Institution of Civil Engineers).

cement where the tensile strain capacity can reduce by an order of magnitude (1 to 0.1%) over a period of years of natural weathering or water curing. The movement of the neutral axis towards the compressive surface depends on a high post-cracking strain in tension and if this tensile strain decreases sufficiently, the composite will have a reduced moment of resistance (HANNANT, 1975).

The effect has been demonstrated by LAWS and ALI (1977) and can be seen in Fig. 12. For instance, OAC is a tensile stress–strain curve and OAC' is the associated bending curve. If the tensile strain reduces from C to A at constant stress, the bending strength will reduce from C' to A with an increasing rate of reduction as A is approached and the material becomes essentially elastic.

4.3.3. Flexural strengthening of short, random fibre concrete

For short fibre composites in which the fibres pull out rather than break, the minimum fibre volume, $V_{f(min)}$, required for flexural strengthening depends mainly on the ratio of the tensile cracking strength of the composite σ_t to the bond strength τ and on the inverse of the length/diameter ratio (l/d) of the fibres.

For instance, HANNANT (1978) has shown for three-dimensional fibre orientation that

$$V_{f(min)} \simeq 0.82 \frac{\sigma_t}{\tau} \frac{1}{(l/d)} . \tag{4.14}$$

For steel fibre concrete $\sigma_t/\tau \sim 1$ and $l/d \sim 100$ so that strengthening can occur at fibre volumes in excess of 0.8%. This is of considerable benefit for mass concrete production because normal mixing and compaction techniques are effective at these low volumes.

When pre-cast production is involved it is more common to improve fibre efficiency by arranging the fabrication technique to give two-dimensional fibre orientation (Section 5.2.3) and in this case the flexural strength is approximated by (4.15) (HANNANT, 1978)

$$\sigma_{f\ell} = 1.55 V_f \tau \frac{l}{d} . \tag{4.15}$$

5. Fabrication of fibre reinforced concrete

The main types of fibre which have been used in concrete, as opposed to cement mortars, have been chopped steel (Section 2.3.5) or chopped polypropylene fibres (Section 2.3.4) and this section will be mostly devoted to the technology involved in the production of concrete containing these two materials.

It has been shown in Section 4 that flexural reinforcement can be achieved for random steel fibre concrete with fibre volumes less than 2% and even although 2% by volume may be considered to be a relatively small volume when related to fibre composites in other fields it nevertheless presents considerable problems to the concrete production industry whether steel or polypropylene fibres are used.

Some technical problems in production are the uniform introduction of the fibres into the mixer and the achievement of uniform fibre dispersion within the concrete during mixing. The material must then be transported and must be made to flow into its final position in such a manner that the fibres have the required orientation in relation to the future stress field and a good surface finish must be achieved.

5.1. In-situ steel fibre concrete

Virtually all in-situ fibre concrete has been reinforced with steel fibres and it has been found that normal concrete mixes with a high proportion of aggregate greater than 10 mm maximum size will generally not accept the required fibre volume uniformly and will be difficult to compact. This practical experience has led to the development of mix designs which will easily accept a sufficient quantity of fibres of an appropriate type, which will give acceptable compaction characteristics and which can also provide useful properties in the hardened state.

The parameter which is used to describe the ease with which concrete can be made to flow under vibration is loosely defined as the 'workability' and this property can be assessed for steel fibre concrete by measuring the time in seconds (known as V–B time) required to re-mould a frustum of a cone of concrete into a cylinder when subjected to a standard vibration procedure as described in British Standard 1881 (1970).

5.1.1. Effect of fibre and aggregate parameters on workability of steel fibre concrete

5.1.1.1. Fibre length and diameter. A collection of long thin fibres of length/diameter (l/d) greater than 100 will, if shaken together, tend to interlock in some fashion to form a mat, or a type of bird's nest from which it is very difficult to dislodge them by vibration alone. Short stubby fibres on the other hand of l/d less than 50 are not able to interlock and can easily be dispersed by vibration.

Similar effects are observed when fibres are dispersed in mortar or concrete and the ease with which the fibres can move relative to each other under vibration is shown in Fig. 13 for mortars, with a particle size less than 5 mm.

It can be seen from Fig. 13 that the l/d ratio has a crucial influence on the volume of fibres which can be included in the mix with relatively easy compaction (say V–B < 20 seconds).

The critical fibre volume for strengthening in direct tension may be about

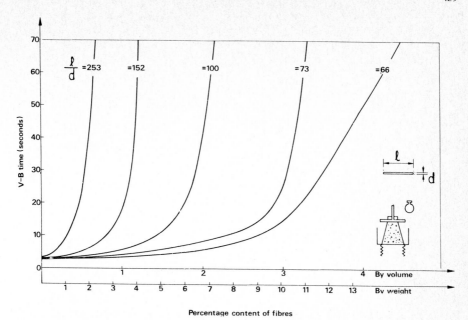

FIG. 13. Effect of fibre aspect ratio on V–B time of fibre reinforced mortar (EDGINGTON et al., 1974, by permission of The Controller, HMSO, Crown Copyright reserved).

2% at an l/d ratio of 100 and Fig. 13 indicates that practically this may only just be achieved with mortars, let alone with concretes. However, the fibre volume required for strengthening in flexure (cf. (4.14)) can be achieved much more easily with acceptable compaction characteristics.

5.1.1.2. Aggregate size and volume. The problem is more complicated when fibres are introduced into a concrete rather than a mortar matrix because they are separated not by a fine grained material which can move easily between them, but by particles which will often be of a larger size than the average fibre spacing if the fibres were uniformly distributed. This leads to bunching and greater interaction of fibres between the large aggregate particles and the effect becomes more pronounced as the volume and maximum size of the particles increase.

The effect of a range of aggregate sizes and volumes on the compaction times of composites made with wires of $l/d = 100$ is shown in Fig. 14.

Fig. 14 indicates that for a V–B time of 20 seconds the 10 mm concrete will only accept about 50% of the fibre volume compared with mortar and the 20 mm concrete will carry less than the 10 mm concrete. These particular results are due to a combination of the effects of aggregate size and aggregate volume because the motar fraction is lowest for the 20 mm mix.

Thus it can be seen that the fibre sizes and volumes which can be calculated from the basic theory in Section 4 for adequate final properties in the hardened state (i.e., high l/d ratios and fibre volumes greater than 2%) may have to be

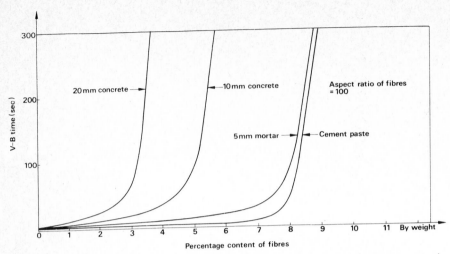

FIG. 14. Compaction time against fibre content for matrices with different maximum aggregate size (EDGINGTON et al., 1974, by permission of The Controller, HMSO, Crown Copyright reserved).

carefully balanced against the mix design for adequate compaction properties which require a low l/d ratio and fibre volume less than 2%.

5.1.2. Mix design for workability of streel fibre concrete
The concrete matrix generally contains a high proportion of sand (more than 50% by weight of the aggregate) together with a maximum aggregate size of 10 mm. However, the fines content may also be effectively increased by the use of pulverised fuel ash or air entraining admixtures. Water reducing admixtures have also been used but trial mixes are essential to determine the handling characteristics and strength properties.

Eq. (5.1) may be used to enable an approximate estimate to be made of the maximum weight of a particular fibre which can be included in a mix containing aggregates of normal density whilst maintaining adequate workability for normal in-situ compaction techniques (HANNANT, 1978):

$$W_f < \frac{600(1 - A_g)}{l/d} \qquad (5.1)$$

where

W_f = weight of fibres, as a percentage of
the concrete matrix, which can be
compacted with normal site techniques,

$A_g = \dfrac{\text{weight of aggregate greater than 5 mm}}{\text{total weight of concrete}}$,

l/d = length/diameter of the fibre.

Very approximately,

$V_f = 0.28 W_f$ for normal density concrete.

5.1.3. *Mixing methods for steel fibre concrete*

A variety of methods is available for introducing the steel fibres into the concrete mixer either to the dry constituents or to the wet mix. These techniques range from charging the aggregate conveyor with fibres, sieving fibres directly into the mixer drum, sieving the fibres and blowing them into the drum or alternating sieves for laboratory use. Also, the development of fibres, glued with water soluble adhesive into units similar to staples, enables the fibres to be dispersed into the mixer as a normal aggregate and they then separate in the mixing process.

The critical factor in whatever techniques is used for single fibre addition is that the fibres should reach the mixer individually and be immediately removed from the point of entry by the mixing action. Mixer characteristics can affect the uniformity of fibre distribution but an excess of fibres can collect into balls in the mixer regardless of mixer type. Also, it is essential that no fibre balls are introduced into the mixer initially as these are unlikely to be broken up by the mixing action.

The fibres are normally delivered to the site in boxes or drums and equipment such as protective gloves, goggles, forks or rods may be useful to assist in transferring the fibres into the dispersing plant.

5.1.4. *Compaction techniques for steel fibre concrete*

Steel fibre concrete may be compacted by poker vibrators, by shutter or table vibration, or by surface vibrating beams as in floor slabs or slip form pavers. However, the type and direction of vibration can have a critical effect on the orientation of the wires relative to the future loading direction and hence on the properties in the hardened state (Section 5.2.3.1). Therefore, prototype

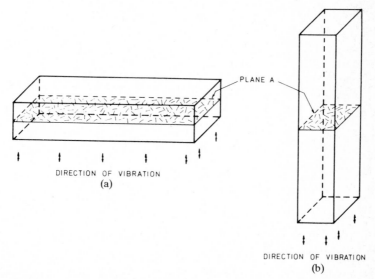

FIG. 15. Effect of table vibration on fibre alignment (EDGINGTON and HANNANT, 1972, by permission of RILEM).

products and laboratory or trial applications should be carried out using the same vibration techniques to be used in the full scale operation otherwise misleading strength performance could result.

Although the fibres may be orientated randomly in three dimensions in the mixer, this is seldom the case after vibration and compaction have been completed and the hardened concrete can then exhibit anisotropic behaviour with strengths up to 50 per cent higher in one direction than another.

Fibre alignment may be achieved accidentally or intentionally in a variety of ways. Table or surface vibration tends to cause the fibres to align in planes at

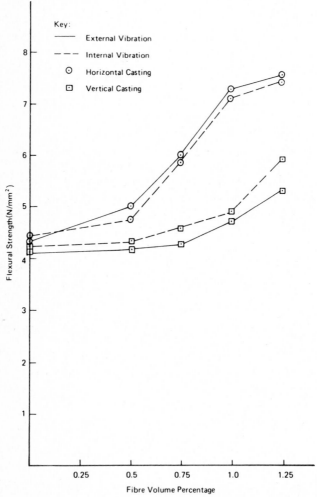

FIG. 16. Influence of method of casting and type of vibration on flexural strength (SWAMY and STAVRIDES, 1975, by permission of Construction Press Ltd.).

right angles to the direction of vibration or gravity (see Fig. 15) and hence horizontal casting is preferable for beams or slabs.

Internal vibration causes a smaller amount of fibre alignment than table vibration and the effect of these two compaction techniques on the flexural strength of $100 \times 100 \times 500$ mm beams has been measured, the results of SWAMY and STAVRIDES (1975) being shown in Fig. 16.

Component thickness is also a factor in fibre alignment and, as the emphasis is generally on thin sections for fibre concrete, this effect can be used to improve fibre efficiency. For instance, sections less than 100 mm thick containing 50 mm long fibres will inevitably result in the fibres tending towards two-dimensional rather than three-dimensional alignment.

From the practical point of view of in-situ concreting, the anisotropic properties could be put to good use by arranging the compaction procedure so that the fibres are aligned in the most beneficial direction relative to the stress field. On the other hand, if the effects of vibration on fibre alignment are not fully appreciated, the strengths of steel fibre reinforced concrete in the field could be much lower than predictions based on laboratory tests using different compaction procedures.

5.1.5. Sprayed steel fibre concrete

The problems associated with mixing and compaction techniques may be partially avoided by spraying the concrete through a gun by means of compressed air. Unreinforced concrete is commonly sprayed using two processes, the dry process and the wet process and both may also be used for fibre concrete. The traditional sprayed concrete machines have been modified so that the fibres can be mixed with the wet or dry constituents before being sprayed with or without additional water being added at the gun. Alternatively, the fibres may be separately projected into the wet or dry concrete as it is sprayed onto the work or they may be blown into the dry materials just before they emerge from the nozzle.

An example of the type of equipment which is available is shown in Fig. 17, this technique being developed by the BESAB Company in Sweden. The principle involves pneumatic conveying of fibres from a rotary 'fibre feeder' to a nozzle via a 75 mm diameter flexible hose. The nozzle has an outlet for surplus air which reduces the speed of the mass and thus reduces the fibre rebound. Fibres with l/d ratios up to 125 have been used.

Sprayed processes are of particular economic advantage in the stabilisation and strengthening or rock slopes after blasting or for the lining of mine shafts and tunnels to provide structural support and a reduction in oxygen and moisture attack on the rock.

Traditional techniques utilised steel mesh pinned to the uneven rock surfaces before the spraying of concrete and, in addition to being a laborious and costly operation, the mesh often had to span large concavities in the uneven surfaces which had to be filled with concrete thus increasing the cost of materials.

The technical and economic advantages therefore are the following:

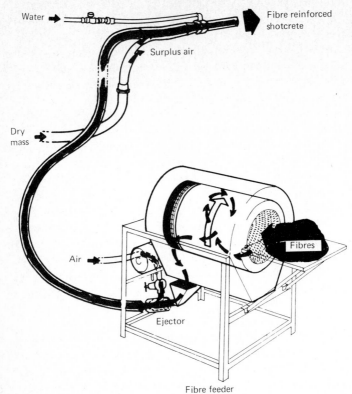

FIG. 17. Fibre spraying equipment developed by BESAB AB (EDGINGTON, 1977, by permission of
The Institution of Civil Engineers).

(a) To provide a thin layer of reinforced concrete on an uneven surface saving part of the labour and material costs associated with conventional methods.

(b) Due to the thinness of the layer and the compaction by impact, the fibres are preferentially aligned in two-dimensional random distribution giving improved properties in the hardened state.

(c) When no mixing is required, wires with l/d ratios in excess of 100 can be utilised thus increasing the efficiency of the expensive fibres.

5.1.6. Examples of in-situ construction of steel fibre concrete
There have been many trial applications but only a few are described to indicate the type of techniques actually used on site.

5.1.6.1. Hydraulic applications.
Steel fibre concrete performs well in situations where cavitation, erosion and abrasion are common in hydraulic structures such as sluiceways and spillways associated with dams. HOFF (1975) has described the construction using steel fibre concrete of spillway deflectors for the Lower Monumental Lock and Dam in the State of Washington, U.S.A.

The fibre concrete was batched at a central plant 45 km from the site and mixed at the plant and during transit in mixer trucks. The composite was discharged from the trucks into the hopper of a double acting positive displacement pump and pumped 37 m before being discharged into the formwork. Some balling of the fibres occurred in the concrete but these were screened out before entering the pump. Mix proportions per m³ were: cement—384 kg; sand—842 kg; 10 mm coarse aggregate—366 kg; water—155 kg; steel fibres—100 kg of dimensions 0.41 mm diameter by 19 mm long; water reducing admixture—1.42 kg; air entraining admixture—0.26 kg; pumping aid—0.04 kg. It can be seen from the mix design that even with fibres of a very low l/d ratio (46) and low volume (about 1.3%) the mix design required several additives to ensure successful placing.

The 28 day flexural strengths were less than 6 MN/m² indicating that the fibres had too low an l/d ratio to contribute greatly to the strength of the composite. Another major application of this type has been the repair of the chutes and stilling basins of the outlet works of the Tarbela dam in Pakistan.

5.1.6.2. Pavement applications. A large highway overlay project (4.8 km long) has been described by LANKARD (1975).

It includes 33 sections of steel fibre concrete, each 122×6.7 m, laid to thicknesses of 50 and 76 mm. In addition, for comparison, there are four sections of continuously reinforced concrete pavement 76 and 100 mm thick and five sections of plain concrete and mesh reinforced concrete at thicknesses of 100 and 125 mm. The experiment also makes provision for mixes in which the binder is shrinkage compensated cement or cement/fly-ash.

Two types of fibre ($0.25 \times 0.56 \times 25.4$ mm and 0.64 mm diameter $\times$ 63.5 mm) were used in amounts giving approximately 0.45, 0.75 and 1.2 per cent by volume. Two mixes were specified (356 and 445 kg/m³ of cement) the maximum aggregate size being 10 mm and the sand content 50%.

A slip-form paver 6.7 m wide was used with minor modification to the vibrators but in the 50 mm thickness small fibre balls caused problems by dragging and catching in front of the screed. Some fibre orientation is inevitable in this process which is basically one of extrusion assisted by vibration and hence anisotropic properties are to be expected in the slab. Also, with thicknesses of less than 76 mm the fibres would be nearer to two-dimensional orientation than to three-dimensional with a consequent improvement in fibre efficiency.

5.1.6.3. Premixed steel fibre concrete. The first country to make steel fibre concrete available through the established premixed concrete industry in a similar manner to conventional concrete has been Australia (MARSDEN, 1980).

This industry has placed emphasis on the need to have easy batching and mixing and for this purpose it was found that the ratio (fibre length/fibre cross-sectional area)$^{1.75}$ must be less than 1200 where the dimensions are in mm. This limitation has resulted in the use of short fibres which are relatively inefficient from the point of view of reinforcement but increased bond has been achieved by the use of enlarged ends in fibres chopped from sheet.

A typical premix with an 80 mm slump containing about 1 per cent by volume of 18 mm long fibres is as follows, the quantities being expressed as kg/m^3: cement—380; pulverised fuel ash—100; 20 mm aggregate—600; 10 mm aggregate—200; sand (8% moisture)—800; water plus water reducing admixture—150. This type of mix is generally used as an overlay to industrial flooring or in new flooring construction for demanding applications at thicknesses up to 230 mm.

5.2. Fabrication techniques for precast applications of fibre concrete

In contrast to in-situ concrete applications, the developments in precast concrete have included chopped polypropylene fibres and continuous glass fibres in addition to chopped steel fibres.

5.2.1. Fabrication of concrete products containing chopped polypropylene twine

In the late 1960's the Shell Chemical Co. carried out trials on the use of chopped polypropylene twine in concrete and gave the material the name Caricrete. The principles behind this early work have been described by ZONSVELD (1970) who attributed the bonding mechanism to the opening up of the fibres in the mixer thus allowing the cement matrix to penetrate between the mesh and form a continuous phase in which the fibre is held firmly by mechanical action. However, the volume of fibre which could be easily included in the concrete mix and compacted in the mould was generally limited to less than 1% due to a combination of the mixing action, the concrete proportions and the fibre length. Because of this relatively low fibre volume the improvements in mechanical properties in the hardened composite were mostly limited to improvements in the impact strength.

5.2.1.1. Mixing techniques using polypropylene twine.

A variety of mixers has been used in practice, some requiring an adjustment to the existing equipment, some none at all. Additional equipment has been installed in some plants to chop and/or to facilitate proportioning the fibres. The type of short fibre chosen is mostly based on film, e.g., a twine of 1400 m/kg, chopped to 50 mm staple length. As the fibres cannot be wetted, the mixing need only achieve a homogeneous dispersion and therefore they are often added shortly before the end of mixing the normal ingredients. A long residence time in any mixer leads to undesirable shredding of the fibres and should be avoided.

Tumbler mixers disperse the fibres without complications. This also applies to ready mix lorries which either carry a pre-weighed bag of fibres, or receive the fibres on site from stock held there. On arrival on site the fibres are dropped in the drum which is kept rotating for two or three minutes before placing.

Pan mixers, slow or high speed, have sometimes needed adjustment to cope with fibres which have, of course, different dimensions from the normal aggregates. The scraper blades may need to be set to a different angle if the fibres are caught and collected on edges or in poorly streamlined corners. Also, the discharge opening may require widening if the fibres have tended to bridge the gate and clog it. Many pan mixers, however, have been found in practice to accept mixes with chopped polypropylene twine without alterations. In a fast pan mixer in one plant the mixing time for normal concrete was about one minute, and it was found that 0.5 per cent by volume of fibres could be added at the beginning without fear of shredding in this short mixing time.

If the continuous twine or filament arrives on spools at the precast factory it is cut to a staple fibre by specially developed equipment. The cutter is placed in line with other batching machinery, and can also combine its task with accurate proportioning.

The mix design of polypropylene concrete will take account of the denier or 'runnage' and the staple length of the fibre that will best suit the aggregate, the workability required and the equipment to be used in making the product. For instance, a thin-walled product would not accommodate the fairly stiff fibres of 700 m/kg because some would lie across the wall and would tend to break out on demoulding. The more flexible twine of 1400 m/kg would therefore be chosen, and would be cut to a shorter staple length. A heavy precast pile on the other hand would accept coarse fibres which would give a higher workability for the same fibre content.

The way of handling fresh composites in the plant is dependent on the equipment available and on the routine which was followed prior to the introduction of fibrous mixes. Changes implemented in practice have included a wider discharge opening for the mixer drum, a conveyor belt in place of wheelbarrows and spades, different vibrators and moulds and other similar adaptations of existing plant.

Polypropylene fibres betray their presence usually on the finished surface, in most cases in an inconspicuous way, and not at all in cladding panels with exposed aggregates. Trowelling of panels and slabs poses problems for in-experienced factory hands, but a skilful operative can avoid the showing of fibres by dexterous use of his float.

5.2.1.2. Workability characteristics for concrete containing polypropylene twine. The workability of a mix containing chopped twine is difficult to measure using standard techniques because the slump of a mix with low fibre content can be zero although the mix flows satisfactorily when kept moving and responds well to vibration with some pressure being required at higher fibre volumes.

The effect of varying fibre dimensions and volume is similar in principle to that of steel fibres (Section 5.1.1) but has not been quantified for polypropylene fibres. However, it can be seen intuitively from a simple calculation that some variability in compaction characteristics can be achieved for a given fibre volume and fibre length in the same mix by varying the diameter of the twine.

For instance, a volume of 0.5% of twine of length 50 mm will result in either 63 or 126 fibres being present in a cube of 100 mm side depending whether twine of 700 or 1400 m/kg is used. It is apparent that the sample containing 126 fibres would have the lower workability.

Fortunately, at these low volumes, polypropylene concrete responds well to conventional vibrating tables, and pokers and presses because the fibres, notwithstanding their low specific gravity do not easily segregate from the mix.

5.2.1.3. Examples of precast fabrication techniques using polypropylene twine.

A very successful application of precasting techniques has been the development in the United Kingdom by Wests Piling and Construction Co. Ltd. of the production of shell pile segments using polypropylene twine as the only reinforcement of the shell. A string of these piles is driven on a steel mandrel, the shells formerly being reinforced with steel mesh. However, since 1969, after extensive impact testing under simulated field conditions (FAIR-WEATHER, 1972) more than half a million shells have been made annually without the steel reinforcement.

The sections of 915 mm length and diameters between 280 and 533 mm are manufactured in a highly automated plant and moulded under pressure and vibration at a high rate of production.

The batching plant for the concrete is standard but the 700 m/kg twine is chopped to 40 mm long staple with a guillotine knife after which the fibre drops on to a running conveyor belt which takes the other components of the mix to the pan mixer. The production rate of the cutter is accurately known so that a time switch starting and stopping its operation adds the exact amount of fibre per batch for a content of 0.44 per cent by volume in the piling shells.

After mixing the fibre composite is transferred to conveyors via the bottom opening mixer drum and is discharged into the shell moulds where simultaneous vibration and pressure are applied. The shells are then extruded from the moulds in the fresh state and transported on pallets to the curing area, the fibres assisting in holding the matrix together during handling. The fibres remain essentially in a random orientation in the completed pile and thus help to resist impact damage which may occur in any direction during handling and driving.

Another example of the use of chopped twine in concrete is the production of buoyancy units to carry jetties and walkways in harbours and marinas. The manufacturing technique developed by Walcon Ltd. in the United Kingdom comprises placing a block of expanded polystyrene with the help of spacers accurately in the centre of a mould and pouring the concrete mixed with polypropylene fibres in the 18 mm wide space between mould and shutter while external vibrators take care of the compaction. Various standard sizes of pontoon are made, mostly 0.9 m deep and with top surfaces between 1 and 2 m by 1.5 m.

Due to the small amount of twine used to keep the costs low and to ensure adequate workability in a confined space, the major benefit to the hardened composite is in impact strength together with the resistance to corrosive attack even after damage has occurred. The tensile stress–strain curves are of the type shown in Fig. 5(a) whereas a typical load–deflection curve for the higher volume composites in bending is shown in Fig. 18.

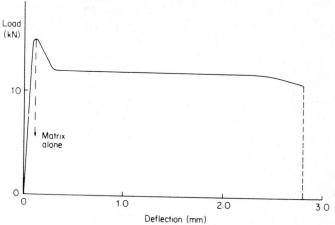

FIG. 18. Load–deflection curve for $100 \times 100 \times 500$ mm beam containing chopped polypropylene twine (1.2 per cent by volume, 700 m/kg, 75 mm long) (HANNANT, 1978, by permission of John Wiley Ltd.).

5.2.2. Fabrication of concrete products containing chopped polypropylene monofilaments

Polypropylene monofilaments do not bond well to cement but nevertheless they have been used in concrete in quantities of about 0.1 to 0.2 per cent by volume to alter the rheological properties of the material rather than to alter the hardened properties. John Laing Research and Development Ltd. in the United Kingdom has developed a fabrication technique in which highly air entrained concretes (up to 40 per cent air by volume) are stabilised by the addition of monofilaments. The main aim was to produce a range of concretes suitable for precast applications with improved thermal properties and with a decorative sculptured finish without the need for moulds and the new material was called Faircrete, which is a shortened version of fibre air concrete.

Variation of the amount of air entrainment or of the type of aggregate used allows a choice of ultimate densities from 700 up to about 2000 kg/m^3. The polypropylene monofilaments have diameters ranging from 0.1 to 0.2 mm and are cut to lengths of 10 to 20 mm by the use of a specially designed rotary high speed cutter. The action of the fibres in the mix has been linkened to that of a three-dimensional sieve, stopping the air passing up through the sieve and holding the aggregate so that it cannot pass down. The resulting properties of the mix, particularly when assisted with very light vibration, are easy flow out of hopper outlets, into restricted areas, and against mould faces. The thixotropic properties enable the concrete after placing to be formed into various shapes and patterns that would not be possible with ordinary concretes. The imprint does not slump back and remains exactly as formed on the hardened concrete.

5.2.3. Fabrication of precast concrete products containing chopped steel fibres

The technical factors which control fabrication in pre-cast applications are basically the same as for in-situ steel fibre concrete (Section 5.1). However, greater control can be obtained of the compaction techniques required to maximise the fibre efficiency in the hardened composite by aligning the fibres in the appropriate direction relative to the applied stress.

5.2.3.1. Magnetic orientation. SKARENDAHL (1980) has described a process

developed by the Institutet för Innovationsteknik in which the fibres can be orientated in the fresh concrete by passing the mould through a magnetic field under simultaneous vibration. This process is used in commercial production by a Scandinavian Company (Precon) to produce units of 20 and 25 mm in thickness. The mould must be of non-magnetic material and when flat sheets are produced the fibre concrete is spread on the mould which may be up to 2 m wide; it is then passed through a water cooled direct current coil with 10–15 kW power requirement. One passage through the coil is normally adequate to obtain sufficient fibre orientation.

5.2.3.2. Thin sections. Beneficial fibre alignment may also be obtained using

conventional casting techniques provided that thin sections are used. SKAREN-DAHL (1980) has described a process in which 50 000 m/annum of edge beams are produced consisting of a 7 mm thick skin of steel fibre concrete surrounding a cellular polymer material. These beams are used as permanent edge formwork for foundation slabs.

5.2.3.3. Spraying techniques. Sprayed steel fibre concrete has been discussed in

relation to in-situ concrete (Section 5.1.5) but specialised techniques have also been developed for precast products (SKARENDAHL, 1980). A process similar to that for sprayed glass reinforced cement can be used in which a continuous wire is deformed, cut and sprayed simultaneously with the matrix from a nozzle. The machine is known as Scanovator and it has the advantage that the fibre length can be up to 250 mm for wire diameters of 0.35 to 0.50 mm and thus very high l/d ratios can be achieved (up to 600) with much greater efficiency in the use of the wire strength in thin walled products. Pipes of 3 m length have been produced with wall thicknesses of 10 to 15 mm and fibre aspect ratios of 400.

5.2.3.4. Other applications. Precast concrete manhole covers and frames have

been produced in a fine concrete mix containing about 2 per cent by volume of wire fibres (HOLLINGTON, 1973). Hydraulic pressing was used at a pressure of 7 to 8 MPa and the elements could be de-moulded immediately. Flexural strengths of about 20 MPa were achieved and the products are an economic replacement for cast iron and can also comply with existing standards for cast iron products.

SZABO (1975) has described the manufacture of pipes up to 1.5 m diameter by die rolling in an Italian Siome machine. The pipes contained about 0.75 per cent by volume of 40 mm long by 0.5 mm diameter fibres and these pass from the fibre dispenser onto a belt conveyor carrying the concrete to the pipe moulds.

5.2.3.5. Effects on the properties of the product of improved fabrication techniques. The benefits of aligning the fibres in the appropriate direction and concentrating them in the most appropriate position have been demonstrated by BERGSTROM (1975). Fig. 19 shows that the bending properties of the composite with only 1.5 per cent by volume of wire can be considerably improved both in strength and strain capacity by suitable magnetic vibration alignment techniques.

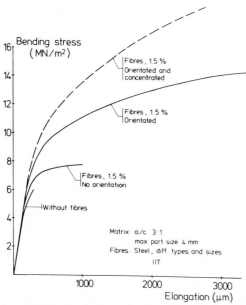

FIG. 19. Effect of fibre orientation on the flexural strength of steel fibre reinforced concrete (BERGSTROM, 1975, by permission of Construction Press Ltd.).

The improvements in bending strength which can be achieved theoretically by increasing the fibre l/d ratio are shown in Fig. 20 for l/d ratios between 50 and 600 for fibres in a two-dimensional orientation (see (4.15)). In Fig. 20 the bond strength is assumed to be 3 MPa and the bending strength of the matrix alone is 4 MPa.

The composite strengths at the highest l/d ratios and fibre volumes may not be achievable in practice because the ultimate tensile strength of the fibres may be exceeded or compression failure of the concrete may occur first but

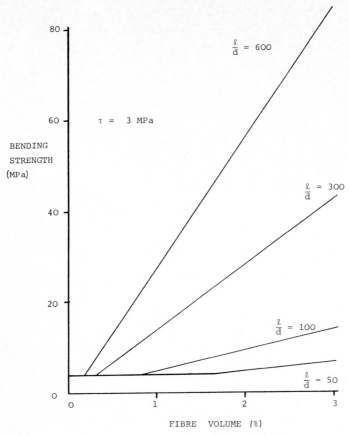

FIG. 20. Theoretical effect of fibre length/diameter ratio (*l/d*) on the bending strength of steel fibre concrete with a two-dimensional fibre orientation.

nevertheless the economic incentives towards achieving such composites are apparent.

5.2.3.6. Castable refractory concretes. One of the most successful fields of application of precast steel fibre concretes has been that of refractories. Refractory concretes are normally made from high alumina cements and heat resistant aggregates. Crushed fire brick is often used for temperatures up to 1400°C but at temperatures between 1400 and 1800°C high quality fused alumina aggregates are necessary. The material retains sufficient strength at these high temperatures by the development of a ceramic bond rather than a hydraulic bond.

In the refractory industry an important requirement is increased resistance to thermal shock in terms of service life. Stainless steels produced by drawing and by the melt extract process have been used in volumes up to about 2% and reliable performance has been obtained at temperatures up to 1600°C

(EDGINGTON, 1977). Three examples of the improvements in performance which can be achieved in comparison with existing products have been described by EDGINGTON (1977) as follows.

STIRRING PADDLES. The service life of the T-shaped paddles, used for stirring 1560°C molten iron, has been extended from an average of 50 to 140 heats following their manufacture from a fibre reinforced refractory castable.

CRUCIBLE LIDS. The replacement of one piece steel lids for aluminium melting furnaces operating at 1800°C by precast fibre reinforced segmental lids has increased service life expectancy from 9 months to 2 years. Elimination of maintenance costs alone have brought about considerable savings.

SUPPORT ARCHES—CARBON RING FURNACE. Precast arches manufactured from a steel fibre castable mix have been used to replace brick arches in a furnace roof operating at temperatures up to 1430°C. Whereas the conventional brick arches would last typically only one year in service and require frequent maintenance, the fibre reinforced refractory arches have been maintenance free for more than $2\frac{1}{2}$ years.

5.2.4. Fabrication of precast concrete products containing glass fibre
There have been relatively few attempts to include chopped glass fibres in concrete, as opposed to cement mortar, because the glass is a difficult material

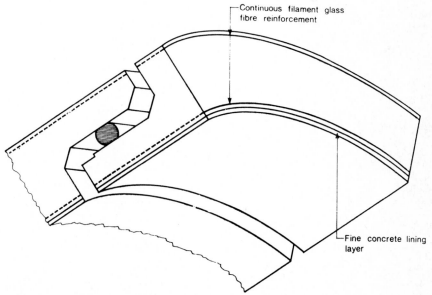

Continuous filament glass
fibre reinforcement

Fine concrete lining
layer

FIG. 21. Section through glass reinforced concrete pipe (Slimline produced by ARC Concrete Ltd., U.K., FARAHAR, 1978, by permission of Precast Concrete).

to handle in fresh concrete, resulting in serious reductions in workability at small percentage additions. Also, Farahar (1978) has shown that the effects on flexural strength are small when random fibre orientation is used. However, by suitable concentration and alignment of fibres the flexural strength can be considerably improved and a major application using this philosophy for precast pipes has been developed by ARC Concrete in the United Kingdom. Continuous glass fibres are used in a relatively low volume (<1%) but the fibres are concentrated at the inner and outer surfaces of the pipes as shown in Fig. 21 to promote maximum efficiency. The main proportion of the pipe consists of unreinforced concrete but, due to the local concentration of glass fibres, the bell end, which is usually a feature of concrete pipes, has been removed and an in-wall joint is possible with consequent advantages when handling and placing. This development of fabrication techniques to enable the production of a price competitive product with improved properties has required heavy investment, but such investment in production machinery is essential when the success of the product is so dependent on the success of the fabrication techniques.

5.2.5. Fabrication of products containing natural fibres

The use of natural vegetable fibres in concrete suffers from the high water absorption of the fibres and reduced workability of the composite in addition to all the other problems which have been described for short man-made fibres. In order to overcome the workability problem, additional water has to added which reduces the strength of the product so that the main benefits result from increased toughness under impact conditions rather than from improvements in tensile or flexural strength. Many of the fibres which have been used in mortar and in concrete have been described by Cook (1980) but the applications have been mostly at trial stage rather than at a commercial level.

6. Fabrication of sheet materials from cements and mortars

6.1. Asbestos-cement

Since 1900 the most important example of a fibre cement composite has been asbestos-cement. The proportion by weight of asbestos fibre is normally between 9 to 12 per cent for flat or corrugated sheet, 11 to 14 per cent for pressure pipes and 20 to 30 per cent for fire resistant boards, and the binder is normally a Portland cement. Fillers such as finely ground silica at about 40 per cent by weight may also be included in autoclaved processes where the temperature may reach 180°C. According to Klos (1975) important fibre parameters to be checked before inclusion in cement are fibre-length distribution, dry and wet density, dust content, specific surface and filtrability, and there are various standard tests to assess these parameters (Asbestos Products Association, 1962).

6.1.1. Production technology

Before the asbestos fibres can be combined with cement, heavy pre-treatment is required to break up the blocks of fibres into thin fibre units of an effective diameter of 1 micron or less. Edge runners or hammer mills are used and because of their crystalline structure, the fibres tend to split into ever thinner fibres or cohesive bundles of parallel fibres without significant shortening of the fibre length. The fibres can then be easily dispersed in a water and cement suspension by continuous dilution and mechanical stirring. It is found that asbestos has a curious affinity for Portland cement which settles on the surface of the fibres and tends to remain there even under high dilution or water extraction. This enables the water content to be reduced from about 90 to 20 per cent without segregation of the cement and leads to a very well distributed and bonded fibre composite (KRENCHEL, 1964).

6.1.1.1. Hatschek process.

The most widely used method of manufacture of asbestos-cement was developed from paper making principles in about 1900 and is known as the Hatschek process. A slurry, or suspension of asbestos-fibre and cement in water at about 6 per cent by weight of solids is continuously agitated and allowed to filter out on a fine screen cylinder. The filtration rate is critical and coarser cement than normal (typically $280 \text{ m}^2/\text{kg}$—compared with the normal value of $320 \text{ m}^2/\text{kg}$) is used to minimise filtration losses. Also, although chrysotile fibres form the bulk of the fibre, most formulations have included a smaller proportion of the amphibole fibre wherever possible. This is because of the de-watering characteristics of crocidolite and amosite. Mixtures of cement and chrysotile drain very slowly and production rates are consequently slow. By using a blend of chrysotile and amphibole fibres (usually between 20 and 40 per cent) considerable improvements in the drainage rate have been achieved, plus the additional bonus of enhanced reinforcement.

Other types of fibre such as cellulose derived from wood pulp or newsprint have also been added to the slurry to produce different effects in the wet or hardened sheet.

COLE (1979) has described the principles behind the operation of the Hatschek machine and the following description is taken from COLE's paper.

In very much simplified terms, the Hatschek machine operates as follows.

Referring to Fig. 22, a dilute slurry pours into the vat. This slurry drains through a porous sieve cylinder depositing the solid contents as a layer on the surface of the sieve. The water passes through the sieve surface and into the cylinder, and then pours out of the open ends of the cylinder to the backwater return circuit. (The ends of the sieve cylinder are sealed against the side walls of the vat.) The sieve cylinder rotates and the layer rises out of the vat.

A continuous felt runs in a loop from the sieve cylinder to an accumulation roll. Surface tension forces the layer to transfer from the top of the sieve cylinder to the underside of the felt. The movement of the felt transfers the layer from the sieve to the accumulation roll. On the way, it is vacuum de-watered. (Note that it is now on the top surface of the felt.) At the

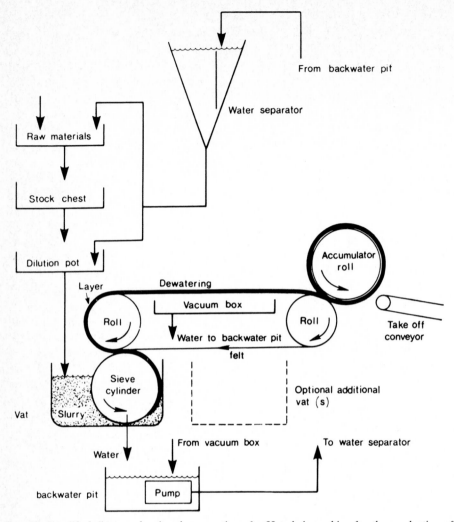

FIG. 22. Simplified diagram showing the operation of a Hatschek machine for the production of asbestos-cement (COLE, 1979, by permission of TAC Ltd.).

accumulation roll, surface tension again transfers the layer, this time to the accumulation roll, where it is laminated up to the desired product thickness and is cut off when this thickness has been achieved.

The length of the sheet is the periphery of the accumulation roll (or a sub-division of it). As the tail of a complete sheet leaves the accumulation roll, accumulation of the next sheet begins. This process is repetitively continuous. After leaving the accumulating roll, the sheet is wet trimmed to size, and the waste trim returned to the process. The sheets are then stacked and matured.

Commercial machines have from two to five vats, all placing a layer on the underside of the continuous felt. They are, of course, much more complex than the simple sketch would indicate.

Typical outputs are one tonne per hour per metre width of vat per vat. Felt speeds range from 40 to 70 metres per minute. A typical three vat machine will make a 6 mm thick sheet in six to nine revolutions and will make one every 20 seconds. Manufacturing costs (other than materials) are therefore very low.

The physical properties of the composite are to a certain extent dependent on this type of fabrication technique because the dispersion of the fibres is essentially in two directions due to the process of forming each thin layer and the layers themselves are laminated to give the product. Also, because of the rotation of the sieve cylinder there is a predominant alignment of the fibres in the direction of rotation.

Curing of the products is often by stacking on pallets and allowing heat built up by the exothermic reaction between cement and water. Temperature in excess of 60°C may be reached in this process but autoclaving may also be used in some countries to accelerate or alter the characteristics of the reaction compounds and increase the rate of turnover of the products. Also, pressure may be applied to flat sheets to increase the density and strength of the composite.

6.1.1.2. Mazza process. RYDER (1975) has described the Mazza process which is used to make asbestos-cement pressure pipes. The de-watered layers on the felt are transferred to a rotating steel mandrel, instead of the accumulator roll and successive layers are consolidated by pressure until the desired wall thickness has been built up. The mandrel plus asbestos-cement is then removed from the machine and the mandrel is withdrawn to leave a pipe which is then matured in a water tank. The fabrication process imparts a considerable degree of orientation to the fibres in the circumferential direction and hence increases the bursting pressure of the pipes.

6.1.1.3. The semi-dry or Magnani process (RYDER, 1975). In the Magnani process, the solids to water ratio is in the neighbourhood of 0.5 and the mix is heated and pumped onto a belt where it is spread and levelled by reciprocating rollers. Both the belt and the rollers may be shaped to form corrugated or profiled sheet, and vacuum boxes under the belt move with it to suck excess water from the hot mix. The de-watered sheet is then transferred to pallets to mature. This process has the advantage for corrugated sheet that it can provide a greater thickness of material at the peaks and troughs of the corrugations and thus increase the resistance to bending of the sheet. The profile also enables lapped sheets to fit more closely together.

6.1.1.4. The Manville extrusion process (RYDER, 1975). The asbestos fibres, cement, fine silica, and a plasticizer such as polyethylene oxide are fed from a hopper into a mixer with just sufficient water to produce a stiff mix. The mix is forced through a steel die by a worm drive to give extruded sections of the desired profile.

The resulting extrusion is cut to lengths which are moved by a take-off belt to pallets on rollers. After drying the extruded sections are autoclaved and finally cut to the required lengths.

6.1.1.5. Injection moulding (RYDER, 1975). This process is now tending to replace the hand moulding of 'wet flat' from the Hatschek process for the manufacture of special fittings, for use with the corrugated sheets. It is also used for other products such as decking tiles. A slurry of asbestos fibre and cement, containing about 45 per cent of solids, is pumped into a permeable mould and then subjected to a hydraulic pressure of about 26 atmospheres via a rubber diaphragm. The mix is de-watered by this process of pressure filtration, and then has sufficient green strength for the product to be demoulded by means of a suction lifting pad, and is transferred to a pallet for curing. The complete cycle from injection to demoulding takes about one minute.

6.1.2. Properties of asbestos-cement
Asbestos-cement is the only fibre composite discussed in this chapter for which there are International Standards requirements for certain properties. These are generally expressed in terms of minimum bending strength, density, impermeability and frost resistance. For instance, the minimum bending strength generally varies between 15 and 23 MPa when tested under defined conditions and depending whether the sheet is semi- or fully compressed. Also, various loading requirements are defined for corrugated sheets such as snow loads up to 1.5 kPa, and point loads to simulate men working on a roof. Water absorption should not exceed 20 to 30 per cent of the dry weight depending on the type of product.

6.1.2.1. Tensile properties. The tensile properties of seven specially fabricated asbestos-cements have been studied by ALLEN (1971). The manufacturing process imparts a certain degree of orientation to the fibres (Section 6.1.1.1) and therefore the properties were measured parallel to and at right angles to the direction of preferential fibre alignment. The fibre volume fraction and the length of the fibres were varied in the seven types and the results are shown in Table 4.

Types 2 to 5 in Table 4 contained similar fibres at an increasing fibre content. The void content increased with fibre content with a resulting decrease in modulus and increase in strain to failure. At high void contents the fibre–matrix bond strength is probably reduced which may account for the relatively low tensile strength of Type 5. Values for bond strength between 0.88 and 3 MPa have been quoted. Types 1, 3 and 7 had short, intermediate and long fibres respectively but the effect of fibre length on the material parameters is not clear due to the interrelated effects due to change in void content.

Types 8 and 9 shown in Table 4 are commercial grades of asbestos-cement and the strong directional effect is again shown in these materials. The moduli and tensile strengths of the fully compressed material are considerably higher

TABLE 4. Properties of seven asbestos-cements (ALLEN, 1971, 1975, by permission of I.P.C. Science & Technology Press Ltd. and CIRIA respectively).

Type	Fibre volume (%)	Void content of matrix (%)	Young's modulus (GPa)		Ultimate tensile strength (MPa)		Strain at failure × 10^{-6}	
			L^a	T^b	L^a	T^b	L^a	T^b
1	5.70	17.9	16.93	16.6	17.8	15.0	1280	1190
2	2.91	14.3	17.29	17.55	14.6	10.8	950	700
3	5.10	26.7	16.09	13.54	20.3	12.3	2320	1540
4	7.32	32.3	14.73	13.82	25.4	18.4	3700	2280
5	14.85	60.4	8.45	8.81	21.3	18.5	5060	4000
6	6.02	12.8	20.41	18.77	14.5	11.3	860	790
7	4.72	32.6	13.05	–	16.1	–	2540	–
8	Commercial semi-compressed		13.6	15.2	16.1	9.5	2110	700
9	Commercial fully compressed		25.6	25.0	27.1	17.2	2130	770

[a] L = load applied parallel to direction of preferential fibre alignment.
[b] T = load applied at right angles to direction of preferential fibre alignment.

than the semi-compressed sheet presumably due to a reduction in the void content. Tensile stress–strain curves for various asbestos-cements have also been measured by ALLEN (1975) and three types of curve are shown in Fig. 23.

Curves A and C are for semi-compressed and fully compressed sheet respectively loaded parallel to the direction of preferential fibre alignment whereas curve B is for fully compressed sheet tested at right angles to that of

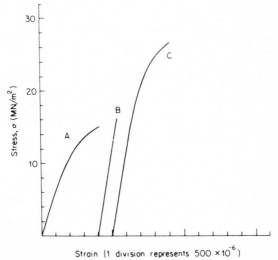

FIG. 23. Tensile stress–strain curves for asbestos-cement (ALLEN, 1975, by permission of CIRIA).

curve C. All the curves show the characteristic feature of smoothness and there was no audible or visual cracking even when examined under load by low-powered microscopy. ALLEN (1975) has suggested that the stress–strain curves are consistent with the progressive development of very fine cracks at strains in excess of the normal ultimate strain of the matrix and this view has been confirmed by LENAIN and BUNSELL (1979) by the use of acoustic emission techniques. It may be the case that the cracks are so fine and so closely spaced that the cement reaction products are still capable of carrying some load on the falling branch of the matrix stress–strain curve and hence, because the crack surfaces have not completely separated, the composite has the appearance of increased strain capacity before cracking when compared with the matrix alone.

6.1.2.2. Flexural properties of commercial asbestos-cement. A range of properties for fully compressed asbestos-cement sheets as quoted in the manufacturers' literature is shown in Table 5. The effect on the tensile and flexural

TABLE 5. Properties of asbestos-cement. Data taken from manufacturers' literature (by permission of John Wiley Ltd.).

Properties	Fully compressed flat sheet autoclaved or steam cured
Density (kg/m^3)	1800–2100
Tensile strength (MPa) { parallel to fibre	19–25
right angles to fibre	12–17
Modulus of rupture (MPa) { parallel to fibre	43–59
right angles to fibre	30–43
Compressive strength (MPa) { parallel to fibre	50–190
right angles to fibre	
Modulus of elasticity (GPa) { parallel to fibre	24–25
right angles to fibre	
Thermal expansion $\times 10^{-6}$ per °C	9–10
Water absorption (% of dryweight)	14–18

strengths of the direction of stress relative to the fibre direction is clearly shown in this table. The compressive strength and modulus of elasticity however are not greatly affected by fibre orientation and these properties are likely to be more dependent on the density, although published data is scarce.

The actual flexural properties for semi-compressed sheets are rarely quoted but the flexural strength is generally required to be in excess of 16 MPa.

6.1.2.3. Impact resistance. The impact resistance of asbestos-cement is notoriously low and this may be due, in part, to the stiffness of the material in the post matrix cracking zone such that imposed deflections caused by shock loads will cause very high local stresses and hence fracture. Contributory factors are short fibres with little energy being absorbed by fibre pull-out.

Fracture energies under impact conditions in an Izod machine of $2 \, kJ/m^2$ have been quoted which is close to that of the matrix alone and to that of 10 year weathered glass reinforced cement (Building Research Establishment, 1979).

However, the fracture energy under impact conditions should not be confused with the work of fracture measured under slow loading, which for the matrix alone has been variously quoted to be between 2 and $10 \, J/m^2$. The work of fracture of asbestos-cement would be expected to be of the same order as this.

6.1.2.4. Durability. Asbestos-cement is known to be very durable under natural weathering conditions and JONES (1947) has reported that no deterioration in flexural properties takes place due to weathering but that the material becomes progressively more brittle.

However, OPOCZKY and PENTEK (1975) have examined asbestos-cement sheets at ages of 2, 16 and 58 years and have shown that asbestos fibres do suffer a certain amount of corrosion which is compensated for, in terms of composite strength, by an increase in bond between the fibre and the cement. The corrosion of the fibre is promoted by the presence of airborne carbon dioxide which causes surface carbonation of the fibre. Also, certain magnesium hydroxides and magnesium carbonates may be formed as reaction products.

6.1.3. Asbestos-cement products
Asbestos-cement products have been used world-wide throughout the past seventy years and they owe their success to their low price and excellent durability under natural weathering conditions.

In the 1970's world asbestos production has exceeded 5 million tonnes per year and about 70 per cent of this has been used in asbestos-cement products. Corrugated sheeting and cladding for low-cost agricultural and industrial buildings has formed a major part of the market and flat sheeting for internal and external applications has also proved very successful.

Pressure pipes have been used for many years for conveying mains water, sewage, gas, sea water, slurries and industrial liquors. Diameters commonly range from 50 to 900 mm with working pressures from 0.75 to 1.25 MPa and, in addition, the crushing loads on buried pipes can be quite severe. An advantage of asbestos-cement pressure pipes is that their smooth uniform bore with freedom from the formation of internal deposits means that the hydraulic resistance is low. Also, non-pressure fluid containers and pipes such as rainwater goods, conduits, troughs, tanks and flue pipes account for a large proportion of the minor applications of asbestos-cement.

The future for asbestos-cement products now depends largely on the attitudes of Government Health and Safety Organisations and whether they follow the trend initiated by Sweden of banning the manufacture of such materials for health reasons.

6.2. Alternatives to asbestos-cement

Due to the high rate of depletion of the world asbestos resources (KREN-CHEL, 1975c) and the increasing concern being expressed about the health hazards of asbestos fibres, there has been a significant interest throughout the 1970's in the production of cement-based materials which can act as alternatives to asbestos-cement.

A number of potentially successful products has been produced from organic and from inorganic fibres but although the mechanical properties have been shown to be satisfactory, the additional criterion for successful commercial exploitation, i.e., that the price of the product should be similar to that of asbestos-cement or at least, only a few per cent higher has not always been satisfied. This commercial limitation is because other alternative products already exist in a slightly higher price bracket. Typical existing alternatives are aluminium sheeting or plastic coated steel cladding, glass fibre reinforced plastic sheets, polyvinyl chloride pipes and other materials with a more attractive visual appearance than asbestos-cement.

Product price is therefore likely to be a factor of equal or greater importance than mechanical properties and therefore automated fabrication procedures with high output rate and a low labour cost element are a vital part of any development programme for alternative fibre cements.

The processes described in this section are mainly still under development and are the result of large capital and research investments often by multi-national companies. The products and processes are protected by Patents and by Trade Names and it is therefore necessary to use these names in parts of the text.

6.2.1. Alternative products based on cellulose fibres

6.2.1.1. Cellulose fibres in cement. Cellulose fibres have been used for many years in the asbestos-cement industry, either as an additive to asbestos in products for internal use or as direct substitutes in wartime in countries which had little or no asbestos of their own. For example, in Finland and in Norway during the first and second world wars alternatives were produced mainly based on cellulose fibres, and PEDERSON (1980) has stated that testing of some of these products has shown that under certain conditions with proper treatment the cellulose fibres could withstand outdoor exposure in harsh climatic conditions over periods of 30–40 years.

The Norwegian company A/S Norcem and the Finnish company OY Partek AB were jointly the first to market successfully (in 1977) replacements for general purpose asbestos-cement flat sheets for internal use (PEDERSON, 1980). The process was based on Hatschek machinery (see Fig. 22) and it was found that cellulose was the only commercially available fibre which could replace asbestos as the basic fibre although additions of fibrillated polypropylene or inorganic fibres could add desirable features to a cellulose based product. The reason for the successful use of cellulose is that the fibre has to pick up or hold the cement in the slurry to prevent the cement and other fillers being removed with the surplus water in the filtration process. Means were found to prepare the cellulose fibres to hold these large quantities of inorganic material and also, under certain circumstances, to have good resistance to alkalis in addition to good strength properties.

A suitable method of preparing the cellulose fibres to give as good a carrying capacity of solids as asbestos fibres, comprises beating the fibres in a cone mill to a degree of 30–70° Shopper Riegler (JUNKKARINEN, 1978). Shopper Riegler is a term used to quantify the degree of fibre disintegration which is relevant to filtration.

Existing Hatschek machinery can therefore be used, with minor modifications and additions, to prepare the cellulosic and other fibre added as well as to mix these components with binder and water. By making use of existing fabrication techniques, together with a low cost fibre, it was therefore possible to produce asbestos-free products at the same price as asbestos-cement counterparts and with a greater profit margin (PEDERSON, 1980). Boards for external use produced by the same process have been under test for some time.

In order that the products could be satisfactorily marketed, performance specifications had to be developed for specific purposes and, in addition, traditional asbestos-cement standards had to be complied with.

6.2.1.2. Cellulose fibres in autoclaved calcium silicate. Asbestos-free building boards for internal purposes or for use in sheltered external situations have been produced in the United Kingdom by Cape Boards and Panels Ltd. using cellulose fibres in an autoclaved calcium silicate binder. The binder consists of a material containing reactive silica, and a calcareous component containing more than 30% of calcium oxide, which may be derived from Portland cement or lime. The fibre is mainly cellulosic produced from wood pulp processes.

The fabrication technique is compatible with commonly used machinery for asbestos-cement production such as Hatschek, Magnani, simple presses and the Fourdrinier process all of which de-water the boards to an extent that makes them into an easily handleable sheet form.

According to BARRABLE (1978) the initial slurry will normally be made up by hydrapulping and dispersing the cellulose and any other fibrous materials in water, followed by the addition of the other powdered materials to form a slurry of approximate water:solids ratio of 5:1 to 10:1. The slurry is then further diluted with water to give a solids:water ratio of about 25:1. Boards are then made from the latter slurry by de-watering on the relevant machine to give a water:solids ratio of approximately 1:1.

The autoclaving process takes place at temperatures up to 200°C at which level the cellulose fibres do not suffer permanent damage in this matrix.

The boards are available in 3 to 12 mm thick sheet of dimension up to 1.2 by 2.4 m and have a density of about 650–875 kg/m². Typical bending strengths are

5 to 10 MPa, respectively. They can be used for most internal panelling and lining purposes and provide excellent fire protection.

Autoclaved systems are said to have the advantage over air cured hydrated cement binders in terms of greater flexural strength at a given density and greater stability in relation to moisture and temperature movements. Because of the absence of free alkalinity the boards can be more easily decorated.

6.2.2. Alternative products based on glass fibres or mineral wool

Glass fibres and mineral wool are, apart from cost, obvious candidates for the production of alternatives to asbestos-cement. Due to the cost factor of glass fibres, product developments in glass reinforced cement have generally been aimed at the higher price bracket of special products where asbestos-cement has not been a competitor. The general technology involved in the production of this wide range of glass reinforced cement products is described in Section 6.3 and this section therefore only covers fabrication technology which is specifically designed to produce alternatives to asbestos-cement by the use of glass in conjunction with other fibres.

The combined use of monofilament glass, alkali resistant strand glass and cellulose fibres in a Hatschek-type process has been described by CAMERON et al. (1979) as a result of developments by Pilkington Brothers Ltd. The advantage of using single filament glass of a mean diameter in the range of 4 to 9 microns at a weight percentage of 2 to 6% of the total solids in the slurry is that the quantity of fine fibrous material is sufficient to reduce the escape of cement fines to an acceptable level whereas chopped glass strand, on its own, would not be satisfactory. Glass wool or mineral wool can be used to provide this filtering action but flocculating agents such as polyacrylamides may also be necessary. As the relatively cheap single filaments are mainly provided to give the correct filtration and drainage characteristics on the Hatschek cylinder, they may not have sufficient alkali resistance and mechanical integrity to give adequate durability to the hardened product and hence the main reinforcing action must be provided by the chopped alkali-resistant glass strands.

A similar type of approach utilising asbestos-cement technology and modified Hatschek machinery has been described by COLE (1979) of TAC Construction Materials Ltd. in the U.K. This company has started mass production of asbestos-free flat boards containing some glass fibre and the material is stated to be suitable for interior and exterior uses.

The modifications required to the standard Hatschek machinery are quite extensive and include the feed stock preparation, de-watering, sheet handling and curing areas as well as to the machine itself. The modifications amount to about 20% of the cost of a new plant (COLE, 1979) and the output which is typically 5 tonnes per hour is not affected. The price of the product is about 25% greater than equivalent asbestos-cement and some typical properties are: flexural strength—17.5 MPa; dry density—1500 kg/m^3; water absorption—25% of dry weight. Standard sheet sizes are made in thicknesses between 3 and 9 mm.

6.2.3. *Alternative products based on polymer fibres and films*

6.2.3.1. *Chopped polymer fibres.*

The use of 6 mm long 'Perlon' fibres of 13 micron diameter was described by KRENCHEL (1975b) as a possible way of achieving adequate properties to replace asbestos-cement sheet. The fibres were made into papers which were then impregnated with cement and built up into a laminate. Bending strengths in excess of 20 MPa at tensile strains of 1 to 2% were achieved with fibre volumes of about 8 to 9% but the manufacturing technique was relatively expensive and labour intensive.

Further developments using chopped polypropylene films have been described by KRENCHEL and JENSEN (1980). The polypropylene in this case was specially stretched and heat treated to give elastic moduli of 9 to 18 GPa with tensile strengths from 500 to 700 MPa and ultimate strains of 5 to 8%. Various surface treatments to improve wetting of the films and increase their bond were carried out before splitting the tape and chopping into lengths between 6 and 24 mm to give fibres with a basically rectangular cross-section but with frayed edges. With the co-operation of Dansk-Eternit-Fabrik A/S in Denmark these chopped and split films have been used mainly on Magnani-type machines to produce alternative products to asbestos-cement but Hatschek and extrusion processes have also been tried with good results.

Some machine modification is necessary but normal production speeds are possible and it is expected that adequate composite properties can be achieved at fibre volumes of 3–5% at a competitive price.

6.2.3.2. *Open networks of continuous film.*

The concept of using layers of opened networks of continuous fibrillated polyolefin films to reinforce a fine grained matrix was developed at the University of Surrey in the United Kingdom (HANNANT et al., 1978). The advantages of this system are that excellent mechanical bonding is achieved by virtue of the micro- and macro-slits in the films (cf. Fig. 3) through which the cement hydration products penetrate. Also, the films are efficiently used because they are continuous and there is a high specific surface which assists in frictional bonding (HANNANT and ZONSVELD, 1980). The resulting composite exhibits very fine multiple cracking with cracks spaced at less than 1 mm intervals, depending on the volume of film, and these cracks may be almost invisible to the naked eye at tensile strains up to 5%.

Polypropylene films have been used to give bending strengths in excess of 40 MPa and tensile strengths above 25 MPa at aligned film volumes of about 9%. Also, composites with adequate two-dimensional strength to replace asbestos-cement can be achieved at a total fibre volume of about 9%, of which 6% and 3% are in orthogonal directions. Fig. 24 shows tensile stress–strain curves in two directions for such a composite in which the flexural strengths in the two directions are in excess of 25 and 17 MPa respectively. A major advantage of these composites is the considerable toughness under impact

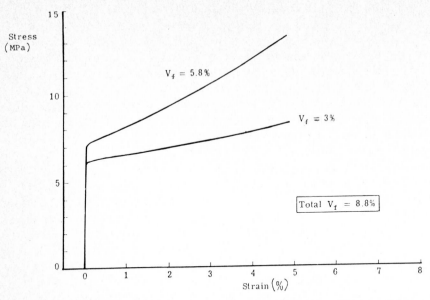

FIG. 24. Tensile stress–strain curves for a composite containing a two-dimensional lay-up of opened networks of polypropylene film.

loading which results from the high strain capacity and large area under the tensile stress–strain curves.

High modulus polyethylenes (CAPACCIO and WARD, 1974) have also been produced as fibrillated film networks and composites made with these films have shown improved stiffness in the post-cracking state when compared with the polypropylenes.

Patent applications for the process using continuous networks of polyolefin films have been taken out world wide (University of Surrey and HANNANT, 1976) and licences to manufacture the composite have been granted to major companies throughout Europe.

BIJEN and GEURTS (1980) have described the policy of DSM (Holland) in their development of fabrication techniques for cement sheets containing polymer films as alternatives to asbestos-cement. The advantages over the Hatscheck process are said to be that the new process is fully continuous, that the sheets are made as monolithic units so that there is little danger of de-lamination, production cost is virtually independent of sheet thickness, the speed of rotation of the machine is equal to the rate at which the product is delivered, no after-compression is needed in the production of the higher grade sheets, production of sandwich panels is possible and only a little water is needed.

Both flat and corrugated sheets will be produced in the initial development and prices are intended to be competitive with equivalent asbestos-cement units.

6.2.4. *Alternative products using natural fibres*

Developing countries often have a need for low cost sheet materials which can be produced by local unskilled labour and without the need for foreign exchange. Natural fibres with adequate properties to replace asbestos fibres may often occur and suitable types of fibre have been reviewed by COOK (1980).

Suitable alternatives have been made by SWIFT and SMITH (1979) using a hand lay-up technique with 500 mm long sisal fibres which had an elastic modulus of about 13 GPa and a tensile strength of 330 MPa.

Also, LEWIS and MIRIHAGALIA (1979) produced sheets from elephant grass which had an elastic modulus of 5 GPa and a tensile strength of 180 MPa. An important characteristic of the sheets was found to be greatly improved resistance to impact loads.

6.3. *Glass reinforced cement*

Much of the credit for the initial development of dispersed glass fibres as reinforcement for cement must undoubtedly be given to the Russians, BIRYUKOVICH et al. (1964) who not only appreciated the technical merits of the material but also developed fabrication techniques. BIRYUKOVICH had previously been involved in the construction of a factory roof in Kiev in 1963 and early work had also been carried out in China since 1958.

The Russian work was mainly concerned with binders of low alkali or high alumina cement whereas research in the Chinese People's Republic had concentrated on Portland cements and high alkali glass fibres with diameters of 18 to 23 microns. The Chinese Cement Research Institute found that the aggressive action of Portland cement on glass fibres reduced the strength of the composite by 50% after 5 months and steam curing reduced strength by over 90%. It was also established by the Chinese that cements which were low in calcium hydroxide such as gypsum slag cements improved the durability of the composites.

This early work was followed by research by MAJUMDAR and NURSE at the Building Research Establishment in the United Kingdom which was directed towards the development of a glass fibre which would resist attack by the highly alkaline ordinary Portland cements commonly used in Europe and America.

The identification of alkali-resistant glasses by the Building Research Establishment and their subsequent development and commercial production by Pilkington Brothers Ltd. led to major research efforts in the early 1970's directed towards understanding the physical properties of the composite material and this resulted in the publication of a large amount of data of value to the construction industry (HANNANT, 1978).

The use of polymers in the matrix to alter the properties of the hardened composite was first investigated by BIRYUKOVICH et al. (1964) using latex cements and later by MAJUMDAR's group at the Building Research Establishment. Also, the desire to use E-glass fibres in the alkaline matrix resulted in the development of a polymer modified matrix (BIJEN, 1979) in which the polymer filled up and surrounded the interfilament spaces thus reducing the possibility of attack on the glass by alkalis. Products have been marketed in Europe using this system.

The use of a material in structural situations must rely on adequate material properties being maintained for periods up to 50 years and therefore much research effort has been devoted to durability testing and to the prediction of long term properties. However, the confirmation of predicted values can only be achieved by tests on the materials or components after continuous exposure to the appropriate weathering conditions and hence the material cannot be proved suitable for continuously loaded structural applications until many years have elapsed. Data for ten years of natural weathering is available for alkali-resistant glass fibre in cement (Building Research Establishment, 1979) and this has been used to predict properties up to 20 years.

6.3.1. Fabrication techniques
In the production of glass reinforced cement the properties of the hardened product are heavily dependent on the fabrication technique adopted because the fabrication method controls the final orientation of the fibres in the composite. Although most of the basic techniques were described by BIRYUKOVICH et al. (1964), more recent refinements have been detailed by MAJUMDAR and NURSE (1974) and HILLS (1975).

The three basic techniques are premixing which gives a random three-dimensional array unless the mix is subsequently extruded or pressed, the spray method which gives a random two-dimensional arrangement, and filament winding which gives a one-dimensional array or a multi-angled oriented array.

6.3.1.1. Premixing.
Premixing is a process in which all the constituents, including short strands of glass fibre are intimately mixed and then further processed to produce a product by casting in open moulds, pumping into closed moulds, extrusion or pressing. The orientation of fibres tends to be three-dimensional random in the mixer but may be altered to a limited extent by the production process.

Glass fibres tend to tangle and matt together if care is not taken in the mixing process and generally pan mixers have been found to give a better result than drum mixers. It has been suggested that it is preferable to disperse the fibres in water containing a thickening additive such as polyethylene oxide or methyl cellulose (0.1 to 1 per cent of total mix water) before adding the solids and mixing in a mixer with driven pan and paddles but stiff fibre strands have also been produced for the premix process which make the use of thickening admixtures unnecessary.

Typically, fibre contents would be between 2 and 5 per cent by weight of the other dry materials using chopped strands about 25 mm long. The highest fibre contents cause some compaction difficulties but the simplest methods of hand tamping in 25 mm layers followed by mould vibration can be effective.

The properties of premixed glass fibre cement have not been as extensively researched as those of the spray de-watered composite but the static strengths are likely to be lower for a given percentage of glass than those quoted for the sprayed material because of the orientation of the fibres will probably be less efficient than two-dimensional random. Fig. 25 shows the measured modulus of rupture for glass contents up to 15 per cent by weight (HILLS, 1975) and if these results are compared with Fig. 28 for the sprayed material at the same age, it can be seen that the strengths of premix are about 50 per cent lower.

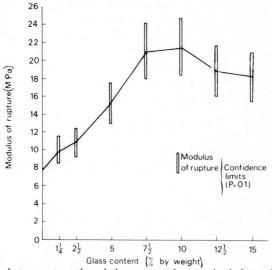

FIG. 25. Relationship between strength and glass content for premixed glass reinforced cement at 28 days. (O.P.C. and alkali-resistant glass) (HILLS, 1975, by permission of the Controller, HMSO, Crown Copyright reserved).

The four main methods of fabrication for premixed glass reinforced cement are outlined below.

GRAVITY MOULDING. Although the casting process is similar to some normal precast concrete production techniques, the glass reinforced premix allows the use of much thinner sections in both open or double sided moulds. Poker vibration is not recommended and generally external mould vibration is sufficient to produce flow, although the mix is less mobile than most concretes. Decorative aggregate finishes can be applied as for normal precast concrete.

PRESSING. Flat sheets between 10 and 20 mm thick have been produced by pressing premix at pressures between 0.15 and 10 MPa. Vacuum assisted de-watering using paper-felt filters may be from either one or both mould faces but a water thickening admixture is required to prevent water expulsion before the fibrous mix has had time to uniformly fill the mould.

A typical mix may contain ordinary Portland cement (O.P.C.), a filler such as Pulverized Fuel Ash (P.F.A.) and sand at a water/cement ratio, before pressing of about 0.8. Proportion by weight of the solids may be O.P.C.: P.F.A.: sand of 1:1:1 with small quantities of polyethylene oxide or methyl cellulose admixture. Mixes of this type can accept between 1.7 and 2.5 per cent (by weight of dry solids) of glass fibres with lengths between 11 and 22 mm.

Simple components with a good surface finish have been produced using this technique and immediate demoulding is possible.

INJECTION MOULDING. Premix containing a thickening admixture and with up to 5 per cent of glass fibre by weight of dry solids has been successfully pumped into closed moulds under pressure. Fibre damage may be increased by the pumping process but window frames, fence posts, and hollow columns have been cast using this process. However, care is required if the presence of small blow holes on the surface of the product are to be avoided.

EXTRUSION. Extruded sections with complex shapes have been produced commercially but careful attention is required to mix design to prevent bleeding through the mix or blocking of the die.

Vibration can be used to assist the flow of the material and it is possible to arrange the vibration and extrusion process to give a beneficial fibre alignment in the finished product.

6.3.1.2. Spray techniques

SPRAY SUCTION—BATCH PROCESS. This technique has been developed from the glass reinforced plastics industry and consists of leading a continuous roving up to a compressed air operated gun which chops the roving into lengths of between 10 and 50 mm and blows the cut lengths at high speed simultaneously with a cement paste spray onto a forming surface containing a filter sheet and the excess water can then be removed by vacuum.

The fibre cement sheet can be built up to the required thickness and demoulded immediately, an additional advantage being that it has sufficient wet strength to be bent round radiused corners to give a variety of product shapes such as corrugated or folded sheets, pipes, ducts, or tubes. Coloured fine aggregate can be trowelled into the surface to produce a decorative finish.

The process can be readily automated for standard sheets as shown in diagrammatic form in Fig. 26.

ALI et al. (1975) have measured stress–strain curves and flexural strengths for glass reinforced cement made using the spray-suction technique and typical results are shown in Figs. 27 and 28. High volumes of the longest fibres gave the best results but volumes of fibre above 6 per cent did not give improved

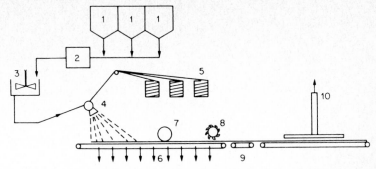

FIG. 26. Main stages in the automated production of glass fibre reinforced flat or corrugated sheets using the spray-suction process. 1. Silos; 2. Weigh batcher; 3. Mixer; 4. Traversing head spraying glass fibre and cement slurry; 5. Bobbins with glass fibre roving; 6. Vacuum de-watering; 7. Roller finisher; 8. Side and cross cut saws; 9. Conveyors; 10. Vacuum lift (HANNANT, 1978, by permission of John Wiley Ltd.).

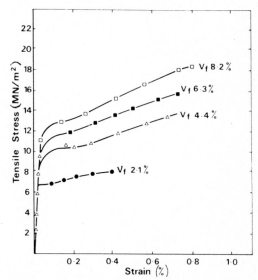

FIG. 27. Tensile stress–strain curves at 28 days of water stored glass reinforced cement composites containing 30 mm long fibres (ALI et al., 1975, by permission of The Controller, HMSO, Crown Copyright reserved).

flexural strengths possibly because of the increase in porosity of the composite. The material parameters relevant to Figs. 27 and 28 are shown in Table 6.

One of the problems with this type of glass reinforced cement has been the reduction in strain capacity with time under water storage or natural weathering conditions. This effect results in the material becoming essentially brittle under these conditions but air storage in a dry atmosphere causes little change in the initial properties even after periods of ten years. The difference in behaviour between wet and dry storage may be a result of continued hydration

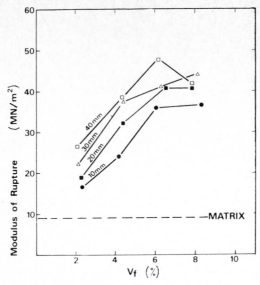

FIG. 28. Relation between fibre volume and modulus of rupture of glass reinforced cement water stored for 28 days for different fibre lengths (ALI et al., 1975, by permission of The Controller, HMSO, Crown Copyright reserved).

TABLE 6. Material parameters applicable to the composites shown in Figs. 27 and 28 (ALI et al., 1975, by permission of The Controller, HMSO, Crown Copyright reserved).

Parameter	Typical value
Fibre cement bond (τ)	3 MPa
Modulus of cement paste (E_m)	26 GPa
Modulus of composite (E_c)	30 GPa
Modulus of fibre (E_f)	76 GPa
Effective area of glass fibre strand (A_f)	0.027 mm^2
Perimeter of glass fibre strand (P_f)	1.42 mm

of the matrix in wet storage the products of which fill all the voids in and around the fibre bundles over a period of time. The matrix becomes very dense with a residual porosity of almost zero and a high contact area at the fibre interface which leads to greatly increased frictional bond and a hard stiff material where the fibres bend across a crack. All these factors contribute to a reduction in tensile strain at failure and hence a reduction in flexural strength (Section 4.3.2).

Time dependent property changes for a variety of storage conditions for periods up to ten years are shown in Table 7 for boards containing five per cent by weight of 34 mm long alkali-resistant glass fibres (Building Research Establishment, 1979). It can be seen from this table that the elastic (Young's)

TABLE 7. Measured mean strength properties of sprayed-dewatered OPC/glass reinforced cement at various ages (5 weight% of glass fibre) (Building Research Establishment, 1979, by permission of The Controller, HMSO, Crown Copyright reserved).

	Total range for air and water storage conditions at 28 days	1 year			5 years			10 years		
		Air [a]	Water [b]	Weathering	Air [a]	Water [b]	Weathering	Air [a]	Water [b]	Weathering
Bending										
ultimate (MPa)	35–50	35–50	22–25	30–36	30–35	21–25	21–23	31–39	17–18	15–19
LOP[d] (MPa)	14–17	9–13	16–19	14–17	10–12	16–19	15–18	14–16	16–17	13–16
Tensile										
ultimate (MPa)	14–17	14–16	9–12	11–14	13–15	9–12	7–8	11–15[c]	6–8[c]	7–8[c]
BOP[e] (MPa)	9–10	7–8	9–11	9–10	7–8	7–9	7–8	9–10[c]	6–8[c]	6–8[c]
Young's modulus (GPa)	20–25	20–25	28–34	20–25	20–25	28–34	25–32	25–33	25–31	27–30[c]
Impact strength (Izod) (Nmm/mm^2)	17–31	18–25	8–10	13–16	18–21	4–6	4–7	15–22	2–3	2–6

[a] At 40 per cent relative humidity and 20°C.
[b] At 18–20°C.
[c] Unreliable due to measurement problems.
[d] LOP—limit of proportionality.
[e] BOP—bend over point.

modulus remains sensibly constant for all curing conditions but the other properties may change with time.

SPRAY-SUCTION—CONTINUOUS PROCESS. In order to minimise production costs a high rate of sheet output is desirable with productivity similar to that of a Hatschek machine for asbestos-cement production. A continuous fabrication technique for the production of spray de-watered glass reinforced cement sheet in Japan has been described by MISHIMA et al. (1979). This fabrication development can produce continuous lengths of de-watered sheet at line speeds of up to 12 m per minute. Mortar is sprayed simultaneously over the whole width of the sheet and the traversing head system has been eliminated. Edge losses of sheet of about 50 mm on each side are still significant in cost terms but the transverse/longitudinal strength can be varied from 50 to 90%. Other fibres such as cellulose and other organic materials, single filament glass, and asbestos can also be premixed into the mortar to give a range of properties.

Typical properties of the product for the standard mix of cement:sand ratio 2:1 and 5% glass of 37 mm fibre length are flexural strengths of 35 MPa with a limit of proportionality of 10 MPa. Bulk density is about 2000–2200 kg/m^3 but no long-term properties for this composite have been published.

DIRECT SPRAY. In principle, this process is similar to the spray-suction technique except that the addition of suitable admixtures to the matrix reduces the water requirement of the cement slurry for satisfactory spraying and therefore filter sheet and suction operations can be dispensed with. Roller compaction is often used to ensure compliance with the mould and to assist in the removal of entrapped air. Water/cement ratios as low as 0.30 to 0.35 may be achieved using this technique for neat Portland cement.

The density may be slightly lower for direct spray materials, lying in the range 1750 to 2000 kg/m^3 as compared with 2000 to 2100 kg/m^3 for spray-suction materials.

The direct spray technique is probably the most used for building components of complex shapes such as permanent formwork or cladding panels.

6.3.1.3. Winding process. The winding process was developed by BIRYUKOVICH (1964) and is very useful for the production of pipes although similar procedures can be adopted for sheets or open sections by cutting and re-forming the freshly made composite.

Continuous glass fibre rovings are impregnated with cement slurry by passing them through a cement bath and they are then wound onto a suitable mandrel at a predetermined angle and pitch. Additional slurry and chopped fibre can be sprayed onto the mandrel during the winding process and roller pressure combined with suction can be used to remove excess cement paste and water. Fibre volumes in excess of 15 per cent have been achieved and hence very high strengths are possible. The process, as for the spray-suction process, can be readily automated for production runs.

A variant of the winding method consists of winding round a cylindrical mould a pre-formed mesh or woven fabric of glass fibre which is sprayed with cement paste and additional chopped strands. Very high strength and accurately controlled composites can be produced using these techniques.

6.3.1.4. Lay-up process. The continuous impregnated rovings used in the winding process can also be laid in moulds in the form of window frames or similar products which can then be vibrated or pressed to improve the penetration by cement paste.

Hand lay-ups using random glass fibre mat or continuous fibre fabric can also be produced on surface moulds on which the impregnation of cement slurry is assisted by surface rollers, pressure and suction. The manufacture of complex shapes of the types produced in glass reinforced plastics is possible using this technique.

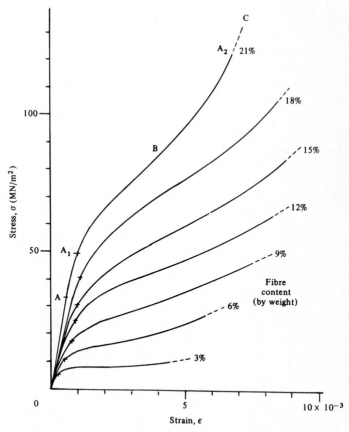

FIG. 29. Stress–strain curves for high-alumina cement with different volumes of continuous parallel glass fibre reinforcement. Tension along the fibres (BIRYUKOVICH et al., 1964, by permission of CIRIA).

The high strengths which could be achieved with lay-up techniques using continuous strands of E-glass aligned parallel to the direction of load application were determined by BIRYUKOVICH (1964) for a matrix of high alumina cement and are shown in Fig. 29. The winding and lay-up techniques enable the highest strengths to be obtained as seen in a comparison with Fig. 27. According to BIRYUKOVICH the increase in properties of the composite in bending for aligned fibre volumes above 9% may be limited by the tendency of the matrix to fail in compression before tensile failure of the fibres occurs.

6.3.1.5. Progressive impregnation of a glass mat by periodically separating its constitutent layers. An ingenious method for producing thin walled structures and thin sheets from glass fibre mats was devised by BIRYUKOVICH utilising the fact that glass fibre cement disintegrates at points where the fibres are not sufficiently impregnated with cement paste.

If a glass mat consisting of an unlimited number of layers of fibre is placed between two planes and immersed in a bath of cement paste, a certain amount of paste will penetrate into the mat. If the planes enclosing the mat are then moved further apart, the cement paste will wet both surfaces of the mat and partially penetrate into it. If now the planes again compress the mat, its external layers impregnated with cement paste will adhere to these planes, so that when the latter again move apart, the mat will segregate along those layers of fibre which the paste has not yet reached. By repeating this procedure many times, the glass mat will become fully impregnated with cement paste, so that it will be converted into a glass reinforced cement sheet.

This principle has been used to make structural units of complex shape in addition to the fabrication of continuous sheeting.

6.3.1.6. Surface coatings. A technique of building blockwork walls by dry stacking the blocks and coating both the surfaces by trowelling on E-glass fibre reinforced ordinary Portland cement was developed in 1967 by the U.S. Department of Agriculture (SIMONS and HAYNES, 1970). Further development work was carried out in the U.S.A. (PECUIL and MARSH, 1974) using 4 per cent by weight of alkali-resistant fibres and also in the U.K., where the material is available in the form of pre-bagged renderings containing about 3 per cent by weight of fibre. Minimum block thickness should be 75 mm and special attention should be paid to stability of the blocks if higher than 3 m. Higher strength and more rapid construction than conventional blockwalls are claimed for the technique and a similar result can be obtained by spraying glass fibres and cement slurry.

6.4. Sheet materials using high cost fibres

Cement composites with very high tensile strengths can be produced with high performance fibres such as carbon and Kevlar.

For instance, AVESTON et al. (1974) have studied the mechanical properties of carbon fibre cement in some detail using two types of composites. In aligned composites, which were made by hand lay-up, a 100 mm wide thin veil (made from 10 000 filament tows) of carbon fibre was used. Pseudo-random composites were prepared from these sheets by rotating the mould through 10° for each layer of carbon. Some tensile stress–strain relationships are shown in Fig. 30 from which it is apparent that very high strengths can be achieved although the cost is likely to rule out bulk commercial applications.

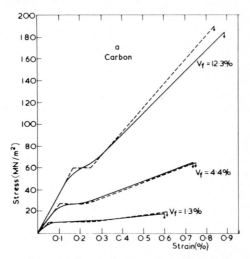

FIG. 30. Tensile stress–strain curves for continuous carbon fibre reinforced cement (—— experimental; – – – – theoretical) (AVESTON et al., 1974, by permission of the National Physical Laboratory, Crown Copyright reserved).

Excellent results can also be achieved in cement by the use of Kevlar fibres (an aromatic amide polymer developed by Dupont in the U.S.A.) (WALTON and MAJUMDAR, 1978) but again cost may rule out significant commercial applications.

7. Concluding remarks

The basic mechanics of the behaviour of fibres in cement and in concrete are now sufficiently well understood that commercial development is not likely to be held back by a lack of theoretical knowledge in designing composites with adequate material properties for a variety of applications.

However, it has not yet been established that fabrication techniques are sufficiently well developed for the new composites to enable the theoretical properties to be achieved on a mass production scale at a low enough price to enable the products to be competitive with the nearest available alternative.

A considerable amount of development work is required before the technical problems can be overcome so that a sufficient quantity of cheap, strong fibres of adequate length can be included in the matrix in the correct angular distribution relative to the directions of applied stresses in the product in service.

References

ALI, M., A.J. MAJUMDAR and B. SINGH (1975), Build. Res. Est., Current Paper CP 94/75, Watford.

ALLEN, H.G. (1971), *Composites* **2**(2), 98.

ALLEN, H.G. (1975), Constr. Ind. Res. and Inf. Assoc. CIRIA, Report 55, London.

American Concrete Institute, E.K. SCHRADER (1978), Formulating guidance for testing of fibre concrete in ACI Committee 544, in: R.N. SWAMY, ed., *Proc. Conf. Testing and Test Methods of Fibre Cement Composites* (Const. Press Ltd., Lancaster) pp. 9–21.

Asbestos Products Association (1962), Manual of testing procedures for chrysotile asbestos-fibre.

AVESTON, J., G.A. COOPER and A. KELLY (1971), Single and multiple fracture, in: *Proc. Conf. on the Properties of Fibre Composites* (IPC Sc. and Tech. Press., Guildford) pp. 15–26.

AVESTON, J., R.A. MERCER and J.M. SILLWOOD (1974), Fibre reinforced cements—scientific foundations for specifications, in: *Proc Conf. Composites Standards Testing and Design* (IPC Sc. and Tech. Press., Guildford) pp. 93–103.

AVESTON, J., R.A. MERCER and J.M. SILLWOOD (1975, 1976), Nat. Phys. Lab. Rept. No. SI No. 90/11/98, Part I, 1975; DMA 228 1976, Part 2, 1976.

BARRABLE, V.E. (1978), U.K. Patent Specification 1498 966.

BERGSTROM, S.G. (1975), A Nordic research project on fibre reinforced, cement based materials, in: A.M. NEVILLE, ed., *Proc. Conf. Fibre Reinforced Cement and Concrete* Vol. 2 (Constr. Press Ltd., Lancaster) pp. 595–600.

BIJEN, J. (1979), Glass fibre reinforced polymer modified cement, in: *Prof. Conf. on the Developing Success of GRC* (International GRC Congress, London).

BIJEN, J. and E. GEURTS (1980), Sheets and pipes incorporating polymer film material in a cement matrix, in: *Proc. Concrete Internat. 1980* (The Construction Press Ltd., London).

BIRYUKOVICH, K.L., YU.L. BIRYUKOVICH and D.L. BIRYUKOVICH (1964), *Glass Fibre Reinforced Cement* (Budivel'nik, Kiev) (English translation, CERA Transl. No. 12, 1965, London).

British Standard 1881, Part 2 (1970), *Methods of Testing Fresh Concrete* (Brit. Stand. Inst., London).

Building Research Establishment (1979), *Properties of GRC: 10 Year Results* IP 36/79 (Build. Res. Stat., Watford).

CAPACCIO, G. and I.M. WARD (1974), *Polymer* **15**, 233.

CAMERON, N.M., K.C. THATCHER, J.P. LOFTUS and P.P. BEMAND (1979), U.K. Patent Specification No. 1543951.

COLE, J.S. (1979), Mass produced GRC—an alternative to asbestos-cement, in: *Proc. Internat. Conf. on the Developing Success of GRC*, London.

COOK, D.J. (1980), Concrete and cement composites reinforced with natural fibres, in: *Proc. Concrete Internat. 1980* (The Construction Press Ltd., London).

Danish Standards Institute (1979), Corrugated sheets of fibre reinforced cement for roofing, Dansk Standardiseringsrad DS/R 1112, 1113, 1114.

EDGINGTON, J. (1977), Economic fibrous concrete, in: *Proc. Conf. Fibre Reinforced Materials* (Inst. Civil Engrg., London) pp. 115–126.

EDGINGTON, J. and D.J. HANNANT (1972), Materiaux et constructions, *RILEM* **5**(25), 41.

EDGINGTON, J., D.J. HANNANT and R.I.T. WILLIAMS (1974), Build. Res. Est. Current Paper CP 69/74, Watford.

FAIRWEATHER, A.D. (1972), The use of polypropylene film fibre to increase impact resistance of concrete, in: *Proc. Conf. Prospects for Fibre Reinforced Building Materials 1971* (Build. Res. Est., Watford) pp. 41–44.

FARAHAR, R.M. (1978), Pre-cast concrete, Publ. Cement and Conc. Ass., London.

GORDON, J.E. and G. JERONIMIDIS (1974), *Nature* **252**, 116.

HANNANT, D.J. (1975), The effect of post cracking ductility on the flexural strength of fibre cement and fibre concrete, in: A.M. NEVILLE, ed., *Proc. Conf. Fibre Reinforced Cement and Concrete* Vol. 2 (Constr. Press Ltd., Lancaster) pp. 499–508.

HANNANT, D.J. (1978), *Fibre Cements and Fibre Concretes* (Wiley, Chichester).

HANNANT, D.J. and J.J. ZONSVELD (1980), Polyolefin fibrous networks in cement matrices for low cost sheeting, *Phil. Trans. Roy. Soc. Ser. A* **294** (1411) pp. 591–597; *Proc. Conf. on New Fibres and their Composites* (Royal Society, London).

HANNANT, D.J., J.J. ZONSVELD and D.C. HUGHES (1978), *Composites* **9**(2), 83.

HILLS, D.L. (1975), Build. Res. Est. Current Paper CP 65/75, Watford.

HODGSON, A.A. (1965), *Fibrous Silicates*, Lecture Series No. 4 (Royal Institute of Chemistry, London).

HOFF, G.C. (1975), The use of fibre reinforced concrete in hydraulic structures and marine environments, in: A.M. NEVILLE, ed., *Proc. Conf. Fibre Reinforced Cement and Concrete* (Constr. Press Ltd., Lancaster) pp. 395–407.

HOLLINGTON, M. (1973), The development of wire fibre reinforced concrete manhole cover and frame, in: L.J. DEN BOER, ed., *Proc. Conf. on Properties and Applications of Fibre Reinforced Concrete* (Stevin Laboratory, Delft) pp. 177–179.

JONES, F.E. (1947), Build. Res. Tech. Paper 29, H.M.S.O. London.

JUNKKARINEN, K.E. (1979), U.K. Patent Application No. GB 2012832 A.

KLOS, H.G. (1975), Properties and testing of asbestos fibre cement, in: A.M. NEVILLE, ed., *Proc. Conf. on Fibre Reinforced Cement and Concrete* Vol. 1 (Constr. Press Ltd., Lancaster) pp. 259–267.

KRENCHEL, H. (1964), *Fibre Reinforcement* (Akademisk Forlag, Copenhagen).

KRENCHEL, H. (1975a), Fibre-spacing and specific fibre surface, in: A.M. NEVILLE, ed., *Proc. Conf. on Fibre Reinforced Cement and Concrete* Vol. 1 (Constr. Press Ltd., Lancaster) pp. 69–79.

KRENCHEL, H. and O. HEJGAARD (1975b), Discussion of paper 7.4, in: A.M. NEVILLE, ed., *Proc. Conf. on Fibre Reinforced Cement and Concrete* Vol. 2 (Constr. Press Ltd., Lancaster) pp. 607–610.

KRENCHEL, H. and O. HEJGAARD (1975c), Can asbestos be completely replaced one day, in: A.M. NEVILLE, ed., *Proc. Conf. on Fibre Reinforced Cement and Concrete* Vol. 1 (Constr. Press Ltd., Lancaster) pp. 335–346.

KRENCHEL, H. and H.W. JENSEN (1980), Organic reinforcing fibres for cement and concrete, in: *Proc. Concrete Internat. 1980* (The Construction Press Ltd., London).

LANKARD, D.R. (1975), Applications of fibre concrete, in: A.M. NEVILLE, ed., *Proc. Conf. Fibre Reinforced Cement and Concrete* Vol. 1 (Constr. Press Ltd., Lancaster) pp. 3–19.

LARNER, L.J., K. SPEAKMAN and A.J. MAJUMDAR (1976), *J. Non-crystall. Solids* **20**, 43.

LAWS, V. (1971), *J. Physics D: Appl. Physics* **4**, 1737.

LAWS, V. and M.A. ALI (1977), The tensile stress–strain curve of brittle matrices reinforced with glass fibre, in: *Proc. Conf. on Fibre Reinforced Materials* (Inst. Civil Engrg., London) pp. 101–109.

LEA, F.M. (1970), *The Chemistry of Cement and Concrete* (Edward Arnold, London).

LENAIN, J.C. and A.R. BUNSELL (1979), *J. Materials Sci.* **14**, 321.

LEWIS, G. and P. MIRIHAGALIA (1979), *Magazine Concr. Res.* **31**(107), 104.

MAJUMDAR, A.J. and R.W. NURSE (1974), Build. Res. Est. Current Paper CP 79/74, Watford.

MAJUMDAR, A.J., J.M. WEST and L.J. LARNER (1977), J. Mater. Sci. **12**, 927.

MARK, R.E. (1971), *J. Polymer Sci. Part C* (36) 393–406.

MARK, R.E., J.L. THORPE and A.J. ANGELLO (1971), *J. Polymer Sci.* (36) 177–195.

MARSDEN, W.A. (1980), Bulk fibrous concrete has arrived in Australia, in: *Proc. Concrete Internat. 1980* (The Construction Press Ltd., London).

MISHIMA, K., M. KOZUKA, T. ISHIZUKA and R. TAKEDA, in: *Proc. Internat. Conf. on the Developing Success of GRC*, London.

NEVILLE, A.M. (1978), *Properties of Concrete* (Pitman, London).

OAKLEY, D.R. and B. PROCTOR (1975), Tensile stress–strain behaviour of glass reinforced cement composites, in: A.M. NEVILLE, ed., *Proc. Conf. on Fibre Reinforced Cement and Concrete* (Constr. Press Ltd., Lancaster) pp. 347–359.

OPOCZKY, L. and L. PENTEK (1975), Investigation of the corrosion of asbestos fibres in asbestos-cement sheets weathered for long times, in: A.M. NEVILLE, ed., *Proc. Conf. on Fibre Reinforced Cement and Concrete* Vol. 1 (Constr. Press Ltd., Lancaster) pp. 269–276.

PAGE, D.H., F. EL-HOSSEINY and K. WINKLER (1971), *Nature* **229**, 252.

PECUIL, T.E. and H.N. MARSH (1975), Fibre glass surface bonding, *A.C.I. Publication S.P.* **44**, 363–374.

PEDERSON, N. (1980), Commercial development of alternatives to asbestos-cement sheet products based on short fibres, in: *Proc. Concrete Internat. 1980* (The Construction Press Ltd., London).

RILEM 19-FRC Technical Committee, Secretariat, 12 Rue Brancion, 75737 Paris.

RILEM Symposium (1978), R.N. SWAMY, ed., *Testing and Test Methods of Fibre Cement Composites* (Constr. Press Ltd., Lancaster).

RYDER, J.F. (1975), Applications of fibre cement, in: A.M. NEVILLE, ed., *Proc. Conf. Fibre Reinforced Cement and Concrete* Vol. 1 (Constr. Press Ltd., Lancaster) pp. 23–35.

SIMONS, J.W. and B.C. HAYNES (1970), Surface bonding. A technique for erecting concrete block walls without mortar joints, Rept. No. CA-42-57, USDA Agricultural Research Service, 6.

SKARENDAHL, A. (1980), Pre-cast and sprayed steel fibre concrete, in: *Proc. Concrete Internat. 1980* (The Construction Press Ltd., London).

SWAMY, R.N. and H. STAVRIDES (1975), Some properties of high workability steel fibre concrete, in: A.M. NEVILLE, ed., *Proc. Conf. on Fibre Reinforced Cement and Concrete* (Constr. Press Ltd., Lancaster) pp. 197–208.

SWIFT, D.G. and R.B.L. SMITH (1979), *Composites* **10**(3), 145.

SZABO, I. (1975), Applications of steel fibre reinforced concrete, in: A.M. NEVILLE, ed., *Proc. Conf. Fibre Reinforced Cement and Concrete* Vol. 2 (Constr. Press Ltd., Lancaster) pp. 483–486.

University of Surrey and D.J. HANNANT (1981), U.K. Patent No. 1582945.

WALTON, P.L. and A.J. MAJUMDAR (1978), *J. Mater. Sci.* **13**, 1075.

ZONSVELD, J.J. (1970), *Plastica* **23**, 474.

CHAPTER IX

Biological Composites

John D. Currey

Department of Biology
University of York
Heslington
York YO1 5DD
United Kingdom

Contents

HANDBOOK OF COMPOSITES, VOL. 4 – Fabrication of Composites
Edited by A. KELLY and S.T. MILEIKO
© 1983, Elsevier Science Publishers B.V.

1. Introduction

In discussing biological composite materials there is a problem about what to include and what to leave out. Virtually all biological materials are composite in the sense of being made of more than one material, and many of them have a mechanical function. The limits I have set myself are as follows.

(a) I shall discuss only passive mechanical materials, and so will not discuss muscle and similar tension-generating tissues.

(b) I shall discuss only tissues that are large enough to be visible to the naked eye. Many cells have composite mechanical structures within them, but these I shall ignore.

(c) In general I shall not discuss materials with a Young's modulus of less than about 2 GPa. This excludes vertebrate tendon, cartilage and blood vessels, for instance. I do this because I suspect that very compliant materials will not be of great interest to readers of this volume. However, I have discussed very compliant materials if they have a chemical constitution very similar to materials of high modulus. Cuticle of insects has materials of similar constitution that show compliances spanning seven orders of magnitude! How such differences are achieved, and what the adaptive reasons for them are, are matters of some interest.

Even within these limits this survey is, of course, not comprehensive. I hope that it will, however, deal with most types of biological composites and include some that most readers of this chapter will have been unaware of.

Almost without exception the skeletal materials of living organisms are composite materials (cf. Table 1). When considering them, it will be helpful to keep in mind some facts about the constraints under which they have been manufactured, and also some of the advantages organisms have over scientists and technologists in their ability to design and manufacture materials.

All biological structures have evolved by a process of trial and error. Alterations in the genetic constitution of organisms that cause them to have a structure which, in some sense, is better than those already existing were likely to be passed on to succeeding generations. The change of *large* random alterations producing a better result are very small, so most structures will have evolved little by little with no innovative leaps. This does not mean that very different composite materials do not evolve, as we shall see, but rather that these different materials have evolved separately from each other over a very long time indeed.

The number of trials that have produced the present range of composite materials we see is huge. Take as an arbitrary starting point the first appearance of calcified shells in the fossil record. This was about 5×10^8 years ago. There are about 2×10^6 species alive now, and we can take the average number alive at any time as not less than 10^6. A very conservative estimate for the mean number of adults in a generation in a species is 10^9. (It will be less for many species of large organisms, but unimaginably more for many small organisms.) We can take the average generation time as a year, and the number of offspring produced by an adult as 100. This rather wild calculation shows that natural selection will have acted on about 10^{25} organisms to produce the variety we now see.

So much for the design process—blind chance operating on an inconceivably large number of gradually improving prototypes. For biological materials there are particular constraints.

TEMPERATURE. The temperature used in fabrication is low. Over the whole range of organisms, virtually no growth takes place outside the range $0°$ and $44°$, and for most the range is much narrower. Some materials, notably wood, have to function at temperatures much lower than their fabrication temperature; at about $-30°$ or so, but they are never fabricated at this temperature. Few materials have to operate at temperatures much higher than their fabrication temperature.

TABLE 1. List of the important groups of animals that have composite skeletal materials (protozoa excluded). The only important animal group that does not have any members with a composite skeleton is the nematodes (roundworms).

Sponges	$CaCO_3$ and SiO_2 spicules in compliant matrix. SiO_2 spicules occasionally form coherent skeleton (*Euplectella*)
Coelenterates	(Sea anemones, Jellyfish, Corals) $CaCO_3$ spicules in compliant matrix. Massive $CaCO_3$ skeletons with tenuous organic matrix (TOM)
Platyhelminthes	(Flatworms) Occasionally $CaCO_3$ spicules in compliant matrix
Annelids	(Segmented worms, earthworms, leeches, fanworms) Tubes made of $CaCO_3$ in TOM, or sandgrains with TOM
Arthropods	(Insects, Crustaceans, spiders, many others) Cuticle of chitin fibrils in protein matrix. Crustaceans and millipedes add $CaCO_3$. $CaCO_3$ with TOM (Barnacles)
Molluscs	(Clams, snails, squids) $CaCO_3$ still with TOM. $CaCO_3$ spicules (chitons' body wall, hinges of clams). Teeth often organic fibres in organic matrix
Brachipods	(Lamp shells; no real common name) $CaCO_3$ with TOM. Feeding arms often have high volume fraction $CaCO_3$ spicules in compliant matrix. Calcium phosphate rich and poor layers alternate (*Lingula*)
Bryozoa	(Tiny, colonial animals) $CaCO_3$ shell with TOM. Spicules in fairly rigid matrix
Echinoderms	(Starfish, sea urchins, many others) Skeletal elements of 'single crystals' of calcite. Whole range from isolated spicules to solid blocks. Sea urchin teeth $CaCO_3$ fibres in $CaCO_3$ matrix
Chordates	(Vertebrates, plus some obscure others) Enamel, bone (and varieties such as dentine). Keratins

AVAILABILITY OF MATERIALS. Most organisms are stationary, few move large distances. They must therefore make use of the raw materials that are available in one place.

PRECISION OF CONTROL. Against the constraints mentioned, organisms have one or more advantages. One is that they are very good at controlling the ionic concentrations in very small volumes. For example, in man the concentration of Ca^{2+} in the body is usually kept between 8.5 and 10.5 mg/100 ml for seventy years or so. Yet, in a volume of the order of $\frac{1}{10} \mu m^3$ the concentration can be increased tenfold in a matter of seconds.

SPEED OF GROWTH. Another advantage organisms have over practical man is that the speed of growth can be, if necessary, very low, of the order of a few microns per day.

These two last effects mean that biological composites are often extremely precise in their construction. Table 1 is a list of the major animal groups that employ composite skeletal materials. For further information one of the multitude of encyclopaedias on animal life will be suitable. VINCENT and CURREY (1980), though concerned with the whole variety of biological skeletal materials, discuss many composites, including very compliant ones.

2. The raw materials

The range of materials used in biology is rather small, although the detailed structure is usually very finely adjusted to the organism's needs. I list the main ones here.

2.1. Ceramics

Calcium carbonate is easily the most widely used. It is found in the crystalline forms of calcite and aragonite. Occasionally it is amorphous. Calcium phosphate, or some variant such as hydroxyapatite is used by the vertebrates and occasionally elsewhere. Silica is much used by plants (though often merely as an abrasive to wear down grazers' teeth), and occasionally by animals, particularly sponges. A short review of the whole range of biological ceramics is given by LOWENSTAM (1981).

2.2. Polysaccharides

Simple sugar molecules, such as glucose, can combine to form very long chain polymers. Particularly important among these are cellulose (polycellobiose) and chitin (poly–N–acetyl–D–glucosamine). They often form side links to neighbouring polysaccharide molecules.

2.3. *Proteins*

Proteins are long chain polymers made of many amino acids bound by peptide bonds. The general formula for an α-amino acid, the types found in proteins, is: $R-NH_2-CH-COOH$. The carboxylic acid group and the amino group are both attached to the α-carbon. 'R' is one of about twenty side groups which range in complexity from the single hydrogen in glycine to the aromatic-ring containing tryptophan. These side groups may be polar or non-polar, acidic, basic or neutral. In the formation of proteins one amino acid is joined to another by a peptide reaction:

$$NH_2-R_1-CH-COOH + NH_2-R_2-CH-COOH \rightarrow$$
$$\rightarrow NH_2-R_1-CH-CO-NH-R_2-CH-COOH + H_2O .$$

This process can be extended indefinitely, resulting in a long polypeptide chain. The characteristics of the R-groups and their infinite permutability allow the polypeptide chains to have a vast array of shapes and functions. Proteins in biological composites tend to be fibrous, in which case the primary structure of the protein (the names of its amino acid residues written in order) tends to be simple, and the amino acid residues are often small, allowing close packing. For instance the primary structure of one of the silks of the silk moth is almost entirely

glycine–alanine–glycine–alanine–glycine–alanine–· · · ·

2.4. *Other organic materials*

Organisms make a fair amount of use of compounds of proteins with polysaccharides as in the mucopolysaccharides. Lignin is an amorphous material, mainly made of long chain polymers of phenylpropane derivatives. Lignin is very widely used as the matrix material in wood.

Water is pervasive in biological composites, very few lacking it. It has a profound effect on the mechanical properties of the materials.

3. High modulus mineralised composites

The composites found in organisms can be classified in many ways: according to the organisms they are found in; according to their function; to their composition; to their mechanical properties; and so on. In what follows I shall deal with them roughly according to their modulus of elasticity. In each class I shall describe the material, what little is known about the mechanical properties, how they are achieved and also a little about function.

3.1. A ceramic–ceramic composite

Sea urchins, like other echinoderms, are remarkable in having skeletons made of sizeable blocks, up to several centimetres in length, made of single crystals of calcite. These blocks do not *look* like single crystals, being very irregularly shaped and pierced by many interconnecting holes. The general skeleton of echinoderms is about the only example in the animal and plant kingdoms of stiff skeletal structures of any size that are not composites. However, the teeth of sea urchins are a delightful exception.

Sea urchin teeth are used for chewing algae on rocks. The usual sea urchin has five teeth supported on an archaic-looking structure called an Aristotle's lantern. Each tooth is ever-growing and has an extremely complex structure of calcite plates cemented to each other by, apparently, small discs of amorphous calcium carbonate (MÄRKEL and GORNY, 1973). The very core of the tooth is called the stone region. In the urchin *Psammechinus* it has a cross section of about 100 by 50 μm. This core region is harder than the cortex of the tooth, and so wears away less. As a result the core stands proud after the tooth has worn down a little, and the tooth is self-sharpening.

This core region is a classical composite. It is made of calcite fibres in a matrix of amorphous calcium carbonate (MÄRKEL and GORNY, 1973; BREAR and

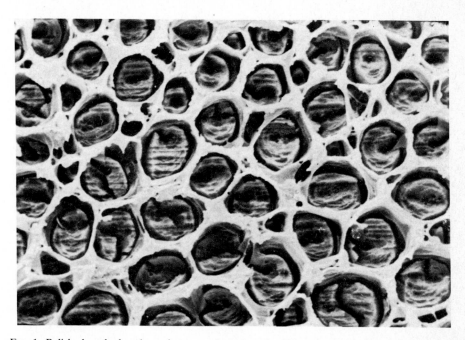

FIG. 1. Polished, etched surface of stone region of tooth of *Psammechinus*. Note the banding on the calcite fibres, showing they all have a similar orientation. The fibres are about 8 μm in diameter.

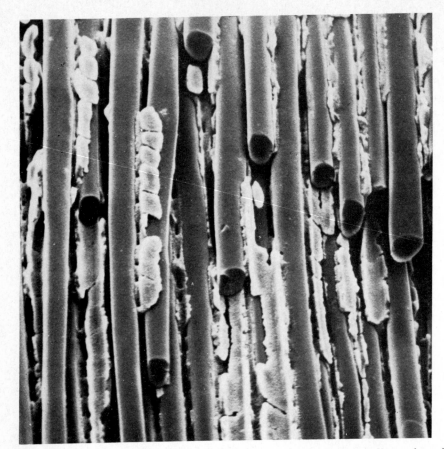

FIG. 2. Fracture surface of stone region of tooth of *Psammechinus*, showing its fibre and matrix structure. The fibres are about 8 μm in diameter.

CURREY, 1976). The fibres have a high volume fraction, about 55%, and can be shown, by etching with dilute acid, to have a rather uniform crystallographic orientation (see Fig. 1). The fibres are very long (millimetres) in relation to their diameter (microns) and have, therefore, a very high aspect ratio. A fracture surface shows characteristic fibres and matrix (see Fig. 2).

Little is known of the mechanical properties of this core region, mainly because it is so small. The hardness of the tooth increases greatly as the stone region is approached. BONFIELD (personal communication) has found the modulus of elasticity of the central part of the tooth (which will include some other material as well as the stone region) to be about 60 GPa.

3.2. Ceramic–organic composites

There is a considerable range of materials and of volume fractions of fibres and matrix in the biological ceramic–organic composites. The mollusc shell and bone are the best known examples and much of the discussion will be about them.

3.2.1. Mollusc shell

THE MOLLUSCS. The snails, clams, squids, etc. usually have a shell. This shell is characteristically made of a composite of calcium carbonate, usually in the crystalline form of calcite or aragonite with a rather tenuous organic matrix, almost entirely protein. The shell is laid down inside an outer organic layer, the periostracum, which is mainly protein. The organic matrix of the shell proper is only about 0.1 to 5% by weight.

A mollusc shell appears in a number of forms which are rather easily distinguished from each other under the scanning electron microscope and sometimes even by the naked eye. These forms have rather different mechanical properties; I list the common types here. They are shown in Fig. 3.

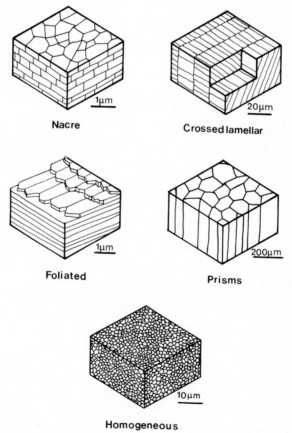

FIG. 3. Block diagrams of fibe molluscan shell types. The lesser dimension of each block is in the thickness of the shell. Note the very different scales of the blocks.

NACRE (mother of pearl) (aragonite). The crystallites are flat tablets of aragonite, arranged in sheets. The sheets are more or less parallel to the surface of the shell. The tablets, and therefore the sheets, are very thin, about 0.3–0.5 μm. Between each pair of sheets is a thin organic layer, which also extends up between the tablets. The tablets in neighbouring sheets are staggered in relation to each other so there is no direct route through the sheets.

CROSSED LAMELLAR (aragonite). This is a lamellate arrangement rather like plywood in which each lamella, which is about 20 μm thick, is built of long aragonite needles all oriented the same way. In adjacent lamellae the predominant direction alters by about 90°. The protein matrix is very tenuous.

FOLIATED (calcite). This is made of long lath-like crystals arranged in overlapping layers, with very variable amounts of change of orientation between the layers. If the change is large, the general appearance is rather like crossed lamellar.

HOMOGENEOUS (usually aragonite). This is made of small granules of aragonite, 0.3–3 μm in diameter with no particular orientation. It is really a very fine-scale rubble. The organic matrix is extremely tenuous.

PRISMS (aragonite or calcite). These are polygonal columnar crystals, of calcite or aragonite, running normally to the surface of the shell. The crystals are quite large, 10–200 μm across, and they may be several millimetres long. The organic sheets surrounding each crystal may be thick, up to 5 μm or so.

The main mechanical properties of the materials are shown in Table 2, derived from CURREY (1980). Nacre is stronger than the others. The values of Young's modulus are not very reliable. The fracture behaviour of nacre and of crossed lamellar structure, which is the most commonly occurring skeletal type in molluscs, have been examined to some extent, and I shall deal with them here.

A tensile test of nacre shows that there is some plastic behaviour. The yield point, occurring at a strain of about 0.002, is rather sharp. The plastic strain as measured over a gauge length of about 10 mm is about 0.01. However, over short regions it is almost certainly greater. When the load–deformation curve flattens out, the optical properties of the specimen change: bands of opacity appear in what was previously translucent. These opaque bands eventually coalesce and the specimen usually fails near where the first band appeared.

Presumably these bands represent the formation of tiny cavities between the aragonite plates of the nacre. It has been suggested by KENDALL (1981) that these bands are cavities that move like dislocations, the structure in any small region healing after the cavity has passed through. Whatever the exact mechanism, it is clear that a considerable amount of energy is absorbed in the plastic region of the curve.

TABLE 2. The properties of the highly mineralised mollusc shell. Median values of species means of various mechanical properties of different mollusc skeletal types (from CURREY, 1980).

Structure	N^a	Tensile strength (MPa)	Compressive strength (MPa)	Bending strength[b] (MPa)	Young's modulus (GPa)
Nacre	8	83	365	196	46
Crossed lamellar	12–15	40	271	85	52
Prisms	1–2	61	252	139	26
Foliated	3–4	40	104	102	32
Homogeneous	1	30	248	60	60

[a] N = number of species determined. For three structures not all four properties were determined for all species, so a range for N is given.

[b] Bending strength is the strength calculated as if for an entirely brittle material (bending strength = bending moment × depth/2 × second moment of area).

FIG. 4. Fracture surface of nacre of the pearl oyster *Pinctada*. The fracture has travelled across the aragonite plates. The plates are about $\frac{1}{3}$ μm thick.

The fracture surface of nacre is highly characteristic (see Fig. 4). When the fracture direction is across the plates the crack always has a rough appearance even if the crack has travelled catastrophically. CURREY (1977) showed that the crack usually travelled round the aragonite plates rather than through them. The plates in the form of nacre he investigated are quite short in relation to their width (the aspect ratio being of the order of 10:1), and so the composite fails because the fibres, in this case the aragonite crystals, pull out rather than breaking. Even so, the fracture process is reasonably energy consuming, presumably because of the shear distortions produced in the protein matrix. The work of fracture, as measured by using a three-point beinding specimen with a deep notch (TATTERSALL and TAPPIN, 1966) is about $1.6 \times 10^3 \, Jm^{-2}$ when the crack travels across the sheets. When the crack travels between the sheets (something unlikely to occur in life) the crack travels very easily (see Fig. 5) and the work of fracture is very low, about $1.5 \times 10^2 \, Jm^{-2}$. See, however, the even lower values for tooth enamel in Section 3.2.3.

This energy consuming ability of nacre is seen well in those shells that have a rather thin sheet of prismatic material on the outside, and nacre on the inside. Cracks tend to travel catastrophically through the prismatic structure, but are brought to a halt in the nacre in a relatively short distance, often a few tens of microns. Readers lucky enough to get the opportunity to eat abalone (*Haliotis* sp.) can demonstrate this by tapping the shell fairly hard against some rigid object.

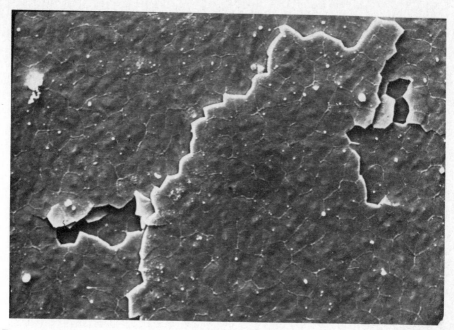

FIG. 5. Fracture surface of nacre in *Pinctada*. The crack travelled between layers of plates. Total width of photograph = 95 μm.

Crossed lamellar structure is not as strong as nacre but is, in fact, much more widely distributed among molluscs. Its plywood-like structure results in there being easy and difficult ways for cracks to travel (see Fig. 6). In the easy direction the crack has to separate two nearly plane surfaces. In the difficult direction the crack is continually having to break across the constituent laths of the lamellae. The orientation of the laths, nearly at right angles to each other, results in the crack that is travelling in the difficult direction continually being led off in two easy directions (see Fig. 7). The fronts cannot travel very far, and soon have to break across the laths to join up again into a coherent crack front.

If a bending specimen is composed of lamellae all oriented in the easy direction, the specimen will fail catastrophically at rather a low load (CURREY and KOHN, 1976). If the lamellae are oriented in the difficult direction, the specimen shows a little plastic deformation and fails at a higher load. If the specimen has the lamellae oriented in the easy direction on the tensile side of the specimen and in the difficult direction on the compressive side of the specimen, the load–deformation curve is saw-toothed; each drop in the load representing a catastrophic crack on the tensile side being brought to a halt where the lamellae change direction (see Fig. 8).

The mechanical properties of other molluscan shell types are less well understood than those of nacre and crossed lamellar structure. Quite remarkable is the so-called homogeneous structure. It is very weak; indeed it is very difficult to machine specimens from it. Prismatic material is not particularly strong, but it seems to have two valuable properties from the animal's point of

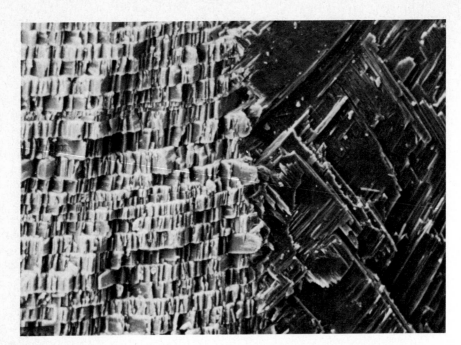

FIG. 6. Fracture surface of crossed lamellar structure in the cone snail *Conus virgo*. The crack travelled right to left. On the right hand it travelled between the laths. In the middle the shell structure is reoriented through a right angle. The crack now has to break across the laths. Cracks frequently stop at this level. The rows of laths are about 25 μm across.

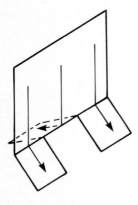

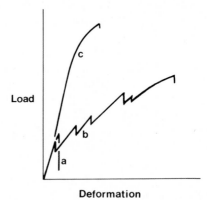

FIG 7. Diagram of the changeover, in crossed lamellar structure, from one orientation of the lamellae to another. A crack, travelling in the direction shown by the arrows, goes first in the 'easy' direction, but is then split up and has to break through lamellae to keep a coherent front.

FIG. 8. Load–deformation curves for crossed lamellar structures loaded in bending. Curve a—Whole thickness of specimen has lamellae in easy direction; Curve b—Lamellae oriented in easy direction on tensile side, difficult direction on compressive side; Curve c—Lamellae oriented in difficult direction on tensile side.

view. One is that it can be laid down very quickly. This makes it useful material to put on the outside of shells; the inner material can be laid down more leisurely and can be a stronger material. The other useful property is that it seems to be rather simple, in evolutionary terms, to alter the modulus of elasticity of prismatic structure by altering the thickness of the protein matrix between the prisms. *Solemya* is a bivalve mollusc living in fine sand. It has a very flexible prismatic shell and this flexibility enables the animal to change the volume of the enclosed contents of the shell, squirting water out in the process. This ejected water turns the sand into a slurry and allows the animal to locomote through it fairly easily (BEEDHAM and OWEN, 1965). Such a flexible shell is probably rather weak, but since the animal spends all its life buried in sand, away from predators, the weakness does not matter.

Pinna, another bivalve mollusc, has a prismatic shell of comparatively low modulus (12 GPa). The animal lives in mud with much of its fan-shaped shell exposed. If the animal is attacked by a fish, the body tissues are withdrawn below the level of the sea bottom. Muscles pull the two values of the shell very hard together. Because the valves are flexible, they become apposed on their inner surfaces. If the fish manages to break the shell, which it is not difficult to do, it gets a mouthful of shell and cannot reach the living tissue (YONGE, 1953). After the still hungry fish has left, the mollusc can build a new shell remarkably quickly: a band a centimetre wide can be produced in less than a day. This represents a volume equivalent to 5% of the body volume per day.

This sequence of events shows rather clearly the pragmatic nature of natural selection. If the feeding mode of the animal makes it very liable to be attacked, it may well be more adaptive to build a structure that can be broken without vital tissues being damaged in the process and which can then be quickly and cheaply repaired, rather than building a metabolically expensive but strong structure, which will also be slow to build but which will withstand attack.

From a biological point of view a very interesting feature of the various molluscan shell structures is their distribution among the different molluscs. The strongest material, nacre, is found in rather primitive molluscs, which have diverged less far from the ancestral types than more advanced molluscs. Most molluscs have, as it were, given up the mechanically superior material in favour of a mechanically inferior one. The selective reason for this is probably that it is difficult to lay down nacre quickly, and that for many molluscs speed of growth is probably more important than strength, particularly high specific strength. Many molluscs grow in highly competitive situations where quick growth is necessary for survival. When, later, time allows, the shell can be made strong simply by adding more of the mechanically inferior material to the thickness. The best example of this is the oyster, which has an extremely weak shell material. This animal grows in a competitive community and is one of the most successful molluscan types to have evolved (CURREY, 1980).

From the materials science point of view, perhaps the most interesting thing about mollusce shell is the strength and toughness that can be achieved by materials made of 99% or so of calcium carbonate.

3.2.2. *Other ceramic–organic shells*

Various other groups have shells made from ceramics with a small amount of protein as matrix. The brachiopods are a group of bivalved animals looking rather like bivalved molluscs but in fact being completely unrelated to them. Most of their shells are calcitic with a tenuous matrix of protein. Just as many molluscs lay down a thin layer of prismatic material on the outside and then put nacre down more slowly on the inside, so these brachiopods have a 'primary layer' of very finely divided calcite on the outside, and on the inside a layer of sheets or fibres of calcite, each bounded by a protein sheath. The general appearance in section is rather like nacre (see Fig. 9) (WILLIAMS, 1970; WILLIAMS and WRIGHT, 1970). However, we know nothing about the mechanical properties of brachiopod shells except that they are quite rigid.

(a)   (b)

FIG. 9. Diagrammatic cross-section of prisms in: (a) human enamel, and (b) brachiopod shell, showing how an interlocking structure is produced in each case. The black lines represent the organic material lining the prisms. The enamel and brachiopod prisms are about 5 μm across.

Another group of brachiopods has the need for a more flexible shell to enable them to pop in and out of their burrows. These shells have calcium phosphate as their mineral (possibly fluorapatite (McCONNELL, 1963)), and the shell has many thin layers of mineral which alternate with rather less heavily mineralized layers (see Fig. 10).

The stony corals, which form the massive coral reefs and coral islands of tropical seas, have skeletons more remarkable for their architectural grandeur than for the properties of the material itself (CHAMBERLAIN, 1978). The mineral

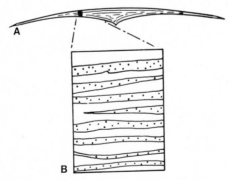

FIG. 10. The flexible shell of the brachiopod *Lingula*. Part A—Cross-section of one valve of the shell. The block is enlarged in Part B. Highly mineralised sheets—dotted; organic regions—plain. The sheets are about 50 μm thick.

is usually aragonite, often arranged in long needles as a result of sphaeritic calcification. There is a tenuous matrix of chitin, or possibly protein in some cases (WAINWRIGHT, 1963; YOUNG, 1971). The organic portion is only about 0.02–0.2% by volume and it seems unlikely that the skeleton derives much mechanical advantage from its composite nature. Nevertheless some evidence is accruing that coral skeleton that is laid down more slowly has more organic material, has more regularly arranged components and is stronger than faster-growing material. Paradoxically the denser corals are more invaded by boring organisms, presumably because the dense coral makes a more secure hiding place (HIGHSMITH, 1981). Mechanical analysis of coral is as yet rather primitive.

3.2.2.1. Birds' eggshells. Unusually for vertebrate mineralized structures the mineral of birds' eggshells is calcite rather than calcium phosphate. The structure, which is reviewed by WILBUR and SIMKISS (1968) and briefly by GILBERT (1979), is rather like the prismatic type of mollusc shell structure. The prisms originate from nucleation centres at the level of the outer organic shell membrane. As the crystals grow they abut, and so growth soon takes place almost entirely normal to the surface of the egg. The region where the crystals have not abutted is called the cone layer; the major part, where the prisms interdigitate, is the palisade layer (see Fig. 11). The prisms are typically 300 μm long, and 50 μm across, though this varies greatly with the size of the shell, of course. The prisms interdigitate in surface view. There are smaller crystallites within the prisms. There is an organic matrix, mainly protein, about 2–4% by volume. It seems to be pervasive right through the mineral. The shell is usually pierced by pores, which contain organic material, but is presumably more permeable than the mineral, and so allows diffusion of gases to and from the actively respiring embryo. The whole outer surface of the egg is covered by the cuticle, which is 5–100 μm thick, and consists of protein with a fairly high proportion of sugars. It probably keeps out bacteria and makes the cuticle waterproof. The variation in structure of various eggshells is reviewed in a long series of papers by TYLER, with references in TYLER (1969a).

The mechanical properties of hen's eggshells are of great interest to poultry

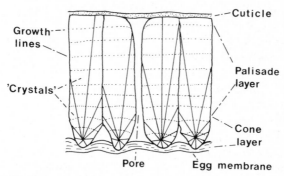

FIG. 11. Diagram of a cross-section of a small part of a bird's eggshell.

farmers, who are concerned that the shells should be strong enough to survive at least until they have been bought by the housewife. However, it is unfortunate that we have little idea of the properties of eggshell in terms that allow us to relate them to other biological composites (TYLER, 1969b). Most eggshells are rather brittle with cracks travelling for considerable distances round the shell. However, some shells, such as those of the quail *Coturnix coturnix*, are much tougher. The hatching chick has to adopt a different strategy to get out of a tough shell. The selective reasons for this difference is mechanical behaviour, and the structural differences that go with them have not yet been worked out (BOND, SCOTT, COOKE and BOARD, 1980).

3.2.3. Enamel

The teeth of vertebrates are capped by enamel, which has a different structure from the underlying dentine. Its function in teeth is to provide a hard-wearing surface for the trituration of food. Its mechanical properties have recently been reviewed by WATERS (1980). Enamel is about 97% mineral, most of which is calcium phosphate in the form of hydroxyapatite. There is about 1% by weight of organic material, mostly protein, and 2% water. Unlike the main other vertebrate hard tissues, bone and dentine, there are no cells or cell bodies in enamel. The enamel mineral crystals are about 25 nm thick, 100 nm wide and 500 nm, or possibly much more, in length. These crystals are bound together in bundles called prisms. There is very little organic material in the prisms. However, at the boundary between prisms there is a concentration of protein. The prisms are keyhole-shaped in cross section, and interlock (see Fig. 9(a)).

Enamel does not usually occur in large blocks and is therefore difficult to test mechanically. The difference in strength reported by different workers is embarrassingly large (cf. Table 3). The values for Young's modulus determined by ultrasonic testing are consistent, averaging about 74 GPa (RICH, BRENDAN and PORTER, 1967; LEES, 1968; GILMORE, POLLACK and KATZ, 1969).

The fracture properties have been determined by RASMUSSEN et al. (1976). They used a Tattersall and Tappin-type three-point bending test in two directions. In one type of specimen the crack could travel between the prisms, in the other it has to break across them. These are like the 'easy' and 'difficult' directions in nacre. The mean value of work of fracture was 13 and 200 Jm^{-2} in these two directions respectively. These are in the same ratio as, but an order of magnitude lower than, the values for the easy and difficult directions in nacre (150 and 1600 Jm^{-2}). That enamel really does have such a low toughness seems a little hard to credit. Possibly the very small size of the specimens that RASMUSSEN et al. (1976) had to use may account for these values, though there is no theoretical reason why it should.

Although static tests show enamel not to be very good at preventing cracks from travelling through it, it is possible that there is a toughening mechanism acting before any crack starts. The late Peter Fox (1980) has suggested that the very narrow pores permeating enamel have an important toughening function. Fox loaded teeth with a ball bearing and observed a small hysteresis loop in the

TABLE 3. Some mechanical properties of enamel (from human teeth, except where otherwise stated) (taken from WATERS, 1980).

Reference	Tooth	Position	Prism orientation in relation to specimen long axis	Compressive strength (MPa)	Tensile strength (MPa)	Young's modulus (GPa)
STANFORD et al. (1960)	Molar	Cusp	Variable	261	–	46.2
	Molar	Side	Across	250	–	32.4
	Molar	Side	Along	94	–	9.6
	Molar	Occlusal	Along	127	–	11.0
CRAIG et al. (1961)	Molar	Cusp	Along	384	–	84.1
	Molar	Side	Along	372	–	77.9
BOWEN and RODRIGUEZ (1962)	?	?	?	–	10.3	–
REICH, BRENDAN and PORTER (1967)	?	?	?	–	–	76.5
LEES (1968)	Bovine	?	?	–	–	73.0
GILMORE et al. (1969)	Bovine	?	?	–	–	74.0

loading–unloading curve. The hysteresis disappeared after repeated rapid loading, but reappeared if the cyclical loading was stopped for a second. Fox claims that the energy absorbed in the hysteresis loop would contribute to the toughness of the enamel, although he was not, because of his experimental set-up, able to quantify this satisfactorily. He concluded the energy absorbed was about 10^4 J per bite. This would be an important contribution to energy absorption, and indeed seems remarkably high. The energy is lost as fluid is forced to migrate through the pores in the enamel. The effective viscosity of the fluid is somewhat increased, as Fox calculated, by the electrokinetic effect. The pores will be lined with electrically charged walls. Associated with these will be layers of ions of the opposite polarity. This electrical double layer will resist disruption.

It is difficult to decide how important these effects are, or how much the toughness of enamel is enhanced by them. What is interesting is the suggestion that important energy absorption mechanisms can act in enamel *before* any cracks start to run through it. These mechanisms are strain-rate dependent, disappearing at very low rates of loading. However, the major threats to enamel are not slowly acting pressures but the sudden biting on a flake of stone or a bone that may crack the tooth before the jaw-opening reflexes can work. In man, enamel lasts for sixty years or more and cannot be repaired. It is important, therefore, that its toughening mechanisms should involve as little microdamage as possible.

In contrast to nacre, enamel prisms are oriented so that cracks travelling through the whole tooth will be able to travel in the easy direction (ANDREASEN, 1972).

Enamel, though apparently not very tough, is quite hard, the Knoop harnesss being about 370 (CALDWELL et al., 1957). This is not much below the value for apatite. It is much harder than dentine, which is a rather bone-like material underlying the enamel in most teeth. Dentine has a Knoop hardness of about 75 (CALDWELL et al., 1957). Many mammals make use of this difference in hardness for self-sharpening. The incisors of rodents grow continually, in a curve. There is enamel on the outer part of the curve, dentine on the inner part. As the tooth wears it remains chisel-sharp at the tip (see Fig. 12) through differential wear. A similar result is obtained in the teeth of cattle, in which dentine, enamel and cement, the last being another rather soft mineralised tissue, are so arranged that curved sharp ridges of enamel always remain proud of the general surface as the tooth wears down (Fig. 12).

The enamel of mammals seems rather uniform in its fine structure. However, there are other enamel-like tissues in other groups of vertebrates. In particular the teeth of sharks are covered by enamel which has been shown by PREUS-CHOFT, REIF and MULLER (1974) to be beautifully adapted to the different functions it has in different teeth.

Sharks' teeth are designed either for crushing or for slashing. In crushing teeth there is a thin layer of enamel lying on a bed of dentine. Typically the enamel layer is 0.2 mm thick, the underlying dentine 15 mm. The function of this

 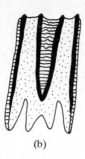

(a)   (b)

FIG. 12. Sections of two mammalian teeth, to show use of material of different hardness to produce sharp edges. (a) Rodent incisor. The enamel (black) on the outside wears less than the dentine (dotted) so producing a chisel edge. The whole tooth is about 3 mm thick. (b) Cow's molar. Differential wear of enamel, dentine and cement (thin lines) produces an accidented surface for chewing tough fibres. The tooth is about 20 mm across.

enamel is to act as a very *hard* cap to the dentine. Being so thin it can contribute very little to the stiffness of the tooth as a whole, which will be determined by the dentine. The dentine has a lower Young's modulus than the enamel. However, presumably the thinness of the enamel layer prevents it from undergoing large strains when it is compressed locally and so loaded in bending. (The crushing teeth are used to deal with snails and crabs and similar hard-shelled animals.)

The enamel consists of hydroxyapatite fibres, about 0.2 μm in diameter and 1–2 μm long, embedded in a rather tenuous matrix. The fibres lie normal to the surface, and are presumably good at resisting compression. The mechanical properties of this enamel are unknown because it is always so thin (see Fig. 13).

In slashing teeth, found in sharks adapted to the more traditional way of life, the enamel layer is thicker, with a mean of 0.6 mm among the various species examined. There are three layers of highly mineralised enamel sheathing the less highly mineralised dentine. On the outside is an extremely thin (roughly 4 μm) 'shiny' layer. This consists of mineral fibres 2.5 μm long and 0.15 μm thick. These lie in the plane of the layer but are at all angles within this plane, forming a two-dimensional random mat. There is no evidence for any organic matrix. The next layer is of 'parallel-fibred' enamel. The mineral in this layer is in the form of high-aspect-ratio needles, 2.5 μm long and about 0.03 μm across. These are embedded in organic fibres of unknown constitution. These fibres, and also their embedded mineral needles, lie in two directions, either normal to the surface of the tooth, or parallel to it (Fig. 13). In any fairly small region the fibres parallel to the surface are also parallel to each other, but this general direction may change over a few millimetres and does indeed vary according to the shape and the mechanical requirements of the tooth.

Underneath the layer of parallel-fibred enamel is a layer of 'tangle-fibred' enamel. The fibres are not completely disordered, as there are fibres that run predominantly in the radial direction. There are also fibres that run in all directions normal to them (Fig. 13).

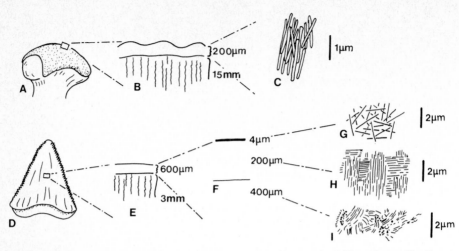

FIG. 13. Diagrams of two tooth enamels in sharks. A—Crushing tooth of *Heterodontus*. B—Cross-section of outer part of tooth. A layer of enamel, about 200 μm thick, overlies 15 mm of dentine. C—The hydroxyapatite needles of the enamel layer. D—Slashing teeth of *Carcharodon*. The teeth have serrated edges. E—Cross-section of outer part of tooth. The enamel is 600 μm thick. F—The three layers of the enamel. The shiny layer = 4 μm thick, the parallel-fibred layer = 200 μm thick, and the tough-fibred layer = 400 μm thick. G—Tangential view of the shiny layer. H—Radial view of parallel-fibred layer. The vertical lines represent apatite needles oriented normal to the surface. I—Tangential view of tangle-fibred layer.

PREUSCHOFT et al. (1974) interpret the design of these parallel-fibred enamels to be principally functioning in relation to tension. It is obvious that the parallel-fibred enamel will be highly anisotropic in tensile strength because, as with crossed-lamellar structure in mollusc shell, there will be an 'easy' way to break this enamel in tension. Although their experimental set-up is not quite clear from their description, this does indeed seem to be the case. They obtained a mean strength of 140 MPa in one direction, 35 MPa in the direction normal to this. Unfortunately, it is also not clear how high the mineralisation is in this enamel. It is almost certainly less than in mammalian enamel. A tensile strength of 140 MPa is quite respectable for a highly mineralised material. There are no measurements of Young's modulus. PREUSCHOFT et al. (1974) produce a series of models that suggests that the direction of the parallel fibres is such that the tensile stresses most likely to be produced in the teeth are always acting along them. The tangle-fibred enamel underneath will, presumably, be fairly isotropic.

They suggest also that the shiny layer on the very outside, with no detectable organic material, acts as a very hard surface, and that any cracks produced in it will not travel through the parallel fibred enamel because of the latter's lower modulus.

Although there is a great deal more to be discovered about shark enamel, it is clear that it is an interesting composite, in which the relationship between loading and structure may be particularly clearly shown.

3.2.4. Bone

Bone is the stiff biological composite we are most familiar with, yet it is remarkably difficult to analyse as a composite, because it has so many levels of organization. Bone is a composite of mineral and organic components, with some water as well. The water is of great importance in controlling the mechanical properties. Water occupies about 12% by weight in ordinary bone (GONG, ARNOLD and COHN, 1964).

3.2.4.1. Structure.

The mineral is mainly hydroxyapatite, though there is a proportion of 'amorphous' calcium phosphate. The mineral occupies about 60% by weight. The organic material is mainly the protein collagen and occupies about 28% by weight. The mineral and the organic constituents of bone occupy about equal volumes. The collagen is arranged in triple helices of helically arranged polypeptides—in a coiled coil. The triple helix is stabilised by hydrogen and other types of bonds (WOODHEAD-GALLOWAY, 1980). Such a coiled coil has a particular length, 260 nm, and is called a tropocollagen molecule. The primary structure of the polypeptide is

glycine–proline–hydroxyproline–glycine–proline–X–glycine.

Glycine has a very small side group (−H) and this allows close packing. Proline and hydroxyproline have larger side chains and make the molecule stiff. X can be any of a variety of amino acids.

The tropocollagen molecules are arranged into fibrils. The exact arrangement of the constituent molecules with respect to each other is still a matter of some debate, but it is almost certain that molecules are laid head to tail, with a gap between each head and tail (see Fig. 14). In bone this gap is important. Neighbouring fibrils are staggered a quarter of their length with respect to neighbours, and as a result there is a 64 nm periodicity in the collagen fibril (Fig. 14).

These collagen fibrils may combine together to form larger units, called fibres. However, in bone the fibrils form such an intimately bound up mass that the idea of a fibre, which is useful in considering tendon, is not really valid.

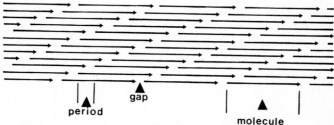

FIG. 14. Diagram of arrangement of tropocollagen molecules in a collagen fibril. There is a gap between each pair of aligned molecules. The periodicity of the whole molecule has a length equal to the distance between homologous points of adjacent molecules. This diagram shows the molecules in one plane. In reality they are in three dimensions.

Collagen is one of the most intensively studied proteins. A simple intro-duction is found in WOODHEAD-GALLOWAY (1980).

Impregnating the collagen molecules is the mineral. There is still uncertainty about the actual form of the mineral. It is most probable that in mature bone it is in the form of plate-like crystals of hydroxyapatite, about 35 nm across and 4 nm thick. The mineral may be virtually amorphous calcium phosphate when first deposited, and become transformed later. There is much uncertainty about this (GLIMCHER, 1976). The mineral seems to be deposited first in the hole region between the heads and tails of the tropocollagen molecules, so that the initial calcification has a 64 nm periodicity. Probably the arrangement of the amino acid residues in the hole region acts as an energetically favourable site, allowing precipitation (BERTHET-COLOMINAS et al., 1979). However, the amount of mineral produced quite quickly fills up the gaps, and it is then deposited within and on the fibrils and between them in the tenuous interfibrillar organic material.

The nomenclature of bony tissue at the next higher level of organization, that visible with the light microscope, is very confused. The reason for this is that there has been a tendency for anthropocentrism, because so much work on bone has been done on mammal and, in particular, human bone. It so happens that human bone has a structure rather atypical of vertebrates in general. I shall not discuss the range of bony tissues seen throughout the vertebrates, mainly because little is known of the mechanical properties of most bone. However, some of the differences have obvious mechanical consequences.

Adult bone substance is in two basic forms: 'fibrous' and 'lamellar'. In fibrous bone the collagen fibrils, although having a preferred orientation, are arranged in fairly coarse bundles which differ considerably in their orientation. It can be laid down more rapidly than lamellar bone, probably accreting at a rate five times that of lamellar bone. Lamellar bone consists of a series of lamellae, about 5–7 μm thick. In these lamellae the collagen fibrils are oriented more or less in the plane of the lamella, and usually have roughly the same orientation, though this may change over many tens of microns (BOYDE, 1972). The orientation of the collagen fibrils in neighbouring lamellae may not be the same; usually it is not. There is some SEM evidence (KATZ, 1980a) that each lamella is itself made of layers of fibrils each layer being one fibril thick. Usually the fibrils in each of these layers are all oriented in the same direction. Between the lamellae there is an interlamellar sheet about 1 μm thick which has a lower concentration of collagen fibres than the lamellae.

In general, the mineral in bone seems to be deposited in parallel to the collagen fibrils, that is to say one of the larger dimensions of the platelets will be along the long axis of the fibril. To a very large extent, therefore, describing the orientation of the collagen is sufficient to describe the orientation of the mineral.

Lamellar bone, apart from its morphological distinction from fibrous bone, also exists in two forms that are different 'historically': it can be either 'primary' or 'secondary'. Primary bone is laid down in space that has never contained bone. Secondary bone forms on surfaces that have resulted from the erosion of previously existing bone. One of the most characteristic kinds of secondary bone, much loved of elementary biology textbooks, is the secondary osteone or Haversian system. This is formed within bone tissue. The bone around a blood channel is eroded away, forming a cylinder of about 200 μm diameter. Lamellar bone is then laid down on the inside of the cylinder, until eventually the blood vessel is narrowly enclosed again (see Fig. 15). When a volume of bone has been extensively remodelled in this way it has a very characteristic appearance (see Fig. 16), and is called 'Haversian bone'. I shall be referring to it frequently in this section.

Haversian systems form round blood vessels; the vessels themselves serve to nourish the bone cells (osteocytes). These osteocytes pervade most, though not all bone. In particular the bone of the most advanced fishes does not contain osteocytes. Strangely, the function of osteocytes is not well understood. Bone seems to survive quite well as a functioning tissue even if the cells within it are dead. The cells lie in small cavities (lacunae) and these tend to be the sites of stress concentrations in the tissue. However it has been shown (CURREY, 1964b)

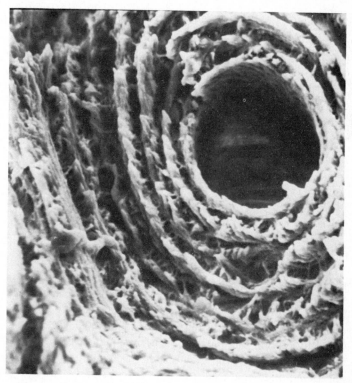

FIG. 15. Scanning electron micrograph of a fracture surface of human bone, showing an Haversian system. Note the concentric rings enclosing the central canal, which is 20 μm in diameter.

FIG. 16. Drawing of a piece of Haversian bone. There are many Haversian systems, each with a central cavity containing a blood channel shown in solid black. It is possible to work out the relative ages of some of the systems by seeing which encroaches on the other. One system is still forming and the central cavity is large. Lamellae with no blood channel are 'interstitial lamellae'. Haversian systems are about 100–200 μm in diameter.

that the shapes of the cavities and their orientation are such that, in lamellar bone at least, the stress-concentrating effect is minimised.

Blood vessels permeate most bone. They have to be incorporated into the bone material during growth, and there are various characteristic ways in which this incorporation is carried out. For a fully comprehensive review see DE RICQLÈS (1977) and previous papers cited therein. The most carefully studied bone type after human Haversian bone is 'laminar', 'plexiform' (ENLOW and BROWN, 1956, 1957, 1958) or, as I think it is best to call it, following DE RICQLÈS (1977), 'fibro-lamellar' bone. This type of bone shows how many mammals, and some dinosaurs, have solved the problem of growing a structure at a greater rate than some of its constituent material can be manufactured. This problem of speed of fabrication is one which is met in a rather different form by engineers.

It is almost certain, though no critical experiments have been performed, that lamellar bone is mechanically superior to fibrous bone. However, fibrous bone can be laid down much more rapidly than lamellar bone. Large mammals and many dinosaurs seem to face (or have faced in the case of the dinosaurs) the problem that, being large, their long bones have to grow in diameter at a greater rate, to keep up with the rest of the body's growth, than lamellar bone, the preferred material, can be deposited. Fig. 17 shows the sequence of growth events used in overcoming this problem. On the outer surface of the whole bone fibrous bone tissue is deposited quickly as a kind of scaffolding. Blood vessels are trapped in the spaces in the scaffolding. Lamellar bone then starts to be laid down on the internal surfaces of these spaces. However, while this is going on at the leisurely pace adopted by lamellar bone, fibrous bone is making another scaffolding, and then another, and so on. Eventually the first space is entirely filled with lamellar bone, trapping the blood channels.

As a result of all this activity a tissue is formed which in essence consists of

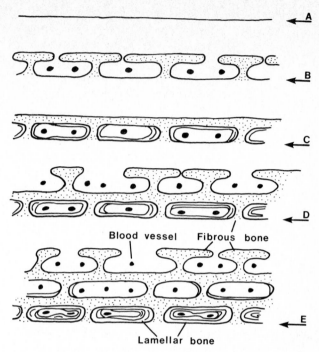

FIG. 17. Development of fibro-lamellar bone in a fast-growing mammal or dinosaur. In each stage the arrow indicates the starting level. Stage A—The original bone surface; Stage B—Fibrous bone scaffolding laid down quickly, trapping blood vessels; Stage C—Fibrous layer complete, lamellar bone starts to appear on internal cavities; Stage D–Second generation of fibrous scaffolding appears. More lamellar bone on inner cavities; Stage E—Third generation of fibrous scaffolding. Lamellar bone appears on completed cavities in second generation fibrous bone. Lamellar bone fills first generation cavities, tightly surrounding blood vessels.

alternating layers of rather disorganised fibrous bone and lamellar bone, the latter having its fibres oriented mainly in the direction of the long axis of the anatomical bone, and containing two-dimensional networks of blood channels. It turns out that the mechanical properties of this structure containing both fibrous and lamellar bone are superior to those of Haversian bone, which contains only lamellar bone. I shall mention possible reasons for this on page 534. It is unfortunate that we have very little information concerning another type of lamellar bone: lamellar-zonal bone (DE RICQLÈS, 1977). This is entirely lamellar and is, unlike Haversian bone, entirely primary. I suspect it would be very strong indeed in one direction at least. It occurs mainly in reptiles.

The level of organization I have been discussing here is that in which the structures are visible with an ordinary light microscope. There is a higher level of organization, which is visible to the naked eye. At this level it is possible to distinguish compact bone, in which the bone is a solid lump (with cavities only for blood vessels and osteocytes) and cancellous bone, in which the bone material is arranged in struts and sheets, with fatty marrow filling the interstices. It is beyond the scope of this chapter, which is concerned with biological *materials*, rather than *structures*, to discuss the various ways in which cancellous bone, which has a low modulus and strength, fulfils various vital roles in bones. I would say, however, that some deep insights into the power of natural selection to produce structures of high mechanical efficiency will come from a study of the interrelationships of compact and cancellous bone.

3.2.4.2. The mechanical properties of bone. The mechanical properties of bone have been the subject of countless studies, and the variation in the results reported is rather alarming. However, in the last twenty years or so the values are settling down, now that some of the more obvious causes of variation have been eliminated. Table 4 shows some representative values, taken from tests carried out on wet bone under reasonably controlled conditions. I shall discuss the properties of compact bone only.

MODULUS OF ELASTICITY. There is a fair agreement between the values obtained for Young's modulus in quasistatic loading and those obtained using ultrasonic techniques. The value for 'ordinary bone' is in the region of 12–25 GPa. There has been considerable interest recently in attempting to determine all the elements of the stiffness tensor of bone. Doing so involves making some assumptions about the symmetry of bone (KATZ, 1980a; LAPPI, KING and LE MAY, 1979; AMBARDAR and FERRIS, 1978; LANG, 1970). (It must be stated that the values of LAPPI et al. are surprisingly low.) Bone is clearly anisotropic in its mechanical properties, as one would expect from its structure; bone tissue has a very definite grain along the length of the anatomical bone. Workers who have considered the stiffness tensor, particularly KATZ and his co-workers, have suggested that bone can be treated, without too much inaccuracy, as transversely isotropic (KATZ, 1980b). This is reasonable for Haversian bone, but is less likely to be true for fibro-lamellar bone, which is found extensively in many mammals.

TENSILE STRENGTH. This has been determined many times in quasistatic loading, and a value between 120 and 200 MPa is usually found. Bone is very anisotropic in tensile strength (see Fig. 18). However, see page 532, this, like all the other mechanical properties, is likely to vary considerably in specialised types of bone material.

COMPRESSIVE STRENGTH. This is in general greater than the tensile strength, though not markedly so. A value of 250 MPa is respresentative.

THE LOAD–DEFORMATION CURVE. Tensile tests on carefully prepared wet specimens of compact bone show a sharp distinction between the elastic part of the load–deformation curve and the later, post-yield part. The curve is almost flat after yield, being little strain hardening there. The strain at yield is about 0.005. The greatest ultimate strain measured is about 0.03 (REILLY, BURSTEIN and

TABLE 4. Some values of mechanical properties of compact bone[a] (taken directly from, or derived from information in REILLY, BURSTEIN and FRANKEL (1974) and REILLY and BURSTEIN (1975)).

Property	Man		Cow			
	Haversian bone		Haversian bone		Fibro-lamellar bone	
	Parallel	Normal	Parallel	Normal	Parallel	Normal
Young's modulus (GPa)	17.0	11.5	22.6	10.2	26.5	11.0
Poisson's ratio	0.46	0.58	0.36	0.51	–	–
Tensile strength (MPa)	148	49	144	46	167	55
Compressive strength (MPa)	193	133	254	146	294	–
Ultimate tensile strain	0.031	0.007	0.016	0.009	0.033	0.007
Ultimate compressive strain	0.026	0.028	0.016	0.031	0.014	–
Yield tensile strain	0.007	0.004	0.006	0.004	0.006	0.005
Yield compressive strain	0.010	0.011	0.011	0.014	0.010	–

[a] The variability of the properties is such that the original papers should be consulted before assertions about differences are made.

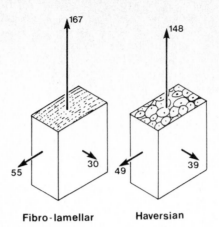

Fibro-lamellar   Haversian

FIG. 18. Block diagram of the strength, in different directions, of cow's cortical bone. The longest dimension of the blocks is in the long axis of the bone, the shortest dimension is in the radial direction. The vectors represent the tensile strength in MPa. Notice in particular that the anisotropy is rather greater in fibro-lamellar bone (5.6:1) than in Haversian bone (3.8:1) (data from REILLY and BURSTEIN, 1974).

FRANKEL, 1974). These measurements of plastic strain were made over quite a large gauge length, so it is probable that the strain locally may be greater, because not all the gauge length will strain to anything like the same extent. The region of plastic deformation seems to spread from the region that first yields to remoter parts of the specimen. This can be observed by the naked eye by viewing a tensile specimen against a strong light. The bone appears unchanged until it yields, when a region of opacity spreads from some part of the specimen. This opaque region is obviously like that which appears in nacre when it yields.

CURREY (1979) has found some extreme values that different mechanical properties of bone can show under the influence of different selective pressures. He compared an earbone of a finwhale (the tympanic bulla), the femur of a cow and the antler of a red deer. Compared with the femur, the bulla needs to be very stiff for acoustic reasons. As it is hidden deep inside the whale it is not exposed to large forces, so it does not need to be strong. The antler of the deer, on the other hand, is exposed to very severe impact loading during the harem-gathering fights the males indulge in. On the other hand, the antler does not need to be particularly stiff. The requirements for the femur are intermediate between those of the others. The mechanical properties of the three bone types are shown in Table 5, together with their mineral contents. It can be seen that the work of fracture is strongly inversely related to the amount of mineralization. The fracture surface of work of fracture specimens is shown in Fig. 19. The difference is striking, and is almost certainly brought about by differences in the volume fraction of the mineral in the three composites.

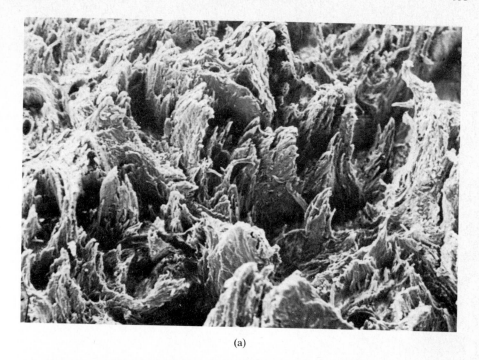

(a)

(b)

FIG. 19. (a) Fracture surface of red deer antler. The crack travelled slowly. Width of photograph about 750 μm; (b) Fracture surface of whale tympanic bulla. The crack travelled slowly. Width of photograph about 750 μm.

TABLE 5. Value (sample size, standard error) of the four physical properties of Antler, Femur and Tympanic Bulla. In alternate rows the values are expressed as a percentage of the highest value. Work of fracture from Tattersall–Tappin-type test. Horn is shown, though it is not bone. Bending strength is calculated as if there were no nonlinear deformation (from CURREY, 1979).

Property	Horn	Antler	Femur	Bulla
Work of fracture	24 582 (9, 528)	6186 (5, 552)	1710 (13, 207)	200 (3, 8)
(Jm^{-2})		100	28	3.2
Bending strength		179.4 (7, 6.3)	246.7 (6, 11.6)	33.0 (3, 6.6)
(MPa)		73	100	13
Young's modulus		7.4 (7, 0.34)	13.5 (6, 0.95)	31.3 (10, 0.97)
(GPa)		24	43	100
Mineral content		59.3 (5, 0.49)	66.7 (25, 0.17)	86.4 (2, 0.05)
(% mass)		69	77	100

Bone and wood seem to be the only two reasonably stiff biological materials on which a fracture mechanics approach more complicated than determining the work of fracture from a Tattersall–Tappin-type test has been tried. The reason for this is that these two substances are of interest to many workers, and also that they come in fairly large lumps. Even so, the number of studies on wet bone is small. Table 6, derived from BEHIRI and BONFIELD (1980), shows the results.

The recent fairly extensive work of BONFIELD and his co-workers has, unfortunately, been performed on specimens oriented so that the crack travels along the grain of the tissue. This is for technical reasons. The values for the fracture mechanics properties are likely to be higher in other directions. BEHIRI and BONFIELD (1980) make the interesting observation that when the crack velocity is controlled (less than about 2.5×10^{-4} ms^{-1}), G_c and K_c both increase with increasing crack velocity. When the crack travels catastrophically, however, the work of fracture drops by an order of magnitude. This change is correlated with a change in the fracture surface, such as was first noted by PIEKARSKI (1970). At low velocities the fracture surface is rough, with many small pull-outs, and the crack tends to be diverted along the grain features of the bone. When fracture is catastrophic, the fracture surface is much smoother. It will be interesting to see what happens if this work can be applied to cracks travelling across the grain, which would be much more like a clinical fracture.

3.2.4.3. Modelling bone as a composite. There have been some attempts to model the mechanical behaviour of bone in terms of its composite structure. My own early attempt (CURREY, 1964a) makes amusing reading, for its primitiveness rather than for hopeless wrongheadedness. Since those early days the subject has advanced a long way. It is probably best to consider the elastic properties and the fracture behaviour separately.

TABLE 6. Fracture mechanics values for bone, mainly derived from BEHIRI and BONFIELD (1980).

Reference	Method	Bone	Crack travel relative to grain	Crack velocity (10^{-5} ms^{-1})	G_c (Jm^{-2})	K_c (MNm$^{-3/2}$)	W (Jm^{-2})
MELVIN and EVANS (1973)	Single edge notch	Bovine femur	Along	–	1970	3.2	–
			Across	–	4330	5.6	–
MARGEL-ROBERTSON (1973)	Three point bending	Bovine femur	Across	–	–	6.6	–
WRIGHT and HAYES (1977)	Compact tension	Bovine femur	Along	–	1200	3.5	–
BONFIELD et al. (1978)	Compact tension	Bovine femur	Along	–	1850	3.8	–
CURREY (1979)	Three point bending	Bovine femur	Across	–	–	–	1710
		Deer Antler	Across	–	–	–	6190
		Whale Bulla	± Isotropic	–	–	–	200
BEHIRI and BONFIELD (1980)		Bovine Tibia	Along	1.75	1740	4.5	760
				3.6	1810	5.0	1340
				12.6	2250	5.2	1900
				23.6	2800	5.4	2120
				Catastrophic	–	–	125

3.2.4.3.1. Elastic properties. KATZ (1980b) has argued that no model that considers bone merely to be mineral fibres in a matrix of collagen will adequately describe the behaviour of bone, even if the various orientations of the two components throughout the bone are taken into account. They assert that such an analysis does produce quite a good fit between theory and experiment at the level of single Haversian systems. However, in life, Haversian systems are separated from each other by the so-called cement lines (cement sheaths would be a better name for these three-dimensional structures). Using experimentally determined values and also some calculated ones KATZ (1980b) fitted the moduli determined from their own ultrasonic data (YOON and KATZ, 1976a, b) and those of BONFIELD and GRYNPAS (1977). Their model is derived from that of HASHIN and ROSEN (1964) for hollow-fibre reinforced composite materials, in which the fibres are assumed to be arranged with hexagonal symmetry. The fit is fair, but by no means exact. This lack of agreement must, furthermore, be considered in light of the fact that several of the model's parameters are derived from the experimental values.

A major problem is that the Haversian systems are by no means all the same size. Furthermore, they have different amounts of interstitial material between them. This interstitial material is partly true matrix material, whose mechanical properties are virtually unknown, and partly the remnants of old Haversian systems which will have quite different properties from the matrix material. Nevertheless the fit is sufficiently good to make it worthwhile refining the model, and this KATZ and co-workers are currently doing.

It must be remembered, of course, that the tissue that KATZ and others are trying to model is only one of the many different types of bone material found in the mammals. It seems that the amount of mineralisation found in bone is at a sensitive volume fraction. In the range of values found in quite ordinary bone there is a very strong dependence of Young's modulus on mineralisation. For instance, CURREY (1975) found that over the range of 63 to 67% mineralisation the Young's modulus of cow's femoral bone increased from 19.9 to 25.3 GPa. In other words a 6% increase in mineralisation produces a 20% increase in Young's modulus. CURREY (1979) showed that compact bone of different types shows an increase in Young's modulus from 7.4 to 31.3 GPa over a range of mineralisations from 59.3 to 86.4%. Any model of bone must be able to account satisfactorily for such sensitivity to mineralisation.

The satisfactory modelling of one type of tissue, Haversian bone, would be a considerable advance. However, it is unlikely that one could extrapolate directly from such a model to other, more common, histological types, such as fibro-lamellar bone. The reason for this is that the mechanical properties of the fibrous bone component are totally unknown. In adult bone fibrous bone is not found on its own but always in association with lamellar bone, so it is difficult to separate the effects of the two.

There have been several suggestions that the presence of mineral may actually alter the behaviour of the collagen molecule itself in some way (HUKINS, 1978; LEES and DAVIDSON, 1977; MCCUTCHEN, 1975). These suggestions are interesting but, although they are worth reading, I personally do not find them convincing.

Nevertheless it is true that reasonable progress is being made towards an adequately quantitative modelling of some part of the elastic behaviour of adult mammalian bone. The situation is not so healthy for fracture behaviour.

3.2.4.3.2. Fracture. Early treatments of the fracture of bone, considering bone as a composite material (Currey, 1964a; Mack, 1964), concentrated on the lowest level: that of collagen as the matrix and mineral as the fibre. This concentration has turned out to be not very fruitful. The problem is that it is difficult, there being such an intimate relationship between the mineral and the protein, to have much idea of how the protein is behaving. In the previous section I have mentioned the ideas of McCutchen (1975), Hukins (1978) and Lees and Davidson (1977), about how the presence of the mineral may alter the elastic behaviour of the collagen. This alteration would also, no doubt, affect the yield and fracture behaviour. Furthermore, at the moment we can hardly talk usefully about the aspect ratio of the mineral, a critical variable, because we know so little about the details of the mineral morphology. Also, at this level we cannot visualise the fracture surface clearly enough to make out, for instance, whether the minute crystal blocks pull-out from the matrix, or not.

However, at the next level of size things are better because the fracture surface becomes usefully resolved by the scanning electron microscope. Before discussing the fracture event we should consider the yield behaviour of bone.

When a bone specimen is loaded into the yield region of its load–deformation curve and then unloaded it shows some residual strain at zero load. On reloading it is more compliant than before. The stress achieved on each loading–unloading–reloading cycle does not change much (see Fig. 20). It seems likely that these aspects of the behaviour of bone could best be explained by assuming that as the bone is loaded into the yield region many tiny cracks form in it. These do not, however, spread and coalesce. Such cracks would account for the increased compliance on reloading. If the cracks were all completely smooth, then on unloading there would be no residual strain. In

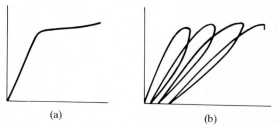

(a)   (b)

Fig. 20. Load–deformation curves for compact bone in tension. (a) A typical curve for a neatly machined, wet specimen; (b) A typical curve for a specimen unloaded and reloaded three times. Note the residual strain after each unloading, and the increased compliance on reloading.

such an irregular material as bone, however, the cracks, once formed, would not all be able to close up again.

The fact that the curve is fairly flat-topped is not explained ipso facto by the formation of cracks. I suggest that what is happening is this: At any level in the specimen we can imagine many elements (without specifying what level of organization we are imagining). Some will be weaker and some stronger. When a weak element cracks the crack will spread a little way, but will then be stopped by some crack arresting mechanism. This event will have two consequences.

(1) There will be an increased tendency for nearby elements in the section to crack because of the *local* high stresses caused by the crack.

(2) The stress in the section *as a whole* will be increased because its effective area has been reduced because of the crack. The effect will be trivial at first.

Consider first the events at the section that is going to break eventually. There is some distribution of strengths of the different elements. Usually the distribution will be a few weak elements, many more somewhat stronger elements, and many many more elements somewhat stronger again. The first element to break is unlikely to have any almost equally weak elements around it, and so the formation of one microcrack is unlikely to cause another. Now the load is increased slightly. Soon another element in the section will crack. The tendency of any element to crack will be related to its strength, the amount of the cross-section that has already cracked (because this will raise the stress generally across the cross-section) and the presence near it of any cracks (because these will modify the stress locally). A small increase in load will induce more and more elements in the section to crack, and sooner or later a cascade will occur and the section will break.

Such a set of events would explain why the stress does not increase much after the yield point has been reached. It does not, however, explain why there is so much strain in the yield region. To help explain this we can make use here of an observation by BURSTEIN and his co-workers (CURREY and BREAR, 1974). When a wet bone specimen is loaded in tension and the curve bends over, showing that yielding is occurring, part of the specimen goes opaque. This opacity does not appear at different places all over the specimen; it starts at one level and spreads along the gauge length. When the specimen breaks, it usually does so very close to the level from which the opacity started to spread.

Presuming the opacity to indicate some kind of microcracking, we can explain the spread of the opacity as being caused by the effect of the section that is beginning to crack on the sections on each side of it. Cracks in the failing section will have stress concentrations at their ends which will locally raise the stress in neighbouring sections. These volumes of greater than average stress will increase the likelihood of cracks appearing in the neighbours. Although the increase in compliance of the whole specimen produced by a single microcrack will be tiny, their cumulative effect will be great.

So far we have merely assumed that microcracks appear in bone when it yields in tension. I shall briefly review the evidence for this assumption.

Optical effects. I have mentioned the opacity that appears in a yielding specimen. This somewhat fleeting effect can be made permanent by loading a specimen in bending and then immersing it in stain (Currey and Brear, 1974). The strain settles on surfaces, particularly rough ones, and the yielded region on the tensile side, which had become opaque, is now picked out in colour. The colouring is not uniform but extends in thin wavy lines part way across the breadth of the specimen. Similar diffuse lines can sometimes be made out at the ends of tensile cracks that have failed to break the specimen in two. The observations do not conclusively demonstrate cracking, but it is difficult to see how the stain could get into the interior of the yielded part of the specimen unless passageways were opened up for it.

Submicroscopic examination. There have been a few reports in the literature of the appearance of microcracks in bone that has yielded but not fractured (Carter and Hayes, 1977; Ascenzi and Bonucci, 1976; Frost, 1960). The observations of Carter and Hayes, who loaded bending specimens six or so times into the yield region, were particularly clear. They reported "*The damage ... consisted primarily of separation (or debonding) at cement lines and interlamellar cement bands .... . Occasional microcracking of interstitial bone was also observed. High magnifications of debonding around an osteon revealed significant fibrous tearing.*" (Carter and Hayes, 1977, p. 269).

It must be added, however, that often it is very difficult to see evidence of microfracture in bone that has yielded. Deer's antler is very tough and a specimen loaded in impact will deform into a bow without breaking. Examination of sections from the part of the specimen that has gone opaque does not reveal any cracks at light microscope level. What cracks there are must be very small indeed.

Acoustic emission. Netz, Eriksson and Stromberg (1980) loaded bones in bending and listened (electronically) to the noise they made. The noise made by the generation or spreading of cracks is, apparently, distinguishable from the noise made by the movement of dislocations (Dunegan and Green, 1972). Essentially all the noises made by bone were characteristic of microcracking. Furthermore they occurred in, or after, the part of the load–deformation curve where the bones started to yield. Knets, Krauya and Vilks (1975) report similar results with an 'avalanche-like rise' in the number of cracks as the bone approached failure. However, as these workers probably tested their specimens dry (the paper is not clear on the point) their results are less relevant to the life situation than those of Netz and his co-workers.

It seems reasonable to conclude that the yield behaviour of bone in tension is the result of the formation of innumerable cracks, which, because of the way various elements of the bone, the Haversian systems, lamellae, fibrils and so on are put together, cannot travel far, and are soon brought to a halt. This ability to crack, yet not fail, accounts for the toughness of bone, such as it is.

4. Arthropod cuticle

ARTHROPODS. The insects, crustaceans, spiders, millipedes and so on are a hugely successful group of animals. Biologists are not prone to attribute success in groups to particular single features but it must surely be the case that the arthropods have been helped by the development of their remarkable cuticle. I shall discuss insect cuticle first, and then describe some variants on the basic theme.

4.1. Structure

The skeleton of insects is an exoskeleton, forming an almost complete covering to the body. To allow movement, relatively stiff sclerites alternate with flexible membraneous areas. However, the chemical compositions of the stiff and compliant areas are not very different (for a review of the chemistry of insect cuticle see ANDERSEN (1979)). A comprehensive discussion of the structure of cuticle is found in NEVILLE (1975).

The main fibrous component in cuticle is chitin, a linear polysaccharide of N–acetyl–D–glucosamine. About 20 molecules come together in parallel forming rods of very high aspect ratio, about 3 nm in diameter. The matrix in insect cuticle is almost entirely protein. Many different protein fractions are extractable from cuticle. Their amino acid composition shows a considerable amount of glycine and valine (both amino acids with small side chains). This is a frequent characteristic of fibrous structural proteins. There is some evidence that the protein close to the chitin fibrils may be oriented in relation to them (HACKMAN, 1960). Some part of the protein seems very firmly attached to the chitin fibrils, while most of the matrix is more loosely attached (HILLERTON, 1980).

The chitin fibrils are arranged in layers. In each layer they are about 6 nm apart and all oriented in the same direction. The orientation of the fibrils usually alters slightly from layer to layer. In the very frequently seen 'helicoidal' arrangement as, for instance, in the water bug *Hydrocirus*, the individual fibrils are about 4.5 nm in diameter and the distance between their centres is about 6.5 nm. This is also the thickness of each layer. The preferred orientation of neighbouring sheets is about 7–8° different and always in the same sense, clockwise or anticlockwise going in the same direction through the cuticle. As a result the preferred direction changes through 180° in about 25 layers; this is a thickness of 160 nm. In a cuticle more than a few microns thick, therefore, the fibres will be pointing in all directions in the plane of the cuticle, which will be isotropic in the plane (see Fig. 21). In low power electron microscopy cuticle appears to be arranged in lamellae, each lamella corresponding to a set of

FIG. 21. Helicoidal structure in arthropod cuticle. The chitin fibres in each layer are parallel, but different in orientation from those in the layers above and below. This exploded diagram is not to scale; for dimensions see text.

layers that change through 180° from top to bottom (see Fig. 22). A rather remarkable feature of many insect cuticles, from the composite point of view, is that the endocuticle laid down at night is different from that laid down during the day. The night cuticle is laid down in the standard helicoidal arrangement; the day cuticle, however, is laid down unidirectionally (NEVILLE, 1975). The mechanical significance of this bizarre alternation is obscure.

There are many variations on this pattern in the insects; the thickness of each lamella (in which the preferred orientation of the fibres rotates through

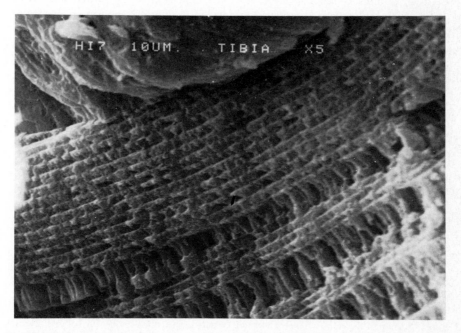

FIG. 22. Freeze-fractured surface of locust tibia, shows the lamellae. Towards the bottom are some unidirectional layers. Width of photograph 25 μm.

180°) may vary, even in the same cuticle, from 0.5 to 0.1 μm or less. A lamella may take twenty seconds to lay down, or six hours.

The helicoidal, isotropic arrangement is common, but frequently there are non-lamellate regions, in which the fibrils are all oriented in the same direction through many layers, not only in 'day layers' but in places where good mechanical reasons prevail, for instance, in tendons where the load is always in the same direction.

Another feature of great mechanical importance is whether the cuticle is sclerotised (stiffened or tanned) or not. Ordinary insect cuticle consists of an outer lipid layer, the epicuticle, which we shall not consider further, and, lying inside this the procuticle, which contains protein and chitin. The outer part of the procuticle may be sclerotised, in which case it is called the exocuticle, and the inner part, which is called the endocuticle, is unsclerotised (see Fig. 23). (As a biologist I apologise for all these names, and can only plead that they correspond to important mechanical differences.) The sclerotisation of the exocuticle results in its becoming harder, as measured by microhardness tests, and almost certainly much stiffer, though this latter fact is difficult to assert because it is very hard to measure separately the mechanical properties of the exo- and endocuticle.

The chemical mechanism involved in this stiffening is still a matter of dispute. It is generally agreed that the process involves the protein matrix and not the chitin component. The most widely accepted story is that during the process of tanning or sclerotisation the proteins become covalently bonded to each other via N–acetyldopamine. ANDERSEN (1974, 1976) suggested that there are two places in which the molecule can be oxidised and that these lead to a transparent or dark cuticle (see Fig. 24). As cuticle becomes sclerotised it loses water. VINCENT (1980b) has recently been championing the view that the loss of water is the critical feature in sclerotisation. He suggests that as water is

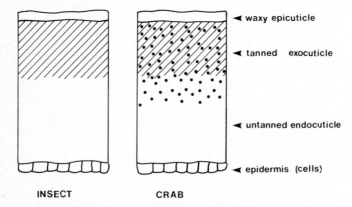

INSECT   CRAB

FIG. 23. Diagram of the cuticle of insect and crab cuticle. Insect cuticle is typically between 20 and 500 μm thick, crab cuticle between 50 and 5000 μm. The sclerotised (tanned) exocuticle is represented by diagonal lines. In the crab the cuticle is partially calcified (represented by spots).

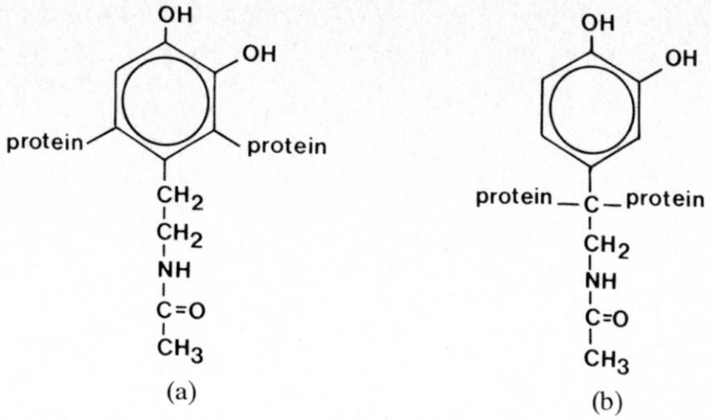

FIG. 24. Two schemes for sclerotisation in arthropod cuticle. (a) Quinone tanning; (b) β sclerotisation. ANDERSEN (1979) suggests that Quinone tanning leads to colourless cuticle, β sclerotisation to dark cuticle.

progressively excluded from the cuticle that protein molecules become more and more able to form secondary bonds with each other rather than with the water. From the biological point of view sclerotisation has two important effects: (i) it alters the mechanical properties of the cuticle, and (ii) it makes the cuticle resistant to degradation by enzymes. As a result of the latter, once the cuticle has been sclerotised it cannot expand, nor can it be resorbed by the animal. Arthropods go through a series of larval stages (instars). At the end of each instar the animal emerges from its exoskeleton through some point of weakness. The cuticle round the newly emerged animal is quite pliant and the animal inflates itself to its new size, stretching the cuticle in doing so. The cuticle then becomes sclerotised and is able to support the animal in day-to-day locomotion and similar activities. Frequently the endocuticle is thicker than the exocuticle. Before the animal moults again much of the endocuticle will be degraded, and the metabolically valuable amino acids in the protein matrix are absorbed into the body fluids to be used again. As I mentioned above the exocuticle cannot be so retrieved.

4.2. Mechanical properties

4.2.1. Young's modulus

Table 7, mainly taken from VINCENT (1980b), gives the values of Young's modulus of various cuticles. The range of values is very large, about seven orders of magnitude. The most compliant—the intersegmental membrane of the female locust—is a remarkable material. Apart from its low Young's modulus it shows extreme stress softening, in that in successive loading–unloading–loading cycles the hysteresis is almost complete, and the load required to extend the cuticle to near the previous strain is extremely small. The amount of extension possible is very great—an engineering strain of 1200% (natural strain of 2.5) being possible without damage (VINCENT, 1975). The molecular mechanism that allows all this is not clear. The chitin fibrils in

Table 7. Values of Young's modulus of different kinds of arthropod cuticle.

Animal	Cuticle type		Young's modulus (Pa)
Locusta (Locust)	Female intersegmental membrane		10^3
Schistocerca (Locust)	Apodeme (tendon)		1.3×10^{10}
	Tibia cuticle		1.7×10^{10}
	Prealar arm (rubber hinge)[a]		2×10^6
Rhodnius (blood-sucking bug)	Abdominal cuticle	{ Unplasticised	6.4×10^8
		{ Plasticised	2.5×10^7
Calliphora (blowfly)	Untanned pupal skin		7.3×10^7
	Tanned pupal skin		2.4×10^8
Phormia (hoverfly)	Wing		6.1×10^9
Carcinus (crab)	Carapace } mineralised		1.3×10^{10}
	Claw }		1.8×10^{10}

[a] The rubber hinge of *Schistocerca* is a true rubber, made of the protein resilin. I do not discuss it further, as it is not a composite (Weis-Fogh, 1961).

the membrane are apparently all oriented normally to the direction of extension, and so presumably act only as a filler, rather than as a reinforcing fibre. The structure of the protein matrix is unknown. The function of this material shows a good structure–function relationship. The female lays eggs in pods in the sand. The posterior end of her abdomen has a self-contained burrowing apparatus which digs down into the sand. The traction on the abdomen causes the intersegmental membrane to stretch (the stiff sclerites, alternating with the membranes, remain of constant length). Eventually a strain of 1000% or so is reached. Periodically the digging apparatus stops digging and turns to packing the walls of the hole. It is here that the complete hysteresis is important. When the digging stops, the membrane does not spring back, being kept extended by the weight of the abdominal contents, about 0.1 N. When the eggs have been laid, the abdomen is withdrawn by muscles inside it. Presumably, somehow the changes that have taken place in the membrane are reversed in the intervals between egg laying.

This composite cuticle is a highly specialised one, showing one extreme that the protein–chitin composites can achieve. Near the other extreme are the apodemes of the locust extensor tibiae muscles. Apodemes are the arthropod versions of vertebrate tendons; they are straps connecting muscles to skeletal structures. The chitin fibrils in apodemes tend to be highly preferentially oriented because the loads are applied only in one direction. Ker (1977) showed that, as a result, the apodeme he investigated was very anisotropic, having a Young's modulus of 11 GPa parallel to the chitin fibrils and 150 MPa normal to them. The highest reliably reported value for Young's modulus is 17 GPa in the cuticle of the locust tibia (Bennet-Clark, 1975). The specific

Young's modulus of cuticle is about $1.4 \times 10^7 \, \mathrm{N \, kg^{-1} \, m}$. This is quite respectable. The corresponding value for mild steel is $2.5 \times 10^7 \, \mathrm{N \, kg^{-1} \, m}$.

KER (1977, unfortunately not otherwise published) was able to analyse the elastic properties of the apodeme as a conventional composite. Usually the cuticle of an insect is effectively isotropic in the plane of the cuticle. This makes it almost impossible to separate the contributions of the matrix and the fibres to the mechanical properties. KER performed standard (though very difficult) experiments, loading the tiny apodeme in different directions. The volume fraction of the chitin component was 17%. The modulus of the matrix was 0.12 GPa, the modulus of the fibres 85 GPa. This is a rather surprisingly large difference in modulus between the two components. KER was able to increase the stiffness of the matrix somewhat by fixing it with glutaraldehyde, and the stiffness of the apodeme increased by the appropriate amount in the different directions. KER also has some, rather less direct, evidence concerning the matrix material in 'standard' sclerotised cuticle. In this it is much stiffer— 1.8 GPa.

The blood-sucking bug *Rhodnius* takes very large infrequent meals of blood. It has been shown that when the insect starts to feed the abdominal cuticle, which is unsclerotised, becomes plasticised and much more viscoelastic. This increase in deformability allows the abdomen to accommodate the blood. The plasticisation is reversible, and seems to be under nervous control. The mechanism producing the change is unclear, but it may be a nervously mediated change in pH (REYNOLDS, 1975a, b; BENNET-CLARK, 1962).

4.2.2. Strength and toughness

The strength of cuticle is shown in Table 8. Most values of tensile strength are about 70–150 MPa. Unfortunately it is difficult to separate physically sclerotised exocuticle from the unsclerotised endocuticle, so it is not at all clear how reliable these results are. Even less is known about the toughness of insect cuticle. It is clear from examination of fracture surfaces that cracks probably have difficulty in running through unsclerotised cuticle. However, as I shall

TABLE 8.[a] Strength of cuticles.

Group	Genus	Type of cuticle	Tensile strength (MPa)
Crustacea	*Carcinus*	Carapace	32
	Cancer	Carapace	35
	Penaeus	Carapace	28
Insecta	*Schistocerca*	Tibia	95
	Schistocerca	Apodeme	600

[a] The very high value for apodeme is from BENNET-CLARK (1975). The chitin of apodeme is very highly oriented, which no doubt accounts for its high strength. HEPBURN (1976) gives a mass of values for strength. Unfortunately they are somewhat confused, some being clearly wrong by an order of magnitude, and I have not included them here.

show in the next section, cuticle is probably designed for stiffness rather than strength.

4.2.3. Relationship of cuticle structure to function

Insect skeletal structures, being 'exoskeletal', and therefore lying outside the living tissues of the animal (unlike our bones, which are endoskeletal) tend to be rather thin-walled tubes. Insects are mostly flying animals, and therefore the premium on lightness is considerable. Thin-walled tubes tend to collapse by buckling, rather than by fracture. Indeed, a crude prodding survey by me shows that this is so.

Buckling is resisted by high stiffness and the specific stiffness of cuticle is high. Very often, however, the structures, being so thin-walled, deform considerably on being loaded and then spring back unharmed when the load is removed. In places this strategy will not work. For instance, the cuticle in the region of joints in the leg of the locust is very highly stressed and must be stiff in order not to deform very much and ruin the kinematics of the joint. The high sclerotisation necessary to produce this stiffness makes the cuticle brittle. If one scratches the cuticle near the 'knee' of a locust, then, when it jumps, the leg snaps, merely from the muscular activity of the animal. The fracture surface is quite conchoidal and glassy-looking (BENNET-CLARK, 1975).

A very characteristic feature of insect cuticle is its division into the stiff exocuticle and the compliant endocuticle. Why is the whole cuticle not sclerotised? We do not know, but there is one explanation that relates to strength, and another to energy storage. The moth *Ephestia kuhniella* has a pupa with a resilient cuticle, which springs back unharmed if pinched by forceps. A genetic mutant form has a cuticle that will crack if pinched. The only apparent difference between the normal and mutant forms is the absence of an endocuticle in the latter (RICHARDS, 1958). Perhaps, then, the endocuticle prevents the stiff exocuticle from being deformed into a curve with a very small radius of curvature, with its concomitant high strains.

KER (1977, quoted in VINCENT (1980b)) has an interesting argument showing that if the function of cuticle is to store energy in bending, either to resist impact loading, or for power amplification of muscles during a jump, the addition of an inner layer or lower Young's modulus can actually produce a structure of lower mass, per unit energy stored, than one in which the material is all of the same modulus.

BARTH (1973) has written a particularly clear and interesting account of the way in which the chitin reinforcing fibrils in spiders' cuticle are oriented in relation to the likely stress direction, and in relation to stress-concentrating holes in the cuticle. He has also investigated the lyriform sense organs of arachnids. These are sets of slits in the cuticle of spiders and similar animals, whose function is to amplify the strain locally so that the strain occurring in the cuticle can be monitored by sense organs connecting with the central nervous system (BARTH and PICKELMANN, 1975).

4.3. Crustacean cuticle

Crustacean cuticle is interesting because it seems to show how, under different selective forces, a structure may be modified in quite different directions. The crustaceans are crabs, lobsters, shrimps, barnacles and so on. Their cuticle has an organization similar to that of insects, except that the protein is replaced to a large extent by calcium carbonate in the form of calcite. In crustacean cuticle the percentages by weight of calcium carbonate, protein and chitin are typically 60:10:30 respectively; in insects 0:40:60 (RICHARDS, 1951). (The use of the word 'replaced' does not imply an evolutionary progression from insects to crustaceans, or vice versa, but merely that in one group one finds protein only, in another less protein and some calcite.) There are few reported values of the mechanical properties of crustacean cuticle but, such as they are, they are inferior to those of insect cuticle, particularly when compared on a per weight basis (cf. Table 9). No attempt has been made, apparently, to analyse the mechanical properties of these three-phase composites.

Crustacean cuticle would be mechanically more efficient, in some senses, if it were more like insect cuticle. However, the selective forces acting on crustaceans are different. First, in the sea, where most crustaceans live, the weight of skeletal materials is less, and so the greater density of crustacean cuticle may be less important. Indeed, having an insect type cuticle, which is less dense than water, might be an embarrassment to a crab. Secondly, the metabolic cost of crustacean cuticle is almost certainly less. Calcium carbonate is cheap to produce, expecially in the sea, whereas protein is expensive. As a rough guess, based on ATP requirements for the synthesis of chitin and protein, one can suppose that if calcium carbonate has a cost of 1, chitin has a cost of 2 and protein a cost of 4. A gram of insect cuticle therefore might cost 1.8 times a gram of crustacean cuticle. Table 9 shows how the costs of producing cuticles of different types relate to their mechanical efficiency. Crustacean cuticle now comes out about the same as insects in Young's modulus. Insects, however, are flying animals and therefore weight saving is of paramount importance. Metabolic cost has to take a second place with respect to specific strength and stiffness, even if the cost is great.

There is a group of shrimp-like crustaceans, in the group Stomatopoda, that have evolved a biological hammer. They are quite small animals, being up to about 15 cm long. One of their pairs of legs, the second thoracic limbs, are inflated into solid bulbs at their main joint (see Fig. 25). This bulb is used to strike the prey (and also other members of their own species in rather ritualised territorial disputes). The blow delivered by the bulb has considerable kinetic energy, 50 J or more. The bulb has the standard crustacean cuticular structure in that there are chitin fibrils arranged helicoidally and the matrix consists of mineral as well as protein. The distribution of this mineral is not uniform. There is an outer layer that is much more highly mineralised than the inner layers. The microhardness is much greater in this outer layer, as might be expected (CURREY, NASH and BONFIELD, 1982). Interestingly, the

TABLE 9. Two mechanical properties of crabs' cuticle and insect cuticle[a] compared in relation to their density and to their metabolic cost.

Species	Structure	Density ($kg\,m^{-3}$)	Property	Value	$\dfrac{\text{Value}}{\text{Density}}$	Value × Relative cost	
						1:1.8	1:3.4
Carcinus	Carapace	1900	Young's modulus	1.3×10^{10}	6.8×10^{6}	6.8×10^{6}	6.8×10^{6}
	Claw	1900	Young's modulus	1.8×10^{10}	9.5×10^{6}	9.5×10^{6}	9.5×10^{6}
Cancer	Carapace	1900	Young's modulus	1.1×10^{10}	5.8×10^{6}	5.8×10^{6}	5.8×10^{6}
Schistocerca	Tibia	1200	Young's modulus	10^{10}	8.3×10^{6}	4.6×10^{6}	2.4×10^{6}
Carcinus	Carapace	1900	Tensile strength	3.2×10^{7}	1.6×10^{4}	1.6×10^{4}	1.6×10^{4}
Cancer	Carapace	1900	Tensile strength	3.5×10^{7}	1.8×10^{4}	1.8×10^{4}	1.8×10^{4}
Schistocerca	Tibia	1200	Tensile strength	10^{8}	8.3×10^{4}	4.6×10^{4}	2.4×10^{4}

[a] I assume, see text, that insect cuticle is 1.8 times more expensive to manufacture than crab cuticle. This is probably conservative. If we assume mineral is free, the ratio of costs is 1:3.4 and is shown in the right-hand column. The values of Young's modulus and tensile strength are in Pascals.

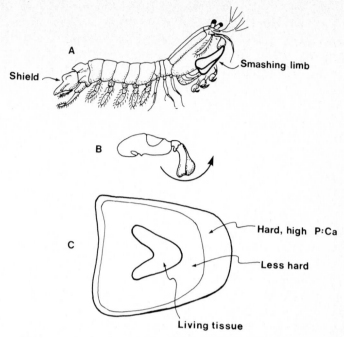

FIG. 25. *Gonodactylus chiragra* a smashing stomatopod. A—General view. The animal is about 10 cm long; B—The smashing action. The rounded bulb strikes the victim; C—Cross-section of the bulb. The harder part has a greater amount of mineralisation as well as more calcium phosphate than the less hard part, whose mineral is nearly all calcium carbonate.

atomic constitution of the mineral also changes with microhardness: the harder the mineral the higher the ratio of phosphorus to calcium. This almost certainly means that the mineral, which is almost exclusively calcite in the less highly mineralised parts, has a significant proportion of calcium phosphate in the harder parts. The fracture surface of the bulb is smooth in the hard region but cracks travelling through the less hard region show a much rougher surface.

The hammer of the animal is designed to break hard objects. It does so by gaining a large amount of kinetic energy and transferring most of it to the other object. The efficiency of this transfer will be greater the greater the stiffness of the bulb. The fact that the bulb is highly mineralised is, therefore, advantageous to the animal. The corollary of this mineralisation, however, is that the material of the other layer is brittle. If one of these bulbs is pressed hard with a metal knife, it will break in a brittle fashion with cracks running in all directions. This outer layer must not, therefore, be deformed significantly or it will break. Therefore it is necessary to have a sufficient thickness of heavily mineralised material for the structure as a whole to be stiff.

This hammer structure contrasts markedly with that of the telson, a shield at the back of the animal. This is used in the territorial disputes mentioned on page 545. Two animals hit the telsons of their opponent alternately until one gives up and goes haltingly away. The telson has a highly calcified layer, but it is only a few microns thick. In the fights, therefore, it can deform considerably without the strain becoming large. The difference in structure between the hammer bulb and the telson is similar to that between the slashing and the crushing teeth of sharks, and the functional implications similar.

Although the precise relationship between the volume fraction of mineral and organic material and the mechanical properties have not been demonstrated, the structures in these shrimps show how detailed are the relationships between structure and function in living organisms.

5. Wood

Wood is perhaps the most obvious natural composite we see around us. A considerable amount is known about the properties of timber (which is dead, with a rather low moisture content); very little is known about living wood, which has a very high moisture content. MARK (1967) published a book on the modelling of wood as a composite that was an extraordinary tour de force. It appeared too soon to be understood by biologists, and to some extent tried to build too much on insufficient foundations. It remains, nevertheless, a remarkable intellectual achievement.

Some representative values for the mechanical properties of wood and timber are given in Table 10. These values are more impressive than they look, when considered on a per weight basis. Wood has a density of about 600 kg m^{-3}, bone of 2000 kg m^{-3}.

TABLE 10. Data are derived from JERONIMIDIS (1980a, b). The last two values are derived from BARRETT (1981).[a]

Type of wood	Property	Direction with respect to grain	Value
'Wood'	Young's modulus	Along	10 GPa
id.	Compressive strength	Along	30 MPa
id.	Tensile strength	Along	100 MPa
id.	Work of fracture	Across	10 kJm^{-2}
id.	id.	Along	0.1 kJm^{-2}
Sitka Spruce	id.	Across	16 kJm^{-2}
Teak	id.	Across	9 kJm^{-2}
Balsa	id.	Across	3 kJm^{-2}
Sitka	K_{IC}	Across	$7.0 \text{ MNm}^{-3/2}$
Balsa	K_{IC}	Across	$2.5 \text{ MNm}^{-3/2}$
Douglas Fir	K_{IC}	Across	$2.5 \text{ MNm}^{-3/2}$
id.	K_{IC}	Along	$0.36 \text{ MNm}^{-3/2}$

[a] Unfortunately these values *appear* to be from air-dried timber, and so their relevance to the living tree is uncertain. BARRETT (1981) has many other values for timber tested in the easy direction.

5.1. Structure

The three main constituents of wood are cellulose, various so-called hemi-celluloses and lignins. Cellulose forms microfibrils, which are about 10–29 nm in diameter, and very much longer. These microfibrils consist of 2000 or so cellulose molecules arranged in parallel, forming a crystalline array. Cellulose and chitin are chemically very similar. About 10^{11} tonnes of these two polysac-charides are synthesised every year. Hemicelluloses are short-chain polysac-charides and do not form crystals. Hemicellulose molecules may branch, and there are many molecular species included in the term hemicellulose. The hemicelluloses are rather hygroscopic, and in the living tree are nearly saturated with water. Lignin is an amorphous material, composed of the oxidation products of phenylpropane derivatives. It is extremely resistant to chemical attack and, together with the hemicelluloses, forms the matrix for the fibrous cellulose.

In the tree, wood has two principal functions: (i) mechanical support, and (ii) the carrying of water up from the earth to the leaves. The two great groups of vascular plants, the gymnosperms and the angiosperms, arrange their woody tissue in rather different ways, and the resulting tissue is called softwood and hardwood respectively, though mechanically this is often a misnomer. However, I shall not discuss these differences further.

The mechanical unit of wood is the cell, which lays woody material around itself. Fig. 26 is an extremely simplified diagram of a woody cell and its walls. The first part to be laid down is the primary wall. When first deposited the primary wall can increase in circumference, but as soon as the secondary wall starts to be deposited, the final size is fixed. The secondary cell wall usually has three layers. S1 and S3 are usually rather thin, and S2 is by far the thickest. The important feature of these layers is that the orientation of the cellulose fibres varies greatly between them. In S1 the cellulose fibres are arranged in helices, about half and half left-handed and right-handed. In S2 the fibrils have an angle with respect to the long axis of the cell of 10° and 40° in different woods; they are usually in right-handed helices. In S3 the fibrils are but at a greater angle to the long axis than those of S2, between 10° and 60°, and usually the sense of the pitch is different from that of S2. The cellulose fibres are laid down initially in a rather porous matrix of hemicellulose. Later the pores become filled with lignin. The central lumen of the cell may become empty or it may become filled with fluid. The fluid will be either cell contents or water being brought up from the earth. The constructional elements formed in this way have various names: tracheids, fibres, vessels and so on, but these need not concern us, because the mechanical properties of the different types are poorly understood. The cells tend to conform to good composite practice in having a high aspect ratio, and they are usually oriented in the direction of the major stresses in the tree. A representative cell might be about 2 mm long, 50 μm across, and the walls take up about 25% of the total cross section.

The cells are bonded together by a matrix called the middle lamella. This

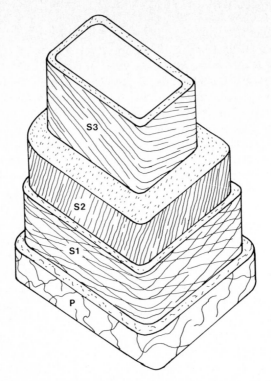

Fig. 26. Diagram of a woody cell with its layered walls. P—Primary; S1, S2, S3—the three layers of the secondary wall. S2 is usually much thicker than S1 and S3. The distribution of the cellulose fibres is shown by the finer lines. The cell is surrounded by the amorphous middle lamella, which acts as a matrix and is not shown. A typical cell is 10–100 μm across.

consists mainly of lignin with some hemicelluloses. It contains no cellulose and is therefore effectively amorphous and isotropic.

5.2. Mechanical properties

The Young's modulus of cellulose is of the order of 250 GPa. The Young's modulus of wood is usually less than 10 GPa. JERONIMIDIS (1980a) has shown how the reduction in modulus is caused. The analysis involves making some assumptions about, for example, the modulus of the matrix, but these are not too far-fetched. In essence, the modulus of wood is less than the modulus of cellulose because: (i) some parts of the molecule are amorphous, not crystal-line, (ii) the matrix has a much lower modulus than the cellulose, (iii) the cellulose fibrils are arranged at an angle to the long axis of the cells, and (iv) the cell material does not occupy the whole volume of the wood. The modulus is also extremely anisotropic being roughly in the ratio of 1:0.05:0.02 in the longitudinal, radial and tangential directions. This is to be expected from the tubular structure of wood.

The strength of wood is greater in tension than in compression (cf. Table 10). This is because the open tubular nature of wood allows it to buckle when compressed along the grain. The buckling seems to start as local buckling on an S2 wall, which then spreads (DINWOODIE, 1968). Trees have to some extent overcome the problems associated with these differences in properties by prestressing (DINWOODIE, 1966; JERONIMIDIS, 1980a). The wood in the centre of a tree trunk may have a compressive stress of 18 MPa when the trunk is not being bent, while the wood underneath the bark may have a tensile stress of 10 MPa. A bending moment that would produce maximum stresses of plus and minus 30 MPa in an unprestressed tree produce stresses of 40 MPa in tension but only 20 MPa in compression (see Fig. 27).

Perhaps the most obvious feature of wood is its toughness (JERONIMIDIS, 1980b). This is, however, an extremely anisotropic property, being almost one hundred times as great when the crack travels across the grain (10 kJm^{-2}) than when it travels with it (0.1–0.2 kJm^{-2}). When wood is fractured along the grain the crack seems to travel through the middle lamellae, which have no fibrous component. When the crack travels across the grain it has to break across the cell walls. JERONIMIDIS (1980a) has shown that the walls of the S2 layer act as extremely efficient energy-absorbing structures. As the strains ahead of the crack increase, the S2 layer, whose fibres are at an angle of 20–30° to the long

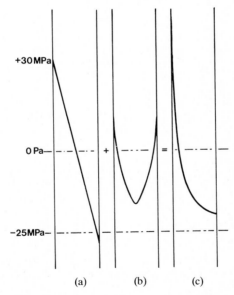

FIG. 27. Prestressing in trees. (a) Some bending loads on a trunk would induce maximum fibre stresses on an unprestressed structure of + and −30 MPa. The compressive strength of many woods is about 25 MPa; (b) The state of stress in an unloaded trunk. The maximum tensile stresses are about 10 MPa, the maximum compressive stresses are about 18 MPa; (c) The addition of applied load and prestress produces a maximum tensile stress of 40 MPa in tension and 20 MPa in compression. Wood can bear such stresses.

axis of the cell, begins to bulge inwards. As this happens cracks appear in the S2 layer, in the direction of the cellulose fibres (see Fig. 28). This results in much energy being used at quite large distances on each side of the crack plane. It would seem that, as usual, a compromise has to be made between the requirement for stiffness, which would be fulfilled with a very low helical angle in S2, and the requirements for toughness, for which a high angle would be good. Working on models similar in their important features to wood cells, GORDON and JERONIMIDIS (1980) have shown that at a fibrillar winding angle of 15–20° there is an optimum relationship between stiffness and toughness. As Jeronimidis points out, the woody system works well at low temperatures. The work of fracture of Sitka Spruce is, if anything, a little higher at −190°C than at 20°C. For trees that have to withstand the bitter cold of a Siberian or Canadian winter, toughness at low temperatures is clearly important.

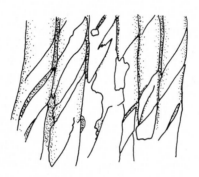

100 μm

FIG. 28. Tensile failure in six tracheids of Sitka Spruce (*Picea sitchensis*). The S2 walls (see Fig. 26) have cracked in the direction of the microfibrils (redrawn from JERONIMIDIS, 1980a).

5.3. Reaction wood

The trunk of a tree is usually upright and therefore not subjected to continuous bending moments from the force of gravity. The boughs springing from the trunk and, if it is not upright, the trunk itself, are subjected to bending moments. The wood developed in these circumstances is different from that found in upright trunks. This special wood is called 'reaction wood'. Interestingly, in conifers and their allies (the gymnosperms) it develops on the compression side, while in angiosperms (the hardwoods) it develops on the tension side.

The main feature of the compression wood of conifers is that the pitch of the helix in S2 is flatter, indeed greater than 45°, whereas usually it is about 25° with respect to the long axis of the cell. Also the S2 layer occupies a greater proportion of the total thickness of the cell. This occludes the lumen to some extent, but will reduce the tendency to buckle. Tension wood, in the angiosperms, shows a different feature. The lumen of the cells, which are not to be confused with cellulose fibrils and which are the main mechanical elements of the wood, is narrowed by the presence of cellulose fibrils forming the so-called 'G layer'. The cellulose has very little matrix and the fibrils are nearly all longitudinally arranged. The mechanical properties of this layer are not understood. However, the longitudinal fibrils presumably increase the stiffness of the wood. (It is a strange fact that the boughs of trees, which are always subjected to bending towards the earth, are usually more or less circular in cross section, rather than being elliptical, which would seem appropriate if the minimum mass for a particular stiffness were being selected.)

6. Keratins

6.1. Structure

Keratin is the name of a class of proteins that are very important as external skeletal structures in the vertebrates. The outer layer of the skin (the *stratum corneum* of the *epidermis*), nails, hair, hooves, claws, beaks, feathers and horns are all keratinous. The major chemical characteristic of keratin proteins is that they have rather large numbers of cysteinyl residues. These are oxidised at a late stage in synthesis, and this results in the chemical and mechanical stabilisation of the protein through sulphur–sulphur cysteine bonds.

A recent review of the structure and mechanical properties of keratin is in Fraser and Macrae (1980).

The keratins usually show, when examined properly, a filament-in-matrix structure. The filaments are narrow, from 3 to 7 nm in diameter, and the volume fraction is high, being about 60%. Despite the standard composite-like appearance of most keratins, there is considerable chemical variation. Table 11 is a simplified classification of keratins. The words 'hard' and 'soft' are embedded deeply in the keratin literature, and I cannot change them, despite their having unfortunate mechanical connotations. The soft keratins are found mainly in the stratum corneum of the epidermis. The hard keratins are found in more rigid structures such as feather hair and claw. Keratins have a characteristic X-ray diffraction pattern: the alpha or beta patterns. The alpha pattern shows the dominant secondary structure of the polypeptide to be helical. The beta pattern is produced by a flattened but pleated sheet of polypeptides. The soft keratins show an alpha pattern. The mammalian hard keratins show an alpha pattern, but the avian and reptilian hard keratins show a beta pattern. The soft keratins have 7 nm diameter filaments, made from three chains in a rope. The alpha helix regions occupy about half the length of the chains. The matrix has two types of protein, one very rich in cysteine, the other rich in histidine.

Mammalian hard keratins, the wools, hair and claws also have an alpha-type X-ray diffraction pattern and a very clear filament matrix structure. The

TABLE 11.

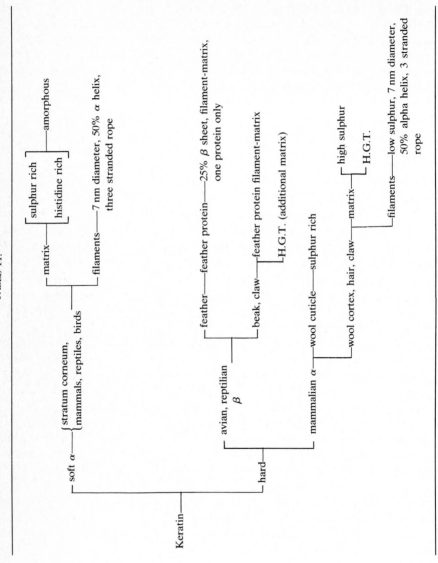

filaments again have about half the chain in the alpha-helical configuration, and the amorphous part is much richer in cysteine than the alpha-helical part. The matrix has two protein types, one called high sulphur because of its many cysteines, the other high glycine tyrosine for obvious reasons. The outer part of the wool has a cuticle of amorphous sulphur-rich protein.

The hard keratins of reptiles and birds are particularly strange from the composite point of view. They show a good filament–matrix structure when seen by electron microscope, but all analyses of the protein show, apparently, no chemical differentiation of filament and matrix. As the protein has well ordered, beta sheet regions, and less well ordered regions it seems most likely that the single protein acts as filament for part of its length, and as matrix for the rest!

6.2. Mechanical properties

The mechanical properties of keratin are profoundly influenced by their state of hydration. Wetting keratinous structures increases their compliance considerably. This is no doubt true for many other biological composites but most of these, like bone, are always wet or, like insect cuticle, in a fairly constant state of hydration that is permitted by the waxy epicuticle covering it. Keratins, by virtue of their position as the outermost part of the body, are likely to be wet some of the time and dry at others. Table 12 shows Young's modulus for various keratins at different humidities. The specific stiffness of keratinous structures is probably a most important mechanical property. It will also be advantageous if they can withstand fairly large strains elastically. This is particularly true of feather, vibrissae (cats' whiskers and similar structures) and fur, the last of which functions, like down feathers, as a heat insulator by trapping a very large volume of air with a small mass of keratin. Keratin can

TABLE 12. Keratins. Young's moduli of various keratinised structures (from FRASER and MACRAE, 1980).

Material	Young's modulus (GPa)		
	Relative humidity		
	0%	Intermediate	100%
Human nail	–	2.6 (70%)	1.8
Human hair	–	2.3 (70%)	1.5
Sheep's wool	5.6	–	2.0
Horse's hair	6.8	5.1 (60%)	2.4
Porcupine quill	–	–	3.5
Feather shaft	–	5.2 (65%)	3.4
Snake scale	–	3.5 (65%)	0.9

strain to about 2–3% elastically, and some types, for instance hair, can show a strain of 40% when wet but much less, only 10% or so, when dry. The density of feather keratin is about $1\,100\,\mathrm{kgm^{-3}}$ so its specific stiffness is $4.7 \times 10^6\,\mathrm{Nkg^{-1}m}$, compared with say $10^7\,\mathrm{Nkg^{-1}m}$ for bone and $1.4 \times 10^7\,\mathrm{Nkg^{-1}m}$ for insect cuticle.

Keratin can be very tough. I have tested dry cow's horn in Tattersall–Tappin-type tests and have found a mean work of fracture of $22\,000\,\mathrm{Jm^{-2}}$.

7. Spicular composites

A form of skeleton often adopted by animals and plants is the spicular skeleton. This consists of little spicules of mineral, often about 50–150 μm long, embedded in an organic matrix. Of course, the mechanical properties of the composite will vary with the volume fraction, aspect ratio and other charac-teristics of the spicules, and as yet little mechanical experimentation has been done. Often, the Young's moduli of the fibres and the matrix may be many orders of magnitude different, unlike the situation in man-made composites except filled rubbers. In filled rubbers the filler particles are very much smaller than the spicules found in animals and plants. There is little help to be gained, therefore, from the technical literature in understanding spicular materials.

Biological tissues with spicules are rather difficult to test mechanically, and there have been virtually no quantitative studies. However, Muzik and Wain-wright (1977) have compared the skeletons of various soft corals (octocorals), and I shall discuss their work to show the variation in the amount of spiculation that can be found, and the kinds of variation in mechanical properties resulting from these variations.

The octocorals are a group of marine animals that live in colonies. Each animal has a battery of stinging cells which they use to kill their very small prey. The octocoral animals in a colony between them produce a skeleton which holds the animals erect in the current. The variation seen in the five species studied by Muzik and Wainwright is described in Table 13 and depicted in Fig. 29.

There is a core which may or may not contain spicules (always of calcite) and often a fairly stiff organic matrix. The outer sheath, the coenenchyme, consists of a very compliant matrix in which are embedded spicules, again of calcite. The spicules of these animals are usually spindle shaped. If so they are arranged predominantly along the axis of the branches. *Melithaea* is the odd one out in this respect, because its branches consist of stiff internodes alternat-ing with flexible nodes. In the internodes the very densely packed spicules are oriented longitudinally, while in the nodes they are oriented in all directions. An interesting feature of the nodal sclerites is that they very rarely form triangles. Since the nodes produce flexibility, it is obviously adaptive to have the sclerites arranged so that they do not form rigid substructures. Triangles are rigid.

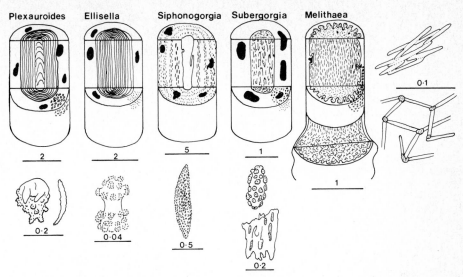

FIG. 29. Schematic diagrams of horny coral structure. The upper row shows the general structure of the core and coenenchyme. Black patches represent cavities into which the animals may withdraw. Lower row (but for *Melithaea* on the right) the shape of characteristic spicules. All scales are in millimetres (see Table 13 for further information; modified from MUZIK and WAINWRIGHT, 1977).

MUZIK and WAINWRIGHT (1977) give no figures for the volume fraction of sclerites, and it is not clear how much the coenenchyme can be stiffened by the spicules. In all except *Siphonogorgia* the mechanical properties of the branch as a whole are mainly controlled by those of the core. The spicules of the coenenchyme have little effect on the flexibility of the branch. However, embedded in the branch are tiny polyps that actually catch the prey. These need a moderate amount of stiffness in themselves and in the coenenchyme in which they are embedded, so that they can stand erect. This stiffness is no doubt produced by the fairly densely packed coenenchymal spicules.

In *Siphonogorgia* there is no mechanically functional central core and the stiffness of the branches is produced by the coenenchymal spicules, which are densely packed but which are not attached to each other.

These five soft corals show the variation that can be seen in animals whose basic requirements are the same: to hold branches out into the sea. The methods vary greatly, and as yet we are not clear as to the reasons for these differences. One area of composite theory that biologists will have to discover much more about is that concerning the behaviour of composites made of large filler particles, up to 0.5 mm long, embedded in an extremely compliant matrix.

Dr. M. KOEHL of the Zoology Department at Berkeley has recently examined the mechanical properties of various spiculated tissues, in particular those of *Alcyonium digitatum* (Dead Man's Fingers). This is a colonial soft coral related to the octocorals. The colonies form bloated, hand-like masses (the

TABLE 13.

Genus	Core		Coenenchyme		
	Radius	Nature	Thickness	Sclerites	Properties
Plexauroides	1 mm	Cross-linked collagen	0.5 mm	Clubs 0.2 × 0.2 mm, or Spindles 0.2 mm × 30 μm	Extremely flexible Tough
Ellisella	0.6 mm	Heavily calcified collagen. Calcifying particles very small	0.6 mm	Knobbly dumbells 60 × 30 μm	Fairly stiff Tough
Siphonorgorgia	0.5 mm	Hollow	3 mm	Densely packed spicules 2 mm × 0.5 μm	Fairly stiff Brittle
Subergorgia	0.5 mm	Irregular, spindly spicules fused to form anastomosing 3-dimensional network in organic matrix	1 mm	Knobbly spindles 0.2 mm × 50 μm	Very stiff Tough
Melithaea Internodes	0.5 mm	Smooth, spindly spicules. Densely packed. 0.15 mm × 10 μm. Fused with CaCO₃	0.1 mm	Knobbly, globular and spindly 50 μm	Very stiff
Nodes	1 mm	Smooth, spindly spicules. 70 × 10 μm. Loosely packed in polyhedra. No triangles. Joined by flexible horny material. Tenuous organic matrix	0.1 mm	As internodes	Flexible

common name is strikingly appropriate) and there is no central core. KOEHL cut conventional tests specimens and loaded them in tension. She also made model tissues by putting various volume fractions of spicules in a matrix (Maid Marion raspberry flavoured jelly) that had properties similar to those of the matrix of the animal. She has found that there is a positive relationship between spicular volume fraction and Young's modulus. Furthermore, this stiffening effect is dependent upon the surface area of the spicules: the greater the surface area of a particular volume fraction, the greater the modulus. Anisodiametric spicules have anisotropic effects on modulus. These results are to be expected, of course, and the moduli are very low, compared with conventional composites, being of the order of 0.5 to 10 MPa. Nevertheless the results are welcome, for, until we have a reasonable body of experimental evidence, it is difficult to make any sensible suggestions as to structure–property–function relationships in such soft-bodied animals.

As can be seen from Table 1 many groups have adopted spicules. We run into tedious semantic difficulties here, in relation to whether we are discussing structures or materials. In some cases, such as the branches of *Siphonogorgia*, the spicules and the compliant matrix obviously make a composite material. Many animals, however, seem merely to use spicules as anti-predator devices. For instance Holothurians (sea cucumbers, sausage-shaped echinoderms) have many intricately-shaped spicules made of single crystals of calcite embedded in the body wall. These function, probably, to deter predators to act as muscle attachment points and to stick out from the skin to act as *points d'appui* during locomotion. What they almost certainly do not do is to alter materially the modulus, strength or toughness of the body wall.

The asteroid echinoderms (the starfish) have a body wall in which there are many spicules. Many of these are quite large and they are called 'ossicles'. These ossicles do have an effect on the general stiffness of the body wall but also, because of the muscles that connect them to each other, can be considered as little skeletal elements.

Finally, consider the siliceous sponges. These have ossicles made of silica, and these probably function both to increase the modulus of the rather compliant sponge body and also to deter predators. In some genera, however, notably Venus' flower basket, *Euplectella*, the spicules are very long and are fused to each other in a complex shape so that they together form a tube with great resistance to both bending and torsion (see Fig. 30). The mechanical properties of the matrix are here irrelevant, all significant stiffness being provided by the silica.

In summary, spicules in animals run the gamut of function from being merely pieces of grit functioning to deter predators, through being the true high modulus moieties of standard composite materials, to being part of skeletal structures, to being the complete rigid structures themselves. Added to this typical biological messiness is the fact that we know very little about the matrices in which they function.

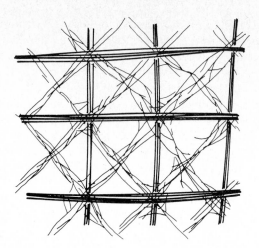

FIG. 30. Diagram of a small part of the skeleton of the glass sponge *Euplectella*. The general shape of the sponge is that of a tube of about 10–15 cm long and 2 cm across. The longitudinal and circumferential glass fibres are much thicker and continuous than those at 45°. The lattice has a periodicity of about 2 mm.

8. Conclusion

This survey of biological composites has no doubt chiefly displayed the biologists' ignorance about the materials they study. It should be borne in mind, however, that biologists work under great difficulties. The pieces of material they have to work with are often very small and irregularly shaped. One obvious result of this is that it is often almost impossible to make test pieces for standard fracture mechanics tests. Furthermore, the materials, being living, tend to be mechanically responsive to small changes in their environment and, indeed, to decay.

A more subtle difficulty is that there is no particular level of size at which one can be said to be dealing with 'the material'. Organisms build their bodies without regard for scientists, and at all levels from the molecular to that of the whole skeleton there are hierarchies of substructures to confuse the issue. Bone and cuticle are particularly good examples of this.

Lastly, the range of materials that organisms have produced is so huge that it is always difficult to have much idea whether what one finds for one species will be true for another. The dangers of extrapolation are great. The example of the soft corals, for instance, shows that it would make no sense to talk about 'the' soft coral skeleton.

On the positive side, for the biologist, is an added delight in trying to find out not only the way in which particular mechanical properties are produced, but also why they have been selected. Natural selection is always performing an optimisation process. The method is crude: survival of the fittest, but the results seem to be, when we can discern them, very finely adjusted. Usually, however, we have only a rather rough idea of what the composite we are investigating is designed to do.

References

AMBARDAR, A. and C.D. FERRIS (1978), *Acta Biol. Acad. Sci. Hung.* **29**, 81–94.

ANDERSEN, S.O. (1974), *Nature* **251**, 507–508.

ANDERSEN, S.O. (1976), Cuticular enzymes and sclerotisation in insects, in: H.R. HEPBURN, ed., *The Insect Integument* (Elsevier, Amsterdam) pp. 121–144.

ANDERSEN, S.O. (1979), *Ann. Rev. Ent.* **24**, 29–61.

ANDREASEN, J.O. (1972), *Traumatic Injuries of the Teeth* (Munksgaard, Copenhagen).

ASCENZI, A. and E. BONUCCI (1976), *J. Biomech.* **9**, 65–71.

BARRETT, J.D. (1981), *Phil. Trans. Roy. Soc. Lond. Ser. A.* **299**, 217–226.

BARTH, F.G. (1973), *Z. Zellforschung* **144**, 409–433.

BARTH, F.G. and P. PICKELMANN (1975), *J. Comp. Physiol.* **103**, 39–54.

BEEDHAM, G.E. and G. OWEN (1965), *Proc. Zool. Soc. Lond.* **145**, 405–430.

BEHIRI, J.C. and W. BONFIELD (1980), *J. Materials Sci.* **15**, 1841–1849.

BENNET-CLARK, H.C. (1962), *J. Insect Physiol.* **8**, 627–733.

BENNET-CLARK, H.C. (1975), *J. Exp. Biol.* **63**, 53–83.

BERTHET-COLOMIAS, C., A. MILLER and S.W. WHITE (1979), *J. Mol. Biol.* **134**, 431–445.

BOND, G.M., V.D. SCOTT, R.G. COOKE and R.G. BOARD (1980), Correlation of hatching techniques in some avian species with the mechanical properties of their eggs, in: J.F.V. VINCENT and J.D. CURREY, eds., *The Mechanical Properties of Biological Materials, Symp. Soc. for Experimental Biology* (Cambridge University Press, London) pp. 459–461.

BONFIELD, W.E. and M.D. GRYNPAS (1977), *Nature* **270**, 453–454.

BONFIELD, W.E., M.D. GRYNPAS and R.J. YOUNG (1978), *J. Biomech.* **11**, 473–479.

BOWEN, R.L. and M.S. RODRIGUEZ (1962), *J. Amer. Dent. Assoc.* **64**, 378–387.

BOYDE, A. (1972), Scanning electron microscope studies of bone, in: G. BOURNE, ed., *The Biochemistry and Physiology of Bone* Vol. 1 (Academic Press, New York) pp. 259–310.

BREAR, K. and J.D. CURREY (1976), *J. Materials Sci.* **11**, 1977–1978.

CALDWELL, R.C., M.L. MUNTZ, R.W. GILMORE and W. PIGMAN (1957), *J. Dental Research* **36**, 732–738.

CARTER, D.R. and W.C. HAYES (1977), *Clinical Orth.* **127**, 265–274.

CHAMBERLAIN, J.A. (1978), *Paleobiology* **4**, 419–435.

CRAIG, R.G., F.A. PEYTON and D.W. JOHNSON (1961), *J. Dent. Res.* **40**, 936–940.

CURREY, J.D. (1964a), *Biorheology* **2**, 1–10.

CURREY, J.D. (1964b), *Quart. J. Microscopical Sci.* **103**, 111–133.

CURREY, J.D. (1975), *J. Biomech.* **8**, 81–86.

CURREY, J.D. (1977), *Proc. Roy. Soc. Lond. Ser. B* **196**, 443–463.

CURREY, J.D. (1979), *J. Biomech.* **12**, 313–319.

CURREY, J.D. (1980), Mechanical properties of mollusc shell, in: J.F.V. VINCENT and J.D. CURREY, eds., *The Mechanical Properties of Biological Materials, Symp. Soc. for Experimental Biology* Vol. 34 (Cambridge University Press, London) pp. 75–97.

CURREY, J.D. and K. BREAR (1974), *Calcified Tissue Research* **15**, 173–179.

CURREY, J.D. and A.J. KOHN (1976), *J. Materials Sci.* **11**, 1615–1623.

CURREY, J.D., A. NASH and W. BONFIELD (1982), *J. Materials Sci.* **17**, 1939–1944.

DINWOODIE, J.M. (1966), *Forestry* **39**, 162–170.

DINWOODIE, J.M. (1968), *J. Inst. Wood Sci.* **21**, 37–53.

DUNEGAN, H.L. and A.T. GREEN (1972), *Amer. Soc. Testing and Materials* **505**, 100–112.

ENLOW, D.H. and S.O. BROWN (1956), *Texas J. Sci.* **8**, 405–443.

ENLOW, D.H. and S.O. BROWN (1957), *Texas J. Sci.* **9**, 186–214.

ENLOW, D.H. and S.O. BROWN (1958), *Texas J. Sci.* **10**, 187–230.

FOX, P.G. (1980), *J. Materials Sci.* **15**, 3113–3121.

FRASER, R.D.B. and T.P. MACRAE (1980), Molecular structure and mechanical properties of keratins, in: J.F.V. VINCENT and J.D. CURREY, eds., *The Mechanical Properties of Biological Materials, Symp. Soc. for Experimental Biology* Vol. 34 (Cambridge University Press, London) pp. 211–246.

FROST, H.L. (1960), *Henry Ford Hospital Bulletin* **8**, 25–35.

GILBERT, A.B. (1979), Female genital organs, in: A.S. KING and J. MCLELLAND, eds., *Form and Function in Birds* Vol. 1 (Academic Press, New York) pp. 237–360.

GILMORE, R.S., R.P. POLLACK and J.L. KATZ (1969), *Archives Oral Biol.* **15**, 787–796.

GLIMCHER, M.J. (1976), Composition, structure and organization of bone and other mineralized tissues and the mechanism of calcification, in: G.D. AURBACH, ed., *Handbook of Physiology Vol. 7*; *Parathyroid Gland* (American Physiol. Soc.) pp. 25–116.

GONG, J.K., J.S. ARNOLD and S.H. COHN (1964), *Anat. Rec.* **149**, 325–332.

GORDON, J.E. and G. JERONIMIDIS (1980), *Phil. Trans. Roy. Soc. Ser. A* **294**, 545–550.

HACKMAN, R.H. (1960), *Austral. J. Sci.* **13**, 560–577.

HASHIN, Z. and B.W. ROSEN (1964), *J. Appl. Mech.* **31**, 223–232.

HEPBURN, H.R. and I. JOFFE (1976), On the material properties of insect exoskeleton, in: H.R. HEPBURN, ed., *The Insect Integument* (Elsevier, Amsterdam) pp. 207–235.

HIGHSMITH, R.C. (1981), *Amer. Nat.* **117**, 193–198.

HILLERTON, J.E. (1980), *J. Materials Sci.* **15**, 3109–3112.

HUKINS, D.W.L. (1978), *J. Theoret. Biol.* **71**, 661–667.

JERONIMIDIS, G. (1980a), Wood, one of nature's challenging composites, in: J.F.V. VINCENT and J.D. CURREY, eds., *The Mechanical Properties of Biological Materials, Symp. Soc. for Experimental Biology* Vol. 34 (Cambridge University Press, London) pp. 169–182.

JERONIMIDIS, G. (1980b), *Proc. Roy. Soc. Land. Ser. B.* **208**, 447–460.

KATZ, J.L. (1980a), The structure and biomechanics of bone, in: J.F.V. VINCENT and J.D. CURREY, eds., *The Mechanical Properties of Biological Materials, Symp. Soc. of Experimental Biology* Vol. 34 (Cambridge University Press, London) pp. 137–168.

KATZ, J.L. (1980b), *Nature* **283**, 106–107.

KENDALL, K. (1981), *Philosophical Magazine A* **43**, 713–729.

KER, R.F. (1977), Some structural and mechanical properties of locust and beetle cuticle, Ph.D. Thesis, Oxford.

KNETS, I.V., V.E. KRAUYA and YU.K. VILKS (1975), *Mechanika Polimerov* **11**, 685–690.

LANG, S.B. (1969), *Science* **165**, 287–288.

LANG, S.B. (1970), *I.E.E.E. Trans. on Bio-medical Engrg.* **17**, 101–105.

LAPPI, V.G., M.S. KING and I. LE MAY (1979), *J. Biomech. Engrg.* **101**, 193–197.

LEES, S. and C.L. DAVIDSON (1977), *J. Biomech.* **10**, 473–486.

LEES, S. (1968), *Arch. Oral Biol.* **13**, 1491–1500.

LOWENSTAM, H.A. (1981), *Science* **211**, 1126–1131.

MCCONNELL, D. (1963), *Bull. Geol. Soc. Amer.* **74**, 363–366.

MCCUTCHEN, L.W. (1975), *J. Theoret. Biol.* **51**, 51–58.

MACK, R.W. (1964), Bone—a natural two-phase material, in: *Technical Memoirs of the Biochemical Lab.* (University of California, Berkeley).

MARGEL-ROBERTSON, D.K. (1973), Studies of fracture in bone, Ph.D. Thesis, Stanford University.

MARK, R.E. (1967), *Cell Wall Mechanics of Tracheids* (Yale University Press, New Haven).

MÄRKEL, K. and P. GORNY (1973), *Z. Morph. Tiere* **75**, 223–242.

MELVIN, J.W. and F.G. EVANS (1973), Crack propagation in bone, *A.S.M.E. Biomaterials Symp.* (American Soc. of Mechanical Engineering, Detroit).

MUZIK, K. and S. WAINWRIGHT (1977), *Bull. Mar. Sci.* **27**(2) 308–337.

NETZ, P., K. ERIKSSON and L. STRÖMBERG (1980), *Acta Orthop. Scandinavica* **51**, 223–229.

NEVILLE, A.C. (1975), *The Biology of Arthropod Cuticle* (Springer, Berlin).

PIEKARSKI, K. (1970), *J. Appl. Physics* **41**, 215–223.

PREUSCHOFT, H., W.-E. REIF and W.H. MÜLLER (1974), *Z. Anat. Entwickl. Gesch.* **143**, 315–344.

RASMUSSEN, S.T., R.E. PATCHIN, D.B. SCOTT and A.H. HENER (1976), *J. Dent. Res.* **55**, 154–164.

REICH, F.R., B.B. BRENDAN and N.S. PORTER (1967), Ulstrasonic imaging of teeth, Tech. Rept, Batelle Memorial Institute, Pacific Northwest Laboratory, Washington.

REILLY, D.T. and A.H. BURSTEIN (1974), *Clin. Orth.* **135**, 192–217.

REILLY, D.T. and A.H. BURSTEIN (1975), *J. Biomech.* **8**, 393–405.

REILLY, D.T., A.H. BURSTEIN and V.H. FRANKEL (1974), *J. Biomech.* **7**, 271–275.

REYNOLDS, S.E. (1975a), *J. Exp. Biol.* **62**, 69–80.

REYNOLDS, S.E. (1975b), *J. Exp. Biol.* **62**, 81–98.

RICHARDS, A.G. (1951), *The Integument of Arthropods* (Minnesota University Press, Minneapolis).

RICHARDS, A.G. (1958), *Z. Naturforschung* **13**, 813–816.

RICQLÈS, A., DE (1977), *Ann. Paleontologie* **63**, 133–160.

SCHNIEWIND, A.P. and J.C. CENTENO (1971), *Engrg. Fracture Mech.* **2**, 223–233.

STANFORD, J.W., K.V. WEIGEL, G.C. PAFFENBERGER and W.T. SWEENEY (1960), *J. Amer. Dent. Assoc.* **60**, 746–751.

TATTERSALL, H.G. and G. TAPPIN (1966), *J. Materials Sci.* **1**, 296–301.

TYLER, C. (1969a), *J. Zool. Lond.* **158**, 395–412.

TYLER, C. (1969b), *J. Zool. Lond.* **159**, 65–77.

VINCENT, J.F.V. (1975), *Proc. Roy. Soc. Ser. B* **188**, 189–201.

VINCENT, J.F.V. (1980), Insect cuticle: a paradigm for natural composites, in: J.F.V. VINCENT and J.D. CURREY, eds., *The Mechanical Properties of Biological Materials, Symp. Soc. for Experimental Biology* Vol. 34 (Cambridge University Press, London) pp. 183–210.

VINCENT, J.F.V. and J.D. CURREY, eds. (1980), *The Mechanical Properties of Biological Materials, Symp. Soc. for Experimental Biology* Vol. 34 (Cambridge University Press, London) pp. 1–513.

WAINWRIGHT, S.A. (1963), *Quart. J. Microscop. Sci.* **104**, 169–183.

WATERS, N.E. (1980), Some mechanical properties of teeth, in: J.F.V. VINCENT and J.D. CURREY, eds., *The Mechanical Properties of Biological Materials, Symp. Soc. for Experimental Biology* Vol. 34 (Cambridge University Press, London) pp. 99–135.

WEIS-FOGH, T. (1961), *J. Mol. Biol.* **3**, 520–531.

WILBUR, K.M. and K. SIMKISS (1968), Calcified shells, in: M. FLORKIN and E.H. STATZ, eds., *Comprehensive Biochemistry* Vol. 26-A (Elsevier, Amsterdam) 229–295.

WILLIAMS, A. (1970), *Smithsonian Contributions to Paleobiology* **3**, 47–67.

WILLIAMS, A. and A.D. WRIGHT (1970), *Special Papers in Palaeontology* **7**, 1–51.

WOODHEAD-GALLOWAY, J. (1980), *Collagen: The Anatomy of a Protein* (Edward Arnold, London).

WRIGHT, T.M. and W.C. HAYES (1977), *J. Biomech.* **10**, 419–430.

YOON, H.S. and J.L. KATZ (1976a), *J. Biomech.* **9**, 407–412.

YOON, H.S. and J.L. KATZ (1976b), *J. Biomech.* **9**, 459–464.

YONGE, C.M. (1953), *Phil. Trans. Roy. Soc. Ser. B* **237**, 335–374.

YOUNG, S.D. (1971), *Comp. Biochem. Physiol.* **40B**, 113–120.

Subject Index